METHODS IN MOLECULAR BIOLOGY™

Series Editor
**John M. Walker
School of Life Sciences
University of Hertfordshire
Hatfield, Hertfordshire, AL10 9AB, UK**

For further volumes:
http://www.springer.com/series/7651

Spermatogenesis

Methods and Protocols

Edited by

Douglas T. Carrell

*University of Utah School of Medicine, Department of Surgery,
Andrology and IVF Laboratories, Department of Obstetrics and Gyneclogy,
Department of Human Genetics, Salt Lake City, UT, USA*

Kenneth I. Aston

*University of Utah School of Medicine, Department of Surgery,
Division of Urology, Andrology & IVF Laboratories, Salt Lake City, UT, USA*

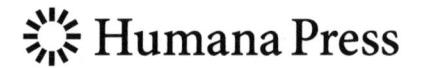

💥 Humana Press

Editors
Douglas T. Carrell
University of Utah School of Medicine
Department of Surgery
Andrology and IVF Laboratories
Department of Obstetrics and Gyneclogy
Department of Human Genetics
Salt Lake City, UT, USA

Kenneth I. Aston
University of Utah School of Medicine
Department of Surgery, Division of Urology
Andrology & IVF Laboratories
Salt Lake City, UT, USA

ISSN 1064-3745 ISSN 1940-6029 (electronic)
ISBN 978-1-62703-037-3 ISBN 978-1-62703-038-0 (eBook)
DOI 10.1007/978-1-62703-038-0
Springer New York Heidelberg Dordrecht London

Library of Congress Control Number: 2012943832

© Springer Science+Business Media, LLC 2013
This work is subject to copyright. All rights are reserved by the Publisher, whether the whole or part of the material is concerned, specifically the rights of translation, reprinting, reuse of illustrations, recitation, broadcasting, reproduction on microfilms or in any other physical way, and transmission or information storage and retrieval, electronic adaptation, computer software, or by similar or dissimilar methodology now known or hereafter developed. Exempted from this legal reservation are brief excerpts in connection with reviews or scholarly analysis or material supplied specifically for the purpose of being entered and executed on a computer system, for exclusive use by the purchaser of the work. Duplication of this publication or parts thereof is permitted only under the provisions of the Copyright Law of the Publisher's location, in its current version, and permission for use must always be obtained from Springer. Permissions for use may be obtained through RightsLink at the Copyright Clearance Center. Violations are liable to prosecution under the respective Copyright Law.
The use of general descriptive names, registered names, trademarks, service marks, etc. in this publication does not imply, even in the absence of a specific statement, that such names are exempt from the relevant protective laws and regulations and therefore free for general use.
While the advice and information in this book are believed to be true and accurate at the date of publication, neither the authors nor the editors nor the publisher can accept any legal responsibility for any errors or omissions that may be made. The publisher makes no warranty, express or implied, with respect to the material contained herein.

Printed on acid-free paper

Humana Press is a brand of Springer
Springer is part of Springer Science+Business Media (www.springer.com)

Dedication

We dedicate this book to our wives and families whose support is the foundation for any contributions we make to the progress of science.

Preface

Spermatogenesis is a biologically complex and essential process. During spermatogenesis, spermatogonia undergo meiotic recombination, reduction of the genome to a haploid state, and extensive cellular modifications that result in a motile cell capable of traversing the female reproductive tract and withstanding various potential assaults to viability. Further, the sperm must be capable of recognizing and binding to the oocyte, undergoing exocytosis of the acrosomal contents, penetrating the oolemma of the oocyte, and then undergoing profound nuclear reorganization events essential for normal syngamy and embryogenesis. Defects in any of these steps, or the many other biological processes needed for successful fertilization, can lead to male infertility, a disease that affects approximately 5–7 % of the population.

Deficiencies in sperm function are usually the result of spermatogenic defects. This edition of Methods in Molecular Biology details protocols used in the study of spermatogenesis and includes basic research tools as well as clinical analytical protocols. We have attempted to provide a comprehensive summary of protocols used in clinical andrology laboratories, as well as common protocols used in the study of spermatogenesis in both the human and animal models. Such protocols include basic techniques, such as obtaining accurate results for a sperm count, and advanced procedures, such as genome-wide genetic study tools and evaluation of nuclear proteins. This volume is unique in its breadth and should be a useful reference for clinicians and researchers alike.

We are very grateful to the chapter authors who have contributed to this book. The authors are leaders in the field, and we appreciate their collegial willingness to assist in the dissemination of good protocols. Accurate, clear protocols are an essential ingredient to both accurate clinical testing and to the advancement of research. It is our hope that this volume will facilitate progress in both of these realms.

Salt Lake City, UT, USA *Douglas T. Carrell*
Kenneth I. Aston

Contents

Preface .... *vii*
Contributors .... *xiii*

PART I   SEMEN ANALYSIS METHODS

1 Methods for Sperm Concentration Determination. 3
   Lars Björndahl
2 Methods of Sperm Vitality Assessment 13
   Sergey I. Moskovtsev and Clifford L. Librach
3 The Hypo-osmotic Swelling Test for Evaluation of Sperm
   Membrane Integrity 21
   Sivakumar Ramu and Rajasingam S. Jeyendran
4 Sperm Morphology Classification: A Rational Method for Schemes
   Adopted by the World Health Organization 27
   Susan A. Rothmann, Anna-Marie Bort, John Quigley, and Robin Pillow
5 Sperm Morphology Assessment Using Strict (Tygerberg) Criteria 39
   Roelof Menkveld
6 Methods for Direct and Indirect Antisperm Antibody Testing 51
   Hiroaki Shibahara and Junko Koriyama
7 Manual Methods for Sperm Motility Assessment 61
   David Mortimer and Sharon T. Mortimer
8 Computer-Aided Sperm Analysis (CASA) of Sperm Motility
   and Hyperactivation 77
   David Mortimer and Sharon T. Mortimer

PART II   TESTS OF SPERM FUNCTION

9 The Hemizona Assay for Assessment of Sperm Function 91
   Sergio Oehninger, Mahmood Morshedi, and Daniel Franken
10 The Sperm Penetration Assay for the Assessment
   of Fertilization Capacity 103
   Kathleen Hwang and Dolores J. Lamb
11 Methods for the Assessment of Sperm Capacitation
   and Acrosome Reaction Excluding the Sperm Penetration Assay 113
   Christopher J. De Jonge and Christopher L.R. Barratt

Part III Sperm DNA Damage Testing

12 Sperm DNA Fragmentation Analysis Using the TUNEL Assay 121
    Rakesh Sharma, Jayson Masaki, and Ashok Agarwal

13 Sperm DNA Damage Measured by Comet Assay. 137
    Luke Simon and Douglas T. Carrell

14 Sperm Chromatin Structure Assay (SCSA®). 147
    Donald P. Evenson

Part IV Advanced Clinical Testing

15 Sperm Aneuploidy Testing Using Fluorescence In Situ Hybridization 167
    Benjamin R. Emery

16 Flow Cytometric Methods for Sperm Assessment 175
    Vanesa Robles and Felipe Martínez-Pastor

17 Human Y Chromosome Microdeletion Analysis by PCR Multiplex
    Protocols Identifying only Clinically Relevant *AZF* Microdeletions 187
    Peter H. Vogt and Ulrike Bender

Part V Sperm Preparation and Selection Techniques

18 Sperm Cryopreservation Methods. 209
    Tiffany Justice and Greg Christensen

19 Density Gradient Separation of Sperm for Artificial Insemination. 217
    David Mortimer and Sharon T. Mortimer

20 Recovery, Isolation, Identification, and Preparation of Spermatozoa
    from Human Testis. 227
    Charles H. Muller and Erin R. Pagel

21 Enhancement of Sperm Motility Using Pentoxifylline
    and Platelet-Activating Factor 241
    Shilo L. Archer and William E. Roudebush

22 Intracytoplasmic Morphology-Selected Sperm Injection 247
    Luke Simon, Aaron Wilcox, and Douglas T. Carrell

23 Sperm Selection for ICSI Using Annexin V. 257
    Sonja Grunewald and Uwe Paasch

24 Sperm Selection for ICSI Using the Hyaluronic Acid Binding Assay 263
    Mohammad Hossein Nasr-Esfahani and Tavalaee Marziyeh

25 Sperm Selection Based on Electrostatic Charge 269
    Luke Simon, Shao-qin Ge, and Douglas T. Carrell

26 Sex-Sorting Sperm Using Flow Cytometry/Cell Sorting. 279
    Duane L. Garner, K. Michael Evans, and George E. Seidel

PART VI STAGING AND HISTOLOGY TECHNIQUES

27 Assessment of Spermatogenesis Through Staging
   of Seminiferous Tubules 299
   Marvin L. Meistrich and Rex A. Hess

28 Immunohistochemical Approaches for the Study of Spermatogenesis 309
   Cathryn A. Hogarth and Michael D. Griswold

29 Ultrastructural Analysis of Testicular Tissue and Sperm
   by Transmission and Scanning Electron Microscopy 321
   Hector E. Chemes

PART VII OXIDATIVE STRESS TESTING

30 Assessment of Oxidative Stress in Sperm and Semen 351
   Anthony H. Kashou, Rakesh Sharma, and Ashok Agarwal

31 Improved Chemiluminescence Assay for Measuring Antioxidant
   Capacity of Seminal Plasma 363
   Charles H. Muller, Tiffany K.Y. Lee, and Michalina A. Montaño

PART VIII ANALYTICAL TOOLS AND METHODS

32 Methods of Sperm DNA Extraction for Genetic and Epigenetic Studies 379
   Jeanine Griffin

33 Isolating mRNA and Small Noncoding RNAs from Human Sperm 385
   Robert J. Goodrich, Ester Anton, and Stephen A. Krawetz

34 A Review of Genome-Wide Approaches to Study the Genetic
   Basis for Spermatogenic Defects 397
   Kenneth I. Aston and Donald F. Conrad

35 Methods for the Analysis of the Sperm Proteome 411
   Sara de Mateo, Josep Maria Estanyol, and Rafael Oliva

PART IX METHODOLOGIES TO EVALUATE SPERM EPIGENETICS

36 Methodology of Aniline Blue Staining of Chromatin and the Assessment
   of the Associated Nuclear and Cytoplasmic Attributes in Human Sperm 425
   Leyla Sati and Gabor Huszar

37 Isolation of Sperm Nuclei and Nuclear Matrices from the Mouse,
   and Other Rodents 437
   W. Steven Ward

38 Protamine Extraction and Analysis of Human Sperm
   Protamine 1/Protamine 2 Ratio Using Acid Gel Electrophoresis 445
   Lihua Liu, Kenneth I. Aston, and Douglas T. Carrell

39	Analysis of Gene-Specific and Genome-Wide Sperm DNA Methylation *Saher Sue Hammoud, Bradley R. Cairns, and Douglas T. Carrell*	451
40	Evaluating the Localization and DNA Binding Complexity of Histones in Mature Sperm *David Miller and Agnieszka Paradowska*	459

Part X In Vitro and Ex Vivo Spermatogenesis Models

41	In Vitro Spermatogenesis Using an Organ Culture Technique *Tetsuhiro Yokonishi, Takuya Sato, Kumiko Katagiri, and Takehiko Ogawa*	479
42	Testicular Tissue Grafting and Male Germ Cell Transplantation *Jose R. Rodriguez-Sosa, Lin Tang, and Ina Dobrinski*	489

Part XI Transgenic Techniques

43	Transgenic Modification of Spermatogonial Stem Cells Using Lentiviral Vectors *Christina Tenenhaus Dann*	503
44	Methods for Sperm-Mediated Gene Transfer *Marialuisa Lavitrano, Roberto Giovannoni, and Maria Grazia Cerrito*	519
45	Phenotypic Assessment of Male Fertility Status in Transgenic Animal Models *David M. de Kretser and Liza O'Donnell*	531
Index		*549*

Contributors

ASHOK AGARWAL • *Center for Reproductive Medicine, Glickman Urological & Kidney Institute, Cleveland Clinic, Cleveland, OH, USA*
ESTER ANTON • *Department of Obstetrics and Gynecology, Center for Molecular Medicine and Genetics, C.S. Mott Center for Human Growth and Development, Detroit, Wayne State University School of Medicine, Michigan, USA; Unitat de Biologia Cellular, Universitat Autònoma de Barcelona, Bellaterra (Cerdanyola del Vallès), Spain*
SHILO L. ARCHER • *Department of Biology, Charleston Southern University, Charleston, SC, USA*
KENNETH I. ASTON • *University of Utah School of Medicine, Department of Surgery, Division of Urology, Andrology & IVF Laboratories, Salt Lake City, UT, USA*
CHRISTOPHER L.R. BARRATT • *Reproductive and Developmental Biology Group, Level 4 MACHS Labs, Division of Medical Sciences, Centre for Oncology and Molecular Medicine, Ninewells Hospital, University of Dundee, Dundee, Scotland, UK*
ULRIKE BENDER • *Molecular Genetics & Infertility Unit, Department of Gynaecological Endocrinology & Reproductive Medicine, University of Heidelberg, Heidelberg, Germany*
LARS BJÖRNDAHL • *Andrology Laboratory, Centre for Andrology and Sexual Medicine, Clinic for Endocrinology, Department of Medicine, Huddinge, Karolinska University Hospital and Karolinska Institutet, Stockholm, Sweden*
ANNA-MARIE BORT • *Fertility Solutions Inc, Cleveland, OH, USA*
BRADLEY R. CAIRNS • *Department of Oncological Sciences, Howard Hughes Medical Institute, and Huntsman Cancer Institute, University of Utah School of Medicine, Salt Lake City, UT, USA*
DOUGLAS T. CARRELL • *University of Utah School of Medicine, Department of Surgery, Andrology and IVF Laboratories, Department of Obstetrics and Gyneclogy, Department of Human Genetics, Salt Lake City, UT, USA*
MARIA GRAZIA CERRITO • *Department of Surgical Sciences, University of Milano-Bicocca, Monza, Italy*
HECTOR E. CHEMES • *Center for Research in Endocrinology (CEDIE), National Research Council (CONICET), Buenos Aires Children's Hospital, Buenos Aires, Argentina*
GREG CHRISTENSEN • *IVF Laboratory, University Women's Healthcare, Louisville, KY, USA*
DONALD F. CONRAD • *Department of Genetics, Washington University School of Medicine, St. Louis, MO, USA*
CHRISTINA TENENHAUS DANN • *Department of Chemistry, Indiana University, Bloomington, IN, USA*

CHRISTOPHER J. DE JONGE • *Department of Obstetrics, Gynecology & Women's Health, University of Minnesota Physicians Reproductive Medicine Center, Minneapolis, MN, USA*

DAVID M. DE KRETSER • *Department of Anatomy and Developmental Biology, Monash Institute of Medical Research, Monash University, Clayton, Melbourne, VIC, Australia*

SARA DE MATEO • *Human Genetics Research Group, IDIBAPS, University of Barcelona, and Biochemistry and Molecular Genetics Service, Hospital Clínic i Provincial, Barcelona, Spain*

INA DOBRINSKI • *Department of Comparative Biology and Experimental Medicine, University of Calgary, Calgary, AB, Canada*

BENJAMIN R. EMERY • *Andrology and IVF Laboratories, Division of Urology, Department of Surgery, University of Utah School of Medicine, Salt Lake City, UT, USA*

JOSEP MARIA ESTANYOL • *Unitat de Proteòmica, Serveis Cientificotècnics (SCT), University of Barcelona, Barcelona, Spain*

K. MICHAEL EVANS • *Sexing Technologies, Navasota, TX, USA*

DONALD P. EVENSON • *SCSA Diagnostics, Volga, SD, USA; Department of Obstetrics and Gynecology, Emeritus, South Dakota State University, Sanford Medical School, and University of South Dakota, Vermillion, SD, USA*

DANIEL FRANKEN • *Reproductive Biology Unit, Tygerberg Hospital, University of Stellenbosch, Cape Town, RSA*

DUANE L. GARNER • *GametoBiology Consulting, Graeagle, CA, USA*

SHAO-QIN GE • *Andrology and IVF Laboratories, Division of Urology, Department of Surgery, University of Utah School of Medicine, Salt Lake City, UT, USA; Basic Medical College, Hebei University, Baoding, Hebei, People's Republic of China*

ROBERTO GIOVANNONI • *Department of Surgical Sciences, University of Milano-Bicocca, Monza, Italy*

ROBERT J. GOODRICH • *Department of Obstetrics and Gynecology, C. S. Mott Center for Human Growth and Development, Center for Molecular Medicine and Genetics, Wayne State University School of Medicine, Detroit, MI, USA*

JEANINE GRIFFIN • *Andrology and IVF Laboratories, Division of Urology, Department of Surgery, University of Utah School of Medicine, Salt Lake City, UT, USA*

MICHAEL D. GRISWOLD • *School of Molecular Biosciences, Washington State University, Pullman, WA, USA*

SONJA GRUNEWALD • *EAA Training Center, Klinik und Poliklinik für Dermatologie, Venerologie und Allergologie, Universitätsklinikum Leipzig AöR, Leipzig, Germany*

SAHER SUE HAMMOUD • *Department of Oncological Sciences, Howard Hughes Medical Institute, Huntsman Cancer Institute, University of Utah, Salt Lake City, UT, USA; IVF and Andrology Laboratories, Departments of Surgery, Obstetrics and Gynecology, and Physiology, University of Utah School of Medicine, Salt Lake City, UT, USA*

REX A. HESS • *Department of Comparative Biosciences, University of Illinois, Urbana, IL, USA*

CATHRYN A. HOGARTH • *School of Molecular Biosciences, Washington State University, Pullman, WA, USA*

GABOR HUSZAR • *Male Fertility Program and Sperm Physiology Laboratory, Department of Obstetrics and Gynecology, Reproductive Sciences, Yale University School of Medicine, New Haven, CT, USA*

KATHLEEN HWANG • *Department of Surgery (Urology), The Alpert Medical School of Brown University, Providence, RI, USA*

RAJASINGAM S. JEYENDRAN • *Andrology Laboratory Services Inc, Chicago, IL, USA*

TIFFANY JUSTICE • *IVF Laboratory, University Women's Healthcare, Louisville, KY, USA*

ANTHONY H. KASHOU • *Center for Reproductive Medicine, Glickman Urological & Kidney Institute and Ob/Gyn & Women's Health Institute, Cleveland Clinic, Cleveland, OH, USA*

KUMIKO KATAGIRI • *Department of Urology, Yokohama City University Graduate School of Medicine, Kanazawa-ku, Yokohama, Japan*

JUNKO KORIYAMA • *Department of Obstetrics & Gynecology, Center for Reproductive Medicine, School of Medicine, Jichi Medical University Hospital, Jichi Medical University, Shimotsuke, Tochigi, Japan*

STEPHEN A. KRAWETZ • *Department of Obstetrics and Gynecology, Center for Molecular Medicine and Genetics, C. S. Mott Center for Human Growth and Development, Wayne State University School of Medicine, Detroit, MI, USA*

DOLORES J. LAMB • *Scott Department of Urology and Molecular and Cellular Biology, Center for Reproductive Medicine, Baylor College of Medicine, Houston, TX, USA*

MARIALUISA LAVITRANO • *Department of Surgical Sciences, University of Milano-Bicocca, Monza, Italy*

TIFFANY K.Y. LEE • *Male Fertility Lab, Department of Urology, University of Washington School of Medicine, Seattle, WA, USA*

CLIFFORD L. LIBRACH • *Division of Reproductive Endocrinology and Infertility, Department of Obstetrics & Gynaecology, CReATe Fertility Center, Sunnybrook Health Sciences Centre and Women's College Hospital, University of Toronto, Toronto, ON, Canada*

LIHUA LIU • *Andrology and IVF Laboratories, Division of Urology, Department of Surgery, University of Utah School of Medicine, Salt Lake City, UT, USA*

FELIPE MARTÍNEZ-PASTOR • *INDEGSAL, University of León, León, Spain*

TAVALAEE MARZIYEH • *Department of Reproduction and Development, Reproductive Biomedicine Center, Royan Institute for Animal Biotechnology, ACECR, Isfahan, Iran*

JAYSON MASAKI • *Center for Reproductive Medicine, Glickman Urological & Kidney Institute, Cleveland Clinic, Cleveland, OH, USA*

MARVIN L. MEISTRICH • *Department of Experimental Radiation Oncology, University of Texas M. D. Anderson Cancer Center, Houston, TX, USA*

ROELOF MENKVELD • *Andrology Laboratory, Department of Obstetrics and Gynaecology, Tygerberg Academic Hospital and University of Stellenbosch, Tygerberg, South Africa*

DAVID MILLER • *Leeds Institute of Genetics, Health and Therapeutics, University of Leeds, Leeds, West Yorkshire, UK*

MICHALINA A. MONTAÑO • *Male Fertility Lab, Department of Urology, University of Washington School of Medicine, Seattle, WA, USA*

MAHMOOD MORSHEDI • *The Jones Institute for Reproductive Medicine, Eastern Virginia Medical School, Norfolk, VA, USA*

DAVID MORTIMER • *Oozoa Biomedical Inc, West Vancouver, BC, Canada*
SHARON T. MORTIMER • *Oozoa Biomedical Inc, West Vancouver, BC, Canada*
SERGEY I. MOSKOVTSEV • *CReATe Fertility Center, Toronto, ON, Canada; Department of Obstetrics & Gynaecology, University of Toronto, Toronto, ON, Canada*
CHARLES H. MULLER • *Male Fertility Lab, Department of Urology, University of Washington School of Medicine, Seattle, WA, USA*
MOHAMMAD HOSSEIN NASR-ESFAHANI • *Department of Reproduction and Development, Reproductive Biomedicine Center, Royan Institute for Animal Biotechnology, ACECR, Isfahan, Iran; Isfahan Fertility and Infertility Center, Isfahan, Iran; Department of Embryology, Reproductive Biomedicine Center, Royan Institute for Reproductive Biomedicine, ACECR, Tehran, Iran*
LIZA O'DONNELL • *Department of Anatomy and Developmental Biology, Monash University and Prince Henrys Institute, Clayton, Melbourne, VIC, Australia*
SERGIO OEHNINGER • *The Jones Institute for Reproductive Medicine, Eastern Virginia Medical School, Norfolk, VA, USA*
TAKEHIKO OGAWA • *Department of Urology, Yokohama City University Graduate School of Medicine, Kanazawa-ku, Yokohama, Japan*
RAFAEL OLIVA • *Human Genetics Research Group, IDIBAPS, University of Barcelona, Barcelona, Spain; Biochemistry and Molecular Genetics Service, Hospital Clinic I Provincial, Barcelona, Spain*
UWE PAASCH • *EAA Training Center, Klinik und Poliklinik für Dermatologie, Venerologie und Allergologie, Universitätsklinikum Leipzig AöR, Leipzig, Germany*
ERIN R. PAGEL • *Male Fertility Lab, Department of Urology, University of Washington School of Medicine, Seattle, WA, USA*
AGNIESZKA PARADOWSKA • *Department of Urology, Pediatric Urology and Andrology, Justus Liebig University of Giessen, Giessen, Germany*
ROBIN PILLOW • *Fertility Solutions Inc, Cleveland, OH, USA*
JOHN QUIGLEY • *Fertility Solutions Inc, Cleveland, OH, USA*
SIVAKUMAR RAMU • *Andrology Laboratory Services Inc, Chicago, IL, USA*
VANESA ROBLES • *INDEGSAL, University of León, León, Spain*
JOSE R. RODRIGUEZ-SOSA • *Department of Comparative Biology and Experimental Medicine, University of Calgary, Calgary, AB, Canada*
SUSAN A. ROTHMANN • *Fertility Solutions Inc, Cleveland, OH, USA*
WILLIAM E. ROUDEBUSH • *Department of Biology, Charleston Southern University, Charleston, SC, USA*
LEYLA SATI • *Department of Histology and Embryology, Akdeniz University School of Medicine, Antalya, Turkey; Sperm Physiology Laboratory, Department of Obstetrics and Gynecology, Reproductive Sciences, Yale University School of Medicine, New Haven, CT, USA*
TAKUYA SATO • *Department of Urology, Graduate School of Medicine, Yokohama City University, Yokohama, Japan*
GEORGE E. SEIDEL • *Animal Reproduction and Biotechnology Laboratory, Colorado State University, Ft. Collins, CO, USA*
RAKESH SHARMA • *Center for Reproductive Medicine, Glickman Urological & Kidney Institute, Cleveland Clinic, Cleveland, OH, USA*

HIROAKI SHIBAHARA • *Department of Obstetrics & Gynecology, School of Medicine, Jichi Medical University, and Center for Reproductive Medicine, Jichi Medical University Hospital, Shimotsuke, Tochigi, Japan*
LUKE SIMON • *Andrology and IVF Laboratories, Division of Urology, Department of Surgery, University of Utah School of Medicine, Salt Lake City, UT, USA*
LIN TANG • *Department of Comparative Biology and Experimental Medicine, University of Calgary, Calgary, AB, Canada*
PETER H. VOGT • *Molecular Genetics & Infertility Unit, Department of Gynaecological Endocrinology & Reproductive Medicine, University of Heidelberg, Heidelberg, Germany*
W. STEVEN WARD • *Institute for Biogenesis Research, John A. Burns School of Medicine, University of Hawaii at Manoa, Honolulu, HI, USA*
AARON WILCOX • *Andrology and IVF Laboratories, Division of Urology, Department of Surgery, University of Utah School of Medicine, Salt Lake City, UT, USA*
TETSUHIRO YOKONISHI • *Department of Urology, Yokohama City University Graduate School of Medicine, Kanazawa-ku, Yokohama, Japan*

Part I

Semen Analysis Methods

Chapter 1

Methods for Sperm Concentration Determination

Lars Björndahl

Abstract

Proper assessment of the number of spermatozoa is essential not only as an initial step in every clinical infertility investigation [Björndahl et al (2010) A practical guide to basic laboratory andrology, 1st edn. Cambridge University Press, Cambridge] but also when attempting to establish the total sperm production in the testis [Amann (Hum Reprod 25:22–28, 2010); Amann (J Androl 30:626–641, 2009); Amann and Chapman (J Androl 30:642–649, 2009)]. Reliable methods combined with an understanding of the specific physiology involved as well as the main sources of errors related to the assessment of sperm concentration are critical for ensuring accurate concentration determination [Björndahl et al (2010) A practical guide to basic laboratory andrology, 1st edn. Cambridge University Press, Cambridge; World Health Organization (2010) WHO laboratory manual for the examination and processing of human semen. WHO, Geneva]. This chapter therefore focuses on these three aspects.

Key words: Sperm, Concentration, Count

1. Introduction

In contrast to most other bodily fluids, semen is not a homogeneous, static fluid. The main constituents are seminal vesicular fluid (approximately 2/3 of the volume) and prostatic fluid (1/3 of the volume). However, these fluids are not mixed until after ejaculation. It is important to acknowledge that most spermatozoa are expelled in the first ejaculate fractions together with prostatic fluid while the last ejaculate fractions contain mainly seminal vesicular fluid with a few spermatozoa. Thus, if the man fails to collect the entire sample, missing one of the first fractions usually means that a substantial number of spermatozoa are missing—the total number of spermatozoa cannot be properly assessed from an incompletely collected sample.

Another important aspect of ejaculate formation is that secretions from the accessory sex glands (prostate and seminal vesicles), as well as transport of spermatozoa from the epididymal storage

Douglas T. Carrell and Kenneth I. Aston (eds.), *Spermatogenesis: Methods and Protocols*, Methods in Molecular Biology, vol. 927, DOI 10.1007/978-1-62703-038-0_1, © Springer Science+Business Media, LLC 2013

(presumed to be mainly in the last part of the epididymis), depend on autonomic nerve stimulation. Although the duration of "sexual abstinence" will affect the total number of spermatozoa in the ejaculate, the duration and quality of sexual stimulation will be important for the number of sperm (transportation) and the concentration of sperm (due to dilution with prostatic and seminal vesicular fluid) in the whole ejaculate.

In the laboratory it is necessary to allow the collected ejaculate to liquefy. Proteins originating from the seminal vesicles cause coagulation, while proteases originating from the prostate cause the coagulum to liquefy. Note that there is no evidence that the ejaculate is mixed in vivo as must happen in vitro, when the whole ejaculate is collected in one container allowing the coagulating proteins to "engulf" the sperm-rich, prostatic fractions. During sexual intercourse it is most likely that the first ejaculated fractions allow the spermatozoa to transfer into the cervical mucus before the spermatozoa have had any substantial contact with the seminal vesicular secretion.

The fact that the ejaculate is made up of two different secretions, one of which forms a coagulum, is an important source of errors. Firstly, even if the ejaculate is macroscopically liquefied (within 30 min if the semen sample is stored at 37°C), microscopically it is common to see that there are still "compartments" containing high numbers of spermatozoa adjacent to other "compartments" with no or very few spermatozoa. In order to ascertain that the aliquot withdrawn for assessment of sperm concentration is representative for the entire ejaculate it is recommended to use at least 50 µL; droplets of 3–10 µL are much more likely to be less representative. Replicate aliquots could also be used to ascertain representativity, but using the larger volume (with the correct pipette) will significantly reduce the risk for sampling errors. The second potential source of error is that all ejaculates—even after liquefaction—have a viscosity that is significantly higher than that of water. Therefore the common type of air-displacement pipette cannot be used: there will always be an error (too low volume withdrawn)—the higher the viscosity of the ejaculate the larger the error. For purposes of pipetting semen, the use of positive displacement pipetters (or PCR-pipetters), which have a piston in the pipette tip, is recommended.

2. Materials

1. Sample collection container (nontoxic plastic).
2. 37°C incubator or heated block (for liquefaction).
3. Pipetters: Air-displacement: 2–20 µL and 100–1,000 µL; positive displacement: 2–50 µL.

4. Microscope slide with coverslips (24×24 mm).
5. Microscope: Positive phase contrast 10× and 20–40× objectives, 10× eyepieces.
6. Small paper wipes.
7. Sperm suspension tubes (plastic 1.5–5 mL with lid, e.g., Eppendorf tubes).
8. Dilution medium:
   (a) 50.0 g $NaHCO_3$.
   (b) 10.0 mL 36–40% formaldehyde solution (a saturated formaldehyde solution).
   (c) Dissolve constituents in distilled water (>10 MΩ/cm), and dilute to 1,000 mL (see Note 3).
   (d) Filter (to eliminate crystals) into a clean bottle.
   (e) Store at +4°C (can be stored for at least 12 months).
   (f) Ensure that the stored medium does not contain solid particles.
9. Vortex mixer.
10. Hemocytometer with improved Neubauer ruling and appropriate cover glass.
11. Humid chamber (e.g., a plastic Petri dish with moistened filter paper and two short glass rods).
12. Bowl to wash reusable counting chambers, detergent.
13. Gloves.

3. Methods

This method has been designed with the aim to reduce the error to less than ±10% related to the counting method. It is essential to acknowledge that due to physiological causes for variations discussed above, one should expect that the variation in results between two ejaculates from the same man will vary much more but not as a result of technical errors.

1. Dilution
   (a) Make sure that the fully liquefied (usually within 30 min at 37°C) ejaculate is macroscopically well mixed.
   (b) Estimate the relevant dilution of the sample by examining a wet preparation: A 10 µL droplet on a microscope slide covered by a 24×24 mm coverslip (see Note 1). Table 1 gives the appropriate dilutions.
   (c) Put the correct volume of diluents into a test tube (see Note 4).

Table 1
Guidelines for the estimation of the appropriate dilution for the assessment of sperm concentration

Spermatozoa per microscope field	Dilution	Semen (µL)	Diluent (µL)
Clean preparation of spermatozoa	1+1 (1:2)	100	100
<15	1+4 (1:5)	100	400
15–40	1+9 (1:10)	50	450
40–200	1+19 (1:20)	50	950
>200	1+49 (1:50)	50	2,450

Clean preparation stands for sperm selected with swim-up or density gradient centrifugation. Under the condition that the diameter of the microscope field (40× objective) is 500 µm, the given numbers of spermatozoa per field can be used to choose the appropriate dilution of the sample (four spermatozoa per field approximately correspond to the concentration of 1 million sperm/mL) (see Note 2)

Fig. 1. Example of a positive displacement pipette with a piston in the pipette tip.

(d) Aspirate the correct volume from the well-mixed semen with a positive displacement pipette (Fig. 1).

(e) Wipe the outside of the pipette tip without affecting the volume inside the tip.

(f) Transfer the exact semen volume into the diluent.

(g) Dip the pipette tip into the diluent and flush it 2–3 times.

(h) Remove the tip from the diluent with the pipetter plunger depressed (see Note 4).

(i) Close the lid and vortex mix immediately (see Note 5).

2. Mounting and loading the counting chamber (see Note 6).

(a) Affix the cover glass on the counting chamber.

(b) Vortex the sperm dilution for at least 15 s (see Note 7).

(c) Aspirate 10 µL (see Note 8), and fill one counting chamber (see Note 9).

(d) Aspirate another 10 µL, and fill the other counting chamber (see Note 10).

(e) Place the filled chamber horizontally in a humid chamber for 15–20 min (see Note 11).

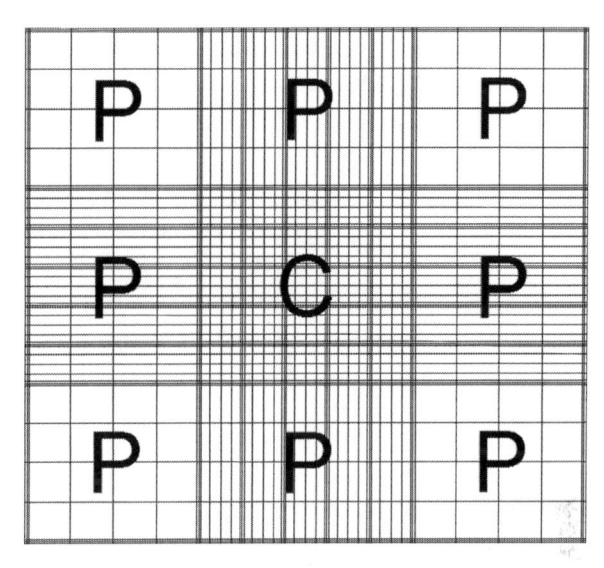

Fig. 2. The layout of the improved Neubauer ruling. The entire field consists of one central area ("C") and eight peripheral areas. The central area ("C") (1 × 1 mm) consists of 25 large squares, each delineated with triple lines. The middle of the triple line is the actual limit of each large square. Each large square is 200 × 200 μm. Within each large square there are 4 × 4 small squares (each 50 × 50 μm). These small squares are helpful when there are many sperm in each large square. The eight peripheral areas ("P") do not consist of large and small squares of the same size as the central area, but each peripheral square is 1 × 1 mm.

Table 2
Guideline for the number of large squares to assess (in a hemocytometer with improved Neubauer ruling) in order to obtain sufficient total numbers of sperm counted

Number of sperm in the upper left large square	Number of large squares to count in each chamber
<10	25
10–40	10
>40	5

3. Counting in the hemocytometer (Fig. 2).
    (a) Count the sperm in the upper left large square in the central area of the counting chamber to estimate how many large squares should be assessed in each of the two chambers. Table 2 shows guidelines for the number of large squares to count (see Note 12).
    (b) Count the number of sperm in the specified number of large squares in the first chamber (see Note 13).

Table 3
Guideline for accepting duplicate assessments

Sum (range)	Value	Sum (range)	Value	Sum (range)	Value	Sum (range)	Value
969–1,000	61	600–624	48	338–356	36	150–162	24
938–968	60	576–599	47	319–337	35	138–149	23
907–937	59	551–575	46	301–318	34	126–137	22
876–906	58	528–550	45	284–300	33	115–125	21
846–875	57	504–527	44	267–283	32	105–114	20
817–845	56	482–503	43	251–266	31	94–104	19
788–816	55	460–481	42	235–250	30	85–93	18
760–787	54	438–459	41	219–234	29	76–84	17
732–759	53	417–437	40	206–218	28	67–75	16
704–731	52	396–416	39	190–205	27	59–66	15
678–703	51	376–395	38	176–189	26	52–58	14
651–677	50	357–375	37	163–175	25	44–51	13
625–650	49						

Choose the row where the obtained sum is included in the range given in the column sum. If the difference between the duplicates is less or equal to the number in the column to the right (value) then the counts are acceptable. The part of the table that has a grey background indicates that the total number of counted sperm is <400 (Fig. 3)

    (c) Count all sperm with head inside the limits of each large square AND all sperm with head on or "touching" the left and upper limits of each large square.

    (d) Count the number of sperm in the same number of large squares in the second chamber.

4. Compare counts, and calculate and present result

    (a) Calculate the sum and the difference of the two counts and go to Table 3.

    (b) If the difference is less than or equal to the value in Table 3 the assessment can be accepted (see Note 14).

    (c) Divide the sum with the factor in Table 4 (or Table 5), based on which dilution was used and how many large squares (or areas) were assessed. The results will be in millions/mL.

    (d) Present the result without decimal places (see Note 15).

    (e) For a full interpretation of results, both the concentration and the total number of spermatozoa should be presented together with duration of abstinence and information

Table 4
Factor by which the sum of counts should be divided to obtain the results in millions/mL

	Number of counted large squares in each of the two chambers		
Dilution	5	10	25
1+1	20	40	100
1+4	8	16	40
1+9	4	8	20
1+19	2	4	10
1+49	0.8	6	4

This table is for occasions when more than the central area (5 × 5 large squares) has been assessed in order to count at least 200 sperm in each counting chamber

Table 5
Factor by which the sum of counts should be divided to obtain the results in millions/mL

Dilution	Number of counted areas in each of the two chambers								
	1	2	3	4	5	6	7	8	9
1+1	100	200	300	400	500	600	700	800	900
1+4	40	80	120	160	200	240	280	320	360
1+9	20	40	60	80	100	120	140	160	180
1+19	10	20	30	40	50	60	70	80	90
1+49	4	8	12	16	20	24	28	32	36

This table is for occasions when more than the central area (5 × 5 large squares) has been assessed in order to count at least 200 sperm in each counting chamber

indicating whether any portion of the sample was lost during collection.

(f) A reference limit is also important; however, the reference ranges provided in the recent WHO manual (5, 6) give the erroneous impression that there is a sharp (and compared to earlier WHO recommendations a lower) limit between fertile and subfertile men. An approach that is more reasonable, taking into account the significant overlap in sperm counts between fertile and subfertile men, is presented in Table 6 (1, 7).

Table 6
Example of experience-based reference limits at two levels to distinguish between truly pathological and truly normal samples as well as acknowledging the significant overlap of results from fertile and infertile men (7) (table after Björndahl et al. (1))

Parameter	Pathological	Borderline	Normal	Unit
Sperm concentration	<10	10–20	20–250	10^6/mL
Volume	<1.5	1.5–1.9	2.0–6.0	mL
Total sperm count	<20	20–79	≥80	10^6/ejaculate

The above data refer to samples collected after 3–5 days of sexual abstinence

4. Notes

1. If 18 × 18 mm coverslips are used, the ejaculate sample volume must be reduced to 6.5 µL.
2. The area of the microscopic field can be calculated by using a microscope slide with a graded scale. Measure the diameter of the field and calculate the field area with the formula Area = πr^2 where r = diameter/2.
3. Dilution of the aliquot withdrawn for assessment of concentration can be done with different solutions. In some clinical settings the use of formaldehyde—even in low concentrations—is not allowed. If this is the case, distilled water can be used, but there is an increased risk for clumping of sperm or adhesion to glass and plastic surfaces. Also some sperm will retain motility for a longer period of time, and bacterial overgrowth can occur within a few hours.
4. If these steps for dilution are followed meticulously, the risk for dilution errors is very low. The sources for errors that must be kept under control by comparison of duplicate assessments are reduced to sampling of diluted spermatozoa and loading of the counting chamber. Under the conditions described above, duplicate dilutions may not be required (1, 8).
5. It is recommended that the dilution tube is shaken on a vortex as soon as possible after addition of the semen aliquot. Even a few seconds' delay increases the risk for large protein precipitate that can be macroscopically visible and that is likely to trap an unknown number of spermatozoa.
6. Hemocytometers with improved Neubauer ruling have a fixed depth of 100 µm if cover glass is affixed appropriately: several

interference lines should be visible (use water, never saliva, to affix the cover glass to the hemocytometer). Cover glass can be affixed several hours before intended use. If disposable Neubauer hemocytometers are used they must be properly validated (9).

7. Vortexing will resolve most sperm aggregates formed after dilution and release most sperm that are adhered to the test tube walls.
8. Flush the pipette tip 2–3 times in the sperm diluent and then aspirate 10 µL.
9. The entire chamber must be filled. Underfilled or overfilled chambers cannot be used for reliable counting.
10. Each hemocytometer has two counting chambers. Sampling from the diluted suspension should be done in duplicate since the sampling and loading procedure can produce errors that cannot be controlled in another way. Vortexing before taking the second aliquot is not necessary if the second sample is taken within 1 min after the first.
11. The focal range of a microscope will not allow the detection of spermatozoa in the entire depth of a 100 µm chamber. Therefore the diluted spermatozoa must sediment to the bottom of the counting chamber. The horizontal position is essential for correct results. The humidity decreases the risk of drying out the preparation during sedimentation—if dried out the chamber cannot be used, and a new one should be prepared to replace it.
12. The aim is to count ideally at least 200 sperm in each counting chamber. If there are fewer than 200 sperm in all 25 large squares in the central area, 1–8 of the peripheral areas can be assessed. The relation between the number of counted spermatozoa and the uncertainty of the result is given in Fig. 3.
13. When counting sperm in the chamber it is advisable to use a form to note which dilution has been used and how large volume (i.e., number of large squares or areas) that has been assessed together with the results of the two replicate assessments. These data can then be used to make the proper comparisons between the replicates to decide if the results should be accepted or discarded.
14. If the difference between duplicate assessments is too large, two new chambers should be prepared and assessed. If a second duplicate count fails to give an acceptable difference, a third duplicate assessment should be done. If the second or third attempt is acceptable, these numbers are used to calculate the final result. If the third attempt fails as well, the final result is calculated from all three attempts, and a note is added to the report form that three attempts for duplicate counting failed and that the results were calculated as the average of all assessments.

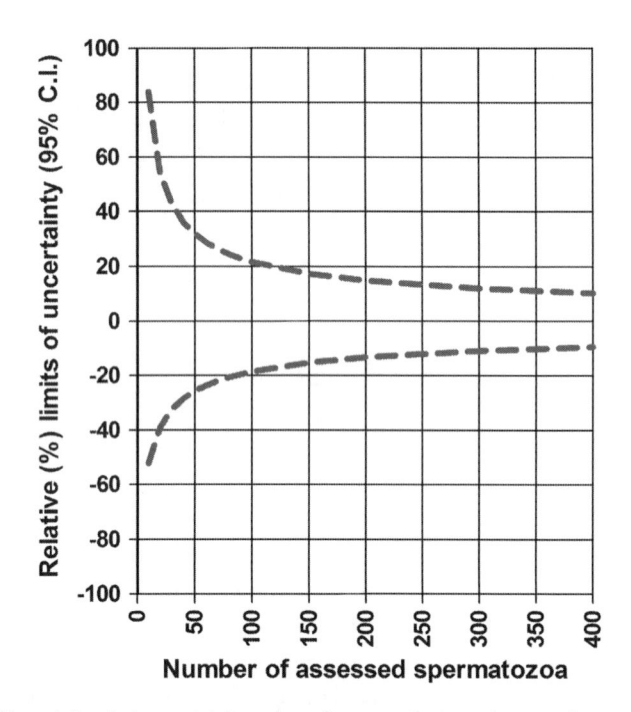

Fig. 3. The relation between total number of assessed spermatozoa and uncertainty in the result (95% confidence interval). Counting a total of 50 sperm to obtain a result of 20 million/mL gives the 95% confidence interval (range of uncertainty) of 15–26 million/mL. If instead 400 sperm had been assessed the range would have been 18–22 million/mL. The width of the range is of importance when interpreting results around a decision limit: the wider the interval, the greater the risk that the result is inaccurate.

15. The level of precision in the assessment does not justify any decimal places. For results less than 1×10^6/mL, one decimal place could be used. Alternatively, results less than 1×10^6/mL could be presented as, for instance, 100 thousands/mL (rounded off to one significant digit) instead of 0.1×10^6/mL.

References

1. Björndahl L et al (2010) A practical guide to basic laboratory andrology, 1st edn. Cambridge University Press, Cambridge
2. Amann RP (2010) Evaluating testis function non-invasively: how epidemiologist–andrologist teams might better approach this task. Hum Reprod 25:22–28
3. Amann RP (2009) Considerations in evaluating human spermatogenesis on the basis of total sperm per ejaculate. J Androl 30:626–641
4. Amann RP, Chapman PL (2009) Total sperm per ejaculate of men: obtaining a meaningful value or a mean value with appropriate precision. J Androl 30:642–649
5. World Health Organization (2010) WHO laboratory manual for the examination and processing of human semen. WHO, Geneva
6. Cooper TG et al (2010) World Health Organization reference values for human semen characteristics. Hum Reprod Update 16:231–245
7. Björndahl L (2011) What is normal semen quality? On the use and abuse of reference limits for the interpretation of semen analysis results. Hum Fertil (Camb) 14(3):179–186
8. Barratt CL, Bjorndahl L, Menkveld R, Mortimer D (2011) ESHRE special interest group for andrology basic semen analysis course: a continued focus on accuracy, quality, efficiency and clinical relevance. Human reproduction (Oxford, England) 26:3207–3212
9. Kirkman-Brown J, Björndahl L (2009) Evaluation of a disposable plastic Neubauer counting chamber for semen analysis. Fertil Steril 91:627–631

Chapter 2

Methods of Sperm Vitality Assessment

Sergey I. Moskovtsev and Clifford L. Librach

Abstract

Sperm vitality is a reflection of the proportion of live, membrane-intact spermatozoa determined by either dye exclusion or osmoregulatory capacity under hypo-osmotic conditions. In this chapter we address the two most common methods of sperm vitality assessment: eosin–nigrosin staining and the hypo-osmotic swelling test, both utilized in clinical Andrology laboratories.

   Key words: Sperm vitality, Eosin–nigrosin staining, Hypo-osmotic swelling

1. Introduction

Vitality is a reflection of the proportion of live spermatozoa determined by the evaluation of cellular and/or membrane integrity. When the percentage of immotile spermatozoa exceeds 40%, it becomes clinically important to verify the proportion of live spermatozoa (1). This assessment can differentiate between necrozoospermia and total absence of motility, indicative of structural defects in the flagellum (2). Accurate motility evaluation can also be verified by sperm vitality tests, as the percentage of immotile spermatozoa should not be higher than the percentage of dead spermatozoa (3).

Most assessments of sperm vitality are based on the ability of the cell membrane to exclude dyes from entering the spermatozoa and permeate into its nucleus. Vital stains are suitable for bright field microscopy (eosin, eosin–nigrosin, trypan blue) or fluorescence microscopy; some fluorochromes can be used for flow-cytometric assessment (propidium iodine, hoechst, ethidium homodimer-1, or Yo-Pro-1) (1, 4).

Sperm vitality can also be assessed by measurement of osmoregulatory capacity under hypo-osmotic conditions (150 mOsm/L), an indication of functional sperm membrane integrity (5). The hypo-osmotic swelling (HOS) test is based on the principle that a living spermatozoon maintains an osmotic gradient by controlled swelling, which results in curling of its tail under hypo-osmotic conditions, whereas a dead spermatozoon exhibits uncontrolled swelling to the degree of membrane rupture, resulting in tail straightening (3). The HOS test identifies live spermatozoa without killing them, allowing utilization of these spermatozoa for therapeutic procedures, such as intracytoplasmic sperm injection (ICSI). However, the test has limitations, in that it is known that some live spermatozoa can have a nonfunctional membrane (6).

While several techniques and dyes are available for sperm vitality evaluation, in this chapter we describe dye exclusion with eosin–nigrosin staining and HOS test, both recommended by the WHO for use in clinical Andrology laboratories (1).

2. Materials

2.1. Eosin–Nigrosin Staining

1. Eosin–nigrosin stain: Dissolve 0.67 g eosin Y (color index 45380) and 0.9 g of sodium chloride in 100 mL of distilled water in a glass beaker placed on a stirring hot plate (see Note 1). Heat gently and add 10.0 g nigrosin (color index 50420) and dissolve it before bringing the stain to a boil. As soon as boiling is observed, remove the beaker from the hot plate and allow it to cool to room temperature. Strain using filter paper, seal, and store at 4 °C in a dark glass bottle (see Note 2).
2. Phosphate-Buffered Saline (PBS): Dissolve one packet of Sigma PBS, pH 7.4 powder in 1,000 mL of distilled water to obtain a 0.01 M PBS solution (see Note 3). Store at room temperature for up to 6 months.
3. Microscope with bright field optics and ×100 oil immersion objective.
4. Stirring hot plate.
5. Microscope slides, 25×75×1 mm.
6. Coverslips, 22×50 mm, #1 thickness.
7. Disposable Pasteur pipettes.
8. Microtubes, 1.5 mL.
9. Immersion oil.
10. Mounting medium.
11. Laboratory counter.
12. Laboratory timer.

2.2. Hypo-osmotic Swelling Test

1. Hypo-osmotic solution: Dissolve 1.375 g D-fructose and 0.75 g of sodium citrate dihydrate in 100 mL of distilled water. Store at 4 °C for up to 6 months. Alternatively the solution can be aliquoted to 1 mL and kept frozen at −20°C for up to 1 year.
2. Microscope with phase contrast optics and ×40 objective.
3. Laboratory hot plate.
4. Microscope slides, 25 × 75 × 1 mm.
5. Coverslips, 22 × 22 mm, #1 thickness.
6. Disposable Pasteur pipettes.
7. Microtubes, 1.5 mL.
8. Laboratory counter.
9. Laboratory timer.

3. Methods

Assess sperm vitality at least 30 min after sample collection to allow for liquefaction of the sample but no longer than 1 h after ejaculation. Mix sample well by swirling before taking aliquots for both types of vitality evaluation. Prior to performing vitality assessment obtain information regarding quality of the semen sample such as sperm concentration and motility. In case of oligozoospermia, when sperm concentrations are below the lower reference limit, the sample can be concentrated by centrifugation (see Note 4).

3.1. Eosin–Nigrosin Staining

1. Add 50 µL of eosin–nigrosin stain to 50 µL liquefied semen in a microtube, and mix well (see Note 5).
2. After 30 s, transfer one drop of the mixture onto a microscope slide and smear (see Note 6).
3. Allow the slide to air-dry.
4. Place one drop of mounting medium and cover with a 22 × 50 mm coverslip (see Note 7).
5. Examine the slide under ×100 oil immersion with a bright field microscope (see Note 8). Live spermatozoa are left unstained (membrane-intact) and dead spermatozoa stain pink or red (membrane-damaged) (see Note 9) (Fig. 1).
6. Evaluate at least 200 spermatozoa, tally the numbers of live and dead spermatozoa using a laboratory counter, and calculate the percentage of live spermatozoa (see Note 10).
7. The lower reference limit for vitality assessed by eosin–nigrosin stain (membrane-intact spermatozoa) is 58% (5th percentile, 95% CI 55–63) (1).

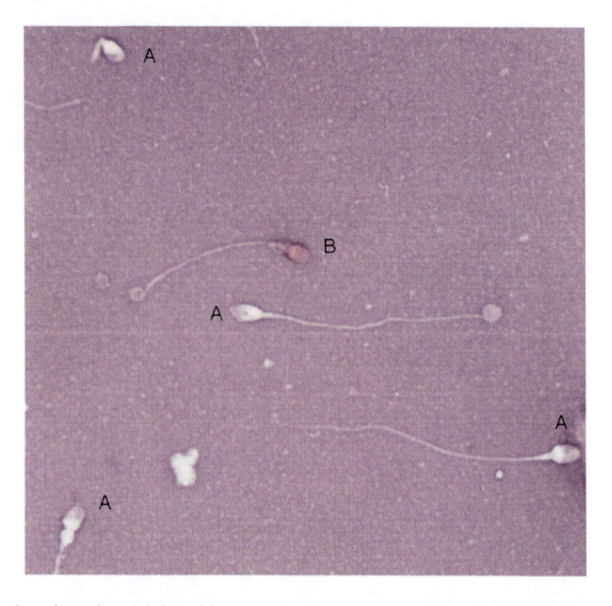

Fig. 1. Eosin–nigrosin staining: Live spermatozoa are unstained (A); dead spermatozoa are stained pink or red (B).

3.2. Hypo-osmotic Swelling Test

1. Place a 1 mL aliquot of hypo-osmotic solution into a 1.5 mL microtube, and warm it up to 37°C.
2. Add 0.1 mL of liquefied, well-mixed semen (see Note 11).
3. Incubate the tube at 37 °C for 30 min (see Note 12).
4. Mix the contents of the tube with a Pasteur pipette.
5. Place one drop of this mixture onto a microscope slide and cover with a 22 × 22 mm coverslip.
6. Allow spermatozoa to settle for 1 min.
7. Examine the slide with ×40 objective utilizing a phase-contrast microscope. Live spermatozoa show controlled swelling visualized by the curling of their tail (membrane-intact) and dead spermatozoa have straight, non-coiled tails (membrane damaged) (see Notes 13 and 14) (Fig. 2).
8. Calculate the HOS test score by evaluating at least 200 spermatozoa. Tally the numbers of live (swollen) and dead (straight tail) spermatozoa using a laboratory counter and calculate the percentage of vital spermatozoa (see Note 15).
9. The lower reference limit for vitality assessed by the HOS test (membrane-intact spermatozoa) is 58% (5th percentile, 95% CI 55–63) (1).

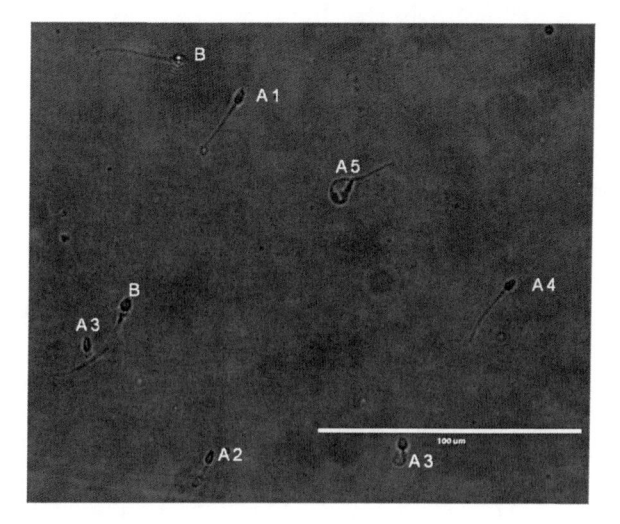

Fig. 2. HOS test: Live spermatozoa show swelling with various changes of the tail region (A1–A5): tail tip enlargement (A1), different degrees of tail curling (A2, A3), shortening and thickening of the tail (A4), partial tail swelling (A5). Dead spermatozoa have a straight tail (B).

4. Notes

1. Use a magnetic stir bar to dissolve ingredients for eosin–nigrosin staining.
2. Warm the stain to room temperature before use.
3. PBS can be used if the specimen is very viscous. Combine sample 1:2 with 0.01 M PBS solution, and mix well with a Pasteur pipette by drawing it in and out of the pipette. Spin the sample down by centrifugation at $300 \times g$ for 10 min. Remove the seminal plasma, and resuspend the pellet with PBS, and then proceed to follow the protocol as for a sample with normal viscosity. If viscosity remains, repeat the process with a second PBS wash.
4. In cases of severe oligozoospermia (sperm concentration less than five million/mL), spin the sample down by centrifugation at $300 \times g$ for 10 min. Remove the seminal plasma, resuspend the pellet in PBS, and then perform the viability testing on a drop from the re-suspended sample. After concentrating sperm follow the protocol as for a sample with a normal sperm concentration. In some cases of severe oligozoospermia, several slides are necessary to prepare for evaluation of a sufficient number of spermatozoa.
5. Use a disposable Pasteur pipette to mix the sample and stain. Avoid making air bubbles.

6. Avoid making the smear too thick, as the background could be too dark to visualize the spermatozoa.
7. Coverslipping using mounting medium is optional and required only for long-term storage of slides. Depending on the mounting medium used, it may take up to several days for the coverslip to be permanently glued to the slide. While slides can be evaluated shortly after coverslipping, wait for mounting medium to dry completely before long-term storage. The slides can be stored indefinitely at room temperature.
8. The nigrosin stain provides a dark purple background that makes it easier to visualize unstained spermatozoa.
9. Spermatozoa with staining restricted to the neck region but an unstained head area are referred to as "leaky necks" and are considered alive.
10. If the percentage of live cells counted is less than the sperm motility, reexamine the sample for motility and perform the stain again.
11. In cases of mild oligozoospermia, an additional 0.1–0.3 mL of semen can be added to the hypo-osmotic solution to increase sperm concentration. In cases of severe oligozoospermia, see Note 4.
12. Swelling of spermatozoa with intact membranes in hypo-osmotic solution will start within 5 min; however the shapes will stabilize after 30 min of the incubation (7). If spermatozoa are going to be utilized for a therapeutic procedure, such as ICSI, incubation should be limited to 5 min.
13. Live spermatozoa with poor osmoregularity, such as senescent spermatozoa, will show uncontrolled swelling similar to dead spermatozoa to the degree of membrane lysis and tail straightening (3).
14. Various types of tail changes can be seen as the evidence of spermatozoa swelling, including only tail tip enlargement, a hairpin curvature at the lower part of the tail, different degrees of tail curling, coiling or swelling, shortening and thickening of the tail, and total swelling of tail (Fig. 2).
15. Prior to the HOS test, neat (unprocessed) semen samples should be examined to determine if any spermatozoa have undergone spontaneous swelling or have spermatozoa with coiled tails. If any affected spermatozoa (tail defects or swollen) are detected, the neat sample should be scored in the same manner as a sample after incubation in hypo-osmotic solution. The number of affected spermatozoa in neat samples should be subtracted from the HOS test score in order to obtain an accurate assessment.

References

1. WHO (2010) World Health Organization laboratory manual for the examination and processing of human semen. WHO Press, Geneva
2. Chemes EH, Rawe YV (2003) Sperm pathology: a step beyond descriptive morphology. Origin, characterization and fertility potential of abnormal sperm phenotypes in infertile men. Hum Reprod Update 9:405–428
3. Mortimer D (1994) Practical laboratory andrology. Oxford University Press, New York
4. Gillan L et al (2005) Flow cytometric evaluation of sperm parameters in relation to fertility potential. Theriogenology 63:445–457
5. Jeyendran RS et al (1984) Development of an assay to assess the functional integrity of the human sperm membrane and its relationship to other semen characteristics. J Reprod Fertil 70:219–228
6. Munuce MJ et al (2000) Does the hypoosmotic swelling test predict human sperm viability? Arch Androl 44:207–212
7. Hossain AM et al (1998) Time course of hypo-osmotic swellings of human spermatozoa: evidence of ordered transition between swelling subtypes. Hum Reprod 13:1578–1583

Chapter 3

The Hypo-osmotic Swelling Test for Evaluation of Sperm Membrane Integrity

Sivakumar Ramu and Rajasingam S. Jeyendran

Abstract

A functional membrane is requisite for the fertilizing ability of spermatozoa, as it plays an integral role in sperm capacitation, acrosome reaction, and binding of the spermatozoon to the egg surface. The hypo-osmotic swelling (HOS) test evaluates the functional integrity of the sperm's plasma membrane and also serves as a useful indicator of fertility potential of sperm. The HOS test predicts membrane integrity by determining the ability of the sperm membrane to maintain equilibrium between the sperm cell and its environment. Influx of the fluid due to hypo-osmotic stress causes the sperm tail to coil and balloon or "swell." A higher percentage of swollen sperm indicates the presence of sperm having a functional and intact plasma membrane. Here, we present the detailed protocol for performing the HOS test and explain the results for interpretation.

Key words: Sperm membrane integrity, Fertility, Hypotonic solution, HOS test, Sperm function

1. Introduction

In the diagnosis of male infertility, a number of sperm characteristics are evaluated. A test for assessing the functional integrity of the sperm membrane is one important measure. Not only sperm membrane integrity is important for its metabolism, but also a correct change in the dynamics of the membrane is required for successful union of the male and female gametes, i.e., for sperm capacitation, acrosome reaction, and binding of the spermatozoon to the egg surface. Thus, the integrity and functional activity of the sperm membrane are of fundamental importance in the fertilization process, and assessment of membrane function may be a useful indicator of the fertilizing potential of spermatozoa.

When exposed to hypo-osmotic stress, water and small-molecular-weight compounds and elements will attempt to enter

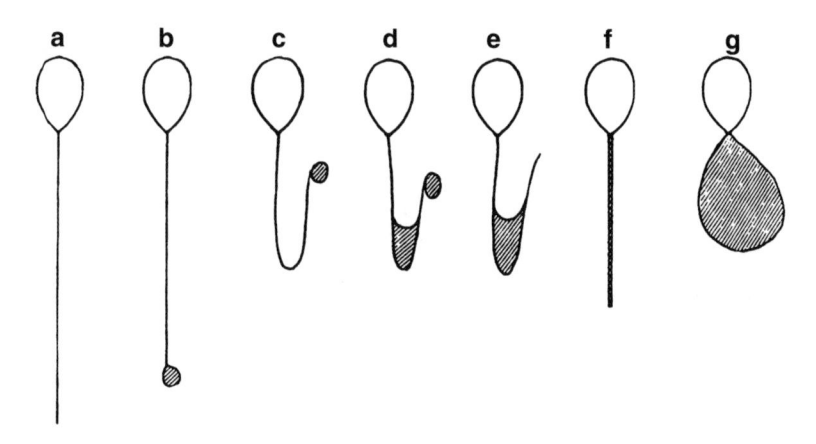

Fig. 1. Schematic representation of various morphological changes of human spermatozoa exposed to hypo-osmotic stress. (**a**) Sperm with unaltered morphology. (**b–g**) Sperm with different types of tail swelling indicated by hatched area. Figure originally published in ref. 1 reproduced with permission.

into the sperm to reach osmotic equilibrium. This inflow of fluids will increase sperm volume, and the plasma membrane will bulge (balloon) to achieve a minimum surface-to-volume ratio (1). The sperm tail appears to be particularly susceptible to such hypo-osmotic stress, and based on the vigor of the sperm, different patterns of tail swelling are observed (Fig. 1). These induced changes to sperm tail morphology are visible with phase-contrast microscopy. This ballooning effect is often referred to as "swelling." It can be assumed that the ability of the sperm tail to swell in the presence of a hypo-osmotic stress is a sign that transport of fluid across the membrane occurs normally, i.e., a sign of both structural and functional integrity of the membrane as opposed to only structural integrity of the cell membrane, which is the basis for eosin Y exclusion in the vital staining test (2). The hypo-osmotic swelling (HOS) test is useful when performing a semen analysis in the diagnosis of the infertile male (1, 3–5). In addition to providing valuable information on sperm viability, this test is widely used to select sperm for intracytoplasmic sperm injection (ICSI) in cases of asthenozoospermia (6). Here we provide in detail the protocol for performing the HOS test and the interpretation of the results.

2. Materials

Prepare all solutions using distilled water and analytical grade reagents. Prepare and store all reagents at the recommended temperature.

2.1. Hypo-osmotic Solution Preparation

1. Solution A: 1.5 mM fructose.

    Weigh 2.7 g fructose (MW 180.16) in a sterile beaker.

    Add 100 ml distilled H_2O.

3 The Hypo-osmotic Swelling Test for Evaluation of Sperm Membrane Integrity

2. Solution B: 1.5 mM sodium citrate.

   Weigh 1.47 g sodium citrate (MW 294.1) in a sterile beaker. Add 100 ml distilled H_2O.

3. Mix equal volume of Solutions A and B, and store in 1 ml aliquots using disposable, polypropylene tubes at −20°C until use or up to 1 year (see Note 1).

2.2. Semen Sample

Collect the semen samples by masturbation or other recommended procedure and allow the sample to liquefy completely (~30 min) (see Note 2).

2.3. Additional Supplies Needed

1. Distilled H_2O.
2. Glass slide with coverslip.
3. 37°C incubator (optional).
4. Laboratory cell counter.
5. Measuring cylinder.
6. Phase contrast microscope.
7. Pipette.
8. Polypropylene tubes.
9. Sterile beaker.
10. Weighing balance.

3. Methods

3.1. HOS Test

1. Incubate tube containing 1.0 ml of hypo-osmotic (HOS) solution at 37°C for 10 min.
2. Add 0.1 ml of thoroughly mixed, liquefied semen to tube containing HOS solution, and mix gently.
3. Incubate semen/HOS solution mixture for at least 30 min, but not longer than 3–4 h, at 37°C (see Note 3).
4. After incubation at 37°C mix the tube gently, place a drop onto a microscopic slide, and cover the drop with coverslip (see Note 4).
5. Place prepared slide on microscope and observe under phase contrast (400×) for the spermatozoa with swollen tails as shown in Fig. 1 (see Note 5).

3.2. Sperm Swelling Pattern

Hypo-osmotic stress will induce several distinct categories of swelling in the sperm tail region as described below (Figs. 1 and 2):

1. *Tip*: Very tip of the tail is swollen; rest of the tail is normal.
2. *Hairpin swelling*: Tail swells at mid piece and main piece junction with tip swelling or without tip swelling.

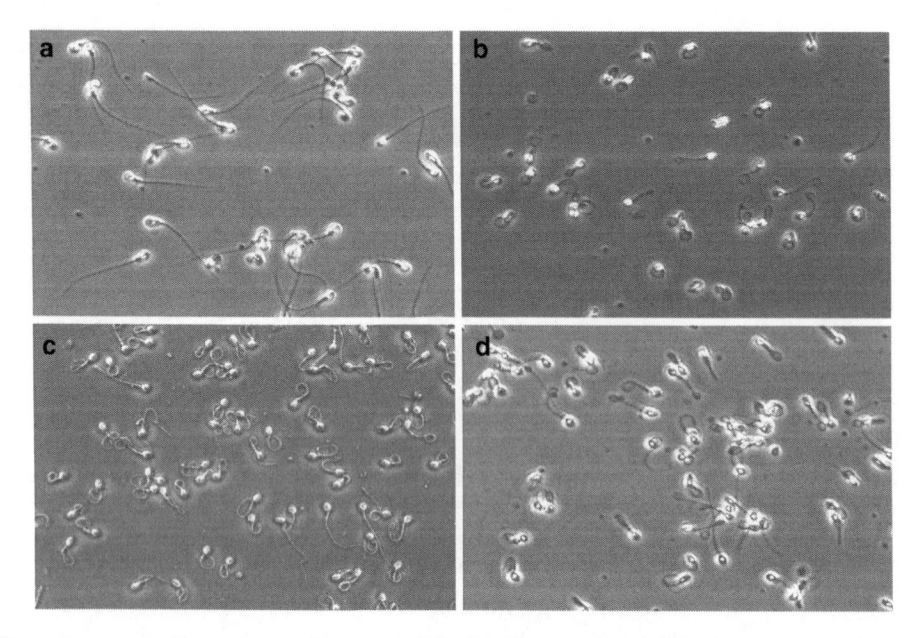

Fig. 2. Spermatozoa were either unexposed (**a**) or exposed (**b–d**) to hypo-osmotic stress for 30 min at 37°C. A representative slide prepared and observed under phase contrast microscope displaying various forms of tail swelling. The pictures were taken using a Nikon microscope with digital imaging system (magnification, ×650).

3. *Shortened and thickened tail*: Tail swells, constricting surface, causing shortening.
4. *Partly or completely enveloped sperm tail*: Tail "balloons" from swelling.
5. Count a total of 200 sperm (at least 100) per sample, including both swollen and non-swollen sperm.
6. Calculate swollen sperm as a percentage of total sperm number counted.
7. Report as percentage of swollen sperm.

3.3. Results and Interpretation

Measured as a percentage of sperm tails observed as swollen: Record the results in percentage swollen on an observation sheet like one shown below:

Sample ID	Result (% swollen)	Interpretation		
		Normal	Equivocal	Abnormal
		≥60 %	50–59 %	<50 %

Since only a single variable (membrane function) is measured, caution is advised. For example, the HOS test is "normal" but some other sperm parameter may be defective rendering the sperm infertile. The abnormal result is more important than the normal result.

4. Notes

1. The HOS test solution should be clear. If you notice any turbidity, discard and prepare a fresh stock.
2. For highly viscous specimens, ejaculate may be diluted with an equal volume of medium or forced in and out of a 3 ml syringe attached to an 18-gauge needle until the sample becomes more pliable.
3. The HOS test can be performed at ambient temperature for the same period without much change in the expected results.
4. Using a Kim wipe, gently pat down the coverslip to remove the excess fluid and form a thin film of one layer of sperm.
5. Count cells under 400× magnification, and appropriate phase for accurate assessment of the tail swelling. If bright field is used, then count coiled tails prior to HOS test and subtract the percent of coiled tails obtained prior to the HOS test from the percent of swollen sperm following the HOS test.

References

1. Jeyendran RS et al (1984) Development of an assay to assess the functional integrity of the human sperm membrane and its relationship to other semen characteristics. J Reprod Fertil 70:219–228
2. Schrader SM et al (1986) Sperm viability: a comparison of analytical methods. Andrologia 18:530–538
3. Van der Ven HH et al (1986) Correlation between human sperm swelling in hypoosmotic medium (hypoosmotic swelling test) and in vitro fertilization. J Androl 7:190–196
4. Jeyendran RS et al (1992) The hypoosmotic swelling test: an update. Arch Androl 29:105–116
5. Hossain A et al (2010) Spontaneously developed tail swellings (SDTS) influence the accuracy of the hypo-osmotic swelling test (HOS-test) in determining membrane integrity and viability of human spermatozoa. J Assist Reprod Genet 27:83–86
6. Verheyen G et al (1997) Comparison of different hypo-osmotic swelling solutions to select viable immotile spermatozoa for potential use in intracytoplasmic sperm injection. Hum Reprod Update 3:195–203

Chapter 4

Sperm Morphology Classification: A Rational Method for Schemes Adopted by the World Health Organization

Susan A. Rothmann, Anna-Marie Bort, John Quigley, and Robin Pillow

Abstract

Sperm morphology is an important measure of testicular health, spermiation, and fertility potential. The World Health Organization (WHO) Semen Manuals advocate different sperm morphology schemes, but, like the schemes themselves, do not describe classification sequence or rules that can be unambiguously applied in a standard method. Our novel dichotomous key provides a rational decision framework for a sperm morphology classification algorithm. Classification order hierarchy is standardized and sperm characteristics are defined. Normal morphology is derived after eliminating abnormal and borderline normal forms. By defining and categorizing borderline normal forms separately from either normal or abnormal, the method can simultaneously produce results for Strict and traditional morphology schemes as adopted by different versions of the WHO Semen Manuals. The algorithm can be used for "recalibration" to a less stringent and potentially more relevant standard of normal, while reducing shift, drift, and variation in classification within and among analysts.

Key words: Sperm morphology, Sperm classification, Dichotomous key, Algorithm, World Health Organization, WHO, Borderline normal sperm, Strict morphology, Semen analysis

1. Introduction

The World Health Organization (WHO) has sponsored the publication of five laboratory manuals on human semen analysis over a 30-year interval, each with different authors and consequently, different emphasis and variations in title and content (1–5). The WHO Semen Manuals are not "standards," but are recommendations for semen analysis methods formulated by a panel of editors whose opinions sometimes oppose each other, and other practitioners of semen analysis (6, 7). Originally conceived as practice guides to facilitate male contraceptive studies, the WHO Semen Manuals can serve as useful references for semen analysis but are neither the "gold standard" nor an incontrovertible source of information (8).

Few areas of the WHO Semen Manuals have changed more than the sections about sperm morphology. Classification methods and reference values differ widely in the five editions, reflecting the plethora of opinions about the definition of a normal sperm. The latest edition, *WHO Laboratory Manual For Examination And Processing Of Human Semen*, 5th *Edition* (WHO5), was published in 2010 (5). Like the 4th Edition, WHO5 adopted the Strict morphology classification (9, 10) validated primarily using in vitro fertilization data (11). The Strict classification scheme assumes that sperm cells recovered from the upper endocervical mucus post coitus are normal and those that look different from this mostly homogeneous population are abnormal. Borderline normal forms with minor variation from the paradigm normal forms are considered to be abnormal (9–11).

The WHO Semen Manuals' adoption of the Strict scheme over others where borderline normal sperm are classified as normal is controversial and arguably appropriate (7, 12). By definition, borderline forms have neither definitively normal nor abnormal morphology; therefore, they have uncertain functionality. Furthermore, the Semen Manuals primarily emphasize analysis of sperm in ejaculated semen. The morphology of ejaculated sperm measures the quality of sperm at spermiation, whereas the morphology of sperm in the endocervical mucus provides insights into potential fertility (12). The relevance of using the morphology of postcoital sperm with sufficient function to penetrate the endocervical mucus as the normal standard for sperm in the ejaculate is questionable. When using semen analysis as a screening tool for male fertility, classifying borderline forms as abnormal because fewer were observed in cervical mucus seems unreasonably conservative.

The original Strict description of normal was either a perfectly smooth oval head or an oval form with a smooth or regular contour but slightly tapered at the postacrosomal end; minor variations from these characteristics were regarded as normal (13). As the scheme moved into clinical practice and into the WHO Sperm Manuals, the reference spermatozoon head became interpreted as perfect in all respects and the original definition of normal was lost (14). Many analysts now have difficulty identifying any normal sperm, even in the semen of fertile men, and percent normal forms above 5–10% are rare in many laboratories. A loss of morphology's predictive value for fertility treatment outcomes, especially in vivo, ensued in many centers (15, 16).

Interpretation of the WHO4/WHO5/Strict scheme differs among similarly trained technologists (17) and changes in classification can occur over time without intervention or awareness (14, 18).

Grouping borderline normal forms with abnormal, as advocated by recent WHO Sperm Manuals, appears to cause inconsistent results often with downward classification shift or drift until normal cannot be detected (13, 14). In our competency/proficiency testing program (19), the first in the USA, and in others (20), challenge results typically include much of the possible analytic range of values and indicate the diversity of sperm classification. Our recent survey of morphology classification practice revealed how differently experts classify the same cell: for 35 sperm, one-third of analysts classified them as normal, one-third as borderline normal, and one-third as abnormal (unpublished data). Clearly the threshold for acceptable variation differs among analysts because they have different concepts and paradigms of normal morphology. The latest WHO Semen Manual included a study defining new reference values and population distributions for sperm morphology (5, 21). The values of percent normal forms from four reference populations based on fertility status showed higher medians and wider, more normal distribution of values rarely observed in many assisted reproductive technologies (ART) laboratories, confirming the lack of uniform application of sperm morphology concepts.

This significant problem is likely due to the absence of rules for the precise process of classification. The WHO Sperm Manuals and published morphology classification schemes describe and illustrate various sperm shapes but do not specify the order and priority of decision logic. Knowing that an investigator classified sperm "according to WHO" is like hearing that a traveler visited a continent without specifying the itinerary. You would have a general idea of where the journey took place but would be unlikely to replicate the trip. Unlike other areas of laboratory medicine where clear cell typing rules exist, sperm morphology classification was put into practice without a defined decision hierarchy. Sperm visual attributes alone are not sufficient for consistent analysis across time. Without a standardized approach to classification, teaching and learning the same system become virtually impossible.

Our solution is a novel algorithm based on a dichotomous key, a method commonly used in taxonomy (22). The Morphology Algorithm (Fig. 1) uses three distinct and essential components to determine spermatozoon morphology: (1) a series of binary choices based on describable and readily recognizable visual attributes for sperm shapes and anatomical characteristics; (2) a defined order of classification that uses a process of eliminating sperm abnormalities to identify normal forms, in contrast to typical approaches where normal is assigned a priori through subjective recall of the analyst's concept of what normal looks like; and (3) definition and separate categorization of borderline normal forms as described by Menkveld

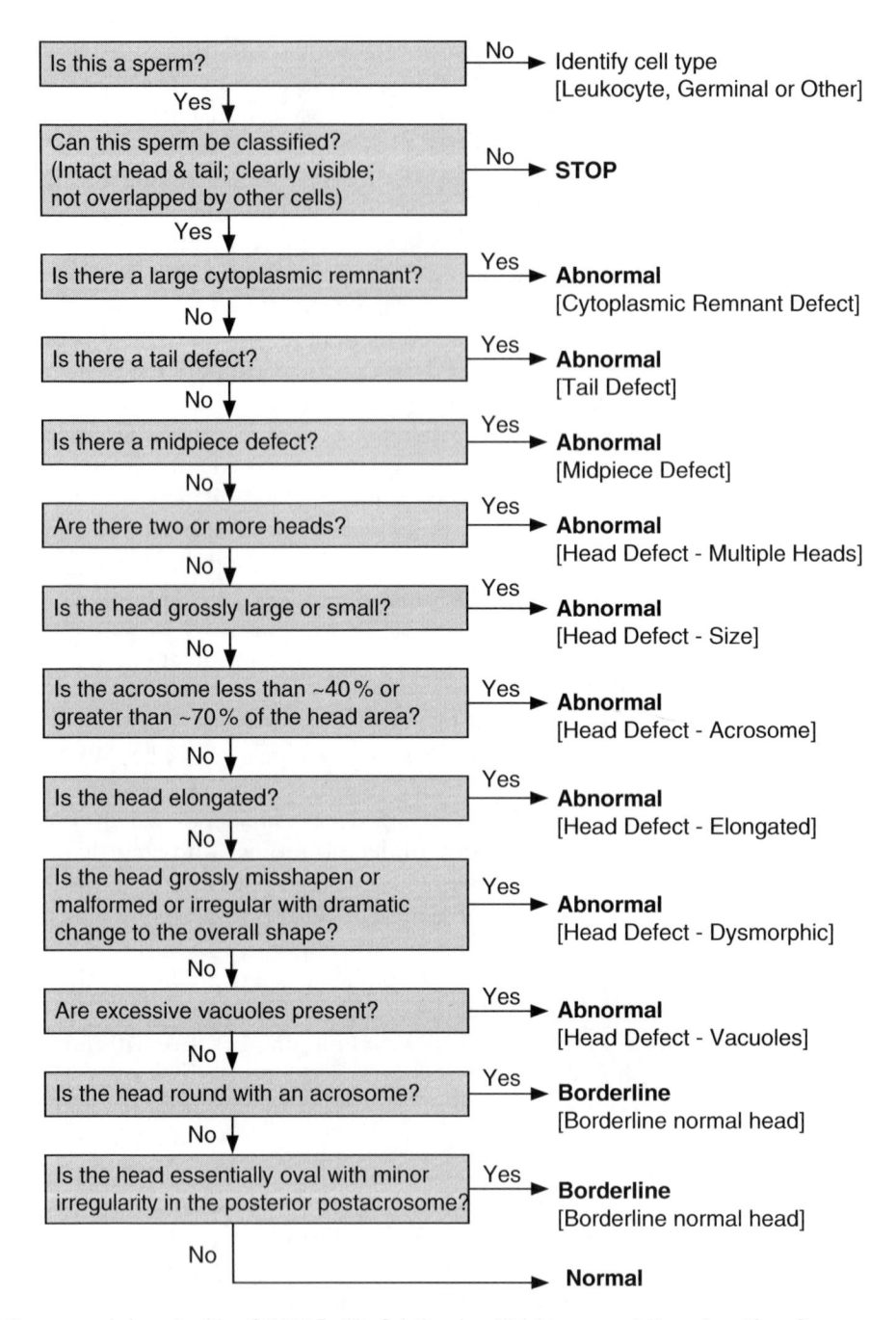

Fig. 1. Sperm morphology algorithm © 2011 Fertility Solutions Inc. All rights reserved. Reproduced from Sperm morphology: a rational approach to classification with permission of Fertility Solutions Inc.

(9) and Kruger (23), from which Strict or traditional percent normal forms can be calculated (24). The Morphology Algorithm provides a framework and method for sperm classification that is rational, repeatable, and defined (24).

2. Materials

1. Glass microscope slides, pre-cleaned, premium quality (Superfrost®).
2. Spray cytology fixative (e.g., PROTOCOL®).
3. Modified Papanicolaou (PAP) stain (www.fertilitysolutions.com).
4. Limonene clearing agent (MasterClear™).
5. Clinical grade microscope with high-wattage Kohler illumination and high quality, low numeric aperture 100× oil brightfield objective for analysis, and low power (10× or 20× objective for locating focal plane and analytic areas of smear).
6. Immersion oil.

3. Methods

3.1. Semen Smear Preparation and Staining

Sperm morphology analysis relies on the quality of the semen smear and its stain. As soon as the semen is smeared, it should be sprayed while wet using a spray cytology fixative (e.g., PROTOCOL®) to prevent artifacts created by air-drying. Modified PAP procedure yields the clearest identification of sperm anatomy. Commercial kits eliminate the lengthy preparation of staining reagents (e.g., www.fertilitysolutions.com). Modern clearing agents containing inorganic agents such as limonene (MasterClear™) eliminate toxic solvents such as xylene that required a fume hood. Most quick stains intended for blood smears mask the fine structural detail of sperm and stain the seminal plasma, leaving little contrast between sperm and background. In our experience, prestained slides intended for hematology do not provide adequate quality for sperm morphology and are difficult to analyze. The time spent applying good-quality PAP stain typically lessens analytic time by providing a smear that is easier to interpret.

3.2. Microscopy

Compared to other body fluid cells, sperm are small and are best observed with bright field microscopic examination using a high-quality, low numeric aperture 100× oil objective with high-wattage illumination. The quality of the light and the objective can affect the appearance of the sperm, especially when optimal staining methods are not used. With inferior optics, the details of sperm anatomy may be difficult to see, especially in the midpiece and tail. Köhler illumination provides high contrast for evaluating sperm morphology by properly aligning the image planes and defocusing the light waves. Poor alignment can cause blurred, washed out images and problems focusing the image plane. Numerous online tutorials illustrate the technique. Thorough objective cleaning after daily use prevents oil accumulation. Annual professional cleaning of the entire optical path is highly recommended.

3.3. Classification Algorithm General Approach

The Morphology Algorithm is a defined sequence of 12 dichotomous questions that identify whether specific visual attributes of the spermatozoon are present or absent. If the specific attribute is not present, the analysis proceeds to the next question. The Morphology Algorithm evaluates readily discernable sperm anatomy in the following order: cytoplasmic remnants, tail, midpiece, and head. The head classifiers are positioned at the end of the pathway because sperm heads exhibit considerable pleiomorphism, often subtly (25). After abnormal forms are eliminated, the attribute classifiers for borderline normal forms are evaluated. Through this process of elimination, a spermatozoon that does not have the describable visual attributes of abnormal or borderline normal forms is classified as normal.

Separate categorization of borderline forms has several distinct advantages. First, the analyst can calculate percent normal forms for Strict and traditional schemes at the same time. In the Strict scheme, the borderline forms are added to the abnormal sperm; in traditional schemes, they are combined with the normal forms. Second, the borderline category conveys the classification uncertainty that inherently exists in subjective morphology analysis, readily apparent not only to the analyst, but also to anyone who must interpret the results. Since it is unknown whether borderline normal sperm are functionally normal or abnormal, separate classification is prudent.

3.4. Algorithm Sequence

Figure 1 shows the algorithm sequence for the most basic differential analysis of abnormal, borderline, and normal. Specific defect classifications for a teratozoospermia index or complete differential are italicized below and shown in brackets in the Figure.

Two essential queries begin the analysis.

Is this a sperm? If not a sperm, identify the cell type, e.g., polymorphonuclear (PMN) leukocyte, immature germ cell, epithelial cell, debris, etc. (see Note 1).

Can this sperm be classified? Classify only intact spermatozoa with complete head, midpiece, and tail. All regions of the cell must be clearly visible and not obscured by other cells or debris (see Note 2).

Next, classify each spermatozoon using the following queries:

1. Is a large cytoplasmic remnant present? If the quantity of cytoplasm is greater than one-third of the area of a normal sperm head, classify as abnormal (*cytoplasmic remnant*) (see Note 3).
2. Is there a tail defect? Tail defects include broken, coiled, hairpin, irregular width, multiple, or short tails (see Note 4). If present, classify the cell as abnormal (*tail defect*).

3. Is there a midpiece defect? Midpiece defects include bent necks or midpieces, and abnormally thick, irregular, abnormally thin, or abaxial midpieces (see Note 5). If present, classify as abnormal (*midpiece defect*).

4. Are there two or more heads? Sperm with two or more heads are usually joined at the neck or midpiece. If present, classify as abnormal (*head defect—multiple head*).

5. Is the head grossly large or small? The head must be at least two times or less than half the size of a normal head to classify this abnormality (see Note 6). If yes, classify as abnormal (*head defect—size*).

6. Is the acrosome less than ~40% or greater than ~70% of the head? This category includes missing, very small, or very large acrosomes (see Note 7). If present, classify as abnormal (*head defect—acrosome*).

7. Is the head elongated? The length-to-width ratio is significantly greater than 2:1, as seen in tapering sperm (cigar-shaped or dumbbell forms) and in pyriform sperm where the postacrosome is elongated and narrows toward the neck (see Note 8). If present, classify as abnormal (*head defect—elongated*).

8. Is the head misshapen or malformed, or does it have an irregularity that dramatically changes the overall shape? This category includes bizarre head shapes, or heads with multiple or severe irregularities, such as large indentations, perimeter roughness, ruffling, or waviness that significantly change the shape. Some forms commonly seen resemble acorns, balloons, bullets, dumbbells, mushrooms, etc. Older schemes call these forms amorphous. Since they obviously have a shape a more appropriate term is dysmorphic (see Note 9). If present, classify as abnormal (*head defect—dysmorphic*).

9. Are excessive vacuoles present? One or more vacuoles that occupy more than 20% of the head area or a large, prominent vacuole in the postacrosomal region are abnormal (see Note 10). If present, classify the cell as abnormal (*head defect—vacuoles*).

10. Is the head essentially round with an acrosome present? Does the sperm head have an overall round shape (i.e., have a length-to-width ratio of 1:1) and an acrosome of normal size and shape? (see Note 11). If yes, classify as a borderline normal form (*borderline normal head*).

11. Is the head essentially oval with minor irregularity in the posterior postacrosome? Included in this group are three variants: the posterior postacrosome segment has straight sides that narrow to a point like a V (see Note 12), ends in a pinched projection (see Note 13), or has minor contour irregularity that does not dramatically change the shape of the head from oval (see Note 14). If any of these are present, classify the cell as a borderline normal form (*borderline normal head*).

12. *Conclusion*: If no abnormal or borderline normal forms were identified after the last query, classify the sperm morphology as normal (see Note 15) (*normal head, midpiece, tail*).

3.5. Calculations

A minimum of 200 sperm should be analyzed. For a simple differential, total the sperm in each category of normal, borderline normal, and abnormal. Calculate the percentage of each type by dividing by the number of spermatozoa analyzed. Calculations for more complex differentials are described in the WHO5 in Chapter 3 (5).

4. Notes

1. Typically the PMN leukocytes and immature germ cells are tallied separately as they are observed within the analysis and each is expressed as number per 100 sperm. If the total sperm count is known, the number in the ejaculate can be calculated.

2. Avoid the temptation to classify a partially obscured sperm with an obvious abnormality to prevent bias toward abnormal scoring. Since the normal sperm cannot be classified unless all anatomy is visible, proportionally more abnormal sperm can be classified if partially visible abnormal forms are included in the differential. For example, when the sperm tail looks normal but the head is obscured, the cell cannot be analyzed in its entirety; however, when the tail is broken but the head is obscured, the sperm still could be tallied as abnormal. Categorizing only those sperm that can be seen in their entirety constitutes the best practice.

3. The WHO5 changed the standard nomenclature of cytoplasmic droplet to excess residual cytoplasm (ERC), but a better term for abnormally large cytoplasm is cytoplasmic remnant (12). We disagree that cytoplasmic remnants disappear during staining. When a smear has been appropriately sprayed with fixative when wet, cytoplasmic remnants are readily identified in a PAP-stained preparation.

4. Distinguish normal tails that are curved or looped from abnormal tails that are either tightly coiled (often a sign of hypo-osmotic stress) or have a hairpin bend, a tight 180° turn that folds the tail back toward the head. Classify a broken tail only when the break appears in the principal piece of the tail, not the neck or midpiece, with either an obvious physical separation or an angle greater than 90° with a sharp point.

5. We found the nomenclature for midpiece defects of bent or broken tails confusing. To reduce ambiguity, we have assigned the following definitions. When a sharp, greater than 90°, bend

is located in the neck area just below the head, we categorize this midpiece defect as a bent neck. When the sharp bend is located anywhere else on the midpiece or at the junction of the midpiece with the tail, the classification is a bent midpiece. Thick midpieces should be clearly wider than twice the tail width. A common form gradually widens until it appears to cup the posterior postacrosome, in a shape similar to a golf tee. Abnormally thin midpieces are the result of the mitochondrial sheath loosening (10). Once loosened, the sheath can either slide up or down giving a "fallen sock" appearance, easily misclassified as a small cytoplasmic remnant, or it can come off completely exposing the midpiece, that is thinner than the width of the tail. If the attachment of the midpiece is not at the end of the central axis of the head, the midpiece is abaxial.

6. Size of sperm is influenced by fixation and type of stain. Measurements can be found in WHO5 (5). Using the classification rule described here allows for these differences and the variation in measurement. Most analysts cannot discern a micron difference and our rule reduces or removes the need to measure each cell.

7. Acrosomes >70% of the head area are rarely observed but are considered abnormal in almost all schemes. Before classifying an acrosome as too small, consider that angle, orientation, and staining have a large impact on the apparent acrosome size. Detecting minor differences in acrosome size can be difficult, but focusing through multiple planes reveals the actual size. This query also identifies round-headed sperm with no visible acrosome. When most of the sperm have this defect, a condition known as globozoospermia is present.

8. The posterior postacrosome of a pyriform head has acute narrowing that can make evaluation of the length-to-width ratio difficult. To determine if the ratio is greater than 2:1, compare the head length to the equatorial width. Pyriform sperm often resemble ice cream cones with the extended "cone"-shaped base.

9. Do not include minor variation in head shape here. This category of dysmorphic head shapes only includes obviously gross, major malformations.

10. We disagree with the Kruger and WHO5 definition that more than two small acrosomal vacuoles is an abnormality (5, 23) and instead, use previous definitions from WHO3, WHO4, and Menkveld (3, 4, 9). Prominent, solitary postacrosome vacuoles as described in WHO5 (5) and Kruger (23) also are abnormal since these appear to have a negative impact on embryo development (26). Multiple small vacuoles observed in most or all sperm on a smear often are artifacts induced by

inadequate fixation and sample aging. Do not misidentify indentation along the edge of the head as a vacuole.

11. Round-headed sperm with no acrosome and sperm with acrosomes of abnormal shape and size should have been identified in previous queries.

12. Although the posterior postacrosome segment narrows, it is not elongated like the pyriform shape. The sides of the V must be straight and lack the curvature associated with the perimeter of a U, oval, or egg shape. The straight sides form an angle between 60° and 135°. If there is any curvature in the postacrosomal segment, the head should not be classified as a borderline normal form. There should be no bulging at the equatorial region.

13. The pinched projection at the posterior post acrosome resembles the keyboard symbol "}" for curly bracket. This form should be distinguished from grossly abnormal dysmorphic shapes that have a very flat posterior postacrosome that resemble a gardening spade.

14. Do not overuse this category for sperm that are not perfectly oval. If the contour of the posterior postacrosome is slightly irregular with either minor indentation or a slightly ruffled, ridged, or wavy appearance, but the overall head shape is essentially oval, the sperm fits into this category. Focusing at different depths helps distinguish a slight indentation caused by smear preparation from a vacuole viewed laterally along the edge.

15. Descriptive sperm morphology must be interpreted with caution because cells that are not abnormal in appearance may or may not have normal function (12).

References

1. World Health Organisation (1980) Laboratory manual for the examination of human semen and sperm-cervical mucus interactions. Press Concern, Singapore
2. World Health Organisation (1987) Laboratory manual for the examination of human semen and sperm-cervical mucus interactions, 2nd edn. The Press Syndicate of the University of Cambridge, Cambridge, UK
3. World Health Organisation (1992) Laboratory manual for the examination of human semen and sperm-cervical mucus interactions, 3rd edn. Cambridge University Press, Cambridge, UK
4. World Health Organisation (1999) Laboratory manual for the examination of human semen and sperm-cervical mucus interactions, 4th edn. Cambridge University Press, Cambridge, UK
5. World Health Organisation (2010) Laboratory manual for the examination and processing of human semen, 5th edn. Cambridge University Press, Cambridge, UK
6. Mortimer D, Menkveld R (2001) Sperm morphology assessment—historical perspectives and current opinions. J Androl 22:192–205
7. Eliasson R (2003) Basic semen analysis. In: Matson P (ed) Current topics in andrology. Ladybrook Publishing, Perth, pp 35–89
8. Handelsman DJ, Cooper TG (2010) Afterword to semen analysis in 21st century medicine special issue. Asian J Androl 12:118–123
9. Menkveld R et al (1990) The evaluation of morphological characteristics of human spermatozoa according to stricter criteria. Hum Reprod 5:586–592

10. Menkveld R et al (1991) Atlas of human sperm morphology. Williams and Wilkins, Baltimore, MD
11. Kruger TF et al (1988) Predictive value of abnormal sperm morphology in in vitro fertilization. Fertil Steril 49:112–117
12. Amann RP (2010) Tests to measure the quality of spermatozoa at spermiation. Asian J Androl 12:71–78
13. Menkveld R (2010) Clinical significance of the low normal sperm morphology value as proposed in the fifth edition of the WHO laboratory manual for the examination and processing of human semen. Asian J Androl 12:47–58
14. Brazil C (2010) Practical semen analysis: from A to Z. Asian J Androl 12:14–20
15. Horte A et al (2001) Reassessment of sperm morphology of archival semen smears from the period 1980–1994. Int J Androl 24:120–124
16. Hughes PM et al (2009) Sperm morphology and intrauterine insemination (IUI) outcome then and now: when morphology mattered. Fertil Steril 92(3):S73
17. Morbeck DE et al (2011) Sperm morphology: classification drift over time and clinical implications. Fertil Steril 96:1350–1354
18. Prinosilova P et al (2009) Selectivity of hyaluronic acid binding for spermatozoa with normal Tygerberg strict morphology. Reprod Biomed Online 18:177–183
19. Kinzer DR et al (1998) Sperm morphology analysis problems as demonstrated by proficiency testing. J Androl 19:54
20. Keel BA et al (2000) Results of the American Association of Bioanalysts national proficiency testing programme in andrology. Hum Reprod 15:680–686
21. Cooper TG et al (2010) World Health Organization reference values for human semen characteristics. Hum Reprod Update 16: 231–245
22. Winston J (2009) Describing species. Columbia University Press, New York, NY
23. Kruger TF, Franken DR (2004) Atlas of sperm morphology. Taylor and Francis, London
24. Rothmann SA et al (2012) Sperm morphology: a rational method for classification. Fertility Solutions, Cleveland, OH
25. Katz DF (1991) Human sperm as biomarkers of toxic risk and reproductive health. J NIH Res 3:63–67
26. Vanderzwalmen P et al (2008) Blastocyst development after sperm selection at high magnification is associated with size and number of nuclear vacuoles. Reprod Biomed Online 17: 617–627

Chapter 5

Sperm Morphology Assessment Using Strict (Tygerberg) Criteria

Roelof Menkveld

Abstract

Although sperm morphology evaluation is one of the most important aspects of the semen analysis if done correctly and accurately, a trend is developing in which many laboratories or clinicians no longer regard sperm morphology as relevant due to the very low normal reference value of only 4% morphological normal spermatozoa given in the newest (2010) WHO-5 semen analysis manual. However, to maintain its relevance, sperm morphology evaluation, like the rest of the standard semen analysis, should be performed according to well-defined procedures. If performed correctly and according to high standards, morphology data are of high predictive value for male fertility potential and are critical in selecting the assisted reproductive techniques (ART) treatment. With the new low normal value it is becoming even more important that not only percentage normal and abnormal be reported but also that an in-depth report be given on the types of abnormalities present which will be important in selecting the type of clinical procedure to be adopted in ART, being, IUI, IVF, or ICSI. The methods to properly evaluate human sperm morphology are described in this chapter.

Key words: Semen analysis, Sperm morphology

1. Introduction

Sperm morphology is one of the routine basic semen parameters to be evaluated according to the WHO manual guidelines (1) and if done correctly is of strong prognostic value (2). One of the most important aspects for performing sperm morphology evaluation is that it should be performed according to very strict guidelines with a clear definition for a morphologically normal spermatozoon. For this purpose the guidelines of the WHO manuals have adopted the strict criteria principles (3), partially in the 1992 edition (4) and in full in the 1999 edition (5) as well as in the 5th and latest edition (WHO-5) published in 2010 (1).

Accurate evaluation of sperm morphology requires adherence to the following principles which are discussed in detail in the current chapter:

1. Correct preparation of slides and smears.
2. Correct fixation of smears.
3. Correct staining of the smears.
4. Correct mounting of stained smears.
5. Correct microscopic equipment, optics, and setup.
6. Correct examination procedure of the semen smear.
7. Clear concept and definition for morphologically normal spermatozoa.
8. Correct identification of the four main abnormal spermatozoa classes.
9. Correct identification and calculation of the teratozoospermia index (TZI) and identification of sperm morphology patterns.
10. Correct use of a schedule (algorithm) for the identification of morphologically normal spermatozoa.

2. Materials

2.1. Preparation of Slides and Smears

1. Air-displacement pipette—0–20 µl.
2. Microscope slides with frosted ends.
3. 95% Ethanol.
4. Detergent solution.
5. Marking pencil.

2.2. Fixation of Smears

1. Staining dishes with tight fitting lids or Coplin jars.
2. Absolute ethanol or methanol.

2.3. Staining of Smears

1. Staining dishes.
2. Stainless steel staining racks with ten slots to hold 10 or 20 slides configured such that slides can be placed lengthwise in a horizontal position.
3. Distilled water.
4. Analytical grade ethanol: Prepare 95, 80, 70, and 50% ethanol with distilled water.
5. Harris hematoxylin.
6. 0.5% HCl: Add 1 ml concentrated HCl to 200 ml distilled water.
7. Scott's water: Add 20 g $MgSO_4 \cdot 7H_2O$ and 2 g $NaHCO_3$ to tap water.

8. Orange G6 solution.
9. Polychrome EA50 solution.
10. Absolute propanol.
11. Xylene solutions.
12. 1-Propanol and xylene solution: Mix 1 part 1-propanol with 1 part xylene.

2.4. Mounting of Stained Smears

1. Heating tray or low-temperature oven.
2. Tweezers (stainless steel or plastic).
3. Mountant such as DPX.

2.5. Microscopic Equipment, Optics, and Setup

1. High-quality microscope with 10×, 12.5×, or 15× magnification eyepieces and at least three objectives: A low-power magnification (LPM) objective of 10×, 15×, or 20×, a high-power magnification (HPM) objective of 40×, and a 100× oil immersion objective of the highest quality available such as planachromat objectives (see Notes 12 and 13 and WHO-5, page 235 for detail (1)).
2. Tally counter with at least five buttons for differential counts.
3. Lens immersion oil.
4. Lens cleaning paper.

3. Methods

3.1. Slide and Smear Preparation

1. Slides for the morphology smears must be clean and grease free. If they are not, the slides should be washed with a detergent solution, rinsed with clean water and then rinsed with alcohol, and dried.
2. Label the slide with patient or sample identification.
3. Place a drop of semen on the microscope slide immediately adjacent to the label (see Note 1).
4. Place a microscope slide in front of the semen drop, pull back to make contact with the semen drop, and move the slide forward drawing the drop behind (see Note 2).
5. Prepare smears in duplicate for backup purposes in case staining was not successful or for later reference purposes.

3.2. Fixation of Smears

1. As soon as the semen smears show a change in appearance, indicating that they are air dried, but not completely dried-out, place the smears in the container with the alcohol for fixation.
2. Fix slides for a minimum of 10 min (see Note 3).

3.3. Staining of Smears

There are many methods available for the staining of semen smears, which can be used for specific staining results or reasons. However, for routine semen analysis a modified Papanicolaou method is the only method recommended because it gives good differential staining of spermatozoa and round cells without causing too many alterations in sperm structures (form) and size (6).

1. Transfer slides from fixation solution or storage box to slots of staining racks.
2. Set up sequential staining jars containing the following solutions with one jar for each step (see Notes 4 and 5).
3. Dip slides ten times in 80% ethanol.
4. Dip slides ten times in 70% ethanol.
5. Dip slides ten times in 50% ethanol.
6. Dip slides ten times in distilled water.
7. Transfer slides to Harris Hematoxylin for 15 min.
8. Rinse slides in running water for 3–5 min (until slides are clear).
9. Dip slides two times in 0.5% HCl.
10. Rinse slides in running water for 5 min.
11. Transfer slides to Scott's water for 1 min.
12. Rinse slides in running water for 1–5 min.
13. Dip slides in 50% ethanol ten times.
14. Dip slides in 70% ethanol ten times.
15. Dip slides in 80% ethanol ten times.
16. Dip slides in 95% ethanol ten times.
17. Transfer slides to Orange G6 solution for 5 min.
18. Dip slides through two sequential washes of 95% ethanol five times each.
19. Transfer slides to polychrome EA50 solution for 5 min.
20. Dip slides through three sequential washes of 95% ethanol five times each.
21. Dip slides in 1-propanol five times.
22. Dip slides in 1-propanol:xylene solution five times.
23. Dip slides through two sequential washes of xylene ten times each (see Note 6).
24. Transfer slides to a third jar containing xylene for 10 min.
25. Mount the slides immediately after staining with an appropriate mounting medium like DPX (see Notes 7 and 8).

3.4. Mounting of Stained Smears

1. Apply a thin streak of mounting medium over the length of the coverslips to be used, start and ending about 5 mm from the edges of the coverslip (see Notes 9–11).

2. Take the slide with the stained semen smear directly out of the Xylene solution and place it with the semen smear side directly on the mounting medium. Allow the weight of the microscope slide to distribute the mounting medium evenly over the entire coverslip area.

3. With a tweezers maneuver the coverslip carefully to a central position. If any air bubbles are present carefully maneuver them to the side of coverslip until they disappear with the aid of the tweezers.

4. Allow slides to air-dry overnight or place them on a heating tray or low-temperature heating oven with moderate heat to accelerate the drying process.

3.5. Microscopic Equipment, Optics, and Setup

1. For optimal bright field microscope performance adjust the light path according to the Köhler illumination system.

2. For ease of operational procedure place the LPM objective in the middle of the HPM and oil immersion 100× objectives. This will allow for easy alternation between the different objectives, especially when screening the semen smears (see Notes 12 and 13).

3.6. Examination Procedure of the Semen Smear

1. Following microscope setup, scan the stained semen smear with the LPM objective to observe the spreading of the spermatozoa in the smear, the staining quality, and the presence of round cells.

2. If round cells are observed move to the HPM to enable better identification of the type of round cell.

3. Identify a suitable area under the LPM to perform the sperm morphology evaluation process. An ideal area should contain between 5 and 10 spermatozoa per oil immersion field.

4. When the right area is identified place a very small drop of immersion oil on the spot to be investigated, not larger than the light spot that can be observed on the objective glass (see Note 14), and then bring the oil immersion objective into place.

5. Assess at least 200 spermatozoa in a single series (see Notes 15 and 16).

6. Record the morphological appearance of each spermatozoon as normal or abnormal as discussed in Subheading 3.7 (see Note 17).

3.7. Concept or Definition for Morphologically Normal Spermatozoa

The definition for a morphologically normal spermatozoon is based on the modal form of spermatozoa seen on smears of cervical mucus obtained from the endocervical canal, after intercourse (3). For the evaluation of the morphological normality of a spermatozoon the whole spermatozoon must be considered. A morphologically normal spermatozoon is defined as follows:

1. The head of the spermatozoon must have an oval form with smooth contours.
2. A clearly visible and well-defined acrosome must be present exhibiting a homogeneous light-blue staining.
3. The tail should be apically inserted to the post-acrosomal end.
4. No abnormalities should be present at the neck insertion or within the neck/midpiece or of the tail.
5. No cytoplasmic residues may be present at the neck/midpiece region or on the tail.
6. The acrosome should cover about 30–60% of the anterior part of the sperm head.
7. A normal spermatozoon should measure 3–5 μm in length and 2–3 μm in width (see Note 18).
8. The midpiece should be about 1.5 times the length of a normal spermatozoon and about 1 μm thick.
9. The tail should be about 45–50 μm long and without any sharp bends.
10. For a spermatozoon to be considered normal all of the above aspects must be normal.
11. Borderline or slightly abnormal spermatozoa are considered as abnormal.

3.8. Identification of the Four Abnormal Sperm Morphology Classes

In the literature four main abnormal sperm morphology classes are: head abnormalities, neck/midpiece abnormalities, tail abnormalities, and the presence of cytoplasmic residues sometimes also called cytoplasmic droplets.

1.1. *Head abnormalities*: Head defects include abnormalities in size, form, and/or structure. Size abnormalities occur when the head is too large or too small based on the actual staining method used, but still presents with an overall oval form. Form defects are present when the spermatozoa do not present with the classical oval form. This category includes spermatozoa that are described as elongated forms, tapering forms, which according to Eliasson (1970) may be smaller or larger than the normal oval form (7), and the so-called pyri- or pear-shape forms, which are usually larger than the normal oval form. Spermatozoa with a V-shaped distal (post-acrosomal) end and which are not elongated are regarded as morphologically normal.

1.2. *Acrosome abnormalities*: Acrosomal defects include size defects, staining defects, and structural defects. Size defects occur when the acrosome is >60% or <30% of the normal sized head. Acrosomal defects can also include staining defects if the acrosomes do not show the homogeneous light-blue staining when stained according to the Papanicolaou method, or if the acrosome staining is only present in certain areas of the acrosome while other areas do not show any staining. Structural defects include the presence of large vacuoles or cysts (see Note 19).

1.3. *Duplicate heads*: Duplications generally occur when two heads are joined together at the neck/midpiece, but may also occur within any other part of the sperm structure. When a head duplication is present this is classified as the primarily head abnormality.

1.4. *Amorphous heads*: Amorphous head classification includes all other head abnormalities that cannot be classified under any of the named abnormal head categories (see Note 20).

2. *Neck/midpiece defects*: As specific neck defects per se are difficult to identify, neck and midpiece defects are combined as one defect type. Neck/midpiece defects include bent necks, where the neck/midpiece forms a definite angle with the sperm head, asymmetrical implantation of the neck/midpiece into posterior region of the sperm head, thickened neck/midpieces, asymmetric bent midpieces, and cases where the mitochondrial material has shifted to the neck or towards the principal tail region, giving the appearance of a very thin midpiece (see Note 21).

3. *Tail defects* include cases where a definite bend (some definitions specify >90°) is observed at any part of the principal tail piece (see Note 22) and shorter than normal tails (see Note 23). Other tail defects include duplicate tails, with two or more extruding from a single sperm head or midpiece, and coiled or irregular tails (see Note 24).

4. *Cytoplasmic residues*: Although no cytoplasmic material should be present, a small amount of cytoplasmic material of <30% of a normal sized sperm head can be regarded as normal (see Note 25).

3.9. Additional Structures and Cells in Semen Smear

1. *Non-spermatozoal structures and cells*: Loose heads are not counted as abnormal spermatozoa or included in the sperm morphology evaluation schedule. The presence of >20% loose heads should be recorded on the sperm morphology report form. Pinpoint heads are not included in the sperm morphology evaluation schedule. The presence of more that 20% pinpoint heads should also be reported on the sperm morphology report form (see Note 26).

2. *Round cells and other cells*: Round cells as seen on a semen smear can be of two sources: cells originating from the germinal epithelium or the so-called precursors, and inflammatory or white blood cells. Other cells that may be present include epithelial cells and cells originating from the male accessory organs such as the prostatic cells.

3.10. Identification and Calculation of TZI and Identification of Sperm Morphology Patterns

1. Due to the very low percentage of morphologically normal spermatozoa counted in today's setting it is becoming of increasing importance to identify the type of abnormality present and not just to report normal or abnormal. Two criteria can be used for this purpose: the TZI or specific abnormal sperm morphology patterns (8).

2. *TZI*: Each morphologically abnormal spermatozoon can have from 1–4 types of abnormality classes as described above, in any number or combination. To reflect this, the TZI was introduced. If the spermatozoon is normal, only press the normal button on the five-button tally. If the spermatozoon is abnormal, press the button indicating the abnormality. If a second abnormality is present keep the first abnormality button pressed down, and press the second button indicating the specific abnormality and release that button. If a third abnormality is present keep the first button pressed down, and press the button indicating the third abnormality. Repeat this procedure if a fourth abnormality is present. Continue in this manner until 200 spermatozoa have been scored. Add all the numbers of the four abnormal classes together and divide by 200 minus the number of normal heads scored. The result will be the TZI. The numerical value of TZI should always be between 1.0 and 4.0.

3. *Sperm morphology patterns*: The use of sperm morphology patterns is a more clinical approach, as the primary type of sperm head abnormality present can be identified and a more clinical decision can be made to direct assisted reproductive technique (ART) treatments. Some morphological abnormalities may be due to environmental factors, which may be reversible with lifestyle changes, and other abnormalities are likely due to genetically determined factors. Abnormalities that may arise due to lifestyle factors include sperm elongation or tapered forms due to male genital infections, excessive exercise or mental stress, and large headed spermatozoa, which can be caused by medication such as in the treatment of ulcerative colitis or Crohn's disease with Sulfasalazine (2). Genetically determined sperm abnormalities include small headed spermatozoa with small acrosomes, and sperm sterilizing defects which are more extreme cases of genetic defects. The best characterized are short tail syndrome and globozoospermia defects, although other types have also been described (8). Short tail syndrome is characterized by short thick tails and dysplasia of the fibrous sheath. Globozoospermia is characterized by the absence of the acrosome and small round sperm heads. Men with short tail syndrome and complete globozoospermia are considered sterile as far as in vivo conception is concerned. These conditions should be treated with ICSI only. It is therefore important that clear notes of these defects be made on the sperm morphology report.

3.11. Algorithm for the Identification of Morphologically Normal Spermatozoa

The following simple algorithm can be used to categorize the spermatozoa as morphologically normal or abnormal:

1. Is the sperm an oval form with smooth contours? If no = abnormal; if yes—continue.
2. Is the sperm size normal? If no = abnormal; if yes—continue.
3. Is an acrosome visible? If no = abnormal; if yes—continue.
4. Is the acrosome normal in size? If no = abnormal; if yes—continue.
5. Is the tail inserted correctly? If no = abnormal; if yes—continue.
6. Is there another neck/midpiece defect? If yes = abnormal; if no—continue.
7. Is there a tail defect? If yes = abnormal; if no—continue.
8. Is there a cytoplasmic residue present? If yes = abnormal; if no—a morphological normal spermatozoon is present.

4. Notes

1. The size of the semen drop will depend on the estimated sperm concentration. Adjust the drop size such as to provide a thin smear with about 5–10 spermatozoa per oil immersion field. When the sperm concentration is high a small drop of ≥5 µl is used and with decreasing sperm concentration the drop size is increased to a maximum of 15 µl. With drops lager than 15 µl there is a strong possibility that part of the smear may dislodge from the slide during fixation or staining. With low sperm concentrations an aliquot of the sample can be concentrated by centrifugation and the pellet resuspended in part of the supernatant so as to increase the sperm concentration.

2. By altering the angle between the two slides and controlling the speed of movement of the slide with which the feathering is performed the thickness of the smear can also be controlled.

3. Smears can be kept in fixative for a prolonged period of time, until staining, or they can be stored after air drying for later use; however best results are obtained if smears are stained immediately after removal from the fixative.

4. It is important to make sure that staining dishes are adequately filled so that slides are completely submerged under the staining solutions.

5. Staining solutions should be replaced regularly, especially the Xylene solutions. Hematoxylin solution should be filtered from time to time if not replaced completely.

6. Xylene is toxic and is prohibited in many laboratories/countries. It can be replaced by other products such as Masterclear™ (American MasterTech, California, USA), propanol, or Neoclear® (Merck, Germany) (9). In cases where substitutes for xylene are used, complementing ethanol mounting mediums are available such as Clearmount™ (American MasterTech, California, USA).

7. When semen material dislodges from the slide, the smears were probably made too "thick" or the running water flow was too strong.

8. It is important to know the reasons for the different staining steps and the effect of every stain and other solutions employed through the staining procedure so that adjustments can be made to obtain the best staining effects possible (see WHO-5 (1), Björndahl et al. (9), and Kvist and Björndahl (10)).

9. It is preferable to use coverslips of 50×24 mm, as this will allow the coverslip to be placed some distance from non-frosted end of the slide. This then will allow two slides to be placed back to back in one slot (opening) of a slide-storing box, allowing double the normal amount of slides being stored in a slide-storing box.

10. Make sure that coverslips are clean. If not, clean them by washing with detergent and rinse with ethanol.

11. Use as little mountant as possible, as thickly mounted coverslips will negatively affect the optics of the microscope, especially with the 40× but also with the 100× objectives.

12. It is advisable that one of the eyepieces can be exchanged for an eyepiece with a built-in micrometer in order to make sperm head measurements after calibration with the use of a stage micrometer.

13. The ideal total magnification for the sperm morphology evaluation is a total magnification of 1,200×.

14. To place the drop of oil on the microscope slide, do not remove the slide from the stage but turn the nosepiece so that the light spot on the microscope slide can be seen halfway between the LPM and oil immersion objectives.

15. The sperm morphology evaluations should ideally be performed in more than one area to improve the accuracy of the evaluation as experience has shown that in some cases nonrandom distribution of normal and abnormal spermatozoa may be present, especially when evaluations are performed on the edges of the microscope slides.

16. According to the WHO-5 (2010) manual, evaluating one series of 200 spermatozoa is better than evaluating 100 spermatozoa twice (1). To increase accuracy two times 200

spermatozoa can be evaluated, but this is not always possible due to time restraints. However, the WHO-5 (2010) manual also recommends that in cases where the treatment of the patients/couple crucially depends on the percentage of morphologically normal spermatozoa present in the sample, 200 spermatozoa should be assessed twice (1).

17. Depending on the type of evaluation to be performed use can be made of the extra buttons to record the type of sperm abnormality present. If the TZI is to be calculated the class of abnormalities can be recorded, or if sperm pattern abnormalities are to be recorded the type of head abnormalities can be recorded. An abnormal spermatozoon can have between one and four abnormalities (see Subheading 3.7). If sperm abnormality patterns are to be recorded the button for an abnormal spermatozoon is pressed and kept down and the type of head abnormality is recorded by pressing down one of the other buttons indicating a specific abnormality as well. In this case only once as there can only be one type of head abnormality present (see next session). If any uncertainty about the size of a spermatozoon exists, as too small but especially as too large, the head can be measured by means of the built-in micrometer in the eyepiece.

18. Spermatozoa measurements are staining specific. Measurements provided here are according to the Papanicolaou staining technique. The generally given measurements for a normal sized spermatozoon of between 3–5 μm in length and 2–3 μm in width are not evidence based and are probably too wide. New measurements provided in the WHO-5 as measured by a computerized sperm morphology analysis system (CAMA) are the following: median length 4.1 μm (95% CI 3.7–3.7 μm) and median width 2.8 μm (95% CI 2.5–3.2 μm) (1). Means (SD) for CAMA obtained by Maree et al. (6) are 4.28(0.27) μm for length and 2.65(0.19) μm for width; and dimensions published by Menkveld et al. (2) are 4.07(0.19) μm for length and 2.98(0.14) μm for width.

19. A vacuole is defined as a lighter stained area with well-defined round borders. A cyst is defined as a lighter stained area with well-defined round borders with an extruding appearance. When the extrusion is present at the tip of the acrosome this is called "a nipple" defect.

20. Spermatozoa with slight but with a definite alternation from the morphological normal form is considered as borderline and classifies as an abnormal head form.

21. The mitochondrial material present at the neck region with the presence of a thin midpiece should not be confused with excessive cytoplasmic material.

22. A bend between the end of the midpiece and beginning of the principal part of the tail is regarded as a tail abnormality and not a neck/midpiece abnormality. A bent tail is not to be confused with a curved tail.

23. Short tails are shorter tails than the normal length specified and not to be confused with abnormal tails due to the short tail syndrome.

24. Coiled tails are in most cases not due to artifacts resulting from smearing the slides, hypo-osmotic stress, or aging, but are usually due to definite abnormalities such as the DAG defect in bulls where the tails are enclosed in a membrane structure (9).

25. The presence of cytoplasmic material, sometimes called cytoplasmic droplets, are better described as cytoplasmic residues to distinguish them from the normal structures present during the last stages of spermiation (9). There is some controversy around this topic as according to some publications cytoplasmic droplets are present on spermatozoa seen in the seminal plasma. However, cytoplasmic material is seldom seen in stained semen smears and is usually associated with other sperm defects such as bent necks and elongated spermatozoa (9). The presence of cytoplasmic material on spermatozoa is associated with the excessive production of reactive oxygen species (9).

26. Pinpoint heads are not to be confused with small heads as they usually do not contain any DNA material or any part of the head structure.

References

1. World Health Organisation (2010) WHO laboratory manual for the examination and processing of human semen, 5th edn. World Health Organisation Press, Geneva
2. Menkveld R, Holleboom CAG, Rhemrev JPT (2011) Measurement and significance of sperm morphology. Asian J Androl 13(1):59–68
3. Menkveld R, Stander FSH, Kotze TJW, Kruger TF, van Zyl JA (1990) The evaluation of morphological characteristics of human. Speomatozoa according to stricter criteria. Hum Reprod 5(5):586–592
4. World Health Organization (1992) WHO laboratory manual for the examination of human semen and sperm-cervical mucus interaction, 3rd edn. Cambridge University Press, Cambridge
5. World Health Organization (1999) WHO laboratory manual for the examination of human semen and sperm-cervical mucus interaction, 4th edn. Cambridge University Press, Cambridge
6. Maree L, Du Plessis SS, Menkveld R, Van der Horst G (2010) Morphometric dimensions of the human sperm head depend on the staining method used. Hum Reprod 25(6): 1369–1382
7. Eliasson R (1971) Standards for investigation of human semen. Andrologie 3:49–64
8. Menkveld R (2010) Clinical significance of the low normal sperm morphology value as proposed in the 5th WHO laboratory manual for the examination and processing of Human Semen. Asian J Androl 12(1):47–58
9. Björndahl L, Mortimer D, Barratt CLR, Castilla JA, Menkveld R, Kvist U, Alvarez JG, Haugen TB (2010) A practical guide to basic laboratory andrology. Cambridge University Press, Cambridge
10. Kvist U, Björndahl L (2002) Manual on basic semen analysis. ESHRE Monographs Oxford University Press, Oxford

Chapter 6

Methods for Direct and Indirect Antisperm Antibody Testing

Hiroaki Shibahara and Junko Koriyama

Abstract

Antisperm antibodies (ASA) are one well-known cause of refractory infertility in both males and females. In females, a sperm immobilization test, which detects sperm-immobilizing antibodies indirectly in the patient's serum, requires complement for the reaction and thus seems to be a more specific immunological reaction. In males, an immunobead test or a mixed antiglobulin reaction test, which detects ASA directly on the sperm surface, is a screening test because of the nonspecific reaction.

Key words: Antisperm antibody, Infertility, Sperm immobilization test, Immunobead test, Mixed antiglobulin reaction test

1. Introduction

The female genital tract undergoes periodic inoculation with hundreds of millions of highly immunologically alien sperm introduced during sexual activity. However, only rarely do women develop an immune response to these cells. Similarly, the presence of sperm and sperm-associated antigens in the male reproductive tract only occasionally invokes autoimmune stimulation in infertile men. Several mechanisms have been postulated to account for the formation of antisperm antibodies (ASA), which are one well-documented cause of refractory infertility in both males (1) and females (2). ASA adversely affects sperm function in a number of ways, including inhibitory effects on sperm migration through cervical mucus (3, 4), uterine cavity, and fallopian tubes (5), or blocking effects on fertilization (6–11).

There are two ways to test for ASA in infertile couples. One is through assays that directly detect ASA on the sperm membrane. For this purpose, the mixed antiglobulin reaction (MAR) test and direct-immunobead test (D-IBT) are now recommended by the World Health Organization (WHO) (12). D-IBT is widely used as a screening test for ASA (13), and the inhibitory effects by ASA on sperm motion and fertilizing ability are evaluated to make a decision for the strategy of infertility treatment (1).

The other ASA tests involve assays that indirectly detect ASA in serum, cervical mucus, follicular fluid, or seminal plasma. For indirect ASA testing, serological tests are usually performed for clinical purposes. In females, a sperm immobilization test (SIT) (14), which detects sperm-immobilizing antibody indirectly in a patient's serum, requires complement for the reaction and thus seems to be a more specific immunological reaction.

Serological tests developed to detect and quantitate isoantibodies and autoantibodies to sperm may be categorized into three groups based on the nature of the antigen source: (a) "sperm extract" assays such as immunodiffusion or immunoelectrophoresis; (b) "fixed sperm" assays such as immunofluorescence, mixed agglutination tests, enzyme-linked immunoassays, and radioimmunoassay; and (c) "live sperm" assays such as macroagglutination, microagglutination, cytotoxicity, immobilization, or sperm/cervical mucus interaction tests. All of the assays can detect ASA; however, not all antibodies bound to sperm are necessarily relevant to infertility. The ASA that cause infertility exert some adverse effects on sperm biological functions necessary for the process of fertilization. Therefore, methods to detect the sperm-binding antibody can be used clinically as a first screen for infertile patients to determine whether they possess any ASA, but biological tests for ASA such as sperm agglutination (15), sperm immobilization (14, 16), or blocking of fertilization (7, 8) are necessary as secondary tests to determine whether antibodies are relevant to immunological infertility.

2. Materials

2.1. Mixed Antiglobulin Reaction Test

1. The SpermMar-Test (FertiPro N.V., Beernem, Belgium) is available for the detection of immunoglobulin G (IgG) and immunoglobulin A (IgA) (see Note 1).

2.2. Immunobead Test

1. Tyrode solution or Dulbecco's phosphate-buffered saline can be used as the buffer medium for the IBT. Both are available commercially from Gibco, Grand Island, New York, USA.

2. Dried, unsterile bovine serum albumin (BSA; see Note 2).

3. Tyrode solution containing 0.3% BSA (TBSA; prepare on the day of use): To prepare 300 ml of TBSA, dissolve 0.9 g of BSA in approximately 20 ml of Tyrode solution, then filter (0.45 µM Millipore) the solution into a flask, and make up to 300 ml with Tyrode solution. Warm the mixture to 30–37°C before use.

4. Tyrode solution containing 5% BSA (TBSA-5): Dissolve 5 g BSA in 100 ml Tyrode solution, filter through a 0.45 μM Millipore filter, then aliquot in 2–5 ml aliquots. Store aliquots below 30°C. Thaw and warm individual vials to 30–37°C before use.

5. The beads are initially in dry form. Reconstitute 50 mg immunobeads with 10 ml of plain Tyrode solution (pre-filtered, 0.45 μM) to obtain a 5 mg/ml working solution (see Note 3). After reconstitution, the beads can be used for up to 2 months if kept at 4°C (see Note 4).

2.3. Sperm Immobilization Test

1. Guinea pig serum (see Note 5).
2. Terasaki plates (Greiner, Frickenhausen, Germany).

3. Methods

3.1. Testing for ASA Coating of Sperm (Direct Method) by the MAR Test

Advantages of the MAR test include low cost, quick turnaround time, and sensitivity, but it is less informative than D-IBT (12). The MAR test employs a "bridging" antibody (anti-IgG or anti-IgA) to bind the antibody-coated beads with unwashed IgG- or IgA-bearing sperm. The direct MAR tests are performed by mixing fresh, untreated semen with latex beads or treated red blood cells coated with human IgG or IgA. A monospecific anti-human-IgG or anti-human-IgA is added to the suspensions. The binding or agglutination of sperm and particles indicates the presence of IgG or IgA antibodies in the sperm. Agglutination between beads indicates successful antibody–antigen recognition.

1. Mix the semen sample well.
2. Place two 3.5 μL aliquots of semen on separate microscope slides.
3. Prepare similar slides with ASA-positive semen and ASA-negative semen as controls in each direct test (see Note 6).
4. Add 3.5 μl of IgG-coated beads to each slide containing test and control semen, and mix by stirring with a pipette tip.
5. Add 3.5 μl of human IgG antiserum to each slide containing the semen–bead mixture, and mix by stirring with a pipette tip.
6. Cover the suspension with a coverslip (22 mm × 22 mm), and store the slides on a flat surface in a humid chamber at room temperature for 3 min (see Note 7).
7. After the 3-min incubation examine the wet preparation under phase-contrast at 200× or 400× magnification and repeat the examination after 10 min.
8. Repeat the procedure with IgA-coated beads and IgA antiserum.

3.2. Scoring the MAR Test

1. Evaluate the samples for motile sperm with beads attached. With increased time the agglutinates will become very large, and sperm motility will diminish significantly. Sperm not expressing antibodies will swim freely between the beads.
2. Determine the percentage of motile sperm with beads attached to them (see Note 8).
3. Score only motile sperm that have two or more latex particles attached as positive. Ignore tail-tip binding.
4. Evaluate a minimum of 200 motile sperm in each replicate.
5. Calculate the percentage of motile sperm with particles attached.
6. Record the class (IgG or IgA) and the binding site (head, midpiece, principal piece) of the latex particles to the sperm. Sperm displaying only tail-tip binding should not be counted.
7. The current threshold value for MAR test binding recommended by WHO is 50% (12).

3.3. Testing for ASA Coating of Sperm (Direct Method) by D-IBT

The IBT allows for the detection of membrane-bound antibodies indicative of local immunity to sperm antigens (13). The D-IBT assay utilizes beads coated with covalently bound rabbit anti-human immunoglobulins against IgG or IgA. These beads are mixed directly with washed sperm, and the binding of beads to motile sperm indicates the presence of IgG or IgA antibodies on the surface of sperm (Fig. 1). D-IBT is relatively simple to perform, and requires minimal equipment and commercially available reagents. While the assay is slightly more time-consuming than the MAR test, the initial wash step is advantageous in that it may remove possible masking components in seminal plasma that may otherwise result in a false negative result.

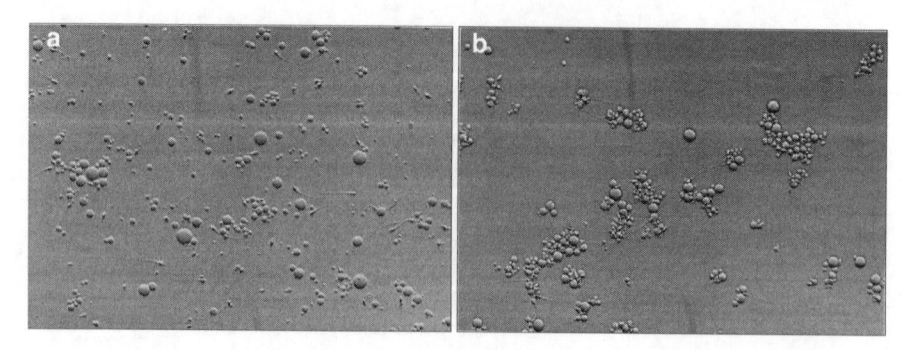

Fig. 1. Assessment of the results of the direct-immunobead test. Test examples that were (**a**) negative or (**b**) positive in the immunobead test. Immunobeads (IB) adhere to the motile sperm that have surface-bound antibodies (**b**). The percentage of motile sperm with surface antibodies is determined, the pattern of binding is noted, and the Ig class of these antibodies can be identified using different sets of IB.

1. Following semen liquefaction wash sperm in at least 2 volumes of a suitable medium, centrifuge for 10 min at 300×g, and pour off supernatant.

2. Resuspend the washed sperm pellet, and place 5 μl of the suspension on a microscope slide. Likewise prepare control slides with ASA-positive and ASA-negative sperm.

3. Add 5 μl of IgG immunobead suspension to each slide and mix by stirring with a pipette tip.

4. Place a 22 mm × 22 mm coverslip over the mixed droplet and store the slides on a flat surface in a humid chamber at room temperature for 3 min (see Note 7).

5. Assess binding within 10 min of preparing the slides, as immunobead binding decreases significantly with increased incubation time.

6. Examine the slides under phase-contrast at 200× or 400× magnification.

7. Repeat the procedure using anti-IgA immunobeads.

3.4. Scoring the D-IBT Test

1. Determine the percentage of motile sperm with beads attached to them (see Note 8).

2. Score only motile sperm that have one or more beads attached as positive (Fig. 1). Ignore tail-tip binding. Record the class (IgG or IgA) and the binding site (head, midpiece, principal piece) of the beads to the sperm. Sperm displaying only tail-tip binding should not be counted.

3. Evaluate a minimum of 200 motile sperm in each replicate.

4. As with the MAR test, the WHO recommends a threshold value of 50% motile sperm binding (12). However, we have been utilizing the cutoff value of 20%, demonstrated in the original manuscript by Bronson et al. (13) because this test is only a screening for immunologically infertile men and secondary tests, such as postcoital test and hemizona assay, are required to make a decision for appropriate treatments (1).

3.5. Testing for Antisperm Antibodies Secreted in Body Fluid (Indirect Method): Sperm Immobilization Test Preparation

The SIT was developed by Isojima et al. (14). Later, a modified assay for SIT was devised to minimize the volume of test samples through the use of Terasaki plates. For samples determined to be positive with the semiquantitative assay, a quantitative assay is performed to evaluate the exact value (SI_{50}) of sperm-immobilizing antibody.

1. Heat inactivate fresh or frozen-thawed patient serum at 56°C for 30 min.

2. Centrifuge serum at 1,350×g for 10 min. The supernatant is used as test specimens. Control sera are collected from women

without ASA and tested in advance for sperm immobilizing and sperm agglutinating activities in the presence or absence of complement.

3. Allow freshly ejaculated semen from healthy donors to liquefy for 20 min, then perform a count, and centrifuge ($1,350 \times g$, 5 min).

4. Wash with 5 ml of a medium, repeat centrifugation, and overlay the sedimented sperm with 2 ml of the medium. Let stand for 5–10 min. Pipette the supernatant, containing actively motile sperm into a small test tube, and adjust the sperm concentration to 40×10^6/ml with the medium before use.

5. Guinea pig serum is used as the source of complement. Test the serum from each guinea pig for toxicity to human sperm, and use only the serum lacking sperm toxicity for assay.

3.6. Semiquantitative Assay for Sperm Immobilization Test

1. To a Terasaki plate add 10 µl of heat-inactivated test serum, 1 µl of human sperm suspension (40×10^6/ml), and 2 µl of guinea pig serum (10 C'H$_{50}$ units) as complement, or heat-inactivated guinea pig serum as control to each well in a Terasaki plate under paraffin oil.

2. Incubate at 32°C for 60 min on a shaking mixer.

3. Also include two other controls, 10 µl of a standard serum containing sperm-immobilizing antibodies with complement and 10 µl of an SIT-negative serum, which does not cause any sperm agglutination and sperm immobilization, with complement for the SIT.

4. Count the sperm motility directly from the plate using a phase contrast microscope (200×). Count 50 sperm in each of the four microscopic fields for a total of 200 sperm, and calculate the percentage of motile sperm.

5. Calculate the percentage of motile sperm for the test specimens with complement (T%). Likewise, calculate the percentage of motile sperm for the control specimens with inactivated complement (C%). The ratio of C/T is designated as the sperm immobilization value (SIV).

6. When the SIV is 2 or more, the test serum is judged as positive for sperm-immobilizing antibody. As the antibody activity increases, the SIV becomes larger but it is only semiquantitative. For the sample positive in this semiquantitative assay, a quantitative assay is performed to evaluate the SI$_{50}$ titer.

3.7. Quantitative Assay for Sperm Immobilization Test

1. For the quantitative assay of sperm-immobilizing antibodies, serially dilute each test serum twofold with the control serum eight times.

2. Mix 10 μl of each diluted serum, 1 μl of sperm suspension (40×10^6/ml), and 2 μl of guinea pig serum (10 C'H$_{50}$ units), or heat-inactivated guinea pig serum as control, to each well of a Terasaki plate under paraffin oil, and incubate at 32°C for 60 min on a shaking mixer.

3. Determine the percentage of motile sperm under the microscope (200×).

4. Determine the percentage of motile sperm in each diluted patient's serum (T%) and in the control serum (C%). Calculate the antibody activity for sperm immobilization using the formula $C - T/C \times 100$.

5. Plot the value of the antibody activity against the dilutions of the test serum on a semilogarithmic chart to obtain a dose–response line with a sigmoid curve (Fig. 2).

6. Determine the dilution of the test serum at which the sigmoid curve crosses the 50% line for sperm immobilization activity ($C - T/C \times 100$). This value is designated as 50% sperm immobilization (SI) unit (SI$_{50}$ unit) (Fig. 2). If SI$_{50}$ units cannot be determined with the applied serum dilutions, the test serum

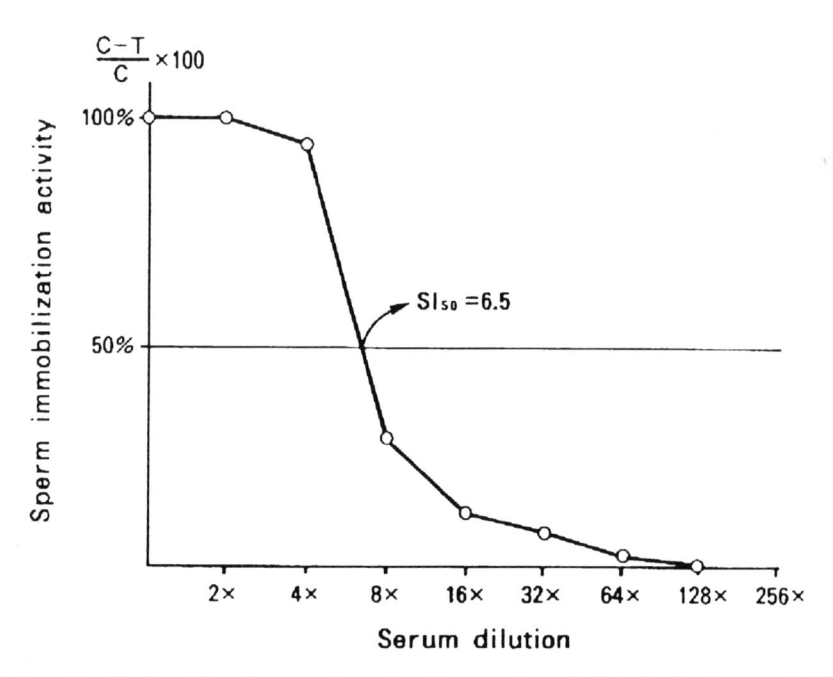

Fig. 2. Calculation of the SI$_{50}$ titer. When the percentage of motile sperm in each diluted patient's serum is T% and that in the control serum is C%, then the antibody activity for sperm immobilization is calculated by the formula of $C - T/C \times 100$. The value of the antibody activity is plotted against the dilutions of the test serum on a semilogarithmic chart to obtain a dose–response line with a sigmoid curve. The dilution of the test serum at which the sigmoid curve crosses the value of 50 for the antibody activity ($C - T/C \times 100$) is determined on the dose–response line and is designated as 50% sperm immobilization (SI) unit (SI$_{50}$ unit).

should be further diluted until SI_{50} units are obtained on the straight portion of the sigmoid curve.

7. To compare the sensitivity of the reaction system in different assays, use a human serum with relatively high sperm-immobilizing antibody activity as a standard serum and determine the SI_{50} units of the standard serum for each assay performed.

8. The value of antibody activity in various dilutions of proposed standard serum was plotted and SI_{50} titer of the serum was calculated as 6.5 units (Fig. 2).

9. Reference values for SIT have not been determined; however the SI_{50} titer may be clinically relevant (see Note 9).

4. Notes

1. The direct SpermMar test should be performed using untreated patient's sperm. One of the reasons is that some washing media contain human serum albumin (HSA). HSA tends to form a "shield" around the sperm, which hides the antibodies so that they cannot be detected by the test method. The other reason is that washing the sperm sample will remove most of the antibodies that might be present in the seminal plasma and might thus render false negative results.

2. The source or purity of the BSA is unlikely to affect the performance of the IBT. However, it is important to obtain the BSA in dried or powder form without preservatives and to filter the dissolved BSA before use. Preservatives such as sodium azide can affect sperm viability, and some bacterial contaminants of crude BSA products can cause false positive interactions between immunobeads and sperm.

3. The immunobead test is available for the detection of IgG and IgA from Irvine Scientific (Santa Ana, CA, USA) and consists of polyacrylamide beads of 2–10 µM diameter with covalently bound rabbit antibodies directed against human IgG and IgA.

4. Because the beads contain sodium azide as preservative they must be washed once in TBSA immediately prior to use. This can usually be done at the same time as the second sperm wash.

5. Rat and rabbit sera can also be used as complement for SIT but guinea pig serum is recommended as the best source of complement to obtain the highest sensitivity of reaction (18).

6. This semen can be obtained from men with and without ASA, respectively, or positive sperm can be produced by incubation in antibody-positive serum.

7. A humid chamber can easily be fashioned by placing water-saturated filter paper in a covered Petri dish. This is important to prevent slides from drying out.
 8. It may be difficult to distinguish non-progressively motile sperm that are close to beads from progressively motile sperm that have beads attached. Lightly tapping the coverslip with a small pipette tip can help to verify that the sperm in question are indeed bound to beads.
 9. Kobayashi et al. (17) proposed a strategy for the treatment of infertile women with sperm-immobilizing antibodies according to the patterns of variation of SI_{50} titers in individual patients. They divided patients with sperm-immobilizing antibodies into three groups according to their follow-up SI_{50} titers. Group A, which consisted of patients with continuously high SI_{50} titers (>10 units), did not conceive by ordinary or repeated IUI; however a satisfactory pregnancy rate was obtained by IVF and embryo transfer. Group B, in which the patients had intermediate SI_{50} titer patterns around 10, showed low rates of success with IUI. In Group C, the patients with continuously low SI_{50} titers (<10 units), conception by repeated or ordinary IUI was achieved, although the success rates were lower than those by IVF-ET.

References

1. Shibahara H et al (2005) Diagnosis and treatment of immunologically infertile males with anti-sperm antibodies. Reprod Med Biol 4:133–141
2. Shibahara H et al (2009) Diagnosis and treatment of immunologically infertile women with sperm-immobilizing antibodies in their sera. J Reprod Immunol 83:139–144
3. Koyama K et al (1979) Effects of antisperm antibodies on sperm migration through cervical mucus. Excerpta Med Int Congr Series 512:705–708
4. Shibahara H et al (2007) Relationship between level of serum sperm immobilizing antibody and its inhibitory effect on sperm migration through cervical mucus in immunologically infertile women. Am J Reprod Immunol 57:142–146
5. Shibahara H et al (1995) Sperm immobilizing antibodies interfere with sperm migration from the uterine cavity through the fallopian tubes. Am J Reprod Immunol 34:120–124
6. Shibahara H et al (1991) Inhibition of sperm-zona pellucida tight binding by sperm immobilizing antibodies as assessed by the hemizona assay (HZA). Nihon Sanka Fujinka Gakkai Zasshi 43:237–238
7. Shibahara H et al (1993) Effects of sperm-immobilizing antibodies on sperm-zona pellucida tight binding. Fertil Steril 60:533–539
8. Shibahara H et al (1996) Diversity of the blocking effects of antisperm antibodies on fertilization in human and mouse. Hum Reprod 11:2595–2599
9. Shibahara H et al (2003) Diversity of the inhibitory effects on fertilization by anti-sperm antibodies bound to the surface of ejaculated human sperm. Hum Reprod 18:1469–1473
10. Taneichi A et al (2002) Sperm immobilizing antibodies in the sera of infertile women cause low fertilization rates and poor embryo quality in vitro. Am J Reprod Immunol 47:46–51
11. Taneichi A et al (2003) Effects of sera from infertile women with sperm immobilizing antibodies on fertilization and embryo development in vitro in mice. Am J Reprod Immunol 50:146–151
12. World Health Organization (ed) (2010) WHO laboratory manual for the examination and processing of human semen, 5th edn. WHO, Switzerland

13. Bronson R et al (1982) Detection of sperm specific antibodies on the spermatozoa surface by immunobead binding. Arch Androl 9:61
14. Isojima S et al (1968) Immunologic analysis of sperm immobilizing factor found in the sera of women with unexplained sterility. Am J Obstet Gynecol 101:677–683
15. Franklin RR, Dukes CD (1964) Antispermatozoal antibody and unexplained infertility. Am J Obstet Gynecol 89:6–9
16. Isojima S, Koyama K (1974) Quantitative estimation of sperm immobilizing antibody in the sera of women with sterility of unknown etiology: the 50% sperm immobilization unit (SI_{50}). Excerpta Med Int Congr Series 370:10–15
17. Kobayashi S et al (1990) Correlation between quantitative antibody titers of sperm immobilizing antibodies and pregnancy rates by treatments. Fertil Steril 54:1107–1113
18. Isojima S, Koyama K (1989) Techniques for sperm immobilization test. Arch Androl 23:185–199

Chapter 7

Manual Methods for Sperm Motility Assessment

David Mortimer and Sharon T. Mortimer

Abstract

Progressive motility is a vital functional characteristic of ejaculated human spermatozoa that governs their ability to penetrate into, and migrate through, both cervical mucus and the oocyte vestments, and ultimately fertilize the oocyte. A detailed protocol, based on traditional manual/visual methods, is provided for performing an accurate four-category differential count including the reliable identification of rapid progressive (grade "*a*") spermatozoa—the most biologically, and hence clinically, important subpopulation. Thorough prior training and the use of a microscope fitted with a heated stage are both essential requirements for achieving accuracy and an acceptable uncertainty of measurement of no more than ±10%.

Key words: Sperm motility, Progressive motility, Temperature

1. Introduction

Progressive motility is a vital functional characteristic of ejaculated human spermatozoa that governs their ability to penetrate into, and migrate through, both cervical mucus and the oocyte vestments (cumulus–corona complex and zona pellucida) (1). The quality of sperm progression is of far greater importance in the prediction of functional competence than either sperm concentration or percentage motility (2). When using traditional manual/visual methods, assessment of sperm progression is best achieved using a four-category differential count (3, 4), allowing the identification of the proportion of rapid progressive (grade "*a*") spermatozoa, i.e., those with a progression velocity of ≥25 μm/s at 37°C. The importance of this is supported by evidence relating to the ability of more rapidly progressive spermatozoa to penetrate cervical mucus in vitro (5, 6), achieve in vivo conceptions (7, 8), and give higher clinical outcomes when using donor insemination (9), intrauterine insemination (IUI) (10), in vitro fertilization (IVF) (10, 11), and even intracytoplasmic sperm injection (ICSI) (12).

That the World Health Organization (WHO) has eliminated the differentiation of rapid and slow progressive motility in the latest edition of its laboratory manual "WHO5" (13) has been considered—based on the evidence cited above—an unjustified oversimplification which ignores the biologically and clinically important information on the quality of sperm progression (14–16). The WHO's reason for the simplification appears to have been based on the observation that poorly trained technicians cannot reliably distinguish between the two categories, although this should be obvious, since without proper training technicians cannot do any sperm assessments (3). Following proper training, four-category differential sperm motility counts can, indeed, be performed accurately and reproducibly (3, 17) (see Note 1), and the method described in this chapter is optimized—especially if a video monitor is used in conjunction with an acetate overlay—for this purpose (see Note 2).

The same method can also be used for washed (e.g., by density gradient centrifugation) spermatozoa that are suspended in either capacitating (bicarbonate-buffered) or non-capacitating (e.g., HEPES-buffered) culture medium. However, in this case the most rapid spermatozoa will swim at velocities far exceeding anything seen for ejaculated spermatozoa in seminal plasma, and some spermatozoa will likely show the hyperactivated pattern of movement (1), and these cells should be counted in a separate, fifth, category.

Finally, since sperm velocity is highly temperature dependent (18, 19), it is essential that sperm motility assessments be made as close as possible to 37°C in order to obtain the most physiologically relevant information on sperm functional potential.

2. Materials

1. *170 mM Saline*: Dissolve 0.993 g NaCl in 100 mL of sterile water for injection. Filter sterilize through a 0.22 μm Millipore Millex-GV filter into labeled sterile containers (borosilicate glass or medical grade polystyrene) and cap tightly. Store at +4°C for up to 6 months. Discard if there are any signs of precipitation or cloudiness.
2. *Sperm Medium*: This can be any bicarbonate-buffered culture medium designed to sustain human sperm metabolism and capacitation in vitro. It must be equilibrated to 37°C prior to use, and then used under an appropriate CO_2-enriched atmosphere (e.g., 6% CO_2-in-air for most such media containing 25 mEq/L bicarbonate ions when used at 37°C close to sea level) to maintain its pH.
3. *Sperm Buffer*: This is any HEPES-buffered medium designed to sustain sperm metabolism in vitro under an air atmosphere;

it will not support sperm capacitation. An equivalent medium buffered using MOPS is also acceptable. Sperm Buffer is used whenever spermatozoa need to be maintained outside a CO_2 incubator (e.g., during sperm washing procedures), or when they are to be prevented from capacitating in vitro. Ideally a Sperm Buffer should be matched in its general formulation to the Sperm Medium when the two are used in combination.

3. Methods

3.1. Specimen

1. Ensure the sample is thoroughly mixed. For a semen sample, swirl it around the specimen container for 10–20 s, for a washed sperm suspension pipette it up and down gently using a (sterile if need be) glass Pasteur pipette. In all cases, take great care to avoid frothing.
2. For semen samples, check for the completion of liquefaction. If the specimen is not fully liquefied (normal specimens should liquefy within 20 min after ejaculation) it can be returned to the incubator for a few minutes longer, but only within the time frame allowable according to any biological degradation or protocol timings. Incomplete liquefaction results in a heterogeneous sample consisting of gelatinous material in a fluid base. See Note 4 regarding unliquefied semen samples.
3. For semen samples, also note the viscosity of the semen, being careful not to confuse abnormalities of liquefaction with abnormal viscosity (see Note 3). Viscosity refers to the fluid nature of the whole specimen, which can be normal even though liquefaction might be incomplete.

3.2. Making a Wet Preparation

All slides or counting chambers prepared from semen or a washed specimen should be labeled to identify unambiguously the source of the sample, as well as any treatments or time points. Ideally, a specimen should be labeled with at least two unique identifiers. Various types of wet preparations can be made, according to the type of specimen and the intended purpose of the assessment (20). Spermatozoa in seminal plasma require a preparation depth of at least 10 µm for unimpeded motion, whereas washed spermatozoa in culture medium (Sperm Medium or Sperm Buffer) require at least 20 µm; hyperactivation assessments usually require at least 30 µm. Microscope slides and coverslips or various special counting chambers can be used (Fig. 1).

1. *Microscope slide*
   (a) Place 10 µL of semen on a labeled clean microscope slide and cover with a 22 × 22 mm #2 or #1½ thickness coverslip;

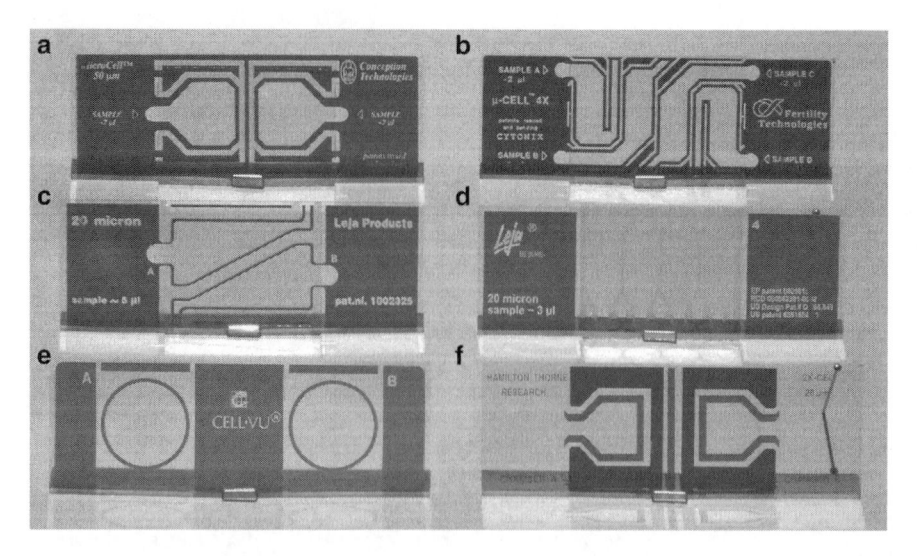

Fig. 1. Examples of fixed-depth chambers and slides. (**a**) 2-Chamber MicroCell; (**b**) 4-chamber MicroCell; (**c**) double Leja chamber; (**d**) quadruple Leja Standard Count chamber; (**e**) double Cell-Vu slide; (**f**) double 2X-CEL slide. Types **a–d** have fixed coverslips and are filled by capillary loading while types **e** and **f** have separate cover glasses and are filled by "drop loading"; see Subheading 3.2 in the text for more details.

take care to avoid trapping any air bubbles. This will give a preparation of approximately 20 µm depth.

(b) Examine the preparation as soon as any "flow" in the preparation has stopped. If this does not occur within 60 s then discard the slide and prepare another.

2. *Makler Chamber*

   This device was designed as a rapid, simple method for determining the concentration and motility of spermatozoa without the need for prior dilution (Sefi-Medical Instruments, Haifa, Israel) (21). It has a chamber depth of 10 µm and a ruled area of 1 mm² divided into 100 squares, hence the number of spermatozoa seen in 10 squares corresponds to the concentration in millions per mL. Although such chambers are convenient, they are prone to substantial errors when used in routine practice for sperm concentration determination and are not recommended for general semen analysis purposes except with very low concentration samples, e.g., $<5 \times 10^6$/mL (3, 4). The grid, however, does facilitate sperm motility assessments.

   (a) Since the Makler Chamber must not be over-filled, use only about 4.0 µL of semen or sperm suspension. The aliquot does not need to be dispensed exactly, so an air displacement pipette can be used.

   (b) To avoid creating small bubbles on top of the dispensed droplet (bubbles that are trapped under the cover glass often obscure the counting grid), draw up the specimen aliquot by depressing the pipetter plunger to the second ("blow out") stop so that additional sample is taken into the tip. Then, to dispense the 4.0 μL, depress the plunger only to the first stop.

(c) Gently lower the cover glass onto the four pillars of the Makler Chamber base unit and rotate it by up to a quarter turn to ensure that it is seated firmly on top of the pillars. When observed under suitable illumination, refraction patterns can be seen on top of the four pillars.

(d) Assess the preparation as described in Subheading 3.3 below.

(e) After use wash and dry the Makler Chamber and its cover glass carefully, and allow them to rewarm to 37°C before using them with another specimen.

3. *Fixed-depth chambers with fixed coverslips using "capillary loading"*

Although there is a problem of significant underestimation of sperm concentration with many chambers of this type, caused mainly by the Segre–Silberberg Effect (22), they can be used for assessments of sperm motility. Examples of these chambers include the MicroCell (Conception Technologies, San Diego, CA, USA) and Leja (www.leja.nl) chambers.

(a) Ensure that the chambers are pre-warmed to 37°C prior to use.

(b) Load the chamber with the required volume of semen as per the manufacturer's instructions for use; since the depth of the chamber controls the vertical dimension of the preparation an exact volume is not required, and hence an air displacement pipette can be used.

(c) Assess the preparation as described in Subheading 3.3 below.

4. *Fixed-depth chambers with non-fixed coverslips using "drop loading"*

Because these chambers (e.g., the 2X-CEL chamber, Hamilton-Thorne, Beverly, MA, USA) or the CellVU chamber (www.cellvu.com) are not filled by capillary flow they are not subject to the Segre–Silberberg Effect (22).

(a) Ensure that the chambers and coverslips are pre-warmed to 37°C prior to use.

(b) Place the volume of sperm suspension stated in the manufacturer's instructions in the center of each chamber area. To avoid air bubbles on the top of these drops use the same pipetting technique as described for the Makler Chamber (see 2(b), above).

(c) Carefully place the coverslip over the specimen area and allow the sperm suspension to spread across the chamber area. Turn the chamber upside down on a clean paper tissue and press down on the bottom surface of the slide to ensure complete spreading of the sperm suspension and establishment of the fixed depth of the preparation, which is determined by the thickness of the painted outlines on the slide surface. Use another paper tissue to avoid fingerprints on the slide surface, which could degrade the image quality.

(d) Assess the preparation as described in Subheading 3.3 below.

(e) Although intended for one-time disposable use, these chambers and coverslips can be reused in accordance with a due risk assessment (i.e., consideration of finger-stick injury resulting from broken coverslips or slides and necessary decontamination). To reuse, wash and dry the chamber slide and its coverslip carefully, and allow them to rewarm to 37°C before using them with another specimen.

5. *Hemocytometers*

Because these chambers have depth of at least 100 μm they are too deep to visualize all the freely swimming spermatozoa at the same time. Therefore, they must not be used for assessing sperm motility.

3.3. Assessing the Preparation

Phase contrast optics are essential.

1. In order to assess sperm progression correctly, the preparation must be kept as close to 37°C as possible. Slides, chambers, and coverslips or glasses (and pipette tips) must be kept at 37°C and the microscope fitted with a heated stage set to maintain the wet preparation at 37°C.

2. Examine the preparation under the microscope either:

   (a) By direct vision through the microscope using a 20× or 40× objective or

   (b) Using a 10× objective and a video camera (with intermediate magnification) to project the image onto a small monochrome video monitor (see Note 2).

   The method is highly recommended as it provides a simple calibration of the field of view, thereby permitting easy training and facilitating effective intra- and inter-observer quality control.

3. Quickly scan the preparation for any artifacts that will adversely affect the assessment, e.g., large air bubbles. If any such artifacts are present then discard the preparation and make another.

4. Qualitative sperm motility, i.e., the overall perceived quality of sperm progression seen in the preparation, can be assessed subjectively by grading the degree of forward progression exhibited by the largest proportion of the motile spermatozoa

(not just the progressive spermatozoa). This form of categorical assessment is more prone to observer variability and is not recommended for expert or research use; the following categories are most often used:

0 = Nonprogressive motility only (abnormal/pathological finding)

1 = Poor, non-directed progression (abnormal/pathological finding)

2 = Moderate progression

3 = Good progression (normal finding)

4 = Excellent progression (normal finding)

5. Quantitative sperm motility (differential count) is determined by counting the numbers of motile and immotile spermatozoa in several randomly selected fields that are not near the coverslip edge (where drying-out artifacts can affect sperm motility).

   (a) Count at least 200 spermatozoa sampled from at least five different fields of view and classify each spermatozoon according to the principles described in Fig. 2.

   (b) Within each field, first count the number of progressively motile spermatozoa. Count only those spermatozoa that are in the field at one moment in time (do not include tailless heads or "pinhead" forms). Progressive motility is defined as achieving a space gain of at least 5 µm/s

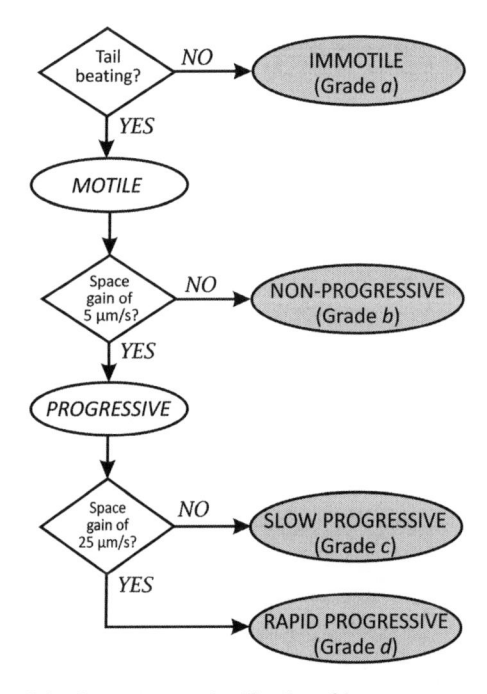

Fig. 2. Principles of the four-category classification of human sperm motility in semen (Redrawn from ref. 26).

(i.e., about one sperm head length per second). The cutoff between rapid progressive motility (grade *a* motility) and slow or sluggish progressive motility (grade *b* motility) is 25 µm/s progression; this is equivalent to approximately half a spermatozoon length of space gain per second.

Note: If the number of progressively motile spermatozoa in the entire field is too great for rapid visual counting then a small area of the field should be defined using an eyepiece graticule (see Note 5).

(c) After counting the progressive spermatozoa, count the nonprogressively motile and immotile spermatozoa (grades *c* and *d* motility, respectively) that are present within the same area. Nonprogressive motility is defined as having flagellar activity but with <5 µm space gain per second; immotility is defined as no evidence of any flagellar beating.

(d) Duplicate counting is essential to minimize sampling error. Therefore motility must be assessed on a second wet prep. Provided that there is no significant variation between the two counts they can be averaged. The acceptability of counts is determined using Table 1 to confirm adequately low variability between the two counts (4).

Note: A simpler routine means of verifying a discrepancy of <10% between replicate counts, either for motile cells (i.e., grades *a*+*b*+*c*) or immotile cells (grade *d*) for samples with <50% motility, is to check that the difference between the two percentages is <1/20th of their sum. While this method is quick and easy to apply it is more strict and will result in some additional assessments having to be repeated that are actually within the 95% confidence interval for the numbers of cells counted.

(e) Accept the assessment if the difference between the duplicates is less than the limit value, otherwise reject the data and prepare and assess two new preparations.

Note: It is recommended that a Sperm Motility Worksheet (form) be used to record the counts, low discrepancy validation, and calculations (Table 2).

3.4. Calculating the Results

1. Results are expressed as the percentages of rapid progressive (grade *a*), slow progressive (grade *b*), nonprogressive (grade *c*), and immotile spermatozoa (grade *d*).

2. Present the results as integers (whole numbers) only, ensuring that the four figures add up to 100%, with any correction for rounding-off errors being made in the largest class(es).

Note: For compliance with WHO5 (13), which considers only three categories of motility (i.e., progressive, nonprogressive,

Table 1
Limit values for the maximum difference between replicate assessments of sperm motility (based on ref. 4)

Average (%)	Limit
0	1
1	2
2–3	3
4–6	4
7–9	5
10–13	6
14–19	7
20–27	8
28–44	9
45–55	10
56–72	9
73–80	8
81–86	7
87–90	6
91–93	5
94–96	4
97–98	3
99	2
100	1

The values were calculated using a formula based on the binomial distribution required to determine asymmetrical confidence intervals for proportions (which have absolute minimum and maximum values of 0 and 100%). These limit values can only be used for duplicate counts of 200 spermatozoa (2×200); the table is not valid for replicate counts of 100 spermatozoa (i.e., 2×100). *Using the table*: Calculate the average percentage (rounded to an integer value) for the replicate counts as well as the difference between them. The difference between the two assessments must be less than or equal to the limit difference given in the right column of the row for that average percentage (left column), otherwise the assessments must be rejected

and immotile), the proportion of progressively motile spermatozoa is derived simply by adding grade *a* and grade *b* before the rounding-off stage.

3. Some labs also report a Motility Index that integrates the proportions with weighting for the grade(s) of progression (see Note 6).

Table 2
Example worksheet for recording sperm motility counts

#	Lab Ref	Wet prep	Counts (a, b, c, d, Sum)	Percentages (a%, b%, c%, d%, a+b, a+b+c)	Checks (A+B, A−B, OK?)	Results (means) (Rapid, Prog, Motile)
1		A / B				
			Comments:			Initials:
2		A / B				
			Comments:			Initials:
3		A / B				
			Comments:			Initials:
4		A / B				
			Comments:			Initials:
5		A / B				
			Comments:			Initials:
6		A / B				
			Comments:			Initials:
7		A / B				
			Comments:			Initials:
8		A / B				
			Comments:			Initials:
9		A / B				
			Comments:			Initials:
10		A / B				
			Comments:			Initials:

SPERM MOTILITY COUNTING WORKSHEET   Date: __/__/201__

3.5. Interpreting the Results

1. Semen samples: A normal result is ≥50% progressive spermatozoa, including ≥25% rapidly progressive (grade *a*) spermatozoa. An abnormal result is <25% progressive spermatozoa with <15% rapid progressive. Values between these two limits are considered to be of uncertain clinical significance.

2. Washed sperm populations: Following preparation by discontinuous density gradient centrifugation (or direct swim-up from liquefied semen) a sperm population should show ≥95% progressive motility. The presence of >15% nonprogressive or immotile spermatozoa is a poor preparation, likely due to either a technical error in the sperm washing technique or an abnormal response of the man's spermatozoa to the washing technique. In diagnostic terms the latter situation could reflect a problem with the man's spermatozoa that could contribute to subfertility, and in therapeutic terms could be taken as an indication of likely sperm dysfunction, perhaps precluding the use of IUI or IVF, necessitating the use of ICSI. In a research setting, one should question the validity of using such preparations unless they are known to come from subfertile men (*cf.* control or research donors).

4. Notes

1. *Quality control (QC) and quality assurance (QA)*
   Many laboratories—even andrology laboratories—continue to ignore the issue of QC in semen analysis, even though it is well established that internal standardization and QC is quite possible, although it depends heavily on proper operator training (3, 4, 17). Accurate and repeatable assessments of sperm motility can be achieved and their reproducibility confirmed by internal repeat evaluations of samples between operators, ideally from reference video recordings. At the least, all staff members in a laboratory should carry out "group assessments" of, for example, all samples received for analysis on 1 day each month as an internal QC exercise.

   In addition, External Quality Assurance Programs (EQAPs) distribute test specimens to participating laboratories on a regular basis. Probably the best known international EQA scheme is the one operated by the Special Interest Group in Andrology of the European Society of Human Reproduction and Embryology, see: www.eshre.eu/ESHRE/English/Specialty-Groups/SIG/Andrology/External-Quality-Control/page.aspx/104.

2. *Calibration*
   While this "manual/visual" method for assessing sperm motility does not require any special calibration when performed directly on the microscope, if a video monitor is used then a

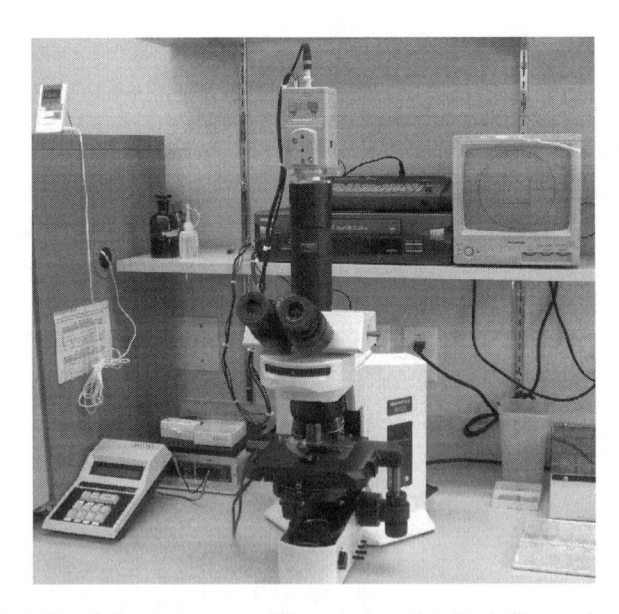

Fig. 3. Illustration of a human sperm motility analysis workstation built around an Olympus BX51 phase contrast microscope with an integral heated stage. Note the Minitüb HT-200 heated stage controller (with boxes of slides and coverslips on it) in the *lower left-hand corner*, behind the multichannel counter. A further stage warmer just visible in the *lower right-hand corner* is used for keeping fixed-depth 2X-CEL chambers warm prior to use. On the shelf behind the microscope next to the 9 in. monochrome CCTV monitor is a video recorder and titler. An acetate overlay is attached to the front of the monitor, the 25-μm-equivalent ruling can be seen within the circular field used for counting, as well as a central smaller box used with samples having a high sperm concentration.

ruled grid with spacing equivalent to 25 μm should be fixed to the screen to facilitate the discrimination of "rapid" spermatozoa from "slow progressive" spermatozoa (Fig. 3). This grid also acts as a reticle to subdivide the field of view and aid counting in cases where the sperm concentration is high.

For example, with the Olympus BX series microscopes, a video camera having a 1/3-in. CCD chip used in conjunction with a 10× objective, 5× camera ocular, and 1.25× setting on the magnification changer module provides an image well suited for assessing sperm motility. A camera with a 1/2-in. CCD will require less intermediate magnification.

3. *Dealing with viscous semen samples*

This remains an extremely difficult problem for which there is no perfect solution. Contrary to common opinion, the use of "needling" to reduce semen viscosity (forcible, often repeated, expulsion of the semen through an 18 G needle) causes significant damage to the spermatozoa and must not be used (23). For semen analysis purposes the best option with moderate to high viscosity samples is simply to do one's best. For extremely viscous samples, Sperm Buffer can be added and the sample mixed gently using a Pasteur pipette (note the volume of Sperm Buffer added so that the dilution factor can be calculated when determining sperm concentration).

4. *Dealing with unliquefied semen samples*

   If a semen sample fails to liquefy completely and a sperm motility assessment must be performed—and especially where the sample is to be prepared for ART—the semen can be liquefied enzymatically using chymotrypsin (24).

   Chymotrypsin can be purchased as glass ampoules containing 5 mg of sterile, lyophilized enzyme (Sigma CHY-5S) which is dissolved in 5 mL of Sperm Buffer and warmed to 37°C immediately before use. The enzyme solution can either be added to the unliquefied semen or, if the patient is known to have a serious liquefaction problem, he can collect his sample directly into a specimen jar containing the 5 mL of enzyme solution. Always correct for the dilution of the semen by the volume of enzyme solution when calculating sperm concentration and derived values.

   Alternatively, a special chymotrypsin-coated MARQ™ Liquefaction Cup is available from Embryotech Laboratories (Wilmington, MA, USA).

5. *Working with very high sperm counts*

   (a) Semen samples: If a semen sample has a very high sperm concentration, or a higher proportion of rapid spermatozoa that can easily be assessed with confidence, it can be diluted using 170 mM saline to a concentration that can be counted more easily. However, the proportion of rapid progressive cells will certainly be increased due to the reduced viscosity of the diluted specimen. The approximate percentage of motile cells must also be determined on the original wet preparation, as must any progression rating (see step 3 of Subheading 3.4).

   – Diluting with 170 mM saline is better than using Sperm Buffer because its osmolarity is closer to that of liquefied semen and will therefore avoid osmotic (sperm swelling) artifacts.

   – Due to the high viscosity of semen, a positive displacement pipette must be used to take the semen aliquot for dilution, while an air displacement pipette can be used for the diluent.

   (b) Washed sperm suspensions: For these specimens, high concentrations can be reduced to facilitate sperm motility assessments by simply diluting an aliquot of the specimen using the same medium in which the spermatozoa are already suspended.

   – Because such specimens are essentially aqueous (low viscosity) then air displacement pipettes can be used when preparing the dilution.

6. *Motility Index*

For some purposes a "motility index" (MI) might be used to summarize or combine the various aspects of a sperm motility assessment into a single numeric value.

(a) A well-established Motility Index, having a maximum value of 300, is calculated as $MI = (\%a \times 3) + (\%b \times 2) + \%c$

(b) When using the modal progression rating ("PR") system of grading sperm motility on a 0–4 scale (see step 3 of Subheading 3.4), the following Motility Index can be used (25):

$$MI = (\% \text{ Progressive} \times PR \times 2.5) + \% \text{ Nonprogressive}.$$

The weighting factor of 2.5 is used to emphasize the importance of the quality of progression and gives a range of MI values between 0 and 1,000.

References

1. Mortimer ST (1997) A critical review of the physiological importance and analysis of sperm movement in mammals. Hum Reprod Update 3:403–439
2. MacLeod J, Gold RZ (1951) The male factor in fertility and infertility. III. An analysis of motile activity in the spermatozoa of 1000 fertile men and 1000 men in infertile marriage. Fertil Steril 2:187–204
3. Mortimer D (1994) Practical laboratory andrology. Oxford University Press, New York
4. Björndahl L, Mortimer D, Barratt CLR, Castilla JA, Menkveld R, Kvist U, Alvarez JG, Haugen TB (2010) A practical guide to basic laboratory andrology. Cambridge University Press, Cambridge
5. Aitken RJ, Sutton M, Warner P, Richardson DW (1985) Relationship between the movement characteristics of human spermatozoa and their ability to penetrate cervical mucus and zona-free hamster oocytes. J Reprod Fertil 73:441–449
6. Mortimer D, Pandya IJ, Sawers RS (1986) Relationship between human sperm motility characteristics and sperm penetration into human cervical mucus in vitro. J Reprod Fertil 78:93–102
7. Comhaire FH, Vermeulen L, Hinting A, Schoonjans F (1988) Accuracy of sperm characteristics in predicting the in vitro fertilizing capacity of semen. J In Vitro Fert Embryo Transf 5:326–331
8. Barratt CL, McLeod ID, Dunphy BC, Cooke ID (1992) Prognostic value of two putative sperm function tests: hypo-osmotic swelling and bovine sperm mucus penetration test (Penetrak). Hum Reprod 7:1240–1244
9. Irvine DS, Aitken RJ (1986) Predictive value of in vitro sperm function tests in the context of an AID service. Hum Reprod 1:539–545
10. Bollendorf A, Check JH, Lurie D (1996) Evaluation of the effect of the absence of sperm with rapid and linear progressive motility on subsequent pregnancy rates following intrauterine insemination or in vitro fertilization. J Androl 17:550–557
11. Sifer C, Sasportes T, Barraud V, Poncelet C, Rudant J, Porcher R, Cedrin-Durnerin I, Martin-Pont B, Hugues JN, Wolf JP (2005) World Health Organization grade 'a' motility and zona-binding test accurately predict IVF outcome for mild male factor and unexplained infertilities. Hum Reprod 20:2769–2775
12. Van den Bergh M, Emiliani S, Biramane J, Vannin AS, Englert Y (1998) A first prospective study of the individual straight line velocity of the spermatozoon and its influences on the fertilization rate after intracytoplasmic sperm injection. Hum Reprod 13:3103–3107
13. World Health Organization (2010) WHO laboratory manual for the examination and processing of human semen, 5th edn. World Health Organization, Geneva
14. Eliasson R (2010) Semen analysis with regard to sperm number, sperm morphology and functional aspects. Asian J Androl 12:26–32
15. Björndahl L (2010) The usefulness and significance of assessing rapidly progressive spermatozoa. Asian J Androl 12:33–35
16. Barratt CLR, Björndahl L, Menkveld R, Mortimer D (2011) The ESHRE Special Interest Group for Andrology Basic Semen Analysis Course: a continued focus on accuracy,

quality, efficiency and clinical relevance. Hum Reprod 26:3207–3212
17. Björndahl L, Barratt CL, Fraser LR, Kvist U, Mortimer D (2002) ESHRE basic semen analysis courses 1995–1999: immediate beneficial effects of standardized training. Hum Reprod 17:1299–1305
18. Appell RA, Evans PR (1977) The effect of temperature on sperm motility and viability. Fertil Steril 28:1329–1332
19. Milligan MP, Harris SJ, Dennis KJ (1978) The effect of temperature on the velocity of human spermatozoa as measured by time-lapse photography. Fertil Steril 20:592–594
20. ESHRE Andrology Special Interest Group (1998) Guidelines on the application of CASA technology in the analysis of spermatozoa. Hum Reprod 13:142–145
21. Makler A (1980) The improved ten-micrometer chamber for rapid sperm count and motility evaluation. Fertil Steril 33:337–338
22. Douglas-Hamilton DH, Smith NG, Kuster CE, Vermeiden JP, Althouse GC (2005) Particle distribution in low-volume capillary-loaded chambers. J Androl 26:107–114
23. Knuth UA, Neuwinger J, Nieschlag E (1989) Bias to routine semen analysis by uncontrolled changes in laboratory environment—detection by long-term sampling of monthly means for quality control. Int J Androl 12:375–383
24. Tucker M, Wright G, Bishop F, Wiker S, Cohen J, Chan YM, Sharma R (1990) Chymotrypsin in semen preparation for ARTA. Mol Androl 2:179–186
25. Pandya IJ, Mortimer D, Sawers RS (1986) A standardized approach for evaluating the penetration of human spermatozoa into cervical mucus in vitro. Fertil Steril 45:357–365
26. Mortimer D (1994) Laboratory standards in routine clinical andrology. Reprod Med Rev 3:97–111

Chapter 8

Computer-Aided Sperm Analysis (CASA) of Sperm Motility and Hyperactivation

David Mortimer and Sharon T. Mortimer

Abstract

Progressive motility is a vital functional characteristic of ejaculated human spermatozoa that governs their ability to penetrate into, and migrate through, both cervical mucus and the oocyte vestments, and ultimately fertilize the oocyte. A detailed protocol, based on the most common computer-aided sperm analysis (CASA) system with phase contrast microscope optics, is provided for performing reliable assessments of sperm movement pattern characteristics ("kinematics") in semen. The protocol can also be used with washed sperm suspensions where, in addition, the percentages of motile and progressively motile spermatozoa can also be derived. Using CASA technology it is also possible to identify biologically, and hence clinically, important subpopulations of spermatozoa (e.g., those in semen with good mucus-penetrating characteristics, or those showing hyperactivation when incubated under capacitating conditions) by applying multi-parametric definitions on a cell-by-cell basis.

Key words: Sperm motility, Progressive motility, Kinematics, Hyperactivation, CASA

1. Introduction

Progressive motility is a vital functional characteristic of ejaculated human spermatozoa that governs their ability to penetrate into, and migrate through, both cervical mucus and the oocyte vestments (cumulus–corona complex and zona pellucida) (1). Moreover, it is the quality of sperm progression that is of prime importance (2), especially the identification of the proportion of rapid progressive (grade "*a*") spermatozoa, i.e., those with a progression velocity of ≥25 μm/s at 37°C, since these are better able to penetrate cervical mucus in vitro (3–6).

Studies using timed-exposure and multiple-exposure photomicrography, and microcinematography (later, videomicrography) combined with frame-by-frame reconstruction of sperm tracks, led to the modern era of objective analysis of sperm motility, and of sperm movement characteristics or "kinematics" (1, 7–11). Although dedicated computer-aided sperm analysis (CASA) systems have been available commercially for 25 years (7), considerable technical issues with their operation as automated semen analyzers remain to be resolved, primarily concerning issues that prevent the accurate determination of sperm concentration in semen, which consequently precludes determination of the percentages of motile spermatozoa (12–14). Nonetheless, CASA technology does allow the objective analysis of sperm motility kinematics, as well as sperm concentration and percent motile in washed preparations, provided that specific expert recommendations are followed (15). CASA analysis of sperm subpopulations with particular patterns of motility has proven invaluable in research (16), and can be used in human infertility diagnosis and pre-assisted conception workup to identify sperm subpopulations with either the appropriate kinematics to penetrate into cervical mucus or that show hyperactivated motility under capacitating conditions (17). It must be stressed that population-averaged values of sperm kinematic measures are meaningless; biologically important subpopulations must be identified on a cell-by-cell basis using multi-parametric definitions based on several kinematic measures (15) (see Notes 1 and 2).

Future validation of methods for the accurate identification of spermatozoa (and their accurate differentiation from non-sperm objects) might allow the extension of CASA technology to the automated analysis of human semen, but this will require additional imaging capabilities such as fluorescence because phase contrast optics cannot achieve such object discrimination in human semen, with its high levels of debris and other cellular elements (14, 15).

This protocol is based on the "IVOS" CASA system from Hamilton Thorne (Beverly, MA, USA) as it is the system that the authors have used exclusively for the past 20 years (currently v12 software); the IVOS is also probably the most widely available CASA instrument around the world. If a CASA instrument is used with an external microscope, then the microscope must be fitted with a heated stage—ideally one that is integral to the stage rather than a separate metal or glass plate that will increase the distance between the specimen and the upper lens of the condenser, and thereby degrade the phase contrast optics.

2. Materials

There are no special reagents required for routine CASA, although either the culture medium in which the spermatozoa are suspended must contain sufficient protein to prevent the "sticking-to-glass" phenomenon—at least 10 mg/mL of human serum albumin, and ideally 30 mg/mL to reflect physiology for washed sperm suspensions (18)—or the glass surfaces must be treated to prevent it from occurring (19).

2.1. Sperm Medium

This can be any bicarbonate-buffered culture medium designed to sustain human sperm metabolism and capacitation in vitro. It must be equilibrated to 37°C prior to use, and then used under an appropriate CO_2-enriched atmosphere (e.g., 6% CO_2-in-air for most media containing 25 mEq/L bicarbonate ions when used at 37°C close to sea level) to maintain proper pH.

2.2. Sperm Buffer

This is any HEPES-buffered medium designed to sustain sperm metabolism in vitro under an air atmosphere; it will not support sperm capacitation. An equivalent medium buffered using MOPS is also acceptable. Sperm buffer is used whenever spermatozoa need to be maintained outside a CO_2 incubator (e.g., during sperm washing procedures), or when they are to be prevented from capacitating in vitro. Ideally a sperm buffer should be matched in its general formulation to the sperm medium when the two are used in combination.

3. Methods

3.1. Specimens

1. Well-mixed samples of either semen or washed sperm suspensions, as appropriate for the analyses being performed. The preparation depth must be sufficient so as not to impede sperm movement (15):

   (a) Human spermatozoa in semen: Minimum 10 µm, up to 20 µm.

   (b) Washed human spermatozoa: Minimum 20 µm.

   (c) Hyperactivating human spermatozoa: Ideally a minimum of 30 µm for unrestrained movement; only under certain assay conditions, and if properly validated, might 20 µm be sufficient.

   For special fixed-depth chambers, see chapter 7, "Manual methods for sperm motility assessment." Ensure that the chamber will fit properly into the CASA instrument being used (e.g., Makler chamber); ideally a clamp mechanism should be used with those chambers having separate coverslips to ensure compression of the preparation to achieve the intended specimen depth.

2. For proper assessment of sperm progression, especially of velocity and other kinematic measures, ensure that both the sperm sample and the chamber (including any cover glass) are at 37°C before making the preparation, and then analyze it as quickly as possible.

3.2. The IVOS CASA Instrument

In order to use the system reliably, each user of the IVOS instrument must have a sound understanding of sperm kinematics and be familiar with all aspects of the IVOS User Manual.

3.3. Calibration

The magnification of the internal optics system of the IVOS must be calibrated for each objective used, although almost all analyses use the 10× negative high (NH) phase contrast objective. Calibration is achieved using the software function of the IVOS (see the IVOS Software Manual for specific instructions according to the software version) using a Makler chamber grid as a 100 µm reference scale. Once calibrated, the magnification values are stored in the software for future use, so in routine use the operator must only verify that this value has not been changed inadvertently by another user before commencing an analysis.

3.4. Quality Control

Although some laboratories insist that a standard reference videotape must be analyzed each day before commencing analyses, since the use of a reference video does not involve the internal illumination and optics systems of the IVOS, it has no real value in practical terms. Similarly the use of standard suspensions of beads does not confirm correct setup of the IVOS for identifying and tracking spermatozoa, and is also considered to have no real practical value. We have used the following protocol for routine quality control of the IVOS system in real situations for the last 15 years.

1. At the start of each day, and again after any change in illumination or other setup parameters, the correct identification and tracking of spermatozoa must be verified using the DISPLAY—Play Back and EDIT/SORT—Edit Tracks functions of the IVOS software. Verify that:
   (a) The spermatozoa in the field have been correctly identified.
   (b) Debris particles have not been spuriously identified as spermatozoa.
   (c) Moving spermatozoa have been tracked reliably, with only occasional points being missed and only rare tracks having been fragmented into separate segments.

If this is not the case, the system will need to be adjusted as described in the IVOS User Manual. For this reason a dummy analysis should always be attempted before any time-critical analyses are due each day, so that if adjustment is required it will not cause delays in time-sensitive analyses.

3.5. Start-Up

Note: The following procedure must be completed at the start of each day before commencing analyses.

1. Switch on the IVOS, its monitor, and the attached printer.
2. At the Windows log-on dialogue box, hit <ENTER>.

3. When the Windows desktop is displayed, double-click on the "HTR Motility" icon to run the IVOS motility software.
4. When requested, log on to the IVOS software.
5. Allow the heated stage to warm to 37°C (indicated at the top right-hand corner of the IVOS screen).
6. From reference printouts, verify each of the parameters in the following SETUP screens for each of the User-defined setups (these should be configured by an expert user for each combination of specimen type, optics, magnification, and specimen chamber/depth being used):
   (a) Analysis Setup.
   (b) Calibrate Optics.
   (c) Configure Stage.
7. For each type of analysis that will be performed:
   (a) Select a setup option from the menu that corresponds to the analysis type, in this example a 20 μm 2X-CEL chamber.
   (b) Prepare a specimen of semen or washed sperm suspension in a 20 μm 2X-CEL chamber by "drop loading" (i.e., not by capillary fill) and load it into the IVOS.
   (c) Select a field from the central region of the specimen and "grab" it using the ACQUIRE—Start Scan option.
   (d) Select DISPLAY and look at the field analyzed. Choose Play Back and verify that all the moving cells have been tracked, and that the tracks appear intact (i.e., not fragmented into shorter lengths).
   (e) Select EDIT/SORT and then Edit Tracks; select a variety of individual tracks, and zoom in on them and their data points one by one by holding the mouse pointer over a track and holding the left mouse button down. Verify that only occasional track points have been lost, and that tracks are not fragments of longer tracks (identifiable by the image number for the track points that make up the track, each intact track should contain points from the first image through to the end of the grab). To return to the Edit Tracks screen simply left click somewhere in the middle of the screen.
   (f) If there is a problem with the tracking, and if your operator status (and experience) allows you software access to recalibrate the optics and setup parameters, proceed as described in the IVOS User Manual. Otherwise call an appropriate person to perform the necessary adjustments.
8. The IVOS is now ready for use.

3.6. Analysis

1. From the INFO screen, select the desired setup option for the analysis to be performed (these need to be pre-configured by an expert user).

    (a) Enter the identifying information for the sample that is to be analyzed, i.e., Patient Name or Donor ID#, and the Laboratory Reference Number (Lab#) for the specimen.

    (b) Enter the Dilution factor for the specimen being analyzed.

    (c) On the DETAILS screen enter any of the information listed that is pertinent to the sample to be analyzed.

    (d) On the NOTES screen enter any other information listed that is pertinent to the sample to be analyzed using lines 2–4 only (line 1 is usually reserved for the laboratory's contact information).

    (e) On the PRINT/FILE screen, enter the filename for the data to be stored (see Note 5 for format conventions).

    (f) Return to the ACQUIRE screen.

2. Check that the correct objective has been selected: This is the 10× NH phase contrast objective (the middle one) for all analyses using phase contrast optics.

3. Prepare a sperm specimen in a 20 μm deep 2X-CEL chamber by "drop loading" (i.e., not by capillary fill) and LOAD it into the IVOS at the position of "Chamber A." Be sure to remix the sperm suspension (by flicking the tube) before taking the separate aliquots to load each chamber.

4. Select a field close to the edge of the specimen (but at least three fields away from the actual edge) and grab it using the ACQUIRE—Start Scan option. If you get a "High Concentration" warning, see Note 3; for a "Concentration Too High" warning see Note 4.

5. For the first field of a new specimen, verify the tracking of the motile sperm. Select EDIT/SORT and look at the field analyzed:

    (a) Choose Play Back and verify that all the moving cells have been tracked, and that the tracks appear intact (i.e., not fragmented into shorter lengths).

    (b) If any of the tracks appear broken click on Edit Tracks; select an individual track and zoom in on it by holding the mouse pointer over the track and holding the left mouse button down. Verify that any broken tracks are due to collisions and not fragmented longer tracks. To return to the Edit Tracks screen simply left click somewhere in the window showing the sperm track.

    (c) If there is no problem with the IVOS operation then proceed to Subheading 3.6, step 6. If there is a problem then, if your operator status (and experience) allows you software access to recalibrate the optics and setup parameters, proceed as described in the IVOS User Manual; otherwise call an appropriate person to perform the necessary adjustments.

6. Use JOG OUT to move to the next field, verify that it is correctly focussed, and then add it to the analysis by clicking on the Add Scan button in the ACQUIRE screen. Repeat this step until at least four fields and at least 1,000 sperm have been analyzed (for low-concentration samples you might need to analyze all the available fields and just accept the total number of available cells).

   Note: Do not analyze any fields that are less than three fields away from the chamber edge.

7. Go to the PRINT/FILE screen and then:
   (a) Verify that the Lab# for the sample, including its specimen type letter qualifier as the last character into the last character of the File Name (see Note 6 for further explanation), is correct.
   (b) If you want a printout then click on the File & Print button. If no printout is required then click only on the File button.

   Note: If you do not "File" the analysis then the results will not be saved, and the data will be lost when the next analysis is commenced.

8. If the analysis only requires one side of the chamber to be analyzed, go to Subheading 3.6, step 12; otherwise return to the ACQUIRE screen and JOG OUT until you reach the other chamber on the slide, and select a field near the edge (but at least three fields away from the actual edge). Verify that the IVOS is still focussed correctly.

9. Start the second analysis by clicking on the Start Scan button.

10. JOG OUT to the next field, verify that it is correctly focussed, and then add it to the analysis by clicking on the Add Scan button in the ACQUIRE screen. Repeat until either all the fields available have been analyzed or the required number of motile cells have been analyzed.

11. Go to the PRINT/FILE screen. If you want a printout then click on the File & Print button; if no printout is required then click only on the File button. This will add a second line of data to the specimen's computer file.

Note: If you do not "File" the analysis then the results will not be saved and the data will be lost when the next analysis is commenced.

12. Remove the specimen from the IVOS by pressing the LOAD button. Remember to "load" the stage back into the machine since its heating system only operates when the stage is in the loaded position.

13. Discard the specimen in a contaminated sharps container.

3.7. Calculations and Results

1. All basic calculations and kinematic measures are performed by the IVOS software.

2. Transfer the required IVOS results from the printed reports onto the Laboratory Form or Experiment Datasheet for the specimen being analyzed.

Note: If two or more separate analyses were performed on a specimen (two sides of a 2X-CEL chamber, or multiple 2X-CELs) then calculate the averages of their results, since they are each derived from a relatively large number of spermatozoa. However, for SORT fractions, add the number of cells together and calculate the correct percentage; do not average the two pre-calculated percentages (this ensures the correct weighting of the results for possible differences in the size of the subpopulations present in the two, or more, analyses).

4. Notes

1. Mucus penetrating fraction

   This analysis is performed on spermatozoa in liquefied semen and requires that duplicate assessments be made on at least 200 progressively motile spermatozoa from each of the two replicates. The SORT fraction reported by the IVOS is the percentage of motile spermatozoa that show movement characteristics that are likely to enable them to penetrate into, and migrate within, ovulatory cervical mucus. The Boolean argument for mucus penetrating characteristics analyzed in the IVOS at a frame rate of 60 Hz is (17) as follows:

   Velocity average path (VAP) ≥ 25 µm/s AND Straightness (STR) $\geq 80\%$ AND amplitude of lateral head displacement (ALH) ≥ 2.5 µm AND ALH < 7.0 µm.

   Because neither the sperm concentration nor the actual percentage of total motile spermatozoa is important in determining this fraction it can be derived accurately even in human semen. Remember that diluting semen with anything other than homologous seminal plasma (see Note 6) will change its viscosity, and hence the sperm kinematics and consequently the mucus penetrating fraction.

2. Hyperactivation

   This analysis is performed on washed spermatozoa incubated under capacitating conditions (i.e., in sperm medium at 37°C under a CO_2-in-air atmosphere) and requires that duplicate assessments be made on at least 200 progressively motile spermatozoa from each of the two replicates. The SORT fraction reported by the IVOS is the percentage of motile spermatozoa that show movement characteristics that reflect the hyperactivated pattern of flagellar beating. The optimized Boolean argument for hyperactivated human spermatozoa analyzed in the IVOS at a frame rate of 60 Hz (16, 17) is as follows:

   Curvilinear velocity (VCL) ≥150 μm/s AND linearity (LIN) ≤50% AND ALH ≥7.0 μm.

3. "High Concentration" warning

   There are too many spermatozoa in the field of view for the IVOS to analyze their kinematics reliably. The sample needs to be diluted (see Note 6) for reliable analysis.

4. "Concentration Too High" warning

   There are too many spermatozoa in the field of view for the IVOS to analyze. Follow the procedure described below and repeat the analysis.

   (a) Remove the sample from the IVOS and discard it.

   (b) Prepare a diluted aliquot of the specimen (see Note 6) and mix thoroughly, but do not vortex.

   (c) Make a new preparation using a clean 2X-CEL chamber and load the slide into the IVOS.

   Note: At step #1c of the procedure you must remember to adjust the Dilution factor.

5. File naming convention

   When storing data to the IVOS hard drive, the HDATA software places an 8-character limit on the filename, so great care must be taken when creating file names. As an example, we have used the following format for this purpose: YRnnnnXX

   where YR is the last 2 numbers of the year (e.g., "11" for 2011).

   nnnn is a 4-digit sample accession number that restarts at 0001 for the first sample each year (use the leading zeroes to keep the filename with the correct length).

   XX is a qualifier that identifies the type of specimen being analyzed:

   S = semen.

   SD = semen diluted for CASA analysis.

   SF = diluted semen (i.e., semen + CPM) pre-freeze.

   HA = hyperactivation assay.

   HC = hyperactivation assay control.

   T = thawed semen.

TD = thawed semen diluted for CASA analysis.

W = washed sperm (fresh).

Z = washed thawed sperm.

Hence "110123HA" denotes a hyperactivation analysis on sample 0123 during the year 2011.

6. Diluting samples

   Because sperm kinematics are affected by the viscosity of the specimen, the following criteria for Dilution must be observed:

   (a) Semen: Use homologous seminal plasma. The simplest and cleanest way of getting this is to use the uppermost layer from a sperm preparation density gradient after centrifugation.

   (b) Washed sperm: Add more Sperm Medium or Sperm Buffer, as appropriate.

   – Mix the diluted specimens thoroughly, but do not vortex. Avoid frothing due to the protein content.

   – Remember to adjust the Dilution factor on the INFO screen of the IVOS.

References

1. Mortimer ST (1997) A critical review of the physiological importance and analysis of sperm movement in mammals. Hum Reprod Update 3:403–439
2. MacLeod J, Gold RZ (1951) The male factor in fertility and infertility. III. An analysis of motile activity in the spermatozoa of 1000 fertile men and 1000 men in infertile marriage. Fertil Steril 2:187–204
3. Mortimer D (1994) Practical laboratory andrology. Oxford University Press, New York, NY
4. Björndahl L, Mortimer D, Barratt CLR, Castilla JA, Menkveld R, Kvist U, Alvarez JG, Haugen TB (2010) A practical guide to basic laboratory andrology. Cambridge University Press, Cambridge
5. Aitken RJ, Sutton M, Warner P, Richardson DW (1985) Relationship between the movement characteristics of human spermatozoa and their ability to penetrate cervical mucus and zona-free hamster oocytes. J Reprod Fertil 73:441–449
6. Mortimer D, Pandya IJ, Sawers RS (1986) Relationship between human sperm motility characteristics and sperm penetration into human cervical mucus in vitro. J Reprod Fertil 78:93–102
7. Mortimer D (1990) Objective analysis of sperm motility and kinematics. In: Keel BA, Webster BW (eds) Handbook of the laboratory diagnosis and treatment of infertility. CRC, Boca Raton, FL, pp 97–133
8. David G, Serres C, Jouannet P (1981) Kinematics of human spermatozoa. Gamete Res 4:83–95
9. Katz DF, Davis RO (1987) Automatic analysis of human sperm motion. J Androl 8:170–181
10. Boyers SP, Davis RO, Katz DF (1989) Automated semen analysis. Curr Prob Obstet Gynecol Fertil XII 5:167–200
11. Davis RO, Katz DF (1993) Computer-aided sperm analysis: technology at a crossroads. Fertil Steril 59:953–955
12. Mortimer D, Aitken RJ, Mortimer ST, Pacey AA (1995) Workshop report: clinical CASA—the quest for consensus. Reprod Fertil Dev 7:951–959
13. ESHRE Andrology Special Interest Group (1996) Consensus workshop on advanced diagnostic andrology techniques. Hum Reprod 11:1463–1479
14. Mortimer D, Mortimer ST (1998) Value and reliability of CASA systems. In: Ombelet W, Bosmans E, Vandeput H, Vereecken A, Renier M, Hoomans EH (eds) Modern ART in the 2000s. Andrology in the nineties. Parthenon Publishing, Carnforth, pp 73–89
15. ESHRE Andrology Special Interest Group (1998) Guidelines on the application of CASA

technology in the analysis of spermatozoa. Hum Reprod 13:142–145
16. Mortimer ST (2000) CASA-practical aspects. J Androl 21:515–524
17. Mortimer D, Mortimer ST (2005) Laboratory investigation of the infertile male. In: Brinsden PR (ed) A textbook of in-vitro fertilization and assisted reproduction, 3rd edn. Taylor & Francis, London, pp 61–91
18. Mortimer D (1986) Elaboration of a new culture medium for physiological studies on human sperm motility and capacitation. Hum Reprod 1:247–250
19. Chapeau C, Gagnon C (1987) Nitrocellulose and polyvinyl coatings prevent sperm adhesion to glass without affecting the motility of intact and demembranated human spermatozoa. J Androl 8:34–40

Part II

Tests of Sperm Function

Chapter 9

The Hemizona Assay for Assessment of Sperm Function

Sergio Oehninger, Mahmood Morshedi, and Daniel Franken

Abstract

The hemizona assay (HZA) has been developed as a diagnostic test for the tight binding of human spermatozoa to the human zona pellucida to predict fertilization potential. In this homologous bioassay, the two matching hemizona halves are functionally equal surfaces allowing controlled comparison of binding from a fertile control versus a test sample, with reproducible measurements of sperm binding obtained from a single oocyte. Oocytes from different sources (surplus from IVF treatment or recovered from ovarian tissue) are salt-stored and used after microbisection. Extensive clinical data have demonstrated excellent predictive power of the HZA for the outcomes of intrauterine insemination and IVF, and therefore the assay has relevance in the clinical diagnostic setting in infertility.

Key words: Hemizona assay, Oocyte, Sperm function

1. Introduction

Work derived from the in vitro fertilization (IVF) and embryo transfer arena has demonstrated that an abnormal sperm–zona pellucida interaction is frequently observed in infertile men. Such a finding can be observed in the presence of normal or abnormal "basic" sperm parameters. An impaired sperm–zona pellucida interaction can result in failure of fertilization, thereby decreasing the chances of pregnancy when couples are being subjected to intrauterine insemination (IUI) or conventional IVF therapies. Consequently, sperm–zona pellucida binding assays were developed due to a real need to assess sperm functional competence in the "extended" evaluation of the infertility workup (1, 2). In the era of intracytoplasmic sperm injection (ICSI), results of such functional assays provide valuable information to the clinician to direct management to low-complexity alternatives such as IUI or directly to IVF augmented with assisted micro-fertilization through sperm injection into oocytes (ICSI) (3–5).

The hemizona assay (HZA) has been extensively validated as a diagnostic test for the binding of human spermatozoa to human zona pellucida to predict fertilization potential (6). In the HZA, the two matched zona hemispheres created by microbisection of the human oocyte provide three main advantages: (1) the two halves (hemizonae) are functionally equal surfaces allowing controlled comparison of binding and reproducible measurements of sperm binding from a single egg; (2) the limited number of available human oocytes is amplified because an internally controlled test can be performed on a single oocyte; and (3) because the oocyte is split microsurgically, even fresh oocytes cannot lead to inadvertent fertilization and pre-embryo formation (6, 7).

A highly specific type of binding is necessary for fertilization to ensue, and therefore, the HZA provides a unique homologous (human) bioassay to assess sperm function at the fertilization level. Different sources of human oocytes can be used in the assay, oocytes recovered from surgically removed ovaries or postmortem ovarian tissue, and/or surplus oocytes from the IVF program. Since fresh oocytes are not always available for the test, different alternatives have been implemented for storage. Others have described the storage of human oocytes in dimethylsulfoxide (DMSO) at ultra low temperatures (8). Additionally, Yanagimachi and colleagues showed that highly concentrated salt solutions provided effective storage of hamster and human oocytes such that the sperm-binding characteristics of the zona pellucida were preserved (9). In developing the HZA, we have examined the binding ability of fresh, DMSO and salt-stored (under controlled pH conditions) human oocytes and have concluded that the sperm binding ability of the zona remains intact under all these conditions (10). Subsequently, we assessed the kinetics of sperm binding to the zona, showing maximum binding at 4–5 h of gamete coincubation, with similar binding curves both for fertile and infertile semen samples.

The assay has been validated by a clear-cut definition of the factors affecting data interpretation, i.e., kinetics of binding, egg sources, variability and maturation status, intra-assay variation, and influence of sperm concentration, morphology, motility, and acrosome reaction status (10–18). Over 90 days evaluation, spermatozoa from fertile men do not exhibit a time-dependent change in zona binding potential, thus reassuring their utilization as controls in the bioassay. Within each pool of donors utilized in the assay, a cutoff value or minimal threshold of binding has to be established in order to validate each assay for the purpose of identification of a poor control semen specimen and/or a poor zona. In our control population, this cutoff value is approximately 20 sperm tightly bound to the control hemizona (fertile donor). Therefore, each laboratory should statistically assess its own control data to establish a reasonable lower limit for assay acceptance. If the control hemizona (matching hemizona exposed to fertile sperm) has a good binding capability, that is, tightly binds at least 20 spermatozoa after the 4-h incubation period (information derived from a statistical evaluation of a pool of fertile donors), then a single oocyte will give reliable information about the fertilizing ability of the test spermatozoa.

The inter-egg variability is high not only for oocytes at different stages of maturation (immature versus mature eggs) but also within a certain population of eggs at the same maturational stage. However, this factor is internally controlled in the assay by the utilization of matching hemizona that allows a comparison of binding of a fertile versus an infertile semen sample in the same assay under the same oocyte quality conditions. Incubating matching hemizonae from eggs at the same maturational stage with homologous spermatozoa from the same fertile ejaculate, we have been able to determine a low (<10%) intra-egg (intra-assay) variability both for human and monkey (cynomolgus) oocytes. Additionally, we have shown that full meiotic competence of human and monkey oocytes is associated with an increased binding potential of zona pellucida (17, 19).

The specificity of the interaction between human spermatozoa and the human zona pellucida under HZA conditions is strengthened by the fact that the sperm tightly bound to the zona are acrosome reacted (15, 20). Moreover, results of interspecies experiments performed with human, cynomolgus monkey, and hamster gametes have demonstrated a high species specificity of human sperm/zona pellucida functions under HZA conditions providing further support for the use of this bioassay in infertility and contraception testing (19, 21, 22).

In prospective blinded studies, we have investigated the relationship between sperm binding to the hemizona and conventional IVF outcome (11, 16, 23–26). Results have shown that the HZA can successfully distinguish the population of male-factor patients at risk for failed or poor fertilization. Using either a cutoff value of fertilization rate of 65% (mean minus 2 standard deviations of the overall fertilization rate in our programs for non-male-factor patients) or distinguishing between failed versus successful fertilization (0% vs. 1–100%) the HZA results expressed as HZI (a percentage derived from the ratio of patient or test sperm binding over control sperm binding on hemizonae from the same oocyte) provide a valuable means to separate these categories of patients. Powerful statistical results allow us to use the HZA in the prediction of fertilization rate. Logistic regression analysis provides a robust HZI range predictive of the oocyte's potential to be fertilized. This HZI cutoff value is approximately 35%. Overall, for failed versus successful and poor versus good fertilization rate, the correct predictive ability (discriminative power) of the HZA was 80 and 85%, respectively. The information gained through the use of the HZA may be extremely valuable for counseling patients in the IVF setting (i.e., considering an HZI below 35% the chances of poor fertilization are 90–100%, whereas for the HZI over 35%, the chances of good fertilization are 80–85%).

A meta-analysis was carried out to examine the value of available sperm function assays to predict fertilization outcome in IVF (27). The aim was carried out through a meta-analytical approach that examined the predictive value of four categories of sperm functional assays: computer-aided sperm motion analysis (CASA); induced-acrosome reaction testing; heterologous hamster oocyte-sperm penetration assay (SPA); and sperm–zona pellucida binding assays (including the HZA) for IVF outcome. Results demonstrated the highest predictive power of the sperm–zona pellucida binding and the induced-acrosome reaction assays for fertilization outcome.

In subsequent studies we investigated the predictive value of the HZA assay for pregnancy outcome in patients undergoing controlled ovarian hyperstimulation (COH) with IUI (28). Overall, patients with an HZI of <30 had a significantly lower pregnancy rate compared to patients with an HZI of > or = 30 (relative risk for failure to conceive: 1.5 [confidence interval 1.2–1.9]). The negative and positive predictive values of the HZA for pregnancy were 93% and 69%, respectively. Logistic regression analysis revealed a model where HZI had the highest predictability of conception in male factor couples ($P=0.021$). The European Society of Human Reproduction and Embryology (ESHRE) and the World Health Organization (WHO) have selected sperm binding assays as research tests approaching diagnostic application in the clinic (29, 30). Consequently, results of this sperm function test are useful in counseling couples before allocating them into COH/IUI therapy, IVF, or ICSI.

2. Materials

2.1. Oocyte Collection

1. Oocytes (see Subheadings 3.1 and 3.2).
2. Salt storage medium (for short-term storage) or DMSO/sucrose storage medium (for long-term storage).
3. 100 mm Petri dishes.
4. Size 23 scalpel blades.
5. 0.2 mm bore size gel sequencing pipette.
6. Culture medium: Modified human tubal fluid (HTF; Irvine Scientific) supplemented with 0.5% human serum albumin (HSA; Irvine Scientific).
7. Dissecting microscope.

2.2. Oocyte Storage

1. *Salt Storage solution (short-term storage)*: 1.5 M magnesium chloride hexahydrate, 0.1% polyvinylpyrrolidone (PVP—MW 40,000), 40 mM sodium HEPES: Measure out about 400 mL IVF water in a graduated cylinder and transfer to a beaker. Completely dissolve 5.2 g sodium HEPES and 0.5 g PVP in the water while stirring. Slowly add 152.5 g MgCl$_2$·6H$_2$O and stir until dissolved. Adjust the pH to 7.4 with several milliliters of 12 M HCl. Bring the volume up to 500 mL with IVF water. Filter sterilize with a 0.2 µM bottle top filter. Label bottle with solution name, date made, expiration date, and initials of the person who made the solution. Store at 4°C for a maximum of 3 months.
2. *DMSO/sucrose storage solution (long-term storage)*: 3.5 M DMSO, 0.25 M sucrose, and 20% blood serum in PBS.
3. *Thawing solution*: 0.25 M sucrose and 3% HSA in PBS.

2.3. Oocyte Microbisection

1. Phase contrast inverted microscope.
2. Micromanipulators.
3. Beaver micro-sharp blade holder.
4. Beaver micro-sharp blades, 3.5 mm, depth 300.
5. Culture medium: Modified HTF supplemented with 0.5% HSA.
6. 0.2 mm bore size gel sequencing pipette.
7. Mineral oil.

2.4. Semen Preparation

1. Test and control semen samples (fresh or cryopreserved).
2. Culture medium: Modified HTF supplemented with 0.5% HSA.
3. HTF supplemented with 3% HSA.

2.5. Assay Incubation, Counting, and Interpretation of Results

1. Bisected hemizonae.
2. 0.2 mm bore size gel sequencing pipette.
3. Prepared sperm suspended in HTF supplemented with 3% HSA.
4. Culture medium.

3. Methods

3.1. Oocyte Collection from Surgically Removed Donated Ovaries and Postmortem Donated Ovarian Tissue

Caution: Always wear appropriate eye protection. The ovaries are a resilient tissue, which can contribute to considerable spatter of potentially biohazardous fluid during cutting.

1. Place a single ovary in a 100 mm plastic Petri dish.
2. Cut the ovary with a size 23 surgical blade into approximate 2 mm slices.

3. Cut the 2 mm slices until the entire ovary is macerated to produce a fine mass of ovarian tissue.
4. Remove small amounts of the macerated tissue to an adjacent dish and add 1 mL of culture medium to the mass. It may be helpful to use square Petri dishes with grids to aid in systematically covering the entire dish.
5. Using a dissecting microscope scan the entire suspension for oocytes.
6. It may be necessary to pipette oocytes recovered in this manner several times in order to strip granulosa cells from the oocytes.
7. Collect the retrieved oocytes in a drop of culture medium.

3.2. Collection of Unfertilized IVF Donated Oocytes

1. Collect the unfertilized donated oocytes with no developmental potential in a freshly prepared drop of culture medium.
2. Use a 0.2 mm bore size pipette to vigorously wash each oocyte to dislodge any attached sperm.

3.3. Storage of Oocytes (See Note 1)

1. For short-term storage (several months), harvested oocytes may be stored in the salt storage buffer at 4°C. Before the assay, remove the desired number of oocytes from the salt storage, rinse them five to ten times with culture medium to remove the storage solution, and then proceed to bisection.
2. For long-term storage, oocytes should be cryopreserved. For cryopreservation, incubate 1–3 oocyte(s) for 3 min in 0.5 mL solution of DMSO/sucrose storage solution. Following this incubation transfer oocytes into 0.25 mL cryopreservation straws, and then plunge straws into liquid nitrogen for permanent storage. When ready for use, thaw oocytes at 37°C for 2–3 min and then expel the contents into a Petri dish. Following thaw observe the oocytes under a dissecting microscope. Transfer the thawed oocytes into the thawing solution for 7–10 min, then rinse the oocytes with culture medium, and then proceed to bisection.

3.4. Oocyte Microbisection (Figs. 1, 2, and 3)

1. Setting up the micromanipulation system is a crucial step in the HZA technique. Improper installation will not allow successful bisection of the zona pellucida. Mount micromanipulators on a phase-contrast inverted microscope.
2. Fix the cover of a 100 mm Petri dish (working area) on the stage of a phase contrast, inverted microscope. Use double sticking tape on the bottom of the dish and affix to the microscope stage to restrain dish movement.
3. Align a microsurgical scalpel handle blade to the pipette holder of a micromanipulator. Blades should be mounted so that the cutting edge is parallel to the Petri dish cover. The Z-axis of

9 The Hemizona Assay for Assessment of Sperm Function 97

Step 1: Positioning of microblade

Step 2: Cut a horizontal line in plastic dish

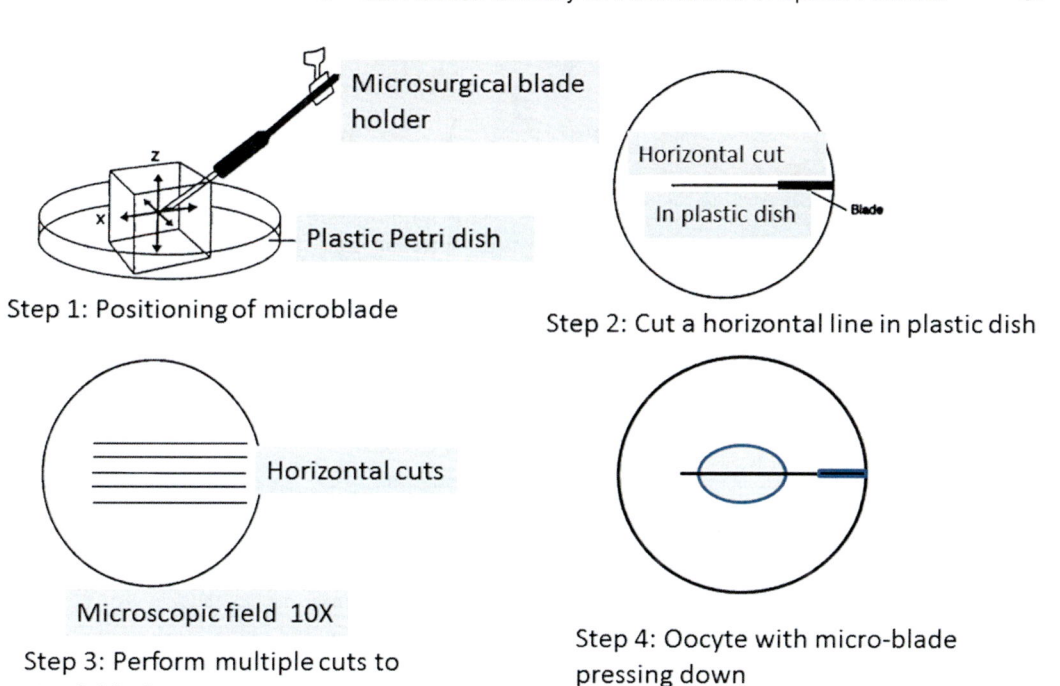

Step 3: Perform multiple cuts to check blade accuracy

Step 4: Oocyte with micro-blade pressing down

Fig. 1. Hemizona assay (HZA) procedure: Setting up the equipment.

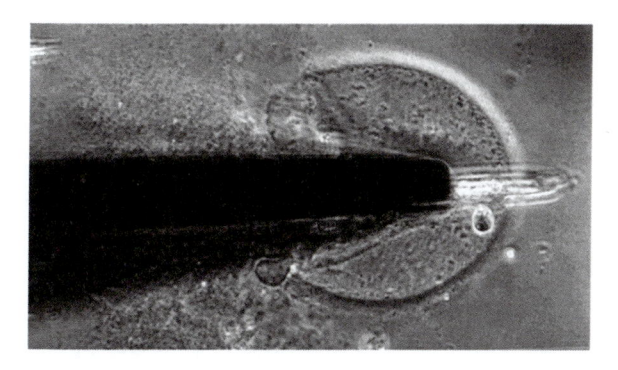

Fig. 2. Photomicrograph of the bisection of a human oocyte to produce two matching hemizonae.

the blade should be straight up and down (perpendicular to the cutting surface).
4. Lower the blade until it touches the bottom of the plastic Petri dish and with the fine manipulator cut a horizontal groove in the dish.
5. Move the stage a few millimeters sideways and repeat step 4 several times until you have produced four or more parallel lines.
6. Following preparation of the dish, proceed with the assay by adding a drop of culture medium to the cutting area.

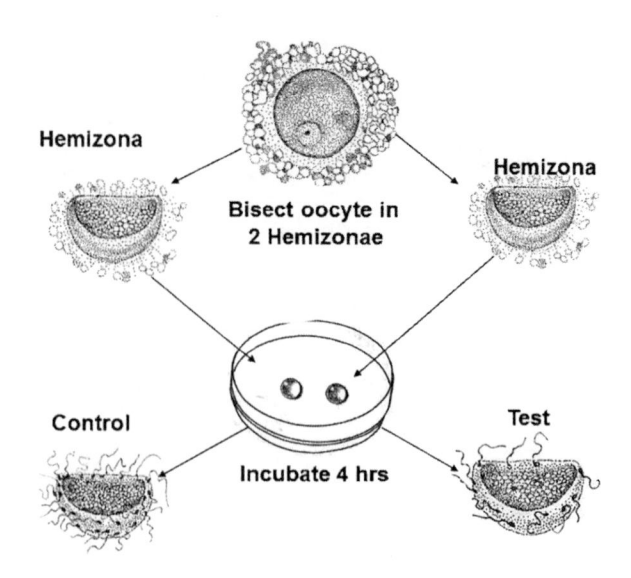

Fig. 3. The Hemizona Assay (HZA): Diagram illustrating the procedures.

7. Remove an oocyte from a working droplet and under microscopic observation, place it in the middle of the microscopic field. Ensure that the oocyte is placed on or between the thin parallel lines previously cut in the plastic dish. This will restrict the oocyte movement during the bisecting process.
8. Carefully lower the blade until it is in the middle of the oocyte. Monitor the position of the blade as it comes down on the oocyte, making minor positioning adjustments as necessary. A micrometer can be used (if necessary) to measure the oocyte diameter and thus to ensure identical hemizonae. This micrometer is similar to the one used to determine sperm dimensions for morphology assessment using strict criteria.
9. Lower the blade until it touches the top surface of the oocyte and by further lowering the oocyte will be flattened.
10. Using side-to-side excursions bisect the oocyte into two identical halves (hemizonae). Be certain that the cuts completely bisect the zona pellucida before releasing the tension on the blade.
11. Keep the hemizona pairs together and pipette each to remove the residual cytoplasm. Aspirate the cut hemizonae about five times or more using a 0.2 mm pipette until the residual cytoplasm is removed. Place the hemizona pairs in fresh drops (about 0.1 mL) of culture medium.
12. Place the matching hemizona pairs in the incubator overnight or for a minimum of several hours before transferring into sperm droplets (see Subheading 3.5). If several eggs are bisected to prepare hemizonae for the entire week of assays, matching hemizonae may be kept in 100 µL droplets of modified culture medium, covered with mineral oil, and refrigerated. In this case, the hemizonae should be incubated at 37°C overnight or

for a minimum of several hours before they are used for the assay. Make sure to remove the oil from the dish to expose the droplets before hemizonae are transferred for incubation and eventual contact with sperm.

13. When sperm are ready for incubation with the hemizonae, one hemizona is transferred into the drop containing control (donor) sperm and the corresponding hemizona is transferred into the drop containing patient's sperm (see Subheading 3.6).

3.5. Semen Preparation

1. Evaluate the test and control semen for basic semen parameters and record.
2. Mix semen with culture medium at a ratio of 0.5 mL semen and 1 mL medium.
3. If a cryopreserved donor sample is used, thaw the sample and check for quality. If not adequate, thaw another sample. Add the medium to the sample at 50–100 µL increments 1–2 min apart until a volume twice that of the semen has been added.
4. Centrifuge each tube at $400 \times g$ for 10 min. Aspirate and discard supernatants.
5. Resuspend the pellet(s) in 1 mL medium and centrifuge/process as in step 4.
6. Gently overlay each pellet with 200 µL of 37°C medium without mixing.
7. Place each tube in a 37°C incubator and allow 1 h for swim up.
8. Remove the sperm suspension (swim up) from the top of each pellet and transfer to properly labeled tubes.
9. Determine the sperm concentration in each suspension.
10. Using fresh 37°C medium to dilute the suspensions, adjust each suspension to make a final concentration of 6×10^6 motile sperm/mL.
11. A total of 60,000 motile sperm per hemizona are needed, and at least three replicates should be performed with each sperm sample (see Note 2). Add 44 µL of the prepared swim-up suspension from step 10 to 352 µL of HTF containing 3% HSA. Mix by gentle pipetting (see Note 3). Make this dilution as close as possible to the time you are ready to add the hemizonae to sperm suspensions for the 4-h incubation. Repeat with other sperm suspensions (i.e., control and other samples from patients, if applicable).

3.6. Assay Incubation Counting and Interpretation of Results (See Note 4)

1. Gently aspirate the bisected hemizonae and place them separately in the patient and control sperm droplets.
2. Cover the droplets with mineral oil.
3. Incubate the droplets for 4 h in a humidified chamber at 37°C and 5% CO_2.

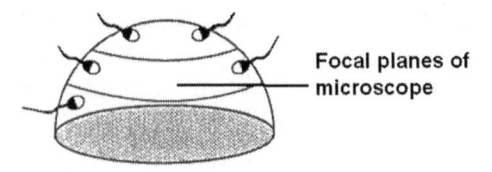

Fig. 4. Counting sperm tightly bound to the hemizona for the HZA.

4. Following the incubation remove each hemizona from the incubation droplets and place it in 50 µL fresh culture medium droplet.
5. Remove loosely attached sperm by aspirating each hemizona separately five times with a 0.2 mm micro-pipette.
6. Remove the hemizona, place it in a fresh culture medium droplet, and count the number of tightly bound sperm on each hemizona. Counting is performed with hemizonae placed with the open end facing the bottom of culture dish. Starting from the bottom of the dish count all spermatozoa that are in focus. Move the microscope stage 4–5 µm upwards and continue counting the in-focus sperm. Proceed until all sperm on the hemizona surface are recorded. Repeat the process by starting at the top end of the hemizona moving incrementally 4–5 µm downwards (Fig. 4).
7. Calculate the average of the two counts to obtain the mean number of bound sperm for the test as well as for the control hemizona (see Note 5).
8. Calculate the Hemizona Index (HZI) based on the formula (see Note 6)

$$\text{Hemizona Index} = \frac{\text{Number of bound sperm from the test sample}}{\text{Number of bound sperm from the control sample}} \times 100$$

4. Notes

1. Oocytes may be stored in a salt solution at refrigeration temperature for short-term storage. Our experience with the salt-stored oocytes shows that they last for several months. There are several ways one can assess if the salt-stored oocytes have maintained their integrity and suitability to be used for the assay. One is the observation of the oocytes during the cutting process. If they have lost their shape and are easily smashed during the cutting process, they should not be used for the assay. Second, if the binding with control (donor) sperm declines

across the board for all oocytes, this may also be indicative of the impact of longer than usual storage in salt solution. The binding pattern may be used as a quality control measure for the assay, particularly if the same control/donor semen sample is used over a period of time. This may be possible when cryopreserved control semen samples are utilized. Nevertheless, for long storage time, it is recommended that oocytes be cryopreserved.

2. In cases where fewer than 500,000 sperm are recovered, use an equal number of sperm in the test and control droplets.

3. Each suspension (396 µL) is sufficient to provide four 90-µL aliquots with 60,000 motile sperm in each aliquot for use in HZA plus a residual 36 µL of the suspension to check sperm motility.

4. Prior to performing the HZA, the zona pellucida binding capacity of several proven fertile donors has to be determined. This is particularly important if cryopreserved semen samples are used. These samples should show binding of >60 sperm/hemizona.

5. In cases where fresh control sperm bind fewer than 20 sperm/hemizona, the assay results for that replicate should not be considered when reporting results.

6. An HZI >30–35 is associated with successful fertilization in IVF and with pregnancy in IUI.

References

1. Oehninger S et al (1997) Clinical significance of human sperm-zona pellucida binding. Fertil Steril 67:1121–1127
2. Oehninger S et al (1997) Approaching the next millennium: how should we manage andrology diagnosis in the intracytoplasmic sperm injection era? Fertil Steril 67:434–436
3. Oehninger S (1995) An update on the laboratory assessment of male fertility. Hum Reprod 10(Suppl 1):38–45
4. Franken DR, Oehninger S (2006) The clinical significance of sperm-zona pellucida binding: 17 years later. Front Biosci 11:1227–1233
5. Oehninger S (2011) Clinical management of male infertility in assisted reproduction: ICSI and beyond. Int J Androl 34:e319–e329
6. Burkman LJ et al (1988) The hemizona assay (HZA): development of a diagnostic test for the binding of human spermatozoa to the human hemizona pellucida to predict fertilization potential. Fertil Steril 49:688–697
7. Hodgen GD et al (1988) The hemizona assay (HZA): finding sperm that have the "right stuff". J In Vitro Fert Embryo Transf 5:311–313
8. Overstreet JW, Hembree WC (1976) Penetration of the zona pellucida of nonliving human oocytes by human spermatozoa in vitro. Fertil Steril 27:815–831
9. Yanagimachi R et al (1979) Retention of biologic characteristics of zona pellucida in highly concentrated salt solution: the use of salt-stored eggs for assessing the fertilizing capacity of spermatozoa. Fertil Steril 31:562–574
10. Franken DR et al (1989) Hemizona assay using salt-stored human oocytes: evaluation of zona pellucida capacity for binding human spermatozoa. Gamete Res 22:15–26
11. Franken DR et al (1989) The hemizona assay (HZA): a predictor of human sperm fertilizing potential in in vitro fertilization (IVF) treatment. J In Vitro Fert Embryo Transf 6:44–50
12. Franken DR et al (1991) Defining the valid hemizona assay: accounting for binding variability within zonae pellucidae and within semen samples from fertile males. Fertil Steril 56:1156–1161
13. Franken DR et al (1991) Comparison of sperm binding potential of uninseminated, inseminated-

unfertilized, and fertilized-noncleaved human oocytes under hemizona assay conditions. Mol Reprod Dev 30:56–61
14. Franken DR et al (1994) Sperm binding capacity of human zona pellucida derived from oocytes obtained from different sources. Andrologia 26:277–281
15. Coddington CC et al (1990) Sperm bound to zona pellucida in hemizona assay demonstrate acrosome reaction when stained with T-6 antibody. Fertil Steril 54:504–508
16. Oehninger S et al (1989) Hemizona assay: assessment of sperm dysfunction and prediction of in vitro fertilization outcome. Fertil Steril 51:665–670
17. Oehninger S et al (1991) Human preovulatory oocytes have a higher sperm-binding ability than immature oocytes under hemizona assay conditions: evidence supporting the concept of "zona maturation". Fertil Steril 55:1165–1170
18. Bastiaan HS et al (2003) Relationship between zona pellucida-induced acrosome reaction, sperm morphology, sperm-zona pellucida binding, and in vitro fertilization. Fertil Steril 79:49–55
19. Oehninger S et al (1989) Validation of the hemizona assay in a monkey model: influence of oocyte maturational stages. Fertil Steril 51:881–885
20. Franken DR et al (1991) Electron microscopic evidence on the acrosomal status of bound sperm and their penetration into human hemizonae pellucida after storage in a buffered salt solution. Andrologia 23:205–208
21. Oehninger S et al (1993) Validation of the hemizona assay (HZA) in a monkey model. II. Kinetics of binding and influence of cryopreserved-thawed spermatozoa. J Assist Reprod Genet 10:292–301
22. Oehninger S et al (1993) The specificity of human spermatozoa/zona pellucida interaction under hemizona assay conditions. Mol Reprod Dev 35:57–61
23. Franken DR et al (1993) The ability of the hemizona assay to predict human fertilization in different and consecutive in-vitro fertilization cycles. Hum Reprod 8:1240–1244
24. Oehninger S et al (1991) Recurrent failure of in vitro fertilization: role of the hemizona assay in the sequential diagnosis of specific sperm-oocyte defects. Am J Obstet Gynecol 164: 1210–1215
25. Oehninger S et al (1991) Hemizona assay: sperm defect analysis, a diagnostic method for assessment of human sperm-oocyte interactions, and predictive value for fertilization outcome. Ann N Y Acad Sci 626:111–124
26. Oehninger S et al (1992) Hemizona assay and its impact on the identification and treatment of human sperm dysfunctions. Andrologia 24:307–321
27. Oehninger S et al (2000) Sperm function assays and their predictive value for fertilization outcome in IVF therapy: a meta-analysis. Hum Reprod Update 6:160–168
28. Arslan M et al (2006) Predictive value of the hemizona assay for pregnancy outcome in patients undergoing controlled ovarian hyperstimulation with intrauterine insemination. Fertil Steril 85:1697–1707
29. (1996) Consensus workshop on advanced diagnostic andrology techniques. ESHRE (European Society of Human Reproduction and Embryology) Andrology Special Interest Group. Hum Reprod 11: 1463–1479
30. World Health Organization (2010) World Health Organization laboratory manual for the examination of human semen and sperm-cervical mucus interaction. Cambridge University Press, Cambridge, UK, pp 31–33

Chapter 10

The Sperm Penetration Assay for the Assessment of Fertilization Capacity

Kathleen Hwang and Dolores J. Lamb

Abstract

The sperm penetration assay, or zona-free hamster oocyte penetration assay, is utilized to measure the ability of sperm to undergo capacitation, fuse with the egg membrane, and decondense the sperm head within the cytoplasm of the oocyte, resulting in the formation of the male pronucleus. The test is scored by calculation of the percentage of ova that are penetrated (the original assay developed) or the average number of sperm penetrations per ovum (the sperm capacitation index of the optimized assay). The sperm penetration assay identifies those couples that will have a high likelihood of success with in vitro fertilization.

Key words: Male infertility, Sperm, Zona pellucida, Gamete interaction, Sperm–oocyte penetration

1. Introduction

Infertility is a difficult and stressful condition for both the patients and the treating physicians. The failure to conceive within 1 year occurs in about 15% of couples (1), and about 50% of problems related to conception are either caused entirely by the male or is a combined problem with the male and his female partner. Male infertility continues to be a clinical challenge of increasing significance. The routine semen analysis remains the cornerstone in the evaluation of male factor infertility; however the diagnostic capabilities currently available to assist in formulating a diagnosis have rapidly evolved to include specialized functional tests.

Even men with normal semen parameters may not be fertile, necessitating the development of tests designed to measure sperm functions such as cervical mucus penetration, capacitation, sperm–zona pellucida binding, sperm–ova binding, acrosome reaction, sperm penetration, and decondensation.

Douglas T. Carrell and Kenneth I. Aston (eds.), *Spermatogenesis: Methods and Protocols*, Methods in Molecular Biology, vol. 927, DOI 10.1007/978-1-62703-038-0_10, © Springer Science+Business Media, LLC 2013

The sperm penetration assay was first developed in the 1970s. In 1976, Yanagimachi and colleagues observed that upon removal of the zona pellucida of hamster ova, the eggs were "promiscuous" and allowed penetration by sperm of other species (2). This test measures the ability of sperm to undergo capacitation, fuse with the egg membrane, and decondense the sperm head resulting in the formation of the male pronucleus. Although a number of studies showed that this test was a useful predictor of fertilization in conventional in vitro fertilization (IVF), the test was plagued by false negatives and assay variability. Johnson, et al. (3, 4) improved the assay by altering the conditions to enhance sperm penetration rates by orders of magnitude and by devising a control method to ensure assay reproducibility, precision, and accuracy. A positive score on this modified test was demonstrated to be highly predictive of a positive outcome in IVF (5, 6). Despite providing useful predictive information, the test rarely is ordered today. A modification of this assay used in combination with intra-cytoplasmic sperm injection (ICSI) into hamster oocytes accurately identified male factor patients who failed to fertilize after IVF/ICSI due to a defect in sperm head decondensation (7).

2. Materials

Prepare and store all reagents at room temperature (unless indicated otherwise). Diligently follow all waste disposal regulations when disposing waste materials.

2.1. Equipment

1. Incubator.
2. Hemocytometer counting chamber.
3. Phase contrast microscope.
4. Microscope slides with coverslips.
5. Pasteur pipette, 9½″.
6. Centrifuge.
7. Glass test tube, 12 × 75 mm.
8. Graduated tube, 15 mL.
9. Micro-test tube, 0.4 mL (Biorad).
10. 26 gauge 3/8″ needle.
11. Culture dish, 35 × 10 mm (Falcon).
12. Dissecting microscope.
13. Mounting micropipette.
14. CO_2 chamber.

2.2. Reagents

1. Sterile water.
2. Sterile saline.
3. Modified Biggers–Whitten–Whittingham (MBWW) media.
4. TEST-yolk buffer (TYB).
5. 37 IU Human chorionic gonadotropin (hCG).
6. 0.5 mg/mL Trypsin (ICN Pharmaceutical).
7. 0.5 mL Hyaluronidase.
8. Mineral oil (Squibb).
9. 8% glycerol.
10. 30 IU Pregnant mare's serum gonadotropin (PMSG): 0.3 mL of 100 IU/mL of MS dissolved in sterile saline.

2.3. Sperm Preparation

1. Semen sample from patient.
2. Semen sample from control (frozen or fresh).

2.4. Ova Preparation

1. Female Hamster: Between 7 and 12 weeks old.

3. Methods

Carry out all procedures at room temperature unless otherwise specified.

3.1. Sperm Preparation

Quality control is maintained where a fertile control, either frozen or fresh, is processed along with the patient sample (see Note 1).

1. Collect the ejaculate in a wide-mouth container provided by the physician and allow to liquefy for 30 min at 37°C.
2. Transfer the specimen to a 15 mL graduated tube to measure the volume.
3. Using a 50 µL pipette, apply a small drop of semen onto a prewarmed (37°C) microscope slide and cover with a #1 coverslip.
4. Aliquot 50 µL of semen directly in 0.95 mL sterile H_2O (1:20 dilution) in a 12 × 75 mm glass test tube. Mix the diluted semen by agitation, and load a small amount into a hemocytometer counting chamber and set aside.
5. Assess sperm quality by viewing the slide of undiluted semen on a 37°C warmed stage at 400× magnification under phase-contrast.
6. Count the sperm once they have all settled to the same focal plane on the hemocytometer counting grid.

7. Multiply the final count by the percent motility and multiply this product by the volume to yield the total number of motile cells per specimen (sperm count ($\times 10^6$ cells/mL) $\times$ % motility $\times$ volume = total number of motile cells).
8. Dilute the semen 1:1 with room-temperature TYB and mix with the aid of a Pasteur pipette. Allow solution to cool gradually, and maintain the sample at 4°C overnight (see Note 2).
9. Remove the samples from the refrigerator at 8:00 a.m. on the day of the assay. A visible sperm pellet should be present at the bottom of the tube.
10. Aspirate the TYB seminal plasma supernatant down to the top of the visible pellet.
11. Immediately add 6 mL of MBWW warmed to 37°C, and thoroughly resuspend the pellet using a transfer pipette (see Note 3).
12. Centrifuge all tubes for 10 min at $600 \times g$.
13. Aspirate the supernatant and resuspend the pellet in 3 mL of MBWW.
14. Divide the resuspension equally into six tubes (12×75 mm) and cap.
15. Centrifuge for 5 min at $600 \times g$ and place the tubes in a 37°C heat block for 60–90 min to allow motile sperm to swim up into the clear supernatant.
16. Using a Pasteur pipette, aspirate the supernatant from each set of tubes, taking care not to disturb the sperm pellets, and pool the supernatants into another 15 mL graduated tube. Centrifuge at $600 \times g$.
17. Determine the sperm concentration using a hemocytometer. Continue counting the sperm until the count reaches 100. Often it is necessary to dilute a 50 µL aliquot into a 200 µL aliquot of distilled water due to high sperm concentrations (see Notes 4 and 5).

Note: If fewer than 250,000 motile sperm are recovered in the swim-up, a micro-assay is performed as follows:

1. Place 0.2 mL MBWW (37°C) in a 0.4 mL micro-test tube.
2. Add the sperm suspension, and centrifuge at $600 \times g$ for 5 min.
3. After centrifugation, place the micro-test tube on a 37°C heater block to await placement of ova into the sperm suspension.

3.2. Ova Preparation

1. At 9:00 a.m. on day 1 of the assay period stimulate each hamster with 30 IU of PMSG injected intraperitoneally (see Note 6).

2. Administer a 37 IU dose of hCG at 3:00 p.m. on day 3 in an identical manner and volume (see Note 7).
3. At 8:00 a.m. on the morning of the 4th day, sacrifice the hamsters in an atmosphere saturated with CO_2.
4. Excise the oviducts and place them in the first of the three ice-cold saline washes after which the oviducts are singularly transferred to a 35 × 10 mm culture dish containing iced saline on a dissecting microscope.
5. At 40× magnification, puncture the bulbous section of the duct with a 26 gauge needle to free the cumulus mass enclosing the ova into the cold saline.
6. Once the entire mass has been teased free transfer it with a Pasteur pipette to a collection tube containing 0.5 mL ice-cold hyaluronidase (see Note 8).
7. Transfer the cumuli to a culture dish containing one drop of room-temperature MBWW. This is done at the time of harvest of the swim-up sperm in order to synchronize the ova and sperm preparations.
8. Stir and allow the ova to settle to remove the majority of the cumulus cells, which will remain suspended in the buffer. Remove most of the supernatant with a Pasteur pipette (see Note 9).
9. Transfer the ova using a Pasteur pipette with an inner bore twice the diameter of the zona-intact ova. By skillful application of suction and pressure, ova can be harvested quickly without any remaining cellular debris.
10. Remove the zona pellucida and the first polar body by placing the ova in a 200 mL droplet of MBWW containing 0.5 mg/mL trypsin for ~2 min.
11. Wash the ova by passage through three consecutive drops of fresh MBWW prior to incubation with sperm (see Note 10).

3.3. Sperm and Ova Incubation

To minimize inter-assay variation in SPA methodology, the period of time after thermal shocking of the sperm to the time the sperm and ova are incubated is held between 2 and 2½ h.

1. Centrifuge all the "swim-up" samples for 5 min at $600 \times g$.
2. Immediately aspirate down to 0.1 mL (or lower when indicated), and then resuspend the sperm to a final concentration of 5×10^6 motile sperm/mL.
3. Put two 50 µL droplets of sperm suspension in a small culture dish and overlay with 37°C silicone oil.

4. Place eight ova into each twin 50 µL droplets containing 250,000 100% motile sperm.

5. Place culture dishes in a tissue culture incubator at 37°C for 2½ to 3 h.

6. When incubation is complete, remove the excess sperm bound to the outer ovum membrane to provide an unobstructed view of those sperm that have penetrated.

Note: If samples with low motile sperm recovery are placed in micro-test tubes, add ova for incubation in the following manner:

1. Carefully draw no more than ten ova into the micropipette, and place the tip just under the surface of the MBWW in the tube. Allow the ova to gently "fall" into the liquid and settle to the bottom of the tube. Be careful not to disrupt the sperm pellet in the bottom of the tube.

2. Place the tube at 37°C, and incubate for 3 h prior to mounting the ova.

3.4. Mounting Ova

1. Use a mounting micropipette with an inner bore the same diameter as the zona-free ova to pass through three individual drops of buffer (see Note 11).

2. Prepare the coverslips just prior to use by placing a very small dot of paraffin–Vaseline (1:1) mixture on all corners.

3. Retrieve the ova so that a sufficient volume of buffer is drawn up in the micropipette after the last ovum.

4. When the ova are placed on the slide, a small droplet will be formed before the ova are expelled, which will prevent the fragile ova from rupturing.

5. The final droplet containing ova should be 1–2 µL. Quickly cover the droplet with the prepared coverslip.

6. Gently press down on one of the corners so that the droplet will spread and gradually draw the coverslip down, expelling air. This flattens the penetrated ova into one or two focal planes when viewed for scoring.

7. Score slides immediately on mounting.

3.5. Scoring Slides

1. Scoring must be performed using a phase-contrast microscope at 400× magnification.

2. Record the number of penetrated sperm in each ovum (see Note 12, Fig. 1).

3. Calculate the sperm capacitation index (SCI), which is simply the average number of penetrations per ovum for all ova counted.

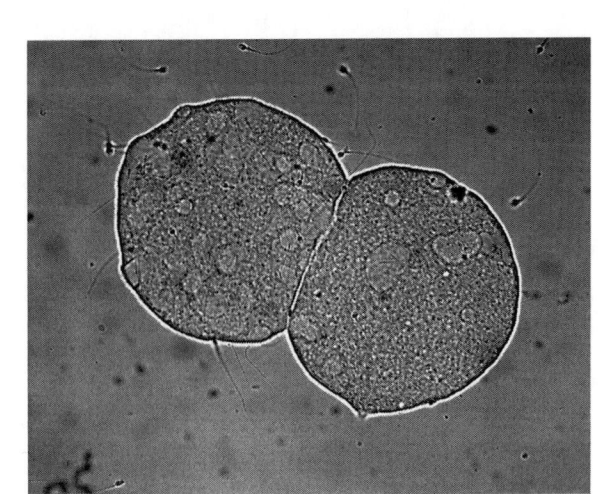

Fig. 1. Photomicrograph of two zona-free hamster oocytes, each displaying multiple sperm penetrations (*round, lighter areas* within the ooplasm).

4. Notes

1. *Frozen control*: A vial from the frozen control is retrieved from liquid nitrogen and submerged into the refrigerated container holding the patient's samples for 14–20 h at 4°C. At 8:00 a.m. on the day of the assay, 6 mL of warm MBWW is added to the frozen control. From this step onwards the control is processed the same way as the patient samples.

   Fresh control: A fresh control (fresh ejaculate from a pregnancy-proven donor) is run with every assay and is processed exactly like the patient samples.

2. It is important to cool the mixture gradually. Immerse the semen/TYB mixtures in a container of water at room temperature, and place this apparatus in the refrigerator. Not more than 2 h should elapse from the time the sample is collected to the start of the cooling process in the refrigerator.

3. The preparation of semen samples from the removal from the refrigerator on the day of the assay to when the warm MBWW is added should be preformed in rapid sequence on individual samples to assure the same temperature shock to all of the samples. This "thermal shock" is an important step for maximizing penetration rates.

4. When using the hemocytometer to count the sperm, if the count of 100 should fall in the middle of a square, finish counting that square. Record the number of sperm counted over the total number of squares on the worksheet. For example: If 109

sperm were counted in three squares, this value would be recorded as 109/3 on the worksheet.

5. The final swim-up volume is calculated as follows:

   Rec volume × (number of sperm/number of squares × 100) = Final volume

6. The intraperitoneal injection is made substernally with downward angulation, using a syringe fitted with a 26 gauge 3/8" needle.

7. Both PMSG and hCG injection solutions can be aliquoted and stored in a –20°C freezer for up to 1 year. It is important to thaw these solutions above room temperature and to use them within 30 min to maximize ova yield. The assay requires that one hamster be injected for every two patients scheduled.

8. The cumuli are collected and kept on ice until it is time to process them for incubation with the sperm.

9. It is necessary to remove the smaller cumulus cells, as they are enzymatically detached from the ova, to prevent interference with the subsequent trypsinization step.

10. It is also important to process the ova (up to 500) within 20 min from the time the cumulus mass is warmed at room temperature with MBWW. The penetration quality of fresh ova is preserved if the ova are kept at a low temperature for up to 24 h. For this reason the oviducts and cumulus masses are stored in iced buffer until needed. This also prevents ova from the animals sacrificed first from having a lower penetration rate than those harvested later.

11. Using the mounting pipette with an inner bore the same diameter as the zona-free ova creates forces in the micropipette tip that shear off the bound sperm. The ova are kept in buffer, while a standard slide is cleaned with ethanol.

12. A penetrated sperm is indicated by the presence of a swollen head associated with a tail. Using this assay, sperm penetrations are numerous, so the clear areas of the heads will often merge together and may become indistinguishable. In this case, it is necessary to count tails without heads. It is very rare that an unpenetrated (normal) sperm bound to the outside of the ova membrane will separate head from tail (Fig. 1).

References

1. Thonneau P et al (1991) Incidence and main causes of infertility in a resident population (1,850,000) of three French regions (1988–1989). Hum Reprod 6:811–816
2. Yanagimachi R et al (1976) The use of zona-free animal ova as a test-system for the assessment of the fertilizing capacity of human spermatozoa. Biol Reprod 15:471–476
3. Johnson A et al (1995) A quality control system for the optimized sperm penetration assay. Fertil Steril 64:832–837

4. Johnson A et al (1991) The microsperm penetration assay: development of a sperm penetration assay suitable for oligospermic males. Fertil Steril 56:528–534
5. Freeman MR et al (2001) Male partner screening before in vitro fertilization: preselecting patients who require intracytoplasmic sperm injection with the sperm penetration assay. Fertil Steril 76:1113–1118
6. Soffer Y et al (1992) Prediction of in vitro fertilization outcome by sperm penetration assay with TEST-yolk buffer preincubation. Fertil Steril 58:556–562
7. Gvakharia MO et al (2000) Human sperm microinjection into hamster oocytes: a new tool for training and evaluation of the technical proficiency of intracytoplasmic sperm injection. Fertil Steril 73:395–401

Chapter 11

Methods for the Assessment of Sperm Capacitation and Acrosome Reaction Excluding the Sperm Penetration Assay

Christopher J. De Jonge and Christopher L.R. Barratt

Abstract

Assessing the ability of human spermatozoa to acquire fertilizing potential (capacitation) by stimulating exocytosis of the contents of the acrosome (acrosome reaction) is thought to have diagnostic potential (De Jonge, Reprod Med Rev 3:159–178, 1994). Calcium-mobilizing agents, such as calcium ionophores (A23187) and progesterone, stimulate the acrosome reaction in vitro (Brucker and Lipford, Hum Reprod Update 1:51–62, 1995). Acrosomal status is easily detected using *Pisum sativum* Agglutinin labeled with fluorescein isothiocyanate (Cross and Meizel, Biol Reprod 41:635–641, 1989). Herein we describe a procedure for assessing capacitation and the acrosome reaction of human spermatozoa in vitro.

Key words: Capacitation, Acrosome reaction, Calcium ionophore, Progesterone, PSA-FITC

1. Introduction

Human sperm must undergo two sequential processes termed capacitation and acrosome reaction in preparation for fertilization in vivo (1, 2). Capacitation involves a complex of changes that include ion flux, sterol removal, redistribution of membrane protein receptors, protein kinase activation, flagellar amplitude, and beat frequency changes—collectively termed "hyperactivation," and other physiological events (1, 3). With the exception of hyperactivation, all of the aforementioned changes are preparatory for the acrosome reaction: an exocytotic event that involves the fusion and vesiculation of the plasma and outer acrosomal membranes and the release of lytic acrosomal enzymes (4). For the fertilizing spermatozoon, the acrosome reaction is induced by an oocyte's vestments—primarily the zona pellucida, and it occurs via typical ligand–receptor binding that stimulates signal transduction pathways and subsequent exocytosis (5).

Measurement of capacitation and acrosome reaction for the purpose of greater basic understanding of sperm cell biology and fertilization or for clinical diagnostics is essential. For the acrosome reaction, the methods for detecting the presence/absence of the acrosome are well defined and straightforward (3). In contrast, assaying capacitation as a process is rather enigmatic. On the one hand, a spermatozoon is deemed to be capacitated based on its induce-ability for the acrosome reaction. Alternatively, capacitation has been correlated with the phosphorylation of proteins containing tyrosine residues (6).

2. Materials

2.1. Culture Conditions for Capacitation and Acrosome Reaction

Capacitation of sperm and responsiveness to stimulators of the acrosome reaction are dependent upon the removal of seminal plasma (1, 7). Capacitation in vitro can be accomplished most easily and reliably by incubation of washed spermatozoa in culture media containing balanced salts, sodium bicarbonate, and protein, e.g., human serum albumin (HSA), at 37°C in 5–6% CO_2 and room air. If room air at 37°C is only possible, then a buffer, e.g., Hepes, should be added to the culture media to control for pH, and the incubation system should be tightly closed.

1. A swim-up (see Note 1) or density gradient prepared (see Note 2) sperm population.
2. Bicarbonate-buffered culture medium, such as Human Tubal Fluid (HTF) or Ham's F-10, with 3.5% HSA (v:v) to induce capacitation (see Note 3).
3. A Hepes-buffered culture medium, such as modified-HTF with 3.5% HSA (v:v) to induce capacitation.
4. Dimethyl sulfoxide (DMSO).
5. Calcium ionophore A23187 (Sigma), 1 mM stock solution in DMSO.
6. Progesterone (Sigma).

2.2. Fluorescence Assessment of Acrosomal Status

The PSA-FITC method was developed by Cross et al. (8) and was subsequently modified (9, 10); while not the only method for assessing acrosomal status, it is a widely used, frequently cited, and reliable technique (see Notes 6 and 7).

1. PSA-FITC: *Pisum sativum* Agglutinin (PSA) labeled with FITC (fluorescein isothiocyanate).
   (a) Stock solution: 2 mg PSA in 4 ml of phosphate-buffered saline (PBS). Store at –20°C.

(b) Working solution: Add 0.5 ml of stock PSA solution to 9.5 ml PBS, and store at 4°C for up to 4 weeks (see Note 8).

2. 0.9% (w/v) NaCl.
3. 95% (v/v) ethanol.

3. Methods

3.1. Procedure for Capacitation and Acrosome Reaction

1. Allow approximately 20–60 min for freshly ejaculated semen to completely liquefy.
2. Use density gradient centrifugation to prepare a highly motile sperm population that is free of leukocytes, germ cells, and dead spermatozoa.
3. Before use, allow test tubes containing bicarbonate-buffered culture medium to warm and equilibrate to 37°C in a 5% CO_2 in room air (v/v) incubator. If a CO_2 incubator is not available, use Hepes-buffered culture medium, and ensure that the tubes are tightly capped. Fresh media should be prepared for each assay.
4. Prepare control and experimental tubes with each tube containing 1×10^6 motile spermatozoa in 1 ml of culture medium (see Note 4).
5. Incubate sperm suspensions for 3 h (see Note 5).
6. Add 10 µl of 1 mM calcium ionophore A23187 to each experimental tube (final concentration 10 µM) to induce the acrosome reaction in capacitated spermatozoa. To control tubes, add an equivalent volume of DMSO (vehicle control). Alternatively, progesterone (3 µM final concentration in experimental tubes) can be used effectively as an inducer of the acrosome reaction.
7. Incubate all tubes for 15 min.
8. Remove a 10 µl aliquot from the control and experimental tubes, and place separately on a slide with coverslip for motility determination.
9. Prepare smears for subsequent PSA-FITC staining.

3.2. Procedure for Fluorescence Assessment of Acrosomal Status (See Note 9)

1. Prepare duplicate sperm smears about 1.5 cm long using aliquots (5–10 µl) from both control and experimental tubes.
2. Inspect the freshly prepared smears for even cell distribution and absence of sperm clumping using phase-contrast microscopy (×400).
3. Air-dry.
4. Fix for 30 min in 95% (v/v) ethanol.

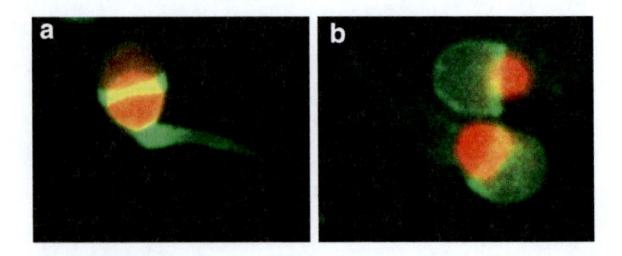

Fig. 1. Sperm stained with PSA and counterstained with propidium iodide. (**a**) demonstrates a sperm that has undergone the acrosome reaction with PSA staining only in the equatorial band. (**b**) demonstrates two sperm that have not undergone the acrosome reaction. Photos courtesy of the Doug Carrell laboratory.

5. Air-dry.
6. Immerse fixed slides into a vertical staining jar containing PSA stain (10 ml) for a minimum of 1 h at 4°C.
7. Wash each slide with distilled water.
8. After the slides have dried add two to three drops of ethanol-soluble mounting medium to the smear area, and apply a coverslip.
9. Using fluorescence optics (450–490 nm) and ×400 magnification view the slide under oil immersion.
10. Sperm are categorized for acrosomal status as follows:
    (a) Acrosome intact: More than half of the sperm head is brightly and uniformly fluorescing (see Fig. 1).
    (b) Acrosome-reacted: A band of fluorescence localized to the equatorial segment or no fluorescence in the acrosomal region.
    (c) Abnormal acrosomes: All other patterns.
11. Count 200 sperm per slide, and categorize the sperm according to acrosomal status.
12. Repeat the assessment on the duplicate slide.
13. Subtract the average percent acrosome reacted (AR) for control from the average %AR for experimental. The difference between the two values reflects the response of capacitated sperm to calcium ionophore A23187 (or progesterone) (see Note 10).

4. Notes

1. The direct swim-up technique is a very simple procedure for obtaining a highly motile sperm population that is free of poorly motile, immotile, and dead spermatozoa, round cells, and other contaminants (11). A drawback of this technique is the relatively low recovery of motile sperm cells.

2. Density gradient centrifugation yields a sperm population that is similar in quality to the direct swim-up technique and with the advantage of greater recovery of motile cells (11). If using this technique to isolate spermatozoa, then it is important to wash sperm completely free of the gradient material. Failure to do so can result in decreased progressive motility due to adherence of the sperm head to the surface of the microscope slide.

3. Selection of culture media should be based on the available resources and understanding that media type may serve as a confounder for acrosome reaction endpoint assessment (12).

4. Whenever possible it is advisable to run a positive control semen sample from a man whose spermatozoa have previously responded well to ionophore-induced acrosome reaction.

5. In the absence of using a live/dead stain, e.g., Hoechst, assessment of sperm motility at the beginning and end of capacitation and after acrosome reaction induction should be performed. If a decrease in sperm motility is recorded then it might be related to toxicity and a subsequent degenerative acrosome reaction that can occur due to cell death. These sperm should not be included as part of the sperm population being evaluated for induced acrosome reaction.

6. Evaluation of acrosomal status after induction of the acrosome reaction can be performed by using other markers and microscopy or flow cytometry (13–19).

7. Personal Communication from Professor R. John Aitken used with permission: "Essentially *Arachis hypogaea* is more sensitive (than *P. sativum*) because the outer acrosomal membrane disperses before the acrosomal contents (detected with *P. sativum*). This is presumably because constituents of the acrosomal matrix remain bound to the inner acrosomal membrane so that they can facilitate secondary zona binding and zona penetration" (13).

8. Perform a crossover test to compare the old and new stain for similarity in staining characteristics. It is suggested that the staining be evaluated on positive control (capacitated, ionophore-treated spermatozoa).

9. Personal Communication from De-Yi Liu used with permission: "Important factors, we wash sperm once using 0.9% NaCl and smear about 1.5 cm on the end of slide, fixed in 95% ethanol for 30 min and then stain in a vertical jar containing 10 ml PSA solution for 1–2 h (no effect if stain >2 h<8 h). Most people use PSA-FITC staining by add one to two drops of stain solution on smear which can dry up or stain uneven. We stain sperm smear in a jar (vertical), the solution can be re-used for extended period (we use for 8–10 weeks). So it is quite economical, and we do not use much PSA chemicals. In the past, many people fix sperm in tube and then make a smear; which often results in sperm stuck together or overlapping,

and making it difficult to read the slide. So it is critical to wash sperm to remove sugar and protein and smear even without overlap. We use this method every day for many years and it is very consistent and reproducible."

10. The %AR in control sperm after a 3-h incubation should, ideally, not be appreciably different from 0 h control sperm. Any significant increase in %AR between controls at the two time points is reflective of spontaneous acrosome reaction or a degenerative AR, the latter of which can be verified by motility determinations at the two time points.

References

1. De Jonge C (2005) Biological basis for human capacitation. Hum Reprod Update 11:205–214
2. Ikawa M et al (2010) Fertilization: a sperm's journey to and interaction with the oocyte. J Clin Invest 120:984–994
3. Visconti PE et al (2011) Ion channels, phosphorylation and mammalian sperm capacitation. Asian J Androl 13:395–405
4. Zanetti N, Mayorga LS (2009) Acrosomal swelling and membrane docking are required for hybrid vesicle formation during the human sperm acrosome reaction. Biol Reprod 81:396–405
5. Yanagimachi R (2011) Mammalian sperm acrosome reaction: where does it begin before fertilization? Biol Reprod 85:4–5
6. Visconti PE (2009) Understanding the molecular basis of sperm capacitation through kinase design. Proc Natl Acad Sci U S A 106:667–668
7. Cross NL (1996) Human seminal plasma prevents sperm from becoming acrosomally responsive to the agonist, progesterone: cholesterol is the major inhibitor. Biol Reprod 54:138–145
8. Cross NL et al (1986) Two simple methods for detecting acrosome-reacted human sperm. Gamete Res 15:213–226
9. Liu DY, Baker HW (1988) The proportion of human sperm with poor morphology but normal intact acrosomes detected with Pisum sativum agglutinin correlates with fertilization in vitro. Fertil Steril 50:288–293
10. Mendoza C et al (1992) Distinction between true acrosome reaction and degenerative acrosome loss by a one-step staining method using Pisum sativum agglutinin. J Reprod Fertil 95:755–763
11. WHO (2010) Laboratory manual for the examination and processing of human semen. World Health Organization, Department of Reproductive Health and Research, Geneva
12. Moseley FL et al (2005) Protein tyrosine phosphorylation, hyperactivation and progesterone-induced acrosome reaction are enhanced in IVF media: an effect that is not associated with an increase in protein kinase A activation. Mol Hum Reprod 11:523–529
13. Aitken RJ, Brindle JP (1993) Analysis of the ability of three probes targeting the outer acrosomal membrane or acrosomal contents to detect the acrosome reaction in human spermatozoa. Hum Reprod 8:1663–1669
14. Cooper TG, Yeung CH (1998) A flow cytometric technique using peanut agglutinin for evaluating acrosomal loss from human spermatozoa. J Androl 19:542–550
15. Cross NL (1995) Methods for evaluating the acrosomal status of human sperm. In: Fenichel P, Parinaud J (eds) Human sperm acrosome reaction. (Colloques INSERM) John Libbey Eurotext, France
16. De Jonge CJ et al (1989) Synchronous assay for human sperm capacitation and the acrosome reaction. J Androl 10:232–239
17. Fenichel P et al (1989) Evaluation of the human sperm acrosome reaction using a monoclonal antibody, GB24, and fluorescence-activated cell sorter. J Reprod Fertil 87:699–706
18. Henley N et al (1994) Flow cytometric evaluation of the acrosome reaction of human spermatozoa: a new method using a photoactivated supravital stain. Int J Androl 17: 78–84
19. Talbot P, Chacon RS (1981) A triple-stain technique for evaluating normal acrosome reactions of human sperm. J Exp Zool 215:201–208

Part III

Sperm DNA Damage Testing

Chapter 12

Sperm DNA Fragmentation Analysis Using the TUNEL Assay

Rakesh Sharma, Jayson Masaki, and Ashok Agarwal

Abstract

Terminal deoxynucleotidyl transferase-mediated deoxyuridine triphosphate-nick end labeling or the TUNEL assay is an important technique in the assessment of DNA damage. Semen samples are routinely assessed microscopically to assess their fertilization ability. In addition to routine semen analysis, the use of the TUNEL assay can provide information on the level of DNA damage present within a sample. This chapter provides a practical walk-through guide aimed at directing a researcher or a clinical facility interested in setting up and using TUNEL and flow cytometry or fluorescence microscopy for sperm DNA analysis.

Key words: TUNEL, DNA damage, Flow cytometry, Chromatin

1. Introduction

This chapter showcases the application of the TUNEL assay in the measurement of sperm DNA damage. The protocol of the TUNEL assay is provided as a model guide for any lab interested in establishing a similar protocol for measuring the DNA damage in ejaculated semen samples with a detailed explanation of the sperm preparation, staining, and inclusion of appropriate controls, instrument specifications, and troubleshooting tips.

1.1. Sperm DNA Damage

The single goal of the spermatozoon is to deliver the paternal counterpart of genetic material to the oocyte. Routine semen analysis is still the cornerstone in the laboratory workup of the infertile male, and the microscopically observable characteristics of the spermatozoa such as concentration, motility, and morphology are important. However, the inclusion of an additional parameter that assesses DNA damage can provide for a more comprehensive description of semen quality. Expanding the parameters of routine semen analysis to include an assessment of genetic integrity is essential in the complete characterization of a semen sample. After all, determining the extent of sperm DNA damage may be of primary importance in predicting the ultimate health of the resulting progeny.

DNA damage may involve single-stranded breaks or "nicks," double-stranded breaks or "fragments," deletions/additions, and base modifications. The term DNA fragmentation is technically associated with endonuclease-mediated double-stranded DNA cleavage as a result of apoptotic programmed cell death. However it has also become interchangeable with the general term "DNA damage" when in the context of TUNEL assay results. This is because TUNEL staining was initially (and still is) used as a marker of DNA fragmentation in distinguishing between apoptosis and necrosis.

The TUNEL staining technique can be used to directly assess DNA damage of a cell in the form of both single- and double-stranded breaks. When coupled with flow-cytometry technology, a quantitative assessment of the proportion of genetically compromised cells within a semen sample can be provided. Understanding the level of DNA integrity of a semen sample can be of diagnostic and predictive value in fertility clinics, as well as in the optimization of experimental methods to aid in treating infertility.

1.2. Etiology of DNA Damage

The causes of sperm DNA damage have been studied extensively. However because of its multifactorial basis and indistinct relationship with infertility, the comprehensive origins of and preventative treatment for DNA damage still remain controversial. Associations of cause and effect are not consistently observed across all studies, and, even more importantly, are confused by various methods of detection (1). In general, the main sources of seminal DNA fragmentation detectable by TUNEL appear to be from free radical damage via reactive oxygen species (ROS) and abortive apoptosis (2, 3). Other factors include radiation, aberrant spermatozoa maturation, or DNA packaging, although these are not specifically attributable to fragmentation as measured by the TUNEL assay.

Oxidative stress occurs when an imbalance of ROS exceeds the locally available antioxidant capacity to prevent oxidative cellular damage. Sperm DNA is especially susceptible to oxidative damage. Spermatozoa have a low cytoplasmic volume, wherein preventive mechanisms against ROS normally exist within a cell. The highly condensed structure of sperm nuclear material rules out enzymatic repair of damaged DNA. Sperm motility is crucial and energy requirements are provided by extremely active mitochondria that generate ROS. Furthermore, processing of semen for various assisted reproductive techniques such as swim-up or density gradient centrifugation removes the seminal plasma that protects the spermatozoa against ROS through its antioxidant properties (4).

Nature provides for a functional deterrent to inhibit spermatozoa exposed to high levels of ROS to reach and fertilize an egg. The sperm plasma membrane contains polyunsaturated fatty acids, which are extremely susceptible to ROS attack (5). Cells exposed to oxidative stress sufficient to cause DNA damage would likely have a compromised plasma membrane and therefore be incapable of natural fertilization. However, this natural process of functional selection is bypassed when techniques such as Intracytoplasmic Sperm Injection (ICSI) are used.

Sources of increased ROS affecting sperm function can be intrinsic (from within the spermatozoa themselves) or extrinsic. The main significant sources of intracellular ROS are superoxide anions produced during oxidative phosphorylation in spermatozoon mitochondria and plasma membrane NAD(P)H oxidase, and nitric oxide produced by nitric oxide synthase within the plasma membrane (6). Extrinsic sources include largely the leukocytes (granulocytes), unknown chemicals/toxins, and environmental pollution, nutritional deficiency, laboratory processing of samples, advanced age, psychological and physical stress, smoking, alcohol, caffeine, cancer and anticancer drugs, or inflammatory processes/cells. Intrinsic sources of ROS appear to be more damaging to sperm DNA than extrinsic ROS. For example, although leukocyte and inflammatory processes can contribute to increased ROS in the ejaculate on the order of 1,000 times that produced intrinsically (7), DNA fragmentation has a stronger correlation with intrinsic ROS than for leukocyte ROS production (8). This may be due to the fact that the antioxidant capacity of the seminal fluid in which spermatozoa are bathed negates the ROS effects from exogenous sources. However in ART procedures where the seminal plasma is removed, the leukocytes come in close proximity to the normal and abnormal spermatozoa. An inadequate antioxidant capacity would therefore greatly contribute to DNA damage by means of oxidative stress. Dietary and in vitro antioxidant supplementation has been shown to be beneficial in reducing some exogenous ROS-induced DNA damage (9).

In the treatment of ROS-mediated DNA damage, it is important to note that eliminating all sources of ROS is not only impossible but also not advisable, as ROS are normal products of metabolism, and their complete elimination may adversely affect sperm function. Certain levels of endogenous ROS produced at specific times are crucial for the normal physiologic maturation, capacitation, and hyperactivation of sperm (10, 11). Oxidative stress has been shown to correlate with apoptosis (12, 13); however, the relationship is not entirely causal; oxidative stress can induce apoptosis, and is also a product resulting from the process of apoptosis.

1.3. How TUNEL Works

TUNEL utilizes a template-independent DNA polymerase called Terminal Deoxynucleotidyl Transferase (TdT) which non-preferentially adds deoxyribonucleotides to 3′ hydroxyl (OH) single- and double-stranded DNA (14). Deoxyuridine triphosphate (dUTP) is the substrate that is added by the TdT

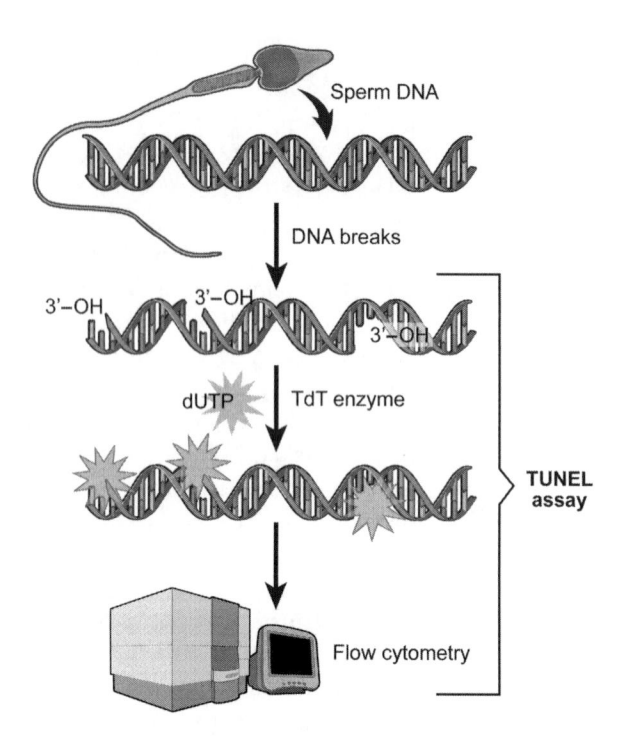

Fig. 1. Schematic of the TUNEL assay.

enzyme to the free 3'-OH break-ends of DNA (see Fig. 1). The added dUTP can be directly labeled and therefore acts as a direct marker of DNA breaks, or the signal can be amplified by the use of a modified dUTP to which labeled anti-dUTP antibody can be adsorbed. The more DNA strand break sites present, the more label is incorporated within a cell. The TUNEL-stained cells can be visualized microscopically; however, for the high-throughput assessment of clinical semen samples it is more practical to use a flow cytometer that detects the intracellular fluorescence of fluorochrome-labeled dUTP within a population of cells. The flow cytometer analyzes one cell at a time, and counts the number of individual events (cells) that it recognizes to have a specific intensity of fluorescence. Fluorescein isothiocyanate (FITC) is the fluorophore label most commonly used in TUNEL staining. Propidium Iodide (PI) is used as a fluorescent counterstain that labels all DNA so that every cell can be counted, and the resultant proportion of those showing DNA damage can be calculated. Because the TdT will bind and incorporate dUTP at the terminal 3' ends of each DNA strand with the same propensity as interior 3' ends (15) irrespective of damage, there is a threshold of labeled dUTP-induced fluorescence which should be considered in order to distinguish between viable, relatively undamaged cells and those with severe, irreparable DNA damage. The parameters used for this threshold are included in the procedure portion of this guide, and are also discussed in Subheading 4.

TUNEL staining kits are readily available in the commercial marketplace. These kits are very cost effective, highly standardized by the manufacturers, and often provide some technical support. Most TUNEL staining protocols are very similar, with only minor variations in centrifugation force, incubation times, temperatures, and reagent concentrations, but all follow the same general principles. The TUNEL kit utilized in the following procedure is the *APO-DIRECT*™ kit (BD Biosciences Pharmingen, San Diego, CA). This kit is designed for use with flow cytometry for the general detection of DNA fragmented apoptotic cells and also provides positive and negative control cells (diploid cell lines) for standardization purposes. When measuring sperm DNA damage it is prudent to include sperm controls for each assay as a precautionary indicator of problems in order to rule out reagent variability between kits, and also because gating is necessary in flow cytometry and must be done on the same cell type.

1.4. Other Assays of DNA Integrity

There are three main groups of assays that are used in the evaluation of cellular genetic integrity and are divided based on what each assay measures. The most obvious and direct method is to assess the actual molecular DNA strand breaks present (in a number of cells in the sample). This snapshot-approach is where the TUNEL assay is categorized, along with COMET using neutral conditions, and in situ nick translation. The next assay group measures the potential susceptibility of the DNA to exogenous damage rather than assessing actual DNA damage directly since these techniques first induce DNA damage by exposure to denaturing conditions. Assays of this type include the Sperm Chromatic Structure Assay (SCSA) and COMET using alkaline conditions. Such assays shed light on associated abnormalities that may affect the genetic integrity such as aberrations in condensation structure/packing, or lack of DNA stabilizing and protective mechanisms. Interestingly, within these two groups of assays there is considerably high correlation between both the measure of DNA damage and the outcome of fertilization using assisted reproductive techniques. The choice of the assay would be based on the ease of the assay, robust nature of the assay (sensitivity and specificity) and if it is cost effective practicality in training, staffing, time, and resources will be the primary determining factors. The third group of assays examines DNA base modification as a means of linking markers of oxidative DNA damage (such as 8-OH-guanosine and 8-OH-2′-deoxyguanosine) with a general loss of DNA integrity. While it is has been shown that these tests can be used to predict male infertility and pregnancy rates (16, 17), the oocyte has the ability to repair such damage, and as a result the focus on the assessment of DNA damage in the clinical

Table 1
Advantages and disadvantages of TUNEL, Comet, and SCSA

Assay	Main advantages	Main disadvantages
TUNEL 1. Adds labeled nucleotides to free DNA ends 2. Template independent 3. Labels SS and DS breaks 4. Measures percent cells with labeled DNA	1. Direct objective 2. Inexpensive 3. Can be performed on few sperm 4. High repeatability 5. Quick and simple (fluorescence microscopy) 6. Objective. High sensitivity (flow cytometry) 7. Fresh or frozen samples 8. Indicative of apoptosis 9. Correlates with semen parameters 10. Associated with fertility 11. Available in commercial kits	1. Thresholds not standardized 2. Variable assay protocols 3. Not designed specifically for spermatozoa 4. Not specific to oxidative damage 5. Need for special equipment (flow cytometer)
Comet For single- and double-stranded DNA 1. Electrophoresis of single sperm cells 2. DNA fragments form tail 3. Intact DNA stays in head Alkaline COMET 1. Alkaline conditions, denatures all DNA 2. Identifies both DS and SS breaks Neutral COMET 1. Does not denature DNA 2. Identifies DS breaks, maybe some SS breaks	1. Indirect assay, subjective 2. Inexpensive 3. Poor repeatability 4. High sensitivity 5. Fresh samples only 6. Correlates with seminal parameters 7. Small number of cells required 8. Versatile (alkaline or neutral)	1. Variable protocols 2. Unclear thresholds 3. Not available in commercial kits 4. Time and labor intensive 5. Small number of cells assayed 6. Subjective 7. Not specific to oxidative damage 8. Lacks correlation with fertility 9. Requires special imaging software
SCSA For single-stranded DNA 1. Mild acid treatment denatures DNA with SS or DS breaks 2. Acridine orange binds to DNA 3. Double-stranded DNA (nondenatured) fluoresces green, single-stranded DNA (denatured) fluoresces red 4. Flow cytometry counts thousands of cells. DNA fragmentation index (DFI)—the percentage of sperm with a ratio of red to (red + green) fluorescence greater than the main cell population	1. Direct objective 2. Inexpensive 3. Established clinical thresholds 4. Many cells rapidly examined 5. High repeatability 6. Fresh or frozen samples 7. Most published studies reproducible	1. Proprietary method 2. Not available in commercial kits 3. Expensive equipment 4. Acid-induced denaturation 5. Small variations in lab conditions affect results 6. Calculations involve qualitative decisions 7. Very few labs conduct this assay

TUNEL terminal deoxynucleotidyl transferase-mediated dUTP nick end-labeling; *SCSA* sperm chromatin structure assay

setting has been mainly on those tests that evaluate double-stranded DNA breaks (fragmentation)—the most genotoxic type of damage (18, 19).

Some of the major advantages and disadvantages of the TUNEL, Comet, and SCSA assays are outlined (see Table 1). The TUNEL assay is by no means the end-all solution, but it is certainly one of the more simple, reliable, objective, and cost-effective methods available for assessing DNA damage. Multiple approaches can be collaboratively used for better evidence of genetic integrity as well as newer, modified, and improved TUNEL methods that also incorporate sperm vitality assessment (20).

2. Materials

2.1. Semen Collection and Handling

1. Sterile open wide mouth specimen cups.
2. Incubator (37°C).
3. Cell counting chambers.
4. Microscope.
5. Centrifuge tubes—12 × 75 mm with caps (polystyrene recommended to limit cell loss).
6. Micropipettes and tips.
7. Phosphate-buffered saline (PBS, pH 7.4).
8. DNase I—1 mg/ml.

2.2. Sperm Fixation

1. Centrifuge—300 × g.
2. Serological pipettes.
3. Transfer/Pasteur pipettes.
4. 3.7% paraformaldehyde.
5. Ice.
6. 70% (v/v) ethanol.

2.3. Staining

1. APO-DIRECT Kit components (store the various components as directed by manufacturer): Reaction buffer (contains coenzyme factors and cacodylic acid), PI/RNase staining buffer (5 μg/ml PI, 200 μg/ml RNase), Rinsing buffer, Wash buffer, TdT enzyme, FITC-dUTP, Positive and Negative control cells.
2. Distilled water.
3. Aluminum foil.

2.4. Flow Cytometry and Data Analysis

1. Flow cytometer FacScan (Becton Dickinson, San Jose, CA).
2. Computer software (FlowJo, Mac Version 8.2.4, LLC, Ashland, OR).

3. Methods

3.1. Semen Collection and Concentration Assessment

1. Collect semen following an abstinence period of 48–72 h. The sample should be collected by means of masturbation into a sterile specimen cup without lubricant. Specimen must be maintained at body temperature and processed within 1 h of collection. Record the sample ID, collection date and time, period of abstinence, and any remarkable conditions. Keep the sample in a 37°C incubator until complete liquefaction has occurred.

2. Following liquefaction perform a sperm count (see Note 1). Approximately $2-3 \times 10^6$ total cells are sufficient to run the assay. To optimize the stains, the sperm concentration should be no more than 5×10^6 (see Note 2). For efficiency samples can be stored at −20°C and batched for TUNEL analysis. Alternatively, the cells can be individually fixed and stored in ethanol until the time of analysis (see fixation step).

3.2. Preparation of Assay Controls (See Note 3)

1. For the positive sperm control, DNA damage is induced by digestion with DNase I. Incubate a sample from a healthy donor after the cell counting step with 100 μl of DNase I (1 mg/ml) for 1 h at 37°C.

2. Similarly, a sample from a healthy donor is used for the negative sperm control, wherein the TDT enzyme is omitted from the staining step as described below.

3.3. Fixation of Sperm

1. Prepare the fixation buffer by adding 10 ml of 37% formaldehyde (100% formalin) to 90 ml of PBS (pH 7.4) to give a 3.7% (v/v) paraformaldehyde solution (see Note 4).

2. Centrifuge the sperm sample at $300 \times g$ for 7 min to pellet and separate the cells from the seminal plasma (see Note 5). All subsequent centrifugation steps are to be done at $300 \times g$ for 5 min.

3. Discard the supernatant by gently aspirating with a transfer or Pasteur pipette (see Note 6).

4. Suspend the cells in the 3.7% paraformaldehyde fixation buffer. Place the suspension on ice for 30–60 min or alternatively refrigerate at 4°C overnight.

5. Discard the supernatant and suspend the pellet in 1 ml of ice-cold 70% (v/v) ethanol at −20°C. At this point the cells can be stored in ethanol at −20°C (see Notes 7 and 8).

3.4. Preparation of Samples for Staining

1. Resuspend all samples (test and controls) by vortexing the tubes since the cells will have settled after prolonged storage in ethanol.

2. For the kit controls, a positive and negative control is provided with the *APO-DIRECT* kit. Remove 2-ml of each of the control cell suspensions (1×10^6 cells/ml, 2×10^6 cells total) and place in 12×75 mm centrifuge tubes.

3. Centrifuge both test and control tubes at $300 \times g$ for 5 min and discard the ethanol supernatant.

4. Resuspend the cells in 1 ml of Wash Buffer (see Note 9), centrifuge at $300 \times g$ for 5 min, and remove the supernatant.

5. Repeat the Wash Buffer treatment a second time.

3.5. Staining

1. Prepare an appropriate volume of the Staining Solution based on the number of samples to be assayed (see Note 10). For each sample to be analyzed, combine 10 µl Reaction Buffer, 0.75 µl TdT Enzyme, 32.25 µl distilled water, and 8 µl FITC-dUTP. The FITC-dUTP reagent should be added last as it is light sensitive (see Note 11).

2. Resuspend the cell pellets in 50 µl of the Staining Solution. For negative semen controls, add staining solution without the TdT enzyme.

3. Vortex to disperse the cells and to allow the staining solution to permeate homogeneously into every cell.

4. Incubate for 60 min at 37°C.

5. After incubation, directly add 1.0 ml of Rinse Buffer to each sample, centrifuge at $300 \times g$ for 5 min, and remove the supernatant by aspiration.

6. Repeat rinse step with an additional 1.0 ml of Rinse Buffer, centrifuge, and again remove the supernatant of each tube.

7. Resuspend in 0.5 ml of the PI/RNase Staining Buffer and incubate tubes at room temperature for 30 min.

8. Analyze the cells in the PI/RNase solution by flow cytometry within 3 h of completing the staining procedure (see Note 12).

3.6. Flow Cytometry, Data Acquisition, and Storage

Care should be taken to use the protocol provided here as only a guide when adjusting the settings of any particular flow cytometry setup. Each machine will be inherently different, but should present similar data when adjusted/calibrated to the same parameters. In our facility an automated flow cytometer is used to analyze the TUNEL-stained sperm. The machine is equipped with a 15 mW argon laser that supplies a filtered 488 nm excitation beam. Green (FITC) fluorescence is measured in the FL-1 channel (480–530 nm) and red (PI) fluorescence is set to be detected in the FL-2 channel (580–630 nm). The steps for setup and data acquisition that we employ for the FacScan flow cytometer is detailed in Subheading 4 (see Note 13).

3.7. Analysis of Data	Data analysis can be performed using the operator's software of choice. The workflow that we employ for the analysis is provided below (see Note 14).
3.8. Clinical Threshold of Infertility	The suggested threshold of infertility used by our lab is 19% DNA fragmentation. This is the cutoff value we have been using to discriminate between a healthy semen sample and one that has positive DNA damage. Every lab should establish its own reference values utilizing appropriate controls (see Note 15).
3.9. Measuring DNA Damage by Fluorescence Microscopy	If flow cytometry is not accessible, the same stained cells can be scored with a fluorescence microscope.

1. Load an aliquot of the stained sample on a slide and cover with a coverslip.
2. Score a minimum of 500 spermatozoa per sample under 40× objective with an epifluorescence microscope (excitation between 460 and 490 nm and an emission >515 nm).
3. First count the number of spermatozoa per field stained with PI (red).
4. Count the number of cells emitting green fluorescence (TUNEL positive) in the same filed.
5. Calculate the percent of TUNEL-positive cells.

4. Notes

1. We recommend that disposable counting chambers specifically made for counting sperm cells (e.g., MicroCell 20 μm 2-chamber) be used for accuracy and operator.
2. The APO-DIRECT kit instructions suggest the use of $1-2 \times 10^6$ cells/ml cell concentration. We suggest that no more than 5×10^6 cells/ml be used, as with each washing and centrifugation step some cells are lost.
3. Rather than making new positive and negative sperm controls every time, previously analyzed samples may be used as standardized/reference control samples with which the flow cytometer analysis software can be calibrated in order to maintain consistent DNA damage percent between batches. Stored aliquots of extra cells from two fixed samples in which DNA fragmentation percentage was previously determined (known positive and negative) should be processed alongside the current batch beginning with the staining procedure. The FC analysis software settings used to gate the new batch of cells can then be adjusted to reflect the previous results of the same samples within ±1%.

While the positive and negative controls provided with commercially available TUNEL kits are adequate for calibration during flow cytometer analysis of somatic cells, they are not suitable as controls for haploid cells such as spermatozoa and will have different nuclear size and level of DNA condensation. Therefore it is important to prepare positive and negative sperm control samples.

4. 3.7% (w/v) paraformaldehyde is used instead of the 1% recommended by the APO-DIRECT kit instructions to ensure that the DNA is completely fixed. We have chosen this concentration out of the convenience of a 1/10 dilution and believe it is justified by the fact that sperm nuclear DNA is the most highly compacted of eukaryotic DNA (21). However, any concentration between 1 and 4% should be sufficient for fixation purposes. Higher concentrations should be avoided as it may inhibit thorough washing of the unwanted formaldehyde from the sample and cause problems in staining. Formaldehyde is carcinogenic and proper handling precaution must be observed.

5. Centrifugation at $300 \times g$ for 7 min instead of 5 min allows for cells to be pelleted more fully in the case of increased seminal plasma viscosity.

6. 50–100 μl of supernatant may be left during aspiration to avoid losing cells after each centrifugation step. It is much more convenient to leave a small amount of fluid rather than to have to recentrifuge the sample due to a disturbed pellet. The repeated washing/rinsing steps should negate any residual supernatant.

7. While the fixed cells can be kept in paraformaldehyde for several days or weeks at 4°C, do not freeze the samples in the paraformaldehyde solution. It is recommended that suspension in 70% ethanol be used for long-term storage of samples at −20°C and that samples not to be processed on the same day. The manufacturer (BD Pharmingen™) also suggests that incubation in ethanol for at least 12–18 h yields better staining results.

8. Always ensure that the quantity of ice is sufficient to last 2–3 days in case of an unexpected delay in delivery.

When shipping samples (individual or batched) from outside labs, cryovials are labeled with the sample information (i.e., date, name, type of sample, volume, etc.). The sample can be either frozen in multiple aliquots soon after liquefaction or processed by suspending the sperm cells ($1–2 \times 10^6$ cells/ml) in 3.7% (w/v) paraformaldehyde prepared in PBS (pH 7.4). The sample is placed on ice for 30–60 min and centrifuged for 5 min at $300 \times g$. The supernatant is discarded and the cells suspended in 70% (v/v) ice-cold ethanol. The cells are stored in 70% (v/v) ethanol at −20°C. At the time of overnight shipping, cryovials are placed in a cryobox and packed with adequate amount of dry ice.

9. Because volume accuracy is not important in the fixation, washing, and rinsing steps, it is easier to use a serological pipette to add the paraformaldehyde, ethanol, wash, and rinse buffers.

10. A special staining solution without the TdT enzyme should also be prepared for the negative semen control. Because the staining solution is only active for approximately 24 h at 4°C after being prepared, mix the reagents together only when required and prepare no more than the amount necessary. The TdT enzyme volume is the only limiting reagent in the kit, as all other reagents are generously supplied. Briefly centrifuge the TdT enzyme tube to make sure that the reagent is at the bottom.

11. All subsequent steps should be conducted in a low-lighted room and the solutions and samples should be covered by a sheet of aluminum foil when exposure to light is expected.

12. The cells should be analyzed as quickly as possible. Prolonged delay >1 h will result in overstaining and degradation of the cells.

13. *CellQuest*™ software is used to adjust the electronics and setup conditions for data acquisition and storage. To optimize the sample, a *CellQuest* template is set up. The forward scatter and side scatter detectors are adjusted, the forward scatter threshold is adjusted, and the population is gated. The FL1 and FL2 channels are adjusted accordingly. Under the '*Acquire*' menu, select '*Connect to cytometer*'. Again under the '*Acquire*' menu, select '*Parameter Description*'. Here you can define your file storage by selecting the storage location. From the '*Parameter Description*', select the folder. Open your assigned folder, create a new folder (enter name of the folder, date created), and select the newly created folder. From the '*Parameter Description*' select '*File*' and under '*File name prefix*' (custom prefix) enter an appropriate file name. Change '*File count*' to 1. Click 'OK' under '*Sample ID*' and '*Patient ID*' insert tube descriptions. Under '*Cytometer*' menu, select '*Instrument Setting*', open the dialogue box and in your folder go to '*Kit*' setting, and hit '*set*' and '*done*'. Now the instrument setting is in place for the assay kit. Go to '*Detectors/Amps*' and only the settings for P4 (FL2 channel) should be changed. Click '*Threshold*'. The threshold for '*FL2-H*' should be fixed at 40 and should not be changed. The first time that the assay kit is run, the settings must be saved. This is saved in a separate folder named "*KIT*". For actual test samples, in the '*Instrument*

setting', open the dialogue box, double click on the 'TUNEL' folder, and then click on '*set*' and '*Done*'. Set the '*FSC-H*' threshold. The threshold for '*FSC-H*' should be fixed at 52 and should not be changed. Similarly, when running the test sample for the first time, once the settings are optimized they can be saved into a folder named "*TUNEL*". When ready to run the sample, change the instrument knob to '*RUN*'. The default is in '*SETUP*' mode where you can preview your cells, adjust instrument setting, and create or modify gates without collecting data. To collect gated data, while in the '*SETUP*', create a gate using the toolbar (Creating Gate/Regions). Under the '*Acquire*' menu, select '*Acquisition & Storage*'. Again select '*Counters*'. This allows monitoring the sample flow rate and event count (see Fig. 2). When ready to collect cells, deselect the '*SETUP*' check box on '*Acquisition Controls*'. A data file will generate when the signal sounds or you can manually save. Once the collecting event is begun, do not alter instrument settings or gates. When finished acquiring the data, exit *CellQuest* and copy/back up your data files to a flash drive.

14. Raw data from the flow cytometer is imported into the *FlowJo* software (Mac version 8.2.4, FlowJo, LLC, Ashland, OR). The '*Workspace*' window is composed of '*Group panel*' and '*Sample*' panel. Transfer the folder into '*Group panel*'. Create subgroups for "*Kit/control*" samples and "*TUNEL/test*" samples. Drag the kit (controls) sample data to the '*Kit*' group and the actual '*Test*' samples to the "*TUNEL*" folder. Next open the sample of interest by double clicking on the sample panel and click on the sample of interest. The *FlowJo* opens a graph window displaying a bivariate dot plot of the cells. The initial graph the *Flow Jo* displays is always a '*Forward vs. Side Scatter plot*' because this view is generally used to gate out dead cells. Select the '*Dot plot*' format. The next step is to create '*Gates*'. To create a polygon gate, click the mouse and add the polygon gate. Once you have closed the gate, give the file name "*Singlet*" to this subpopulation of cells. Next Double click on "*Singlet*" and open a new graph. Display the X-axis options by clicking on the axis label and choose "*FL-W*" from the pull-down menu. Click on the Y-axis label and select "*FL2-A*" (PI channel). To see the gated area without the gate, double click on the plot. Next click on the X-axis and choose "*FL1-H*" (FITC Green channel) and on the Y-axis, click "*Histogram*" from the pull-down menu. Create the '*Negative*' gate and give the file name "*Negative*". Similarly, create the '*Positive*' gate and give the file name "*Positive*". Apply all analysis nodes (gated subpopulations) to other samples. Gating parameters are applied to every sample simultaneously by dragging the gating tree onto the "Kit" and "TUNEL" subgroups. To do this, highlight

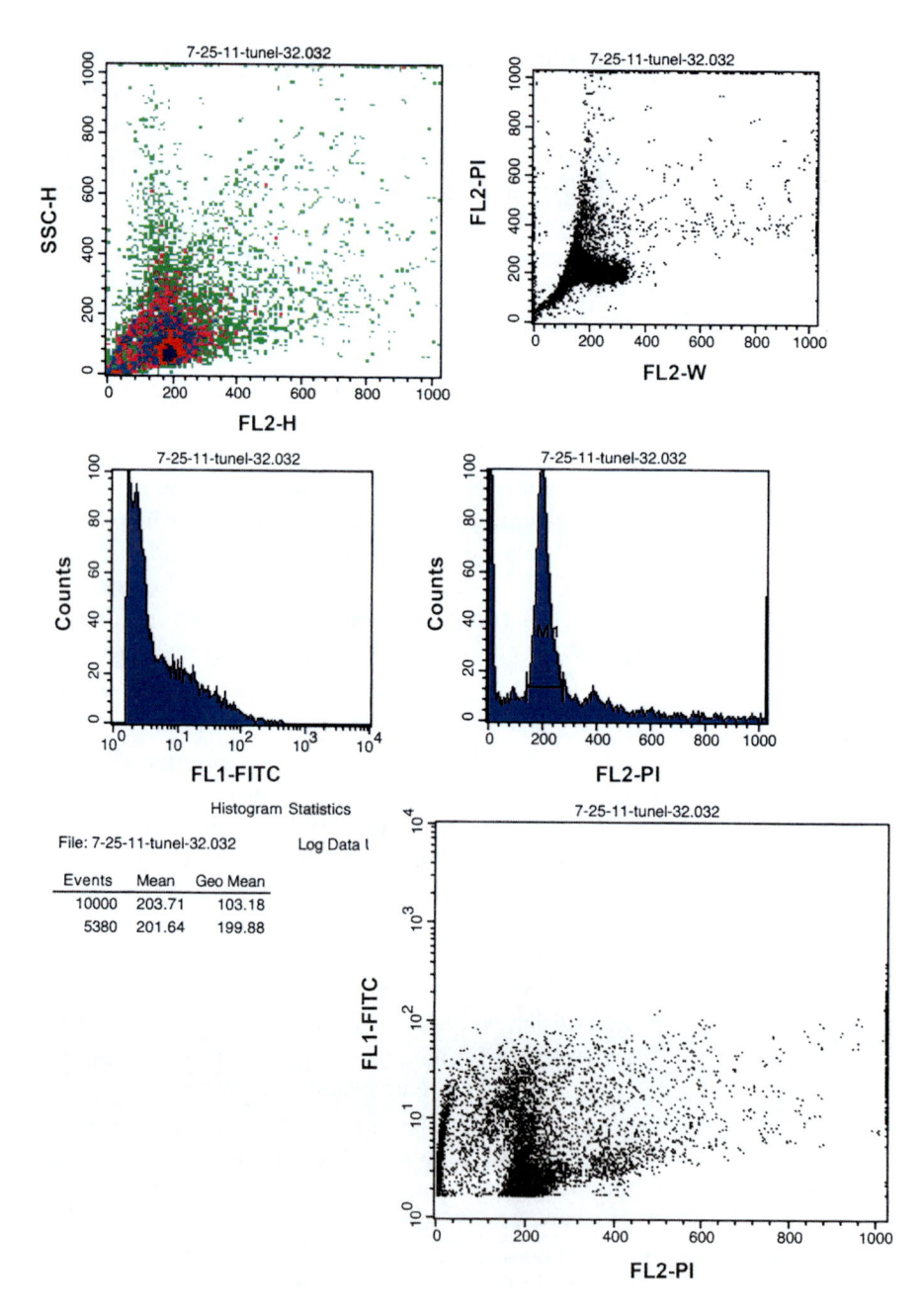

Fig. 2. Representative setup of the data acquisition for a single sample by FACScan before it is analyzed by FlowJo software. FL1 = Green channel; FL2 = Red channel; PI = Propidium iodide; FITC = Fluorescence isothiocynate; SSC = Side scatter count.

the "Singlet", "Negative", and "Positive" subfolders and drag-and-drop to the "KIT" group. Similarly apply the same to the 'TUNEL' group. Note: If the gate needs to be changed/adjusted after the desired gate has been obtained, highlight the "Singlet", "Positive", and "Negative" subfolders and press the

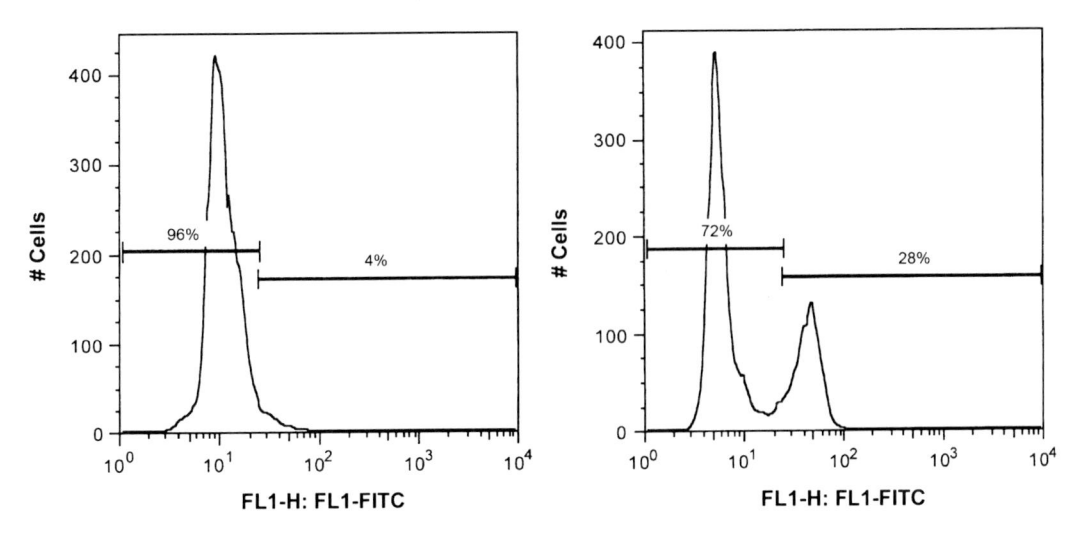

Fig. 3. Representative histogram showing (**a**) TUNEL-negative and (**b**) TUNEL-positive sample.

'Shift' and 'Delete' keys. Redraw the gates and save again. Always save the Workspace frequently.

To prepare graphical reports and generate tables, open the '*Layout Editor*' window by clicking the rightmost function in the Workspace window. The left panel shows the names of the layouts and the right panel is a drawing board. In the drawing board, a graphical display for a single sample can be created. *FlowJo* is able to create a batch report. To do this, click on "*Singlet*" from either the "*Kit*" or the "*TUNEL*" subgroups and drag it to the drawing board. Double click on 'batch' at the upper corner of the 'Batch Report'. Place the report panel according to the number of plots in each row or column (each block represents one printed page). Drag and adjust and print the report accordingly. A representative curve for a sample that is negative and positive for DNA damage is illustrated in Fig. 3. Similarly a table can also be printed with the data statistics.

15. The DNA fragmentation threshold of 19.25% was determined to be the cutoff value which maximized the sum of sensitivity and specificity by receiver operating characteristic (ROC) curve analysis in a recent 2010 study of 194 infertile and 25 healthy male subjects using an identical protocol (22). At this threshold value the study had 100% specificity, 64.9% sensitivity, 97.7% positive and 37% negative predictive values, and an accuracy of 70%. This is very close to the 20% threshold calculated by ROC curve analysis in a 2005 study using a different protocol (yielding 89.4% specificity, 96.9% sensitivity, and 92.8% positive and 95.5% negative predictive values). Several studies have shown a threshold/cutoff range of DNA damage of about 10–40% (23–26).

References

1. Aitken RJ et al (2009) Biological and clinical significance of DNA damage in the male germ line. Int J Androl 32:46–56
2. Sharma RK et al (2004) Sperm DNA damage and its clinical relevance in assessing reproductive outcome. Asian J Androl 6:139–148
3. Tremellen K (2008) Oxidative stress and male infertility—a clinical perspective. Hum Reprod Update 14:243–258
4. Twigg J et al (1998) Iatrogenic DNA damage induced in human spermatozoa during sperm preparation: protective significance of seminal plasma. Mol Hum Reprod 4:439–445
5. Agarwal A, Prabakaran SA (2005) Mechanism, measurement, and prevention of oxidative stress in male reproductive physiology. Indian J Exp Biol 43:963–974
6. Koppers AJ et al (2008) Significance of mitochondrial reactive oxygen species in the generation of oxidative stress in spermatozoa. J Clin Endocrinol Metab 93:3199–3207
7. Plante M et al (1994) Reactive oxygen species released by activated neutrophils, but not by deficient spermatozoa, are sufficient to affect normal sperm motility. Fertil Steril 62:387–393
8. Henkel R et al (2005) Effect of reactive oxygen species produced by spermatozoa and leukocytes on sperm functions in non-leukocytospermic patients. Fertil Steril 83:635–642
9. Zini A et al (2009) Antioxidants and sperm DNA damage: a clinical perspective. J Assist Reprod Genet 26:427–432
10. Makker K et al (2009) Oxidative stress & male infertility. Indian J Med Res 129:357–367
11. Desai N et al (2009) Physiologic and pathologic levels of reactive oxygen species in neat semen of infertile men. Fertil Steril 92:1626–1631
12. Wang X et al (2003) Oxidative stress is associated with increased apoptosis leading to spermatozoa DNA damage in patients with male factor infertility. Fertil Steril 80:531–535
13. Mahfouz R et al (2010) Semen characteristics and sperm DNA fragmentation in infertile men with low and high levels of seminal reactive oxygen species. Fertil Steril 94:2141–2146
14. Robbins DJ, Coleman MS (1988) Initiator role of double stranded DNA in terminal transferase catalyzed polymerization reactions. Nucleic Acids Res 16:2943–2957
15. Robbins DJ et al (1987) Interaction of terminal transferase with single-stranded DNA. J Biol Chem 262:9494–9502
16. Loft S et al (2003) Oxidative DNA damage in human sperm influences time to pregnancy. Hum Reprod 18:1265–1272
17. Shen HM et al (1999) Evaluation of oxidative DNA damage in human sperm and its association with male infertility. J Androl 20:718–723
18. Derijck A et al (2008) DNA double-strand break repair in parental chromatin of mouse zygotes, the first cell cycle as an origin of de novo mutation. Hum Mol Genet 17:1922–1937
19. Lewis SE et al (2008) Sperm DNA tests as useful adjuncts to semen analysis. Syst Biol Reprod Med 54:111–125
20. Mitchell LA et al (2011) The TUNEL assay consistently underestimates DNA damage in human spermatozoa and is influenced by DNA compaction and cell vitality: development of an improved methodology. Int J Androl 34:2–13
21. Ward WS, Coffey DS (1991) DNA packaging and organization in mammalian spermatozoa: comparison with somatic cells. Biol Reprod 44:569–574
22. Sharma RK et al (2010) TUNEL as a test for sperm DNA damage in the evaluation of male infertility. Urology 76:1380–1386
23. Duran EH et al (2002) Sperm DNA quality predicts intrauterine insemination outcome: a prospective cohort study. Hum Reprod 17:3122–3128
24. Benchaib M et al (2003) Sperm DNA fragmentation decreases the pregnancy rate in an assisted reproductive technique. Hum Reprod 18:1023–1028
25. Henkel R et al (2003) DNA fragmentation of spermatozoa and assisted reproduction technology. Reprod Biomed Online 7:477–484
26. Henkel R et al (2004) Influence of deoxyribonucleic acid damage on fertilization and pregnancy. Fertil Steril 81:965–972

Chapter 13

Sperm DNA Damage Measured by Comet Assay

Luke Simon and Douglas T. Carrell

Abstract

Measurement of sperm DNA damage is a useful tool in the evaluation of male infertility, as the sperm nucleus lacks protection against oxidative stress and is vulnerable to oxidation-mediated DNA damage. The Comet assay or single-cell gel electrophoresis is a relatively simple and sensitive method for measuring strand breaks in DNA in individual sperm. During this procedure, sperm cells are embedded in a thin layer of agarose on a microscope slide and lysed with detergent under high salt conditions. This process removes protamines and histones allowing the nucleus to form a nucleoid-like structure containing supercoiled loops of DNA. Alkaline pH conditions result in unwinding of double-stranded DNA, and subsequent electrophoresis results in the migration of broken strands towards the anode, forming a comet tail, when observed under fluorescence microscope. The amount of DNA in the head and tail is reflected by its fluorescent intensity. The relative fluorescence in the tail compared with its head serves as a measure of the level of DNA damage. In this chapter, we describe the alkaline version of the Comet assay, which is highly sensitive for measuring single- and double-strand DNA breaks.

Key words: Alkaline comet assay, Decondensation, Electrophoresis, Sperm DNA damage, Single- and double-strand breaks

1. Introduction

The Comet assay or single-cell gel electrophoresis is one of the simplest methods to measure single- and double-strand breaks in sperm (1). The principle of the assay is based on the separation of broken DNA strands under the influence of an electric field facilitated by the charge and size of the strands. After separation, the intact DNA remains in the comet's head, whereas single- and double-stranded broken DNA fragments migrate into the comet's tail (2). Therefore, the sperm with high levels of DNA strand breaks would show an intense comet tail (3) and increased comet tail length (4). Additional parameters have been used to increase the efficiency of the test such as the diameter of the nucleus, olive tail moment, and comet tail length (5).

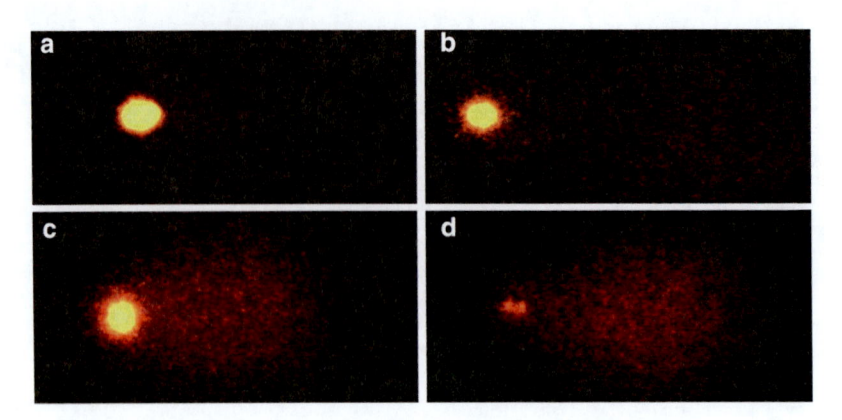

Fig. 1. Photomicrograph of comets with different levels of DNA damage measured by the alkaline Comet assay. Sperm with (**a**) no or 0% damage, (**b**) 12% damage, (**c**) 45% damage, and (**d**) 95% damage.

The Comet assay has been used to assess DNA damage in a number of cell types and a wide range of studies such as evaluation of the effects of UV radiation, carcinogens, toxic substances, and radiotherapy (5). Steps involved in assessment of sperm DNA damage with the comet assay are the following: Sperm cells are mixed with agarose and layered on microscopic slide. Detergents and high salt concentration are used to lyse their cell membranes, and remove nuclear proteins (protamines and histones), which relaxes the DNA into a supercoiled nucleoid structure. The nucleoids are subjected to an electrophoretic field resulting in migration of the broken strands of DNA through the agarose. Cells with DNA damage will resemble comets when visualized using a fluorescent dye, with the comet's tail length and fluorescent intensity proportional to the degree of DNA fragmentation. The non-fragmented DNA remains in the nucleus or comet's head (6, 7). The Comet assay can be performed in a neutral or an alkaline environment: in the neutral buffer only DNA with double-strand damage is measured, while in the alkaline buffer, single- and double-strand damage and alkali-labile sites are detectable due to unwinding of the strands (8). This is the only technique that can directly assess the amount of DNA damage in individual sperm. A photomicrograph of different levels of sperm DNA damage measured by the alkaline Comet assay is shown in Fig. 1.

1.1. Advantages of Comet Assay

The Comet assay is inexpensive and one of the most sensitive techniques available to measure sperm DNA damage. The alkaline Comet assay can be used in all cell types including sperm (9). The assay requires only a few cells; and the level of DNA damage can be obtained from individual cells. In sperm, complete decondensation by the removal of protamines can be obtained only in Comet assay. The significance of Comet assay in assessing male infertility has been demonstrated by a number of authors (10–13). The clinical association between sperm DNA damage measured by the Comet assay and assisted reproductive treatment outcomes was recently established by Simon et al. (14–17).

1.2. Limitations of Comet Assay

The disadvantage of the assay is that it lacks standardized protocols, which makes it difficult to fully understand and relate the results of different authors (8). The assay condition is known to damage the alkaline-labile sites, making it difficult to discriminate between endogenous and induced DNA breaks. In somatic cells, DNA damage can be overestimated by the Comet assay due to the presence of residual RNA which can increase the background intensity during analysis (18). The assay is also criticized for underestimation of DNA damage due to entangling of DNA strands. If chromatin decondensation in sperm is incomplete, DNA will be underestimated. Overlapping comet tails decrease the accuracy of the assay, small tail fragments may be lost, and excessively small fragments are difficult to visualize. It is difficult to differentiate sperm from the contaminant cells present in the semen after the assay. The assay is laborious, has a high level of inter-laboratory variation, and hence is not recommended for clinical use (19, 20).

2. Materials

The laboratory conditions should be maintained for temperature and humidity. The assay should be performed under yellow light (570 nm) to minimize DNA damage.

2.1. Chemicals

1. Normal melting point agarose (NMP).
2. Low melting point agarose (LMP).
3. Phosphate-buffered saline (PBS).
4. Tris(hydroxymethyl)aminomethane Hydrochloride (Tris–HCl).
5. Sodium chloride (NaCl).
6. Sodium salt of Ethylenediaminetetraacetic acid (Na_2EDTA).
7. Triton X-100 (Triton).
8. Dithiothreitol (DTT).
9. Lithium diiodosalicyclate (LIS).
10. Sodium hydroxide (NaOH).
11. Ethidium bromide (EtBr).

2.2. Equipment

The following list of equipment is required:

1. Two water baths.
2. Microwave oven.

3. pH meter.
4. Electronic balance.
5. Magnetic stirrer.
6. Micropipettes.
7. Hemocytometer or Makler chamber.
8. Stopwatch.
9. Refrigerator and freezer (4 and −20°C).
10. Electrophoresis tank with power pack.
11. Epi-fluorescence microscope, camera, and UV light source.
12. Computer with Comet analysis software.

2.3. Consumables and Glass Ware

1. PureSperm® wash (Nidacon International AB, Molndal, Sweden).
2. Fully frosted slides.
3. Coverslips (24 × 50 mm).
4. Coplin jars (25 ml capacity).
5. Eppendorf tubes and stand.
6. General glassware (beakers, measuring cylinders, etc.).
7. Gloves.
8. Micropipette tips (10, 20, 200, 1,000 μl).
9. Syringe (20 ml).
10. Tissue/blotting paper.

3. Methods

3.1. Preparation of Stock Solutions

All stock solutions should be prepared with double-distilled water (ddH$_2$O) and stored at room temperature, unless or otherwise specified.

1. Lysis buffer: Dissolve 146.4 g NaCl, 37.2 g Na$_2$EDTA, and 1.2 g Tris–HCl in 800 ml of ddH$_2$O (final concentrations 2.5 M NaCl, 100 mM Na$_2$EDTA, and 10 mM Tris–HCl), bring the volume up to 950 ml with ddH$_2$O, adjust to pH 10 (with NaOH), then make the volume up to 1,000 ml with ddH$_2$O, and store at 4°C (see Notes 1 and 2).
2. DTT (10 mM): Dissolve 1.58 g of DTT in 50 ml of ddH$_2$O. Prepare 1.25 ml aliquots in 2 ml Eppendorf tubes and store at −20°C (see Note 3).
3. LIS (4 mM): Dissolve 1.54 g of LIS in 50 ml of ddH$_2$O. Prepare 1.25 ml aliquots in 2 ml Eppendorf tubes and store at −20°C.

4. Electrophoresis buffer: Two stock solutions are required to prepare fresh electrophoresis buffer. To prepare fresh electrophoresis buffer for each experiment, add 60 ml of stock solution 1 (10 M NaOH) to 10 ml of stock solution 2 (200 mM EDTA) and make the volume up to 2,000 ml with ddH$_2$O.
   (a) Stock solution 1 (10 M NaOH): Dissolve 400 g of NaOH in ddH$_2$O and make the volume up to 1,000 ml using ddH$_2$O.
   (b) Stock solution 2 (200 mM Na$_2$EDTA): Dissolve 74.45 g of Na$_2$EDTA in ddH$_2$O, adjust to pH 13 (with NaOH), and make the volume up to 1,000 ml using ddH$_2$O (see Note 1).

5. Neutralization buffer (0.4 M Tris): Dissolve 48.5 g of Tris–HCl in ddH$_2$O, adjust to pH 7.5 (with NaOH), and make the volume up to 1,000 ml using ddH$_2$O.

6. Staining solution (200 mg/ml EtBr): Dissolve 10 mg of EtBr in 50 ml of ddH$_2$O in an amber bottle to make a stock of 10× concentration. Wrap the bottle in tin foil and store at room temperature.

7. PBS: Dissolve one tablet of PBS in 200 ml of ddH$_2$O or as per the manufacturer's instructions.

3.2. Experimental Protocol

The protocol is obtained from Hughes et al. (1997) and Agbaje et al. (2008). All steps are performed at room temperature unless otherwise specified.

1. Turn on two water baths: One should be at 37°C and the other at 45°C (see Note 4).

2. Prepare agarose gels by adding 0.250 g of normal-melting-point agarose (0.5% gel) to 25 mL of PBS and 0.125 g of low-melting-point agarose (0.5% gel) to 25 ml of PBS in 50 ml glass flasks. Heat flasks in the microwave to completely melt the agarose (see Notes 5–7).

3. Place the NMP gel in the 45°C water bath and the LMP gel in the 37°C water bath (see Notes 8 and 9).

4. Carefully pipette 200 μl of NMP gel onto the slide and immediately cover with a coverslip (see Notes 10 and 11).

5. Leave the slides on the bench at room temperature for 15 min to allow the agarose to solidify (see Notes 12 and 13).

6. Adjust the concentration of sperm to 6×10^6/ml using PBS or PureSperm® wash (see Note 14).

7. Remove the coverslips very gently, taking care not to remove the layer of agarose (see Note 15).

8. Place 10 μl of the sperm sample (6×10^6/ml) into a 0.5 ml Eppendorf tube, and add 75 μl of LMP gel incubated at 37°C to the sperm. Mix thoroughly and pipette on top of the layer

of NMP agarose gel dropwise. Quickly cover with a coverslip and leave it on the bench at room temperature for 15 min to allow the agarose to solidify (see Note 16).

9. Remove the lysis solution from the refrigerator, add 250 µl of Triton X-100 to 22.5 ml of lysis stock solution, and mix thoroughly in a Coplin jar (see Notes 17–19).
10. Remove the coverslips from the slides and immerse in lysis solution for 1 h at 4°C (see Notes 15 and 20).
11. Take the slides out of the Coplin jar, add 1.25 ml of DTT, invert to ensure mixing, then return the slides to the Coplin jar, and incubate for 30 min at 4°C (see Note 21).
12. Take the slides out of the Coplin jar, add 1.25 ml of LIS, invert to ensure mixing, then return the slides to the Coplin jar, and incubate for 90 min at room temperature (see Note 21).
13. Remove slides and carefully drain off any remaining liquid by standing them vertically on tissue paper against a support.
14. Fill the horizontal gel electrophoresis tank with fresh alkaline electrophoresis solution (see Note 22).
15. Leave the slides in the buffer for 20 min (see Note 23).
16. Electrophorese the slides for 10 min by applying a current at 25 V (0.714 V/cm) adjusted to 300 mA by adding or removing (±1–20 ml) buffer in the tank using a 20 ml syringe (see Notes 24 and 25).
17. After electrophoresis, drain the slides on tissue paper as before, place them on a tray, and flood with three changes of neutralization buffer for 5 min each (see Note 26).
18. Drain the slides thoroughly to remove the neutralization buffer. Prepare fresh EtBr solution by diluting the stock solution ten times in ddH$_2$O. Add 50 µl of EtBr to each slide and cover with a coverslip (see Note 27).
19. View the slides using a fluorescence microscope with appropriate Comet software. Analyze 50 comets per slide (see Notes 28–33).

4. Notes

1. EDTA will not be dissolved in water unless the pH of the solution is increased.
2. Lysis buffer should be prepared fresh every week. Old buffers show reduced ability to lyse cells, so it is better to make lysis buffer in small volumes based on individual requirements.
3. Use a fume hood to prepare DTT to avoid uncomfortable odor.
4. The water bath at 37°C is used to incubate LMA while the water bath at 45°C is used to incubate NMA.

5. If threaded cap-type bottles are used, then the caps should be loosened prior to heating the agarose in the microwave.

6. The remaining agarose in the bottle can be stored for later use. Repeated heating of agarose will result in an increase in concentration of agarose due to evaporation of water from the gel; therefore after 5–7 uses new NMP and LMP agarose should be prepared.

7. Dissolve both NMA and LMA gels in microwave at 50% power for 30 s. Remove the flasks, shake gently, and continue to heat them in the microwave. Repeat the process until the agarose is dissolved completely. Avoid boiling of agarose. Loss of water will increase the concentration of agarose and interfere with the mobility of DNA strands during electrophoresis.

8. Once the NMP and LMP agarose is melted, flasks or bottles should be placed in their respective water baths to cool to the optimal temperature for use. Maintaining a constant temperature will ensure uniform thickness of agarose gel on the slide, which is particularly important for scoring of comets.

9. If only one water bath is available, then the water bath should be used for LMP agarose (37°C) to avoid overheating of sperm in the molten agarose.

10. Only fully frosted microscopic slides should be used for the Comet assay. The agarose should be coated on the frosted surface of the slide as it will not stick to the smooth surface of the slide.

11. The function of the NMP agarose layer is to hold the LMP agarose (containing sperm) to the slide surface. LMP agarose alone will not adhere to the surface of the slide.

12. It is important to leave the slides at room temperature to allow the agarose to set. Some of the earlier procedures recommended refrigeration of the slides at 4°C. However, refrigeration of the slides keeps the agarose moist which allows the gel to slide off at a later stage.

13. If the agarose gels keep sliding off during subsequent experiments, then increase the concentration of the NMP agarose from 0.5 to 1%, but the concentration of the LMP should remain constant.

14. The concentration of sperm and volume of sperm pipetted (Subheading 3.2, step 8) should be kept constant for all the experiments. Increase in the concentration of sperm/volume will lead to overcrowding of sperm comets and overlapping of comet tails, rendering slides impossible for the software interpretation. If the concentration of sperm is too low, then it is difficult to find the comets under the microscope while scoring.

15. Care should be taken when removing the coverslip from the surface of the agarose gel. Gently sliding the coverslip along

the surface of the agarose will minimize cracking of the agarose layer. Even with the gel cracks, the experiment can be continued without any adverse effect. However, cracked gels have a higher probability of sliding off than non-cracked gels.

16. Try to avoid formation of bubbles in the LMP layer. Formation of bubbles in the agarose layers will not interfere with the assay but will make scoring of the comets more difficult.

17. When pipetting Triton X-100, use a 1,000 µl pipette and cut the tip to produce a large opening. Because Triton X-100 is highly viscous it is not easily pipetted.

18. Ensure proper mixing of Triton X-100 in the lysis buffer. Inadequate mixing will result in the detergent floating on the surface of the buffer, leading to improper lysis of sperm membranes. In addition, avoid foam formation by avoiding vigorous shaking of the buffer.

19. If a bigger Coplin jar/container is used then increase the volume of the buffer and reagents accordingly to ensure that the entire surface of the agarose gel is submerged in the buffer.

20. DNA repair enzymes will be released into the buffer when somatic cells are lysed. Hence, incubation of the lysis step under refrigeration conditions will prevent any potential DNA repair.

21. Inadequate mixing of DTT and LIS in the lysis buffer will lead to underestimation of sperm DNA damage and asymmetrical readings between the duplicate slides.

22. Place each slide into the tank making sure that the gels face upwards and the buffer is at a level of 0.25 cm above the slide surface.

23. The sensitivity of the assay will decrease if the electrophoresis buffer is not maintained at pH 13. High alkaline conditions result in the separation of double-strand DNA into single strands, thereby enabling additional measurement of single-strand breaks.

24. Constant voltage and current (milliamps) are required to maintain standard electrophoretic mobility of DNA fragments in all the experiments. The volts should be constant in all the experiments and the milliamps should be adjusted to the required level. At constant voltage, increasing the level of buffer will increase the milliamps and vice versa.

25. In some power packs, the volts and milliamps will fluctuate. In such cases an approximate range should be maintained in all the experiments.

26. Neutralization of slides should be complete; if not the EtBr stain will not properly bind to the DNA.

27. Higher concentrations of EtBr will increase background staining resulting in decreased visibility of migrated tail DNA.

Typically, the background of the comets should be black to brown (Fig. 1).

28. The slides should be analyzed immediately. Storage of slides to analyze the next day will decrease the accuracy of the assay, because the DNA fragments will diffuse in the agarose gel leading to reduced visibility of comet tails or sometimes expanded appearance of comet shapes.

29. The commonly used Comet analysis software packages are Andor Komet software (Andor Technology, UK), MetaSyatems automated imaging (MetaSystems Group Inc., MA, USA), and Comet Assay IV (Perceptive Instruments, UK).

30. During scoring, the first 50 comets observed should be scored. Scoring only good-looking comets will affect the results.

31. Comets with no heads also called hedgehogs should be considered as sperm containing 100% DNA damage.

32. When scoring, comets with overlapping tails should be avoided.

33. Make sure that the background intensity measurement box used by the software is on clear space. If part of the box overlaps any of the comets then the background intensity will be considered high and will affect the experimental results.

Acknowledgement

This work was supported by University of Utah, and Andrology and IVF laboratories.

References

1. McKelvey-Martin VJ et al (1997) Two potential clinical applications of the alkaline single-cell gel electrophoresis assay: (1). Human bladder washings and transitional cell carcinoma of the bladder; and (2). Human sperm and male infertility. Mutat Res 375:93–104
2. Klaude M et al (1996) The comet assay: mechanisms and technical considerations. Mutat Res 363:89–96
3. Hughes CM et al (1996) A comparison of baseline and induced DNA damage in human spermatozoa from fertile and infertile men, using a modified comet assay. Mol Hum Reprod 2:613–619
4. Singh NP, Stephens RE (1998) X-ray induced DNA double-strand breaks in human sperm. Mutagenesis 13:75–79
5. Singh NP et al (1988) A simple technique for quantitation of low levels of DNA damage in individual cells. Exp Cell Res 175:184–191
6. Ostling O, Johanson KJ (1984) Microelectrophoretic study of radiation-induced DNA damages in individual mammalian cells. Biochem Biophys Res Commun 123:291–298
7. Collins AR et al (1997) The comet assay: what can it really tell us? Mutat Res 375:183–193
8. Tarozzi N et al (2007) Clinical relevance of sperm DNA damage in assisted reproduction. Reprod Biomed Online 14:746–757
9. Singh NP et al (1989) Abundant alkali-sensitive sites in DNA of human and mouse sperm. Exp Cell Res 184:461–470
10. Irvine DS et al (2000) DNA integrity in human spermatozoa: relationships with semen quality. J Androl 21:33–44
11. Chan PJ et al (2001) A simple comet assay for archived sperm correlates DNA fragmentation to reduced hyperactivation and penetration of zona-free hamster oocytes. Fertil Steril 75: 186–192

12. Donnelly ET et al (2001) Assessment of DNA integrity and morphology of ejaculated spermatozoa from fertile and infertile men before and after cryopreservation. Hum Reprod 16:1191–1199
13. Lewis SE, Agbaje IM (2008) Using the alkaline comet assay in prognostic tests for male infertility and assisted reproductive technology outcomes. Mutagenesis 23:163–170
14. Simon L et al (2011) Relationships between human sperm protamines, DNA damage and assisted reproduction outcomes. Reprod Biomed Online 23:724–734
15. Simon L, Lewis SE (2011) Sperm DNA damage or progressive motility: which one is the better predictor of fertilization in vitro? Syst Biol Reprod Med 57:133–138
16. Simon L et al (2011) Sperm DNA damage measured by the alkaline Comet assay as an independent predictor of male infertility and in vitro fertilization success. Fertil Steril 95:652–657
17. Simon L et al (2010) Clinical significance of sperm DNA damage in assisted reproduction outcome. Hum Reprod 25:1594–1608
18. Shamsi MB et al (2008) Evaluation of nuclear DNA damage in human spermatozoa in men opting for assisted reproduction. Indian J Med Res 127:115–123
19. Olive PL et al (2001) Analysis of DNA damage in individual cells. Methods Cell Biol 64:235–249
20. Olive PL et al (1992) Factors influencing DNA migration from individual cells subjected to gel electrophoresis. Exp Cell Res 198:259–267

Chapter 14

Sperm Chromatin Structure Assay (SCSA®)

Donald P. Evenson

Abstract

The SCSA® is the pioneering assay for the detection of damaged sperm DNA and altered proteins in sperm nuclei via flow cytometry of acridine orange (AO) stained sperm. The SCSA® is considered to be the most precise and repeatable test providing very unique, dual parameter data (red vs. green fluorescence) on a $1,024 \times 1,024$ channel scale, not only on DNA fragmentation but also on abnormal sperm characterized by lack of normal exchange of histones to protamines. Raw semen/sperm aliquots or purified sperm can be flash frozen, placed in a box with dry ice and shipped by overnight courier to an experienced SCSA® lab. The samples are individually thawed, prepared, and analyzed in ~10 min. Of significance, data on 5,000 individual sperm are recorded on a $1,024 \times 1,024$ dot plot of green (native DNA) and red (broken DNA) fluorescence. Repeat measurements have virtually identical dot plot patterns demonstrating that the low pH treatment that opens up the DNA strands at the sites of breaks and staining by acridine orange (AO) are highly precise and repeatable (CVs of 1–3%) and the same between fresh and frozen samples. SCSAsoft® software transforms the X-Y data to total DNA stainability versus red/red + green fluoresence (DFI) providing a more accurate determination of % DFI as well as the more sensitive value of standard deviation of DFI (SD DFI) as demonstrated by animal fertility and dose–response toxicology studies. The current established clinical threshold is 25% DFI for placing a man into a statistical probability of the following: (a) longer time to natural pregnancy, (b) low odds of IUI pregnancy, (c) more miscarriages, or (d) no pregnancy. Changes in lifestyle as well as medical intervention can lower the %DFI to increase the probability of natural pregnancy. Couples of men with >25% DFI are counseled to try ICSI and when in the >50% range may consider TESE/ICSI. The SCSA® simultaneously determines the % of sperm with high DNA stainability (%HDS) related to retained nuclear histones consistent with immature sperm; high HDS values are predictive of pregnancy failure.

The SCSA® is considered to be the most technician friendly, time- and cost-efficient, precise and repeatable DNA fragmentation assay, with the most data and the only fragmentation assay with an accepted clinical threshold for placing a man at risk for infertility. SCSA® data are more predictive of male factor infertility than classical semen analyses.

Key words: SCSA, DNA fragmentation, ART, IVF

1. Introduction

1.1. Uniqueness of the SCSA®

The SCSA® is considered to be the most precise and repeatable test providing very unique, dual parameter data (red vs. green fluorescence) on a 1,024 × 1,024 channel scale, not only on DNA fragmentation, but also abnormal sperm characterized by a lack of normal exchange of histones to protamines. These abnormal protein–DNA complexes allow greater access of acridine orange dye (AO) to sperm DNA thus producing a High DNA Stainable (green fluorescence) fraction (HDS). HDS data are highly correlated ($r=0.610$, $p<0.001$) with data derived from other tests for excess histones such as the CMA_3 test (1). In one study (2), the HDS fraction contained unprocessed P2 protamines, likely resulting in structural chromatin abnormalities that allowed an increased amount of DNA being available for AO staining.

The SCSA® is ideal for researchers and students since the reagents cost only about $0.10 per test allowing many repeat experimental tests to be done at very little cost, provided that a relatively inexpensive, two-parameter flow cytometer is available.

1.2. Direct Versus Indirect Sperm DNA Fragmentation Tests

Some authors have classified the various sperm DNA fragmentation tests as being either a direct or indirect test. For example, the Tunel test is considered a direct test on the assumption that the DNA strand break marker has "direct" access to the broken strand. The SCSA® has been considered an indirect test since it requires an agent, heat or low pH, to open the DNA double helix at the sites of DNA strand breaks. However, when one considers that the Tunel test requires extensive pelleting and washing of the sperm, permeabilization of the cell membranes, agents (e.g., DTT) in some cases to open up the S-S bonds of chromatin (3), incubation with enzymes of potential varying activity and washing out of nonspecific label, it hardly seems appropriate to consider this a direct test that takes at least 4 h to conduct. In sharp contrast, the SCSA® only requires that fresh or frozen/thawed semen be treated for 30 s with pH 1.2 buffer to open, with extremely high and repeatable precision, the DNA at the sites of strand breaks, followed by staining with AO that stays in equilibrium with the cells as they are measured. Thus, this author considers the SCSA® to be as much or more a "direct test" of sperm DNA fragmentation as others.

1.3. SCSA® Data

Figure 1 illustrates SCSA® raw data for human sperm and the same data converted by SCSAsoft®.

1.4. Characterization of Sperm Populations Identified in SCSA® Analysis

Flow cytometry-sorted SCSA-processed human sperm (4) showed that Feulgen stained moderate-DFI sperm had normal nuclear shape, while high DFI sperm nuclei had a smaller area. The HDS fraction, known to consist of immature sperm, had significantly

Poor DNA Integrity

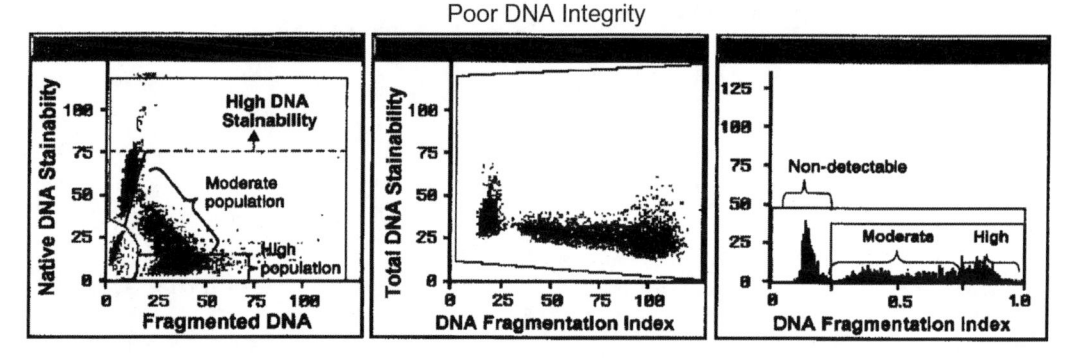

Patient	Date	Measurement	mean DFI	SD DFI	DFI	HDS
7272-113	####	1	563.7	307.0	64.9	6.4
		2	561.4	304.8	64.9	7.2
		mean	562.6	305.9	64.9	6.8
		sd	1.2	1.1	0.0	0.4

Fig. 1. *Left panel*: Green versus red scattergram (cytogram) showing 5,000 dots, each representing a single event with specific green (native DNA) and red (fragmented DNA) coordinates on a scale from 0 to 1,024. The *horizontal dashed line* lays at the top of the highest green fluorescence values for normal sperm. Sperm above this line have "High DNA Stainabilty" (HDS) and are characterized by immature sperm lacking full protamination. *Center panel*: SCSAsoft® software (SCSA Diagnostics, Brookings, SD) converts the data in the *left panel* to total DNA stainability versus the DNA Fragmentation Index (DFI). This reorients the data into a vertical/horizontal pattern of dots. *Right panel*: The data in the middle panel is converted to a frequency histogram of DFI which is divided into (a) nondetectable DNA fragmentation, (b) moderate level of DNA fragmentation, and (c) high level of DNA fragmentation. Total %DFI is Moderate + High level of DNA fragmentation, a parameter that is most frequently used in expressing the extent of sperm DNA fragmentation in a sample. This method, derived from SCSAsoft®, provides a much more accurate calculation of total %DFI due to the difficulties in accurately gating between the populations with no or moderate fragmentation in the left hand panel. Figure 5 is an example of a sample that requires SCSAsoft for correct interpretation.

more nuclear area and roundness as would be expected from immature sperm. Comet analysis showed that approximately 75% of the sperm with moderate and high DNA fragmentation had positive Comets while the population without sperm DNA fragmentation and the population of sperm with high DNA stainability (HDS) only showed a minor degree of background noise level. These results indicate that sperm with fragmented DNA based on SCSA® analysis demonstrate true DNA strand breaks and HDS sperm with an increased ratio of histones to protamines did not have any significant amount of DNA strand breaks.

1.5. Power of the SCSA® Test

Uniquely, biochemical interactions between AO and DNA/chromatin are precisely repeatable within any single sample. This is proven by comparing cytograms (X vs. Y scatter plots) of repeat measures of a single semen sample. The dot pattern from replicate measures is virtually identical on a $1,024 \times 1,024$ scatter plot.

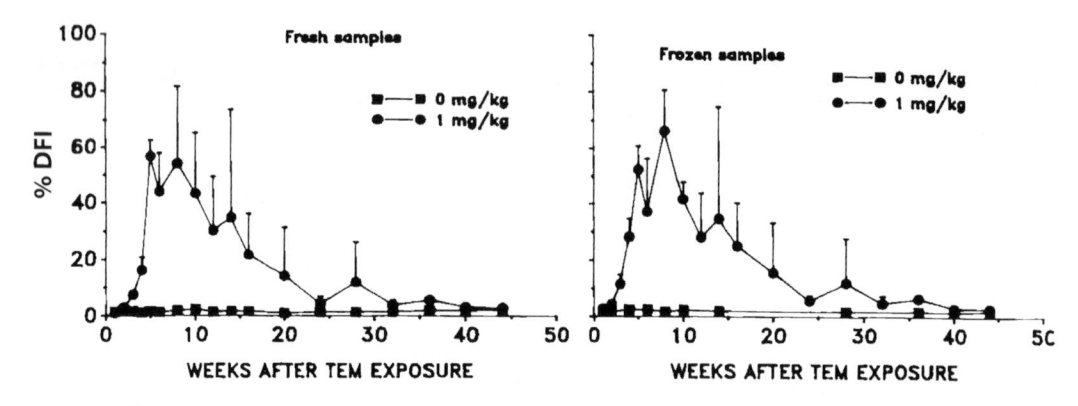

Fig. 2. SCSA® data from a sample with a high frequency of sperm with moderate DNA fragmentation. In this case, it is impossible to gate between sperm with no or moderate DNA fragmentation in the FCM dot-plot (*left panel*). With the SCSAsoft® gating between the two populations is unproblematic (*right panel*).

Thus, both the 30 s low pH induced opening of the DNA strands at sites of DNA breaks and the AO labeling are highly specific and repeatable.

1.6. Precision and Repeatability of SCSA® Data Over Time

The visual patterns of scatter plots of 5,000 sperm/sample are strikingly exact repeat patterns (4) and the CV of the repeat data typically 1–3% as seen for consecutive monthly semen samples from four men (Fig. 2).

The SCSA® values have a very high level of repeatability provided that no significant event has happened between collection times, e.g., hot tubs, medications, fever, illness, etc. One report (5) showed a considerable CV of 30% for human patient semen samples obtained over several years from the same patients. This is, in part, expected as patients presenting with a high %DFI could have factors that cause DFI fragmentation such as varicocele, obesity, poor diet, extreme sports, routine hot tub/sauna use, infections, disease, pesticide/chemical exposure, spinal cord damage, use of medications such as SSRI's. When such factors are removed or treated, the level of %DFI may well have significant decreases. In fact, it can be stated, that unless changes can be made by personal lifestyle and/or medical intervention, the use of SCSA® in clinical settings would have lesser utility.

1.7. Repeatability of SCSA® Data Between Fresh and Frozen/Thawed Epididymal Sperm on Long-Term Recovery from Toxicant Exposure

A fresh, raw semen or sperm sample (e.g., mouse epididymal sperm) can be measured immediately or flash frozen, thawed and analyzed; the data are highly repeatable between the fresh and frozen/stored/thawed samples (6) as seen in Fig. 3.

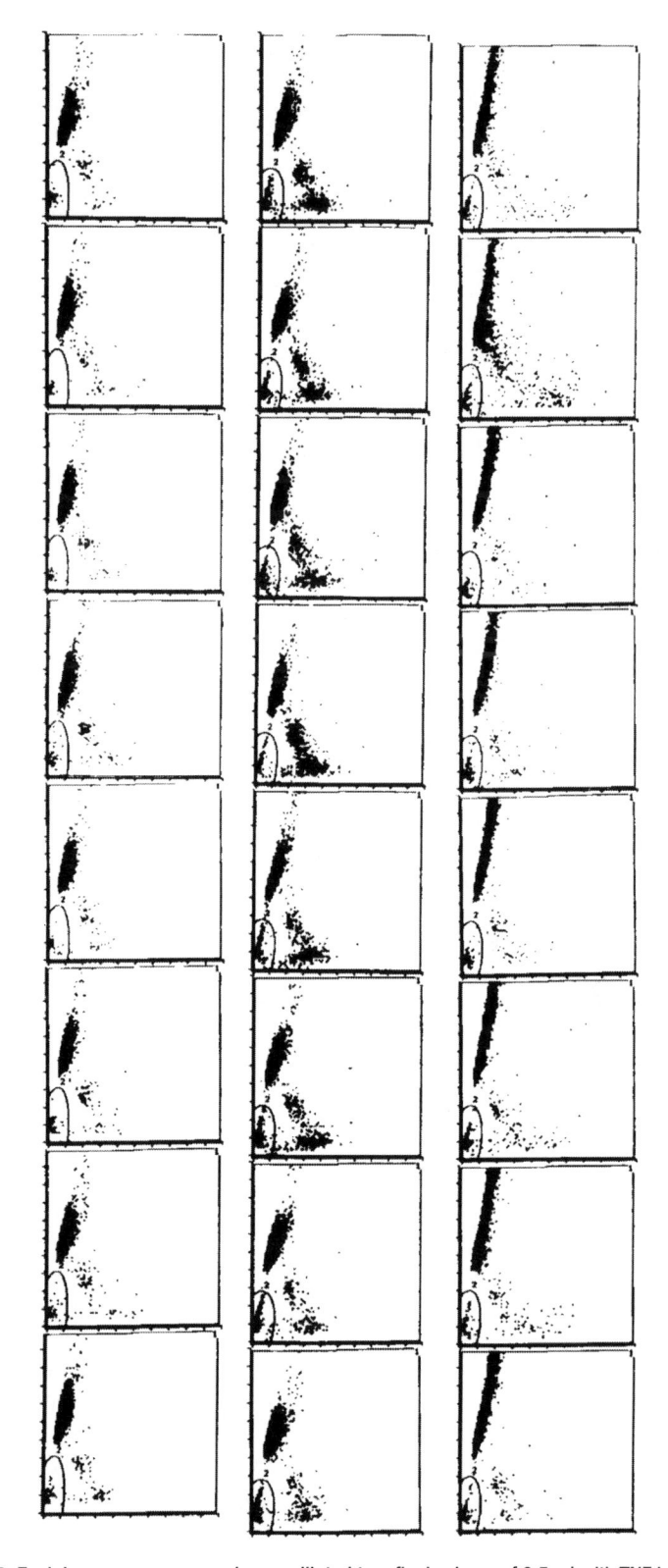

Fig. 3. Each human semen sample was diluted to a final volume of 0.5 ml with TNE buffer to obtain a count of approximately 2×10^6 sperm/ml. The samples in the *right column* were sonicated for 30 s with a Branson 450 Sonifier operating at a power setting of 3 and utilizing 70% of 1 s pulses.

1.8. SCSA Data on Epididymal Sperm Following Exposure to the Mutagen, Methyl Methanesulfonate as Related to Dominant Lethal Mutations/ Embryo Death

Male mice were exposed to methyl methanesulfonate (MMS) then mated, and dominant lethal mutation deaths were scored at 1–4, 5–8, 9–12, 13–16 and 17–20 days post conception. The first deaths were seen at the 5–8 day period while 85% of sperm had extensive DNA damage at 1–3 days post exposure (7). Thus, the SCSA® data likely showed the molecular precursor events (DNA breaks) that led to embryo deaths.

1.9. SCSA Data on Animal Fertility Providing Highly Convincing Argument that SCSA Data Predict Male Sub/Infertility

The SCSA® has established clinically significant thresholds for male factor sub/infertility for both large domestic animals (bulls (8, 9) and stallions (10)) and humans (11–14). When >20–25% of sperm in a semen sample have increased red fluorescence due to (ss) or (ds) DNA breaks, the patient/animal is placed into a statistical category of the following: (a) taking a longer time to pregnancy, (b) more IVF cycles, (c) more miscarriages, or (d) no pregnancy. At the 30% DFI level, the odds ratio (OR) is reduced eight to tenfold for a successful natural or IUI pregnancy. Several studies (15, 16) have shown the exception for boars where pregnancy rates and pigs/litter are reduced when the %DFI is >6%. Sperm with severely damaged DNA fertilize eggs that may lead to embryo death (17).

1.10. Potential RNA Staining Artifacts for SCSA®

Since AO stains both single stranded DNA and RNA the fluorescent color red, it was very important to know if cytoplasmic or nuclear RNA contributed to the red fluorescence that might be erroneously attributed to denatured DNA. We addressed this question (5) by sonicating whole bull, mouse, and human sperm, purifying each sample of nuclei through a sucrose gradient and measuring both the sonicated and nonsonicated sperm by SCSA®. As illustrated in Fig. 4, the unsonicated and sonicated sperm produced cytograms that were virtually identical.

1.11. Clinical Utility of Sperm DNA Fragmentation Tests

1. *Confusion from multiple DNA fragmentation tests.*
   All sperm DNA fragmentation tests report results as %DFI, as was initially coined for the SCSA® test. However, these %DFI numbers may mean different things for different tests. For example, light microscope Tunel data give different %DFI than flow cytometry Tunel. Also, Tunel %DFI may not be the same as for flow cytometry SCSA®. This is most likely due to the requirements of Tunel for large protein enzymes to enter into the highly compact and tangled nuclear chromatin (3) to tag DNA strand breaks. In sharp contrast, the SCSA® molecular tag is a very small acridine orange molecule (MW 265) that can penetrate the maze of nuclear chromatin and, furthermore, remains in equilibrium with the sperm during SCSA® measurement. In attempts to increase access to fragmented DNA sites for the Tunel test, some investigators have used disulfide bond reducing agents in the protocol to open the highly complex

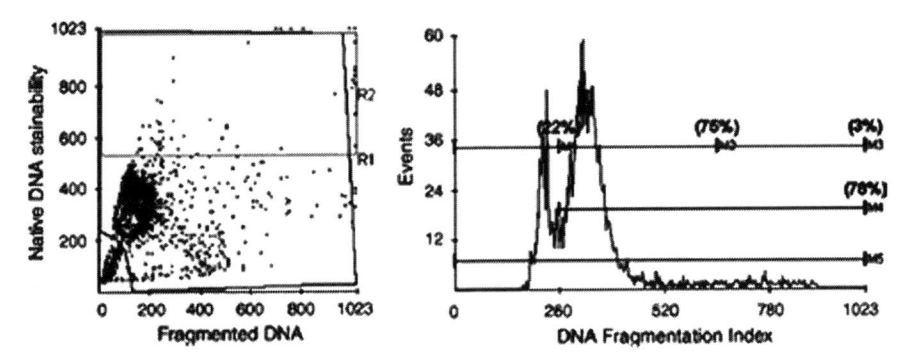

Fig. 4. Green versus red fluorescence cytograms from monthly semen samples provided by three donors. Examples are selected from 45 men to illustrate different types of cytogram patterns (5).

nuclear chromatin with disulfide bond reducing agents. However, since some sperm nuclei of "normal WHO samples" have diffuse chromatin and others highly compact chromatin (3), there would likely be differential effects from a DTT treatment. This was dramatically shown by Evenson et al. (18) when measuring the kinetics of human and mouse sperm chromatin uptake of a DNA stain following DTT and/or protease treatments. The mouse nuclei responded uniformly while the sperm from a fertile man were highly heterogeneous in response to DTT treatment.

2. *Power of SD DFI.*

The use of SCSAsoft® software to reorient the raw green versus red dot plots to green versus red/(red+green) dot plots followed by the generation of a frequency histogram of the converted dot plots provides a more accurate determination of %DFI, especially for those samples that have poor resolution between normal and moderate DFI (Fig. 5) (see Note 1).

1.12. Animal Fertility Studies Related to %DFI, DFI, and SD DFI

Importantly, the value of SD of DFI (SD DFI) is often a more sensitive measure of sperm DNA damage than %DFI alone. One reason for this difference is that DFI is determined on the whole population of sperm measured and is not subject to dividing the population into normal and increased red fluorescence populations by computer gating. This has been solidly shown in animal fertility and animal toxicology studies.

1. Bulls

Semen from individual bulls is often used for hundreds to thousands of cow inseminations. Thus, fertility rankings can be made between bulls in a stud service. Negative correlations were seen between fertility ratings and both SD DFI (-0.58, $P<0.01$) and %DFI (-0.40, $P<0.01$). Inherent in studies as above, and much more so with human studies, are the variables

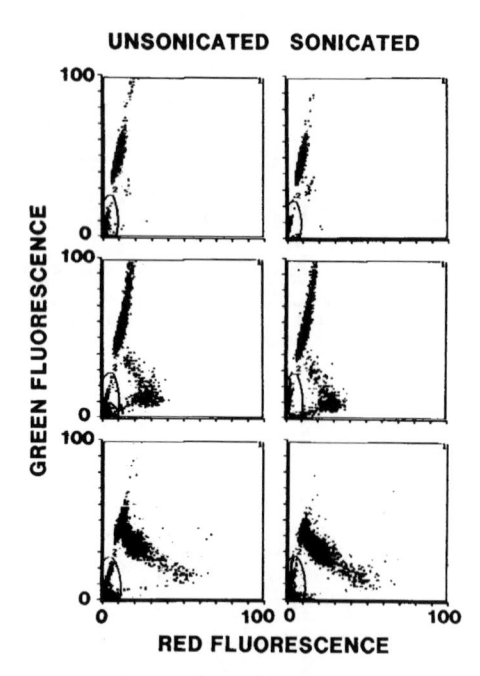

Fig. 5. Effects of 1.0 mg/kg (daily × 5) triethylenemelamine (TEM) on %DFI in epididymal sperm during a 44 week period. *Left*: %DFI on fresh samples. *Right*: Aliquots of the same samples frozen and measured later at a single time period.

in the females and a host of other factors such as experience of the artificial insemination team. To get around this problem, animal studies can use what is known as heterospermic insemination protocols in which equal numbers of motile sperm from two or more phenotypically different bulls are mixed prior to insemination. The parentage of calves resulting from these matings is determined and, based on the number of calves sired with each phenotype, a competitive fertility index is derived for each bull (9). Correlations of SD DFI and %DFI with competitive index were −0.94 ($P<0.01$) and −0.74 ($P<0.05$), respectively.

2. Boars

Multiparous pigs allow for a determination of both fertility rate and number of piglets per litter. The SCSA® was used (15) retrospectively to characterize sperm from 18 sexually mature boars with fertility information. Boar fertility was defined by farrow rate (FR) and average total number of pigs born (ANB) per litter of gilts and sows mated to individual boars. Fertility data were compiled for 1,867 matings across the 18 boars. In contrast to humans and other mammals studied where the threshold for reduced fertility is an approximate 25–30% DFI, the threshold for boars is about 6% DFI. The %DFI and SD

DFI showed the following significant negative correlations with FR and ANB; %DFI versus FR, $r = -0.55$, $P < .01$; SD DFI versus FR, $r = -0.67$; %DFI versus ANB, $r = -0.54$, $P < .01$ and SD DFI versus ANB, $r = -0.54$, $P < 0.02$. The present data suggest that boar sperm possessing fragmented DNA can affect embryonic development corroborating earlier studies in mice showing that fertilization occurs whether the sperm has damaged DNA or not (17) but may cause embryonic death.

1.13. SCSA® Defined Etiologies of Increased DNA Fragmentation

The most likely common factor in causing sperm DNA fragmentation is oxidative stress (ROS) (18). Factors related to increased %DFI include: age (19), genetics (20), air pollution (21), varicoceles (22–25), cancer and cancer and cancer treatment (26, 27), pesticides (29), environmental heat (29, 30), fever (2), medications (31), and diabetes (32–34).

1.14. Human Clinical Utility of the SCSA®

The American Society for Reproductive Medicine (ASRM) consensus committee on "the clinical use of sperm DNA fragmentation assays for clinical patients" has stated that Sperm DNA fragmentation tests are recommended for certain patients but it has not been recommended as a part of the routine semen analysis although some clinics have implemented this approach already. At the time of the 2006/08 ASRM reports (35) there were only two SCSA® reports on the odds ratios (OR) of natural pregnancy. These studies reported an odds ratio of eight to tenfold reduction for pregnancy when the %DFI was >25–30%. Unfortunately, the ASRM consensus report erroneously showed this as OR of 2.08 (35) which implied lesser utility for classifying men's sperm DNA fragmentation levels and reduced fertility. Since that time, the studies of Givercman et al. (14) on natural fertility OR, using the same SCSA® protocol, confirmed an even higher odds ratio when the SCSA® %DFI was >20%.

1.15. Conclusion

The SCSA® is the easiest and most rapid DNA fragmentation test in the laboratory. The SCSA is also the most precise and repeatable with CV of ~1–3%. Due to the low cost of supplies, numerous measurements can be made for both research and clinical applications.

In summary, the SCSA protocol appears offhand rather simple; however, there are numerous very critical points that, unless followed exactly, will give very poor data and serious errors in clinical diagnosis and prognosis.

Basic protocol steps are the following: Fresh or frozen semen/sperm are thawed in a 37°C water bath and diluted to $1-2 \times 10^6$ sperm/ml with TNE buffer. 200 µl of this sperm suspension are added to a test tube followed by adding 400 µl of acid detergent solution. After 30 s 1.2 ml of AO staining solution is added to the sample, and finally flow cytometric measurement commences.

2. Materials

2.1. Basic Laboratory Equipment

1. Oxford adjustable, 0.20–0.80 ml automatic dispenser for the acid detergent solution with glass *amber* bottle.
2. Oxford adjustable, 0.80–3.0 ml automatic dispenser for the AO staining solution glass *amber* bottle (CAT # 13 687 66, Fisher Scientific).
3. Pipettors: adjustable 0–10 µl, 10–100 µl, 100–1,000 µl and a nonadjustable 200 µl.
4. Ice buckets (3) for samples and reagent bottles.
5. Water bath (37°C).
6. Stopwatch.

2.2. Staining Solutions and Buffers

For all solutions, use double distilled water (dd-H_2O). For sterilization use a 0.22 µm filter. Use only the purest grade reagents. All solutions and buffers are stored at 4°C.

1. Acridine Orange (AO) stock solution, 1.0 mg/ml: Dissolve chromatographically purified AO (Polysciences) in dd-H_2O at 1.0 mg/ml. Solution can be stored up to several months (see Note 2).
2. Acid-detergent solution, pH 1.2: Combine 20.0 ml 2.0 N HCl (0.08 N), 4.39 g NaCl (0.15 M), and 0.5 ml Triton X-100 (0.1%) in H_2O for a final volume of 500 ml. Adjust pH to 1.2 with 5N HCl. Store at 4°C up to several months (see Note 3).
3. 0.1 M citric acid buffer: Add 21.01 g/L citric acid monohydrate (FW = 210.14; 0.10 M) to 1.0 L H_2O. Store up to several months at 4°C.
4. 0.2 M Na_2PO_4 buffer: Add 28.4 g sodium phosphate dibasic (FW = 141.96; 0.2 M) to 1.0 L H_2O. Store up to several months at 4°C.
5. Staining buffer, pH 6.0: Combine 370 ml 0.10 M citric acid buffer, 630 ml 0.20 M Na_2PO_4 buffer, 372 mg EDTA (disodium, FW = 372.24; 1 mM), and 8.77 g NaCl (0.15 M). Mix overnight on a stir plate to insure that the EDTA is entirely in solution. Adjust pH to 6.0 with concentrated NaOH pellets. Store up to several months at 4°C (see Note 4).
6. AO staining solution (working solution): Add 600 µl AO stock solution to each 100 ml of staining buffer. Rinse the pipette tip several times into the staining buffer. This solution is kept in the glass amber automatic dispenser for AO staining of samples. Store up to 2 weeks at 4°C.
7. AO equilibration buffer: Combine 400 µl acid-detergent solution with 1.2 ml AO staining solution (see Note 5).

8. TNE buffer, 10×, pH 7.4: Add 9.48 g Tris–HCl (FW = 158; 0.01 M), 52.6 g NaCl (FW = 58.44; 0.15 M), and 2.23 g EDTA (disodium, FW = 372.24; 1 mM) to 600 ml dd H_2O. Adjust pH to 7.4 with 2 N NaOH. Store up to 1 year at 4°C.

9. TNE buffer, 1×, pH 7.4 (working solution): Combine 60 ml 10× TNE and 540 ml H_2O. Check pH (7.4). Store for several months at 4°C.

10. FCM Tubing Cleanser (for unclogging FCM sample lines): 50% ETOH, 50% household bleach (contains ~5% sodium hypochlorite), and 0.5 M NaCl. Store at room temperature (see note 6).

11. 50% household bleach (for eliminating AO from sample lines): Combine 50 ml household bleach (~5% sodium hypochlorite) with 50 ml H_2O. Store at room temperature (see Note 7).

12. FCM Sheath fluid: 2 L 0.45 nm filtered water + 0.1% Triton X-100: (this helps minimize bubbles in the flow channel). Add 0.5 ml Triton X-100 to each liter of filtered dd-H_2O. This is most conveniently prepared by using a gravity flow system containing a 0.45 μ filter unit into a storage bottle. It is NOT necessary to use commercially sold sheath fluid unless FCM sorting of sperm is done with a jet-in-air sorter.

2.3. Major Equipment

1. Ultracold freezer (−70 to −110°C) or, preferably, a LN_2 tank.
2. Biological safety hood.
3. Flow cytometer(s) (see Notes 8 and 9).

3. Methods

3.1. Cell Preparation

1. Human semen samples are typically obtained by masturbation into plastic clinical specimen jars preferably after ~2 days abstinence (see Note 10).

2. Allow ~30 min for semen liquefaction at room temperature, then freeze aliquots of raw or TNE diluted (1–2 × 10⁶ sperm/ml) semen directly without cryoprotectants in an ultracold freezer (−70 to −110°C) in 0.5–1.5 ml snap-cap tubes. Alternatively samples may be frozen in a shipping box with dry ice or placed directly into a LN_2 tank (cryovials) (see Note 11).

3.2. Flow Cytometer Setup

1. Set up the flow cytometry workstation to allow samples to be thawed and processed in the immediate vicinity of the flow cytometer. The following equipment should be handy for quick and easy use: Ice buckets containing wet ice to hold the reagent bottles, sample tubes, TNE buffer, in addition disposable gloves, stopwatch, automatic pipettors and tips and a container with disinfectant for sample disposal are needed.

2. Align cytometer (see Note 13).

3. Prepare a reference sample diluted with cold TNE at a concentration of 1–2×10^6 cells/ml (see notes 14–16).

4. Immerse single frozen samples in a 37°C water bath, just until the last remnant of ice disappears. Dilute sperm to a concentration of 1–2×10^6 cells/ml with TNE buffer (see Note 17).

5. Place a 200 µl aliquot of fresh or frozen/thawed semen sample of known sperm concentration into a 12×75 mm conical plastic test tube.

6. Add 400 µl of the acid-detergent, low pH buffer using an automatic dispenser set deep in the ice bucket (see Note 15). Start a stopwatch *immediately* after the first buffer is dispensed.

7. *Exactly* 30 s later add 1.2 ml of AO staining solution.

8. Place the sample tube into the flow cytometer sample chamber (which varies in design by different instruments), and immediately start the sample flow after placing it in the sample holder.

9. Using the stopwatch that was started with the addition of the acid-detergent solution, start the acquisition of list mode data to computer disk at 3 min (see Note 19).

10. Check the sperm flow rate during this time and if it is too fast, i.e., >250 cells/s, make a new sample at the appropriate dilution (see Note 20).

11. Measure all samples at least twice in succession for statistical considerations and record data on ~5,000 sperm cells (total events recorded are higher due to debris) per measurement.

12. For the second measurement, take the sample from the same thawed aliquot; dilute appropriately, process for the SCSA and measure.

13. After the second measurement of a sample is finished, place a tube of AO equilibration buffer on the instrument to maintain the AO conditions and wash any of the previous sample out of the tubing and start preparing the next sample. There is no need to run this buffer between the duplicate measurements of the same sample, just allow the first one to stay running while preparing the second one.

3.3. Gating and Debris Exclusion

A very important, *but* sometimes difficult point, is deciding where to draw the computer gates to exclude cellular debris signals [signals located at the origin in the red (X) versus green (Y) fluorescence cytograms] from the analysis. This gate is usually best set at a 45° angle, i.e., at the same channel value for both red and green fluorescence values (e.g. 400/1024). Resolution of debris and sperm signal is partly instrument dependent (see Note 21).

4. Notes

1. The variables of DFI other than %DFI (DFI and SD DFI) have been shown to be very useful especially in our studies for toxicology and animal fertility. Future studies will show its importance for human fertility studies. However, a simple determination of the percentage of cells with denatured DNA (%DFI) and the percentage of cells with abnormally high green stainability (%HDS) can be reasonably estimated for the majority of samples without the ratio calculations. %DFI is currently the most used variable of this assay for human fertility assessment.

2. Our laboratory has used only AO obtained from Polysciences and thus we have full confidence in this source. *DO NOT* use a more crude preparation of AO; failure will result. AO is a toxic chemical, and precautions should be taken when handling it. Tare a 15 ml, flat-bottom scintillation vial on a 4-place electronic balance, carefully remove and transfer 3–6 mg AO powder from the stock bottle with a microspatula into the vial. Add an exact equivalent number of milliliters of water. Wrap the capped vial in aluminum foil to protect from light.

3. Use purchased 2.0 N HCl (e.g., Sigma); do not dilute from a more concentrated HCl solution that is likely less pure and may be of questionable strength. The Triton-X stock solution is very viscous. Use a wide-mouth pipette and carefully draw up the exact amount, wipe the outside of the pipette free of Triton-X and then expel with force in and out of the pipette until all is dispensed.

4. Slowly and carefully adjust the pH using very small pieces of concentrated NaOH pellets (cut with a scalpel and handled with a forceps). Note that when the 0.2 M Na_2PO_4 buffer is removed from the refrigerator, salt crystals will be present. Heat in 37°C water bath until the salts are fully dissolved.

5. This is run through the instrument for ≈ 15 min prior to sample measurement to insure that AO is equilibrated with the sample tubing. This is also run through the instrument between different samples to maintain the AO equilibrium and help clean the prior sample out of the lines.

6. This solution is passed through the flow cytometer fluidic components in contact with AO to remove adherent AO. It does not need to be used if the next user of the FCM is also doing AO staining.

7. This solution is used for an occasional thorough cleaning or flushing of flow cytometer sample lines to remove adherent cell debris and any fluorochromes.

8. The flow cytometer must have 488 nm excitation wavelengths and an approximate 15–35 mW laser power. Fluorescence of individual cells is collected through red (630–650 nm long pass) and green (515–530 nm band pass) filters.

9. *Orthogonal flow cytometer configuration and related signal artifacts.* The highly condensed mammalian sperm nucleus has a much higher index of refraction than sample sheath (water) in a flow cytometer. This differential, coupled with the typical nonspherical shape of sperm nuclei and their orientation in the flow channel, produces an optical artifact consisting of an asymmetric, bimodal emission of DNA dye fluorescence when measured in orthogonal configuration flow cytometers where the collection lenses are situated at right angles to both sample flow and excitation source. Since DFI is a computer calculated ratio of red to total (red + green) fluorescence, the optical artifact of AO-stained sperm measured in the orthogonal instruments does not significantly interfere with results and the DFI frequency histogram is very narrow for a normal population of sperm. Although each type of flow cytometer with different configurations of lenses and fluidics produces different cytogram patterns, the DFI data are essentially the same.

10. Of importance is the length of the previous abstinence period; if several days have elapsed, then sperm stored in the epididymis can become apoptotic, in which case such a sample would not be representative of a fresh semen sample. We suggest that a patient ejaculate, wait 2 days and ejaculate again, then the sample for testing be taken after another 2 days, e.g., ejaculate on Monday and Wednesday and collect clinical sample on Friday.

11. Freshly collected semen should be quick frozen as soon as liquefaction has occurred in about a half hour. The majority of semen samples may be kept for up to several hours at room temperature prior to measuring/freezing without significant loss of quality, allowing for collections within a medical institution and transport to the flow cytometry unit. However, we have observed in limited studies that an estimated 10% of samples have an increased DNA fragmentation while setting at room temperature; likely these samples have very low antioxidant capacity. If transport is required outside of a building complex, the sample may be conveyed in an insulated box or jacket pocket to keep from freezing or on liquid ice if the ambient temperature is hot. Once a sample has been diluted in TNE buffer it should be measured or frozen immediately.

12. Samples should be frozen in vials that are approximately ¼ larger in volume than the semen volume in order to reduce the air- surface interface thus minimizing related reactive oxygen damage. Keep the tubes vertical when freezing, as samples frozen at the bottom of a tube are later thawed in a water bath

with greater ease and safety. Cryoprotectants are not needed since quick-frozen cells and those frozen with a cryoprotectant provide equivalent SCSA data. This feature is unique to mammalian sperm cells due to the highly condensed, crystalline nature of the nucleus.

13. *Flow cytometer alignment.* Prior to measuring experimental samples, the instrument must be checked for alignment using standard fluorescent beads. *Very importantly*, an AO equilibration buffer (400 µl acid-detergent solution and 1.20 ml AO staining solution) must be passed through the instrument sample lines for ~15 min prior to establishing instrument settings. This insures that AO is equilibrated with the sample tubing. To save time, this AO buffer can be run through the instrument during its warm up time prior to alignment and again just before measuring samples. Contrary to existing rumors, using AO in a flow cytometer *does not* ruin it for other purposes! The sample lines *do not* need to be replaced after using AO in a flow cytometer! However, the system *does* need to be fully equilibrated with AO, as AO does transiently adhere to the sample tubing by electrostatic force, thus reducing the required AO concentration. After finishing SCSA measurements, AO can easily be cleansed from the lines by rinsing the system for about 10 min with a 50% filtered household bleach solution followed by 10 min of filtered H_2O. Our laboratory has utilized many fluorescent dyes and sample types after measuring AO stained sperm without any associated problems.

14. *Reference samples.* Because SCSA variables are very sensitive to small changes in chromatin structure, studies on sperm using this protocol require very precise, repeat instrument settings for all comparative measurements whether done on the same or different days. These settings are obtained by using aliquots of a single semen sample called the "reference sample" (this is not a "control" sperm from a fertile donor). A semen sample that demonstrates heterogeneity of DNA integrity (e.g., 15% DFI) is chosen as a reference sample and then diluted with cold (4°C) TNE buffer to a working concentration of $1-2 \times 10^6$ cells/ml. CLIA and other licensing agencies, e.g., New York Health, require that for every measurement period that a low %DFI and a high %DFI sample become part of the measurement data. Several hundred 300 µl aliquots of this dilution are *immediately and quickly* placed into 0.5 ml snap-cap vials and frozen at −70 to −100°C in a freezer or, preferably in a LN_2 tank. These reference samples are used to set the red and green photomultiplier tube (PMT) voltage gains to yield the same mean red and green fluorescence levels from day to day. The mean red and green fluorescence values are set at ~125/1,000 and ~475/1,000 channels, respectively. The values established

by a laboratory (preferably the same as above) should be used consistently thereafter. Strict adherence to keeping the reference values in this range must be maintained throughout the measurement period. A *freshly thawed* reference samples is measured after every five to ten experimental samples to insure that instrument drift has not occurred.

15. Very few FCM protocols are as demanding as the SCSA for using a reference sample. Obviously, it would be advantageous to prepare a new batch of reference samples from the same individual donor. However, if a new donor is used, then first set the PMTs for the previous reference sample to be at the same x and y channel positions, then measure the new reference sample and note the red and green mean values and use these values for the next studies.

16. Since reference samples can be stored in LN_2 for years, a donor could provide enough samples for thousands of reference aliquots.

17. When analyzing a series of human samples, it is extremely helpful to obtain the sperm count in advance of SCSA preparation so that time is not lost determining the proper dilution. However, if a sample(s) needs to be measured quickly for a clinical decision, then rather than wait for a sperm count, estimate a dilution, check the flow rate, and if necessary, resample with the proper dilution to attain the required flow rate of ~200 events per second.

18. This dispenser needs to be highly accurate and to have a maximum volume only somewhat more than what is being dispensed. At the beginning of sample measurement and after long breaks in measurement, dispense several volumes from both dispensers before starting with the samples as AO in the delivery tube may have been damaged by light and solutions in the plastic delivery tubes may be warmer than 4°C.

19. This allows ample time for AO equilibration in the sample and hydrodynamic stabilization of the sample within the fluidics, both very important aspects of AO staining.

20. This protocol provides ~2 AO molecules/DNA phosphate group. Thus, to initially set up the proper hydrodynamic conditions, measure several sperm samples that have a predetermined cell count of $\approx 1.5 \times 10^6$ sperm/ml (or known concentration of fluorescent beads) and adjust the flow rate settings (if possible) for ~200 cells/beads per second. On a FACScan, the "low flow" rate setting delivers an approximately correct flow rate. If a sample's flow rate is too high, this same sample cannot be diluted with AO buffer to lower the concentration. Sample and sheath flow valve settings of the instrument are *never changed* during these measurements so the liquid flow rate is constant.

Doing so widens the flow sample stream with consequential loss of resolution. Thus, a change in sperm count rate is a function of sperm cell concentration only. PMT settings should be fairly identical from day to day depending on slight alignment differences between days and sample runs.

21. The real SCSA values of a sample cannot be obtained if the fluorescence from debris (i.e., free cellular components, and other contaminants) is blended in with the sperm fluorescence signal. This can sometimes be eliminated by washing the sperm or processing though gradients. However, there is a risk of losing cell types and the advantage of using whole semen measurements is then compromised. Bacterial debris appears as a straight line to the left of and parallel with the main sperm population in the cytograms; this usually can be gated out, but not in all samples.

References

1. Jaffay SC, Strader LF, Buss RM, et al (2006) Relationship among semen endpoints used as indicators of sperm nuclear integrity. Am Soc Androl Abstract
2. Evenson DP et al (2000) Characteristics of human sperm chromatin structure following an episode of influenza and high fever: a case study. J Androl 21:739–746
3. Evenson DP et al (1980) Comparison of human and mouse sperm chromatin structure by flow cytometry. Chromosoma 78:225–238
4. Evenson D (2011) Sperm Chromatin Structure Assay (SCSA): 30 Years of Experience with the SCSA. In: Zini A, Agarwal A, editors. Sperm Chromatin. Springer. P. 147
5. Evenson DP et al (1991) Individuality of DNA denaturation patterns in human sperm as measured by the sperm chromatin structure assay. Reprod Toxicol 5:115–125
6. Erenpreiss J et al (2006) Intra-individual variation in sperm chromatin structure assay parameters in men from infertile couples: clinical implications. Hum Reprod 21:2061–2064
7. Evenson DP et al (1989) Long-term effects of triethylenemelamine exposure on mouse testis cells and sperm chromatin structure assayed by flow cytometry. Environ Mol Mutagen 14:79–89
8. Evenson DP et al (1993) Effects of methyl methanesulfonate on mouse sperm chromatin structure and testicular cell kinetics. Environ Mol Mutagen 21:144–153
9. Ballachey BE et al (1988) The sperm chromatin structure assay. Relationship with alternate tests of semen quality and heterospermic performance of bulls. J Androl 9:109–115
10. Kenney RM et al (1995) Relationships between sperm chromatin structure, motility, and morphology of ejaculated sperm, and seasonal pregnancy rate. Biol Reprod Mono 1: 647–653
11. Evenson DP et al (1980) Relation of mammalian sperm chromatin heterogeneity to fertility. Science 210:1131–1133
12. Evenson DP et al (1999) Utility of the sperm chromatin structure assay as a diagnostic and prognostic tool in the human fertility clinic. Hum Reprod 14:1039–1049
13. Bungum M et al (2007) Sperm DNA integrity assessment in prediction of assisted reproduction technology outcome. Hum Reprod 22: 174–179
14. Giwercman A et al (2010) Sperm chromatin structure assay as an independent predictor of fertility in vivo: a case-control study. Int J Androl 33:e221–e227
15. Didion BA et al (2009) Boar fertility and sperm chromatin structure status: a retrospective report. J Androl 30:655–660
16. Boe-Hansen GB et al (2008) Sperm chromatin structure integrity in liquid stored boar semen and its relationships with field fertility. Theriogenology 69:728–736
17. Ahmadi A, Ng SC (1999) Fertilizing ability of DNA-damaged spermatozoa. J Exp Zool 284: 696–704
18. Saleh RA et al (2003) Negative effects of increased sperm DNA damage in relation to seminal oxidative stress in men with idiopathic and male factor infertility. Fertil Steril 79 (Suppl 3):1597–1605

19. Wyrobek AJ et al (2006) Advancing age has differential effects on DNA damage, chromatin integrity, gene mutations, and aneuploidies in sperm. In: Annual meeting of the national academy of science, USA
20. Rubes J et al (2007) GSTM1 genotype influences the susceptibility of men to sperm DNA damage associated with exposure to air pollution. Mutat Res 625:20–28
21. Rubes J et al (2005) Episodic air pollution is associated with increased DNA fragmentation in human sperm without other changes in semen quality. Hum Reprod 20:2776–2783
22. Chen SS et al (2004) 8-Hydroxy-2′-deoxyguanosine in leukocyte DNA of spermatic vein as a biomarker of oxidative stress in patients with varicocele. J Urol 172:1418–1421
23. Zini A et al (2005) Beneficial effect of microsurgical varicocelectomy on human sperm DNA integrity. Hum Reprod 20:1018–1021
24. Yamamoto M et al (1994) The effect of varicocele ligation on oocyte fertilization and pregnancy after failure of fertilization in in vitro fertilization-embryo transfer. Hinyokika Kiyo 40:683–687
25. Werthman P et al (2008) Significant decrease in sperm deoxyribonucleic acid fragmentation after varicocelectomy. Fertil Steril 90:1800–1804
26. Evenson DP et al (1984) Flow cytometric evaluation of sperm from patients with testicular carcinoma. J Urol 132:1220–1225
27. Fossa SD et al (1997) Prediction of posttreatment spermatogenesis in patients with testicular cancer by flow cytometric sperm chromatin structure assay. Cytometry 30:192–196
28. Sanchez-Pena LC et al (2004) Organophosphorous pesticide exposure alters sperm chromatin structure in Mexican agricultural workers. Toxicol Appl Pharmacol 196:108–113
29. Karabinus DS et al (1997) Chromatin structural changes in sperm after scrotal insulation of Holstein bulls. J Androl 18:549–555
30. Sailer BL et al (1997) Effects of heat stress on mouse testicular cells and sperm chromatin structure. J Androl 18:294–301
31. Tanrikut C et al (2010) Adverse effect of paroxetine on sperm. Fertil Steril 94:1021–1026
32. Agbaje IM et al (2007) Insulin dependant diabetes mellitus: implications for male reproductive function. Hum Reprod 22:1871–1877
33. Pitteloud N et al (2005) Increasing insulin resistance is associated with a decrease in Leydig cell testosterone secretion in men. J Clin Endocrinol Metab 90:2636–2641
34. Stigsby B (2010) Personal communication
35. (2006) The clinical utility of sperm DNA integrity testing. Fertil Steril 86: S35–S37

Part IV

Advanced Clinical Testing

Chapter 15

Sperm Aneuploidy Testing Using Fluorescence In Situ Hybridization

Benjamin R. Emery

Abstract

Sperm aneuploidy screening has been used as a tool in diagnosis and determining treatment options for male factor infertility since the development of human sperm karyotyping by injection into hamster and mouse oocytes in the 1970s. From these studies and subsequent work with interphase chromosome analysis, at risk populations of men with teratozoospermia, oligozoospermia, and men with translocations, have since been identified. The current technique is an application of fluorescent in situ hybridization (FISH) on interphase sperm nuclei with careful enumeration of the labeled chromosomes to determine sperm ploidy. Typically, five to seven chromosomes are evaluated in individual ejaculates to determine the percent of aneuploid sperm present. This protocol will detail the procedures for: preparation of specimens, exposure of the sperm nuclei to the FISH probes, hybridization, destaining, and scoring criteria.

Key words: Sperm, Aneuploidy, Diagnostic testing, Meiotic errors, Fluorescence in situ hybridization (FISH), Male factor infertility

1. Introduction

Fluorescence in situ hybridization (FISH) allows fluorescent probes to be bound to specific regions of chromosomes thus allowing the enumeration of chromosomes as well as the identification of structural defects such as translocations. By analyzing sperm with FISH prior to an IVF cycle, it is possible to better counsel patients as to the possible risks of transmitting chromosomal aberrations and increase the accuracy of predicting outcomes (1–4). Men with severe sperm morphology defects (2, 5), severely oligozoospermic men, and men with a history of some chemical exposures are at risk of likelihood for increased sperm aneuploidy, and screening should be considered (4, 6).

Analysis of the entire chromosomal complement in each cell is not currently practical due to the labor-intensive nature of the protocol. Therefore, chromosomes with the highest risk for poor outcomes or those that are most prone to aneuploidy during meiosis are typically analyzed. The most commonly tested chromosomes are X, Y, 13, 16, 18, 21, and 22, while those that are most predictive of recurrent miscarriage are 1, 15, 17, 21, and 22 (4, 7).

2. Materials

2.1. General Media and Reagents

1. 0.1 M PBS.
2. 20× SSC: Weigh 175.2 g NaCl and 88.2 g Na citrate. Dissolve NaCl and Na citrate in type I water. Adjust to pH 7.4 with concentrated NaOH solution. Bring the total volume to 1 l with type I water. Store at −5°C for 1 year.
3. 2× SSC/0.1 % NP-40: Combine 100 ml 20× SSC with 850 ml type I water and add 1 ml NP-40 (IGEPAL; Sigma). Add type I water to bring the final volume to 1 l. Adjust the pH to 7.0 ± 0.2 with NaOH. Store at 2–5°C for 6 months.
4. 0.4× SSC/0.3 % NP-40: Combine 20 ml 20× SSC, 950 ml type I water and 3 ml IGEPAL. Add type I water to bring the total volume to 1 l. Adjust the pH to 7.0–7.5 with NaOH. Store at 2–5°C for 6 months.
5. PN Buffer: Solution A: $Na_2HPO_4 \cdot 7H_2O$ solution (2 l). Add 53.614 g $Na_2HPO_4 \cdot 7H_2O$ to 1.8 l distilled water, Adjust to 2 l. Solution B: 0.33 M NaH_2PO_4. Add 39.6 g NaH_2PO_4 to 800 ml distilled water, adjust to 1 liter. To make PN buffer, combine 2 l of solution A and 50–100 ml of solution B (0.1 M). Adjust pH to 8.0 with addition of solution B. Add IGEPAL to the PN buffer for a final IGEPAL concentration of 0.05 %. Store at 2–5°C for 1 year.
6. Antifade slide mounting medium: Add 5 mg p-phenylenediamine to 10 ml PN buffer to dissolve then add 10 ml glycerol. Aliquot into 0.5 ml aliquots in micro centrifuge tubes. Store at −20°C for 90 days or −80°C for 1 year.
7. 1 M dithiothreitol (DTT): Dissolve 3.08 g DTT in 20 ml distilled water. Store at −20°C for 1 year.
8. 10 mM DTT in Tris–HCl Add 400 μl 1 M DTT to 40 ml 0.1 M Tris–HCl pH 8.0. Use within 24 h.
9. Carnoy's Fixative: 1:4 Glacial Acetic Acid in Methanol.
10. Prepared FISH Probes: Commercially available probes are typically available for most applications and are preferred due to the labor-intensive procedure of preparation. Abbott Molecular Vysis probes are the most commonly used, but other manufacturers are available for species variation.

3. Methods

All procedures should be done with care not to contaminate the reagents, samples or prepared slides with sources of exogenous DNA that might hybridize and cause erroneous results.

3.1. Sample Preparation and Sperm Nuclear Decondensation

1. Add a volume of PBS to reach 1:3 ratio of semen to PBS, mix gently (see Notes 1–3).
2. Centrifuge at $500 \times g$ in a swinging bucket centrifuge for 10 min.
3. Resuspend the decanted pellet to a sperm concentration of 100–200 million per ml (see Note 4).
4. Drop the washed sperm sample onto a fluorescence-grade, precleaned microscope slide and smear using a standard sperm smearing technique (Fig. 1) (see Notes 5 and 6).
5. Fix air-dried slides in Carnoy's Fixative for 5 min on ice.
6. Air-dry slides again to prepare for sperm decondensation using DTT (see Note 7).
7. Add slides to a prepared solution of 0.1 M DTT in a Coplin jar.
8. Incubate until sperm heads start to swell and lose their high birefringent property (Fig. 2) (see Note 8). This typically takes 10–20 min.
9. Air-dry the slides once again prior to preparation for hybridization.

3.2. Dehydration, Denaturation and Hybridization

1. Dehydrate slides through a series of 70 %, 80 %, 100 % ethanol washes, 1 min each step and air-dry.
2. Add 10 µl of prepared probe mixture to the center of the slide and coverslip with a 22×22 mm coverslip (Fig. 3a).
3. Remove all air bubbles by gently forcing any trapped air bubbles to the edge of the coverslip (Fig. 3b) (see Note 9).
4. Seal with rubber cement around the entire coverslip (Fig. 3c) (see Note 10).

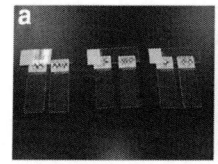

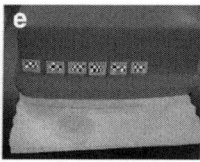

Fig. 1. Step by step slide preparation and slide smears. (a) The slides are precleaned and laid out for dropping sperm suspension. (b) Small drops of the sperm suspension are placed towards the top of the slide. (c) A second slide is slowly and gently moved through the drop at a 45° angle. (d) The slides are then fixed in ice-cold Carnoy's fixative. (e) They are then air-dried on an angle.

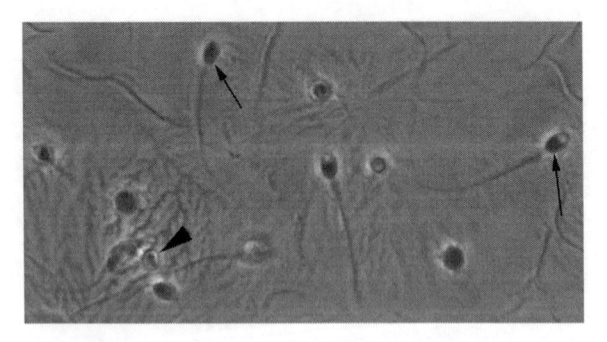

Fig. 2. Decondensed sperm under phase microscopy. *Arrows* indicate appropriately decondensed sperm while the *arrowhead* shows a sperm that has retained its birefringent nucleus.

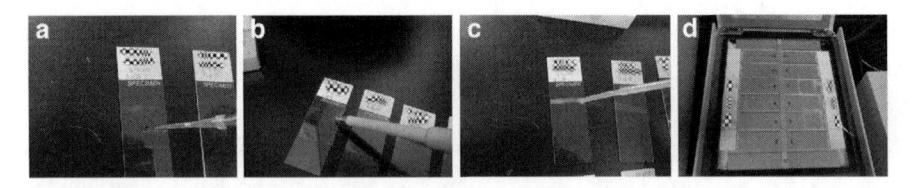

Fig. 3. Step by step probe addition and sealing. (a) The probe mixture is added to the slide in an area of evenly spread sperm nuclei. (b) The coverslip is added and the bubbles are gently pushed to the edge of the coverslip. (c) The rubber cement seal is easily made using a disposable transfer pipette moved along all edges of the coverslip. (d) The slides are added to the incubation chamber or slide PCR machine.

5. Add the slides to a slide PCR machine or denaturation/hybridization chamber (Fig. 3d) (see Note 11).
6. Incubate following the probe manufacturer's specified time and temperatures (see Note 12).

3.3. Destaining and Counterstaining

1. Remove slides from the incubation chamber and carefully remove the coverslip; first remove the rubber cement seal then carefully lift one corner of the slide and slowly pull the coverslip off.
2. The slide should be immediately placed into a preheated Coplin jar containing 0.4× SSC/0.3 % NP-40 wash solution at 74°C (in a water bath) for 5 min (see Notes 12–14).
3. Slides are immediately moved to 2× SSC/0.1 NP-40 wash solution in a Coplin jar at room temperature for 1 min.
4. Remove slides, air-dry and mount with 10–20 μl of antifade mounting media under a 22×22 mm #1.5 glass coverslip (see Note 15).
5. Observe sperm using a standard epifluorescence microscope equipped with the appropriate filters for the direct conjugated fluorophores.

3.4. Visualizing, Scoring and Quality Control

Sperm should be observed for adequate hybridization efficiency, probe tightness, and proper intensity of counterstaining. The slides can be evaluated manually or with an automated spot counting algorithm built into an analysis system (see Note 16).

1. Hybridization efficiency should be greater than or equal to 90 %. This is assessed by counting the number of sperm nuclei with no signal over at least the first 100 sperm.
2. Probe tightness is assessed by looking for scattered probes (Fig. 4a); probes should be a single spot with little or no noise within the nucleus.
3. Probe size is assessed when two signals of a given probe are observed within a single nucleus. The spots need to be equal in size (Fig. 4b).
4. Overlapping spots are assessed by verifying that they are at least 1.5 domains separated from each other. This applies to probes for the same chromosome only (Fig. 4c).
5. Signals must be within the counterstain mask of the nucleus (Fig. 4d).
6. The number of sperm counted per sample has been a point of frequent discussion, but currently the state-of-the-art is still a minimum of 5,000 sperm per sample for each chromosome being assessed (see Note 16).

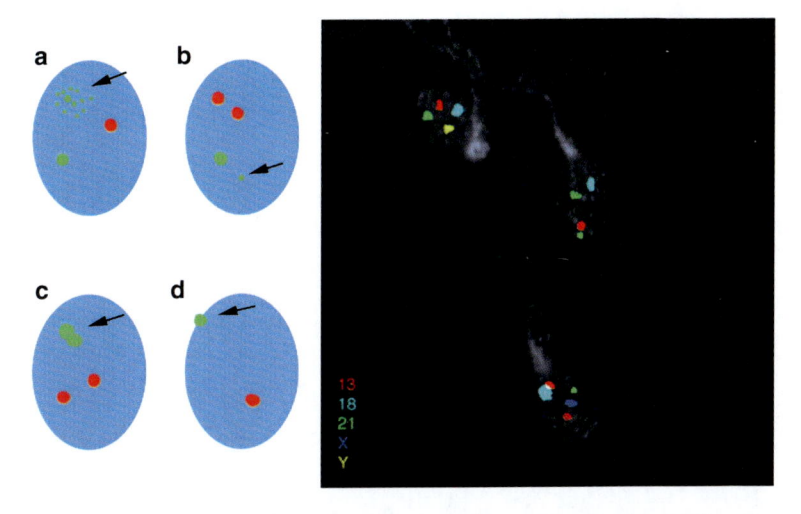

Fig. 4. Scoring criteria. (**a**) The *arrow* points towards a scattered probe. (**b**) The *arrow* shows a probe of the same fluorophore, but unequal in size to the other. (**c**) Here is shown two overlapping probes. (**d**) One probe, at the *arrow*, which is outside the mask of the nucleus.

4. Notes

1. Raw semen samples should be washed whenever possible. Contaminants from raw semen will make the enumeration of the probe counts less reliable and can cause higher levels of background noise.
2. Samples can be prepared using a separating sperm wash, such as a density gradient or swim-up, to remove contaminating cell types (white blood cells and other round cells sloughed from the reproductive tract). But such practices will have an impact on the interpretation of the results since these wash procedures will also remove a subpopulation of the sperm from the sample.
3. Washing also concentrates samples with extremely low sperm counts and will aid in increasing the number of sperm that can be counted.
4. Samples with low concentrations will not reach this count and should be resuspended to a volume just large enough to make the slide smears.
5. You should prepare enough slides to accommodate all the probe mixtures you are using and some additional slides to repeat the assay in case of hybridization failures or overdecondensation of sperm nuclei.
6. The concentration of the sperm on the slide should be such that the sperm concentration is high, but not so high that the majority of sperm will be touching or overlapping after nuclear decondensation.
7. DTT decondensation may be skipped if a high-heat denaturing step is incorporated (8). There is some benefit to this in samples where DTT decondensation causes overly decondensed heads. The high heat option is used exclusively in some labs and is described herein (see Note 11).
8. Overdecondensation will cause scattered probes and the slides will not be able to be read with any degree of confidence.
9. Be very careful not to push too hard, you may rupture the sperm heads.
10. This is easily done using a disposable plastic transfer pipette.
11. If you are using non-DTT treated sperm, you will need to use a high-heat denaturation rather than the temperature specified by the manufacturer. The temperatures for the melt step and hybridization step will be in the range of 80–90°C for 10 min for melting and 40–42°C for hybridization for 4–10 h (8).
12. Plastic Coplin jars are highly recommended. The high heat stresses the glass jars and greatly reduces their life span.

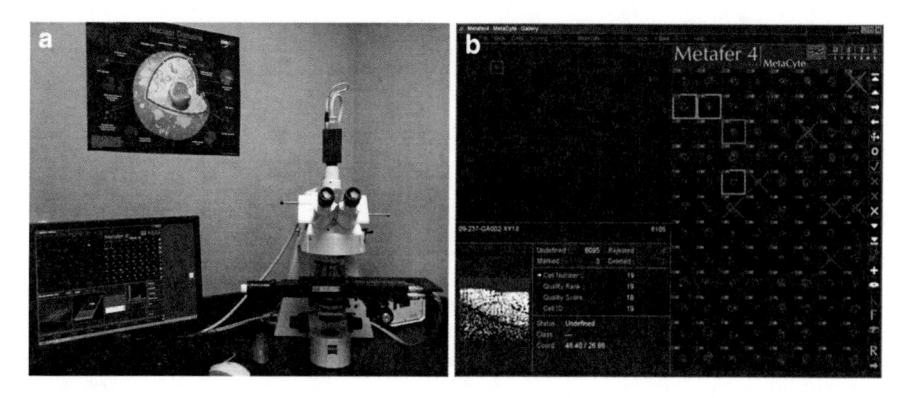

Fig. 5. Automated spot counting. (**a**) An image of the automated spot counting system from MetaSystems Inc. (**b**) A screen capture of the manual review interface that allows selection of miscounts and classification of abnormal cells.

13. If non-DTT treated sperm are used, the distaining time will need to be optimized and is typically 90 s to 2 min.
14. Do not wash more than four slides at a time and check the temperature before starting another batch of four slides.
15. Depending on the probe mixture used, 1 μl DAPI at 100 ng/ml may need to be added to the antifade solution for visualization of the nucleus. Some commercially available probe mixtures contain nonspecific DNA tagged with a mixed fluorophore to give a nuclear counterstain.
16. The efficiency of counting such a high number of sperm accurately can be increased by using an automated system with thorough and careful review of all sperm counted as aneuploid (Fig. 5) (4, 9).

References

1. Shi Q, Martin RH (2001) Aneuploidy in human spermatozoa: FISH analysis in men with constitutional chromosomal abnormalities, and in infertile men. Reproduction 121:655–666
2. Tempest HG et al (2004) The association between male infertility and sperm disomy: evidence for variation in disomy levels among individuals and a correlation between particular semen parameters and disomy of specific chromosome pairs. Reprod Biol Endocrinol 2:82
3. Oliver-Bonet M et al (2006) Studying meiosis: a review of FISH and M-FISH techniques used in the analysis of meiotic processes in humans. Cytogenet Genome Res 114:312–318
4. Carrell DT (2008) The clinical implementation of sperm chromosome aneuploidy testing: pitfalls and promises. J Androl 29:124–133
5. Sun F et al (2006) Is there a relationship between sperm chromosome abnormalities and sperm morphology? Reprod Biol Endocrinol 4:1
6. Tempest HG et al (2008) Sperm aneuploidy frequencies analysed before and after chemotherapy in testicular cancer and Hodgkin's lymphoma patients. Hum Reprod 23:251–258
7. Tempest HG, Martin RH (2009) Cytogenetic risks in chromosomally normal infertile men. Curr Opin Obstet Gynecol 21:223–227
8. Almeida Santos T et al (2002) Analysis of human spermatozoa by fluorescence in situ hybridization with preservation of the head morphology is possible by avoiding a decondensation treatment. J Assist Reprod Genet 19:291–294
9. Carrell DT, Emery BR (2008) Use of automated imaging and analysis technology for the detection of aneuploidy in human sperm. Fertil Steril 90:434–437

Chapter 16

Flow Cytometric Methods for Sperm Assessment

Vanesa Robles and Felipe Martínez-Pastor

Abstract

Flow cytometry allows the assessment of multiple sperm parameters following a diverse number of protocols. Here, we describe three methods for evaluating critical aspects of sperm quality: the double stain SYBR-14 and propidium iodide (PI) for assessing sperm viability; the double stain PNA-FITC (peanut agglutinin-fluorescein isothiocyanate) and PI for assessing acrosomal integrity combined with sperm viability; and JC-1 (5,5′,6,6′-tetrachloro-1,1′,3,3′-tetraethyl-benzimidazolylcarbocyanine iodide) for assessing mitochondrial membrane potential. These three stains are widely used and can be analyzed with basic flow cytometers.

Key words: Flow cytometry, Spermatozoa, Viability, Mitochondria, Acrosome, Propidium iodide, SYBR-14, JC-1, PNA, FITC

1. Introduction

Flow cytometry has gained increased attention in the field of andrology in the last 15 years. There are dozens of flow cytometry applications for sperm analysis, which have been reviewed elsewhere (1–5). We have selected three methods among the myriad of available techniques, considering them the best established and the most interesting for new andrologists, either from the clinical or the research fields. These methods are the viability stain using SYBR-14 and propidium iodide (PI); the acrosomal/viability stain using peanut agglutinin (PNA) conjugated with fluorescein isothiocyanate (FITC) and PI; and the mitochondrial stain using JC-1. These fluorochromes can be excited by the 488 nm line that is standard in flow cytometers. That means that even the most basic cytometers with a single blue laser for excitation and green/red or green/orange/red emission optics, can be utilized to analyze samples stained with these fluorochromes.

The combination of SYBR-14 and PI for assessing sperm viability (i.e., the integrity of the plasma membrane) was first used in 1994 (6), and the manufacturer (Invitrogen) produced a specific kit for assessing sperm viability soon after: the LIVE/DEAD® Sperm Viability Kit. SYBR-14 is a membrane-permeable DNA intercalating agent, with a maximum emission at 516 nm (green). SYBR-14 readily stains all nuclei. PI, another intercalating agent with maximum emission at 617 nm (red), can stain the nucleus only if the plasma membrane is damaged, with costaining resulting in quenching of SYBR-14 fluorescence (7). This dual stain was very convenient for both fluorescence microscopy and flow cytometry analyses, since all nucleated cells are stained, enabling easy discrimination of cells from debris (8).

The integrity of the acrosome is another useful measurement of sperm quality. Some researchers have used the anti-CD46 antibody to target the acrosomal matrix (9). More commonly, lectins targeted to the inner leaflet of the outer acrosomal membrane are used, either *Pisum sativum* (pea) agglutinin (PSA) (10) or *Arachis hypogaea* (peanut) agglutinin (PNA) (11). Here we describe the double stain PNA-FITC (peanut agglutinin conjugated with fluorescein isothiocyanate, maximum emission at 521 nm) with PI, which allows the simultaneous assessment of sperm viability and the condition of the acrosome. PNA is available conjugated with other fluorochromes as well (12), which may be more practical in some applications. In addition, some authors utilize ethidium homodimer in place of PI (13).

Sperm mitochondrial activity is associated with many sperm characteristics, from motility to fertility (14). The JC-1 fluorochrome (5,5′,6,6′-tetrachloro-1,1′,3,3′-tetraethyl-benzimidazolylcarbocyanine iodide; also abbreviated as $CBIC_2(3)$) is one of the most popular fluorochromes for sperm assessment (15–17), but other options do exist (18, 19). JC-1 accumulates in the mitochondria and has a maximum fluorescence emission at 525 nm (green). When JC-1 binds to membranes with a high potential ($\Delta\Psi_m$, 80–100 mV), it forms J-aggregates (20), resulting in a shift in fluorescence emission to ~590 nm (orange-red). Due to the heterogeneity of mitochondrial activity in live cells, it is typical to observe green and orange regions simultaneously, even in the same mitochondrion.

2. Materials

2.1. General Media and Reagents (See Note 1)

1. Phosphate buffer saline (PBS): Prepare a 10× stock solution. (1.4 M NaCl, 0.15 M KCl, 0.07 M Na_2HPO_4, 0.015 M KH_2PO_4) Weigh 8 g NaCl, 1.15 g KCl, 2.4 g Na_2HPO_4 and 0.2 g KH_2PO_4 and add to a 100 mL graduated cylinder containing about 50 mL of water. Seal tightly with lab film and

mix until completely dissolved. Make up to 100 mL with water and filter through a 0.45 μm filter. Store at room temperature. For making up the 1× solution, mix 20 mL of 10× in 180 mL of water in a beaker and add 0.1% BSA. Wait until dissolved and store at 4°C for up to 2 weeks.

2. HEPES buffer: 10 mM HEPES, 150 mM NaCl, pH to 7.4. Weigh 1.19 g HEPES and 4.38 g NaCl and add to about 300 mL of water in a beaker. Stir until dissolved, adjust pH to 7.4 with NaOH and make up to 500 mL. BSA (fraction V) can be added from 0.1% to 10% if required (see Note 2). Store at 4°C, protected from light (see Note 3) for up to 2 weeks.

2.2. Preparation of Fluorochrome Stock Solutions (See Note 4)

Fluorochromes must be manipulated in low light conditions, especially when in solution. DMSO must be at least of analytical grade, and the dyes dissolved with DMSO should be stored with desiccant since moisture is detrimental for fluorochromes. Deionized water should be at the least of milli-rho quality. If PBS is used, it should be 0.2-μm filtered or autoclaved. When preparing aliquots for freezing, it is convenient that each one is sufficient for 1 day of work, thus avoiding refreezing of the solutions.

1. LIVE/DEAD® Sperm Viability Kit: the kit contains both components ready for use, the SYBR-14 solution (1 mM in DMSO) and the PI solution (2.4 mM in water). Prepare a 50-fold dilution of the SYBR-14 solution in DMSO (20 μM), aliquot in convenient volumes to avoid repeated freezing-thawing and store frozen. Do not use aqueous media for preparing stock solutions of SYBR-14 (see Note 5).

2. Preparation of PI (for PNA-FITC/PI stain): PI solutions in a range of 5 mg/mL to 50 μg/mL can be quickly prepared and stored refrigerated or at −20°C. If purchased in powder form, pipet enough water or PBS into the vial to make up the solution, close tightly and vortex until completely dissolved. See Note 4 for advice in case it is necessary to weigh part of the PI before dilution. When using frozen solutions, briefly vortex the vial or sonicate it before use, in order to dissolve any precipitate. It is possible to purchase PI solutions, avoiding the need to prepare it in the lab.

3. Preparation of PNA-FITC: PNA-FITC is usually acquired lyophilized. Add water or PBS to the vial and vortex until completely dissolved. A typical stock solution is prepared at a concentration of 0.1–1.0 mg/mL. The stock should be aliquoted and stored at −20°C. If preferred, 2 mM sodium azide may be added to the solution allowing storage at 2-6°C for several months.

4. Preparation of JC-1: JC-1 is purchased lyophilized and must be diluted with DMSO and stored at −20°C. Prepare a 1–5 mg/mL (1.5–7.7 mM) solution by adding a small quantity of DMSO to the vial, closing it tightly and vortexing for several minutes

until all the dye has dissolved. Transfer the solution to a light-tight tube and rinse the vial with additional DMSO, transferring it to the light-tight tube. Make up to the final volume with DMSO, aliquot and store at −20°C.

3. Methods

3.1. Preparing Sperm Samples

Spermatozoa may be obtained from a variety of sources (fresh ejaculate, frozen semen, epididymal sperm, etc.), and therefore should be submitted to different procedures before analysis. If any component of the sperm medium could interfere with the staining procedure (components of seminal plasma, freezing extenders, incubation media, etc.) samples should be washed or otherwise cleaned up (see Notes 6 and 7). The staining protocols might need to be deeply modified in different species (see Note 8).

3.2. Flow Cytometer Setup

We have used these stains successfully in several flow cytometers: FACScan, FACScalibur, and LSR-I, from Becton Dickinson, and Epics XL, Cytomics FC 500, and Cyan ADP, from Beckman Coulter. Many other machines are cited in the bibliography. Each machine has its own startup procedures, and prior knowledge is necessary for operating them correctly (21). As a precision machine, the flow cytometer should be calibrated daily using fluorescently labeled latex spheres (e.g., Calibrite™ beads). For the methods presented here, only the blue excitation line is required (generally a 488 nm Argon-ion laser). The cytometer software should be set up for acquiring information from the appropriate photodetectors (frequently shown as FL1-H for green, FL2-H for orange, and FL3-H for red; see Note 9). These photodetectors should be set to "logarithmic" scale and to acquire "signal height" (hence the "H"). Fluorescence compensation may be necessary (see Note 10). Samples should be first analyzed for the FSC-H and SSC-H signals (forward and sideward scattering of light; Fig. 1), in order to identify the sperm population, to gate out debris and electronic noise, and to assess the presence of excessive debris. For reading samples, a low flow pressure is always preferable (e.g., 12 µL/s) (see Note 11). From 5,000 to 10,000 events identified as spermatozoa should be acquired for each sample.

3.3. Using the LIVE/DEAD® Sperm Viability Kit for Assessing Sperm Viability

1. Prepare a working solution of the LIVE/DEAD® Sperm Viability Kit by adding 5 µL of 20 µM SYBR-14 and 50 µL of the 2.4 mM PI solution to 10 mL of HEPES buffer (see Note 12) in a 15-mL polypropylene centrifuge tube. The working concentrations of SYBR-14 and PI are 100 nM and 12 µM, respectively. This solution must be stored in the dark and used the same day. This volume is sufficient to assess about 20 samples.

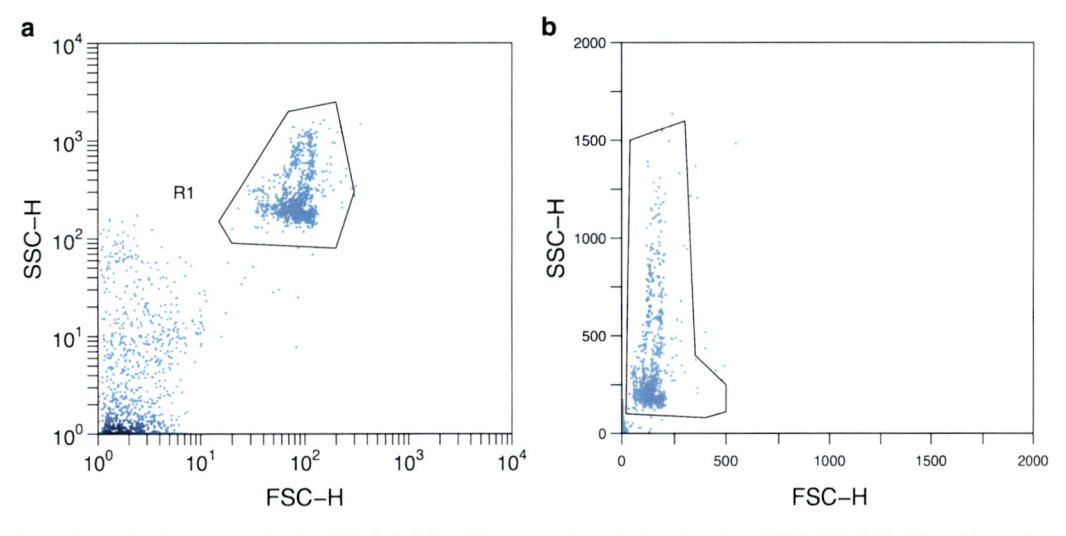

Fig. 1. Example of cytograms for the FSC-H (height of the forward scattering signal) and SSC-H (height of the side scattering signal) produced after running a sperm sample through a flow cytometer. (**a**) Cytogram produced using a logarithmic scale for these two detectors. The sperm population is enclosed in a region, and the events outside (debris and electronic noise) can be gated out of the analysis. (**b**) The sample was acquired using a linear scale. Again, the sperm population has been enclosed in a region, and the debris is barely visible at the origin. Both kinds of configurations are used with spermatozoa, and the choice depends on the experience and the convenience for specific techniques or machines.

2. Aliquot the staining solution in flow cytometer tubes, 0.5 mL each.

3. Dilute the spermatozoa in the staining solution to a concentration of $1-2 \times 10^6$ spermatozoa per mL.

4. Incubate for 10–15 min in the dark. Incubation may be carried out at 37°C if needed to achieve optimal staining.

5. Run the tube in the cytometer. The standard acquisition template consists of a cytogram showing SSC-H versus FSC-H signals, with a region enclosing the area where sperm events can be found. This region is used to establish an "AND" logical gate, which is applied in another cytogram showing FL1-H (green) versus FL3-H (red) fluorescence.

6. Interpretation of flow cytometry results (Figs. 2a and 3a): the FL1-H versus FL3-H cytogram shows two main populations. One of them includes cells with high FL1-H and low FL3-H (SYBR-14 positive and PI negative), which are considered viable spermatozoa. The other population includes cells with low FL1-H and high FL3-H (SYBR-14 negative and PI positive), which are considered dead spermatozoa. Frequently, a transitional population can be detected, composed of spermatozoa with varying levels of FL1-H and FL3-H signals. These cells are sometimes termed "moribund", and they are supposed to have early or minor membrane damage. This transitional

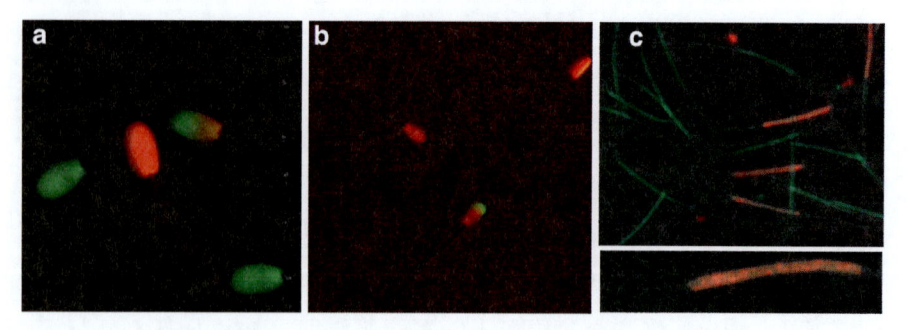

Fig. 2. Ruminant spermatozoa stained with the three techniques and visualized by fluorescent microscopy. (**a**) SYBR-14 and PI. The nuclei of all spermatozoa are clearly visible against the dark background. The three combinations are shown: *green* (SYBR-14 positive, viable), *red* (PI positive, dead), *mixed* (SYBR-14 being quenched by PI, "moribund"). (**b**) PNA-FITC and PI. Some white light allows the distinguishing of a viable spermatozoon with intact acrosome (unstained). Three nonviable spermatozoa are clearly visible with red nuclei (PI positive), one of them showing FITC green fluorescence in the acrosome (PNA positive, damaged acrosome). (**c**) JC-1 stain, with many spermatozoa showing green fluorescence in the midpiece (monomeric JC-1, indicating low $\Delta\Psi_m$). Several spermatozoa show intense orange-red fluorescence in the midpiece (J-aggregates forming at regions of high $\Delta\Psi_m$), indicating active mitochondria. The *inset* shows the granular appearance of a midpiece with active mitochondria, due to the heterogeneity of the $\Delta\Psi_m$ in different regions of the mitochondrial membranes.

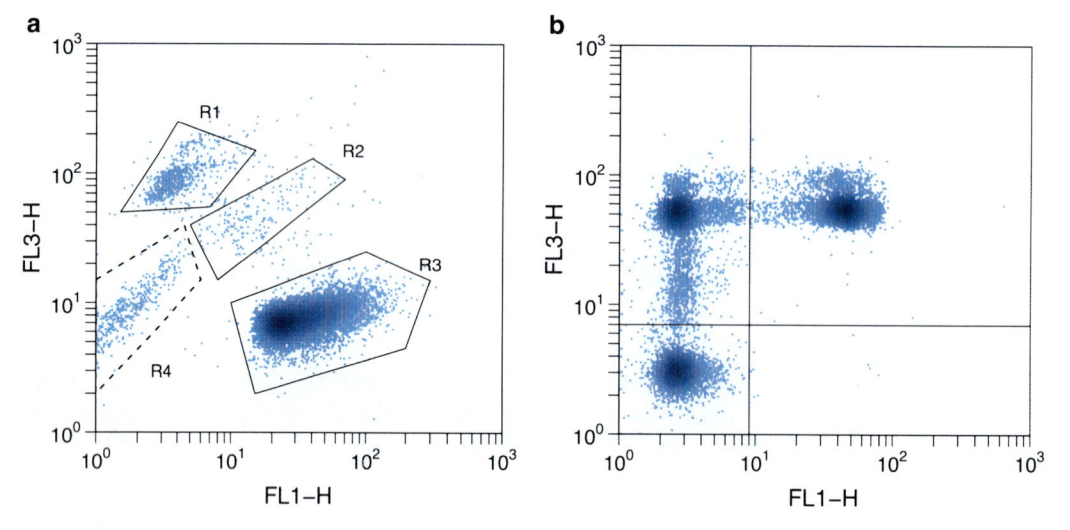

Fig. 3. Example of cytograms for spermatozoa stained with SYBR-14/PI and PNA-FITC/PI (FL1-H: intensity of the green fluorescence; FL3-H: intensity of the red fluorescence). (**a**) SYBR-14/PI yields three populations (enclosed in regions for clarity). R1 encloses the PI positive events (high FL3-H and low FL1-H), corresponding to dead spermatozoa. R2 encloses the events with mixed fluorescence, "moribund" spermatozoa. R3 encloses the SYBR-14 positive events (high FL3-H and low FL1-H), corresponding to viable spermatozoa. Compare with the stain pattern in Fig. 2a. R4 encloses unstained events, which can then be gated out from the analysis. (**b**) PNA-FITC/PI produces a four-quadrant distribution. The unstained population of viable spermatozoa is in the lower-left quadrant. The upper quadrants (high FL3-H) enclose the PI positive (dead) spermatozoa, with acrosome-damaged spermatozoa on the right (high FL1-H). Compare with the stain pattern in Fig. 2b. A fourth population of viable and acrosome-stained spermatozoa may show in the lower-right quadrant. Notice the transitional population from the lower-left quadrant to the population in the upper-left quadrant (see Note 13).

population is often ignored, taking only the percentage of truly SYBR-14 positive spermatozoa as the measure of sperm viability. Nevertheless, some authors have noted that this transitional subpopulation might be informative (7).

3.4. Assessing Acrosomal Status and Sperm Viability Simultaneously (PNA/PI)

1. The following procedures are based on stock solutions of PI and PNA-FITC prepared at 1 mg/mL in water. Thaw a PI aliquot and a PNA-FITC aliquot.
2. Prepare the PI/PNA-FITC working solution at 1 µg/mL PNA-FITC and 1.5 µM PI in PBS: Add 10 mL of PBS to a 15-mL polypropylene centrifuge tube and add 10 µL of the PI dilution and 10 µL of the PNA-FITC dilution. This solution must be stored in the dark and used the same day. This volume is sufficient to assess about 20 samples.
3. Aliquot the staining solution in flow cytometer tubes, 0.5 mL each.
4. Dilute the spermatozoa in the staining solution to a concentration of $1–2 \times 10^6$ spermatozoa per mL.
5. Incubate for 15 min in the dark. Incubation may be carried out at 37°C if needed to achieve optimal staining.
6. Run the tube in the cytometer. Follow indications for the SYBR-14/PI stain (see Subheading 3.3, step 6) (see Note 12).
7. Interpretation of flow cytometry results (Figs. 2b and 3b): the FL1-H versus FL3-H cytogram usually shows four populations. Unstained events (low FL1-H and low FL3-H) are viable spermatozoa with undamaged acrosomes (see Note 13). Events with high FL3-H and low FL1-H correspond to spermatozoa with damaged membranes (PI positive) but undamaged acrosomes (PNA negative) (see Note 14). Events with high FL3-H and FL1-H correspond to spermatozoa with damaged membranes and damaged acrosomes (PI and PNA positive). The population with high FL1-H and low FL3-H (PNA positive and PI negative) is usually very small, although some treatments can increase it considerably (e.g., adding a calcium ionophore to capacitated spermatozoa, which promotes the acrosome reaction in viable cells).

3.5. Assessing Mitochondrial Membrane Potential with JC-1

1. The following procedures are based on the stock solution of JC-1 prepared at 3 mM in DMSO. Prepare the JC-1 working solution at 3 µM JC-1 in PBS: Add 10 mL of PBS to a 15-mL polypropylene centrifuge tube and add 10 µL from a thawed JC-1 stock aliquot. Stir gently to assure a homogenous dilution. This solution must be stored in the dark and used promptly. This volume is sufficient to assess about 20 samples.

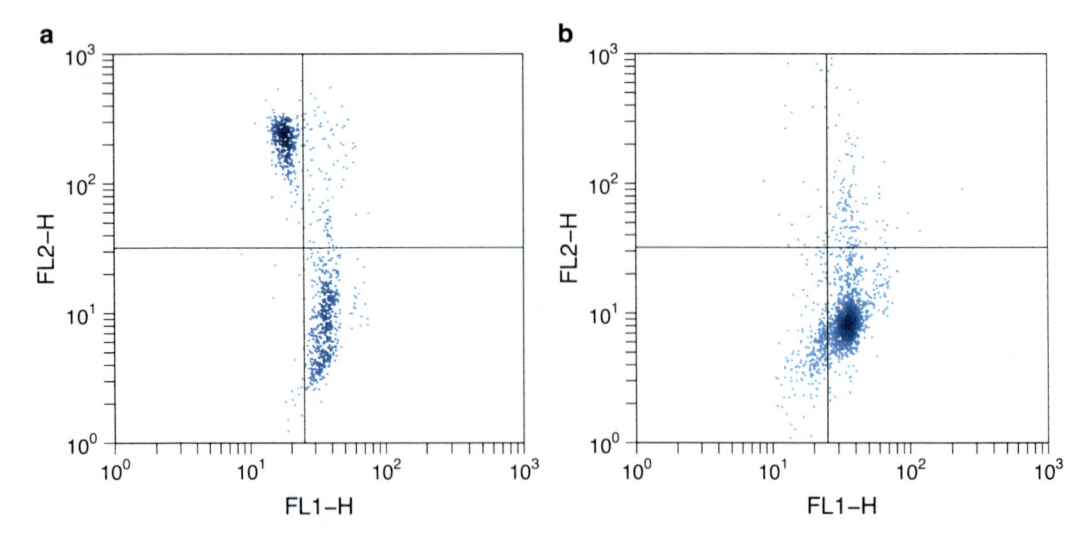

Fig. 4. Cytograms of spermatozoa stained with JC-1 (FL1-H: intensity of the green fluorescence; FL2-H: intensity of the orange fluorescence). (**a**) A good quality sperm sample, with most cells showing high orange fluorescence and low green fluorescence (active mitochondria, upper-left quadrant), some cells showing orange fluorescence and bright green fluorescence (upper-right quadrant) and a population of cells with inactive mitochondria in the lower-right quadrant (green fluorescence). Compare with the stain pattern in Fig. 2c. (**b**) A poor quality sperm sample, with most cells in the lower-right quadrant, and some cells losing green fluorescence (lower-left quadrant), possibly due to mitochondrial disruption.

2. Aliquot the staining solution in flow cytometer tubes, 0.5 mL each (see Note 15).

3. Dilute the spermatozoa in the staining solution to a concentration of $1-2 \times 10^6$ spermatozoa per mL.

4. Incubate for 20 min at 37°C and protected from light (see Note 16).

5. Run the tube in the cytometer. Follow indications for the SYBR-14/PI stain (see Subheading 3.3, step 6), noting that the cytogram for analyzing the fluorescence must show FL1-H (green) versus FL2-H (orange) fluorescence.

6. Interpretation of flow cytometry results (Figs. 2c and 4): the FL1-H versus FL2-H cytogram usually shows three populations. High FL1-H and low FL2-H corresponds with spermatozoa with low $\Delta\Psi_m$, whereas low FL1-H and high FL2-H correspond to spermatozoa with high $\Delta\Psi_m$. Usually, a transition population with high FL1-H and high to moderate FL2-H can be detected, corresponding to spermatozoa that present wide gaps with low $\Delta\Psi_m$ in their mitochondria. Sometimes, a population of low FL1-H and low FL2-H can be detected, possibly corresponding to nonviable spermatozoa with degraded midpiece.

4. Notes

1. Several types of media can be used as staining media for spermatozoa, allowing the adaptation of staining conditions to fit different experiments. PBS buffers are adequate for most applications, and flow cytometer sheath fluid is generally based in phosphate buffers. However, some fluorochromes may be affected by components of the staining solutions, prohibiting the use of these components (see Note 5).

2. Invitrogen recommends 10% BSA. From our experience, lower BSA concentrations can be used, and it can even be omitted.

3. HEPES may degrade with exposure to light, producing hydrogen peroxide (22). Therefore, solutions should be stored in the dark (covering bottles with aluminum foil), exposure of the media with cells to light should be minimized.

4. Many fluorochromes are mutagenic, toxic or potentially toxic, especially in powder form. If possible, always add the diluent to the vial in which the fluorochrome was shipped, dilute it completely, and then carry out further pipetting if required. If it is necessary to weigh a portion of the powder and dilute it in another vial, take precautions according to the manufacturer. Working in a hood or wearing respiratory mask and goggles is advisable when working with any fluorochrome capable of interacting with DNA.

5. According to the SYBR-14 manufacturers (Invitrogen), phosphate-containing buffers may interfere with SYBR-14 staining. Therefore, solutions should be prepared with other buffering systems.

6. Some clean up procedures could be stressful for spermatozoa, reducing the overall quality of the sample or its resilience to follow-up procedures. It is necessary to balance the need for cleaning up the sample against the possible alterations inflicted to the spermatozoa (see Note 7).

7. The presence of debris in the sample can be a problem for some techniques. Freezing extenders for mammal spermatozoa generally include egg yolk or milk. If cleaning up is not possible or incomplete, that debris will remain in the sample. Using dye combinations that are specific for nucleic acids (staining all nucleated events, such as SYBR-14/PI) allows excluding unstained events as debris (see Fig. 3a). If debris cannot be gated out from the FSC/SSC plot, it may overlap unstained cell populations (such as in PNA-FITC/PI). Many fluorochromes are lipophilic and show a high affinity for debris (e.g., JC-1), in which case the debris can be interpreted as part of fluorescent cell populations. In both cases, an overestimation

of one or more sperm populations will occur. It has been proposed that mathematical corrections be applied to the data (23). A very effective approach consists of staining the cells with Hoechst 33342 (24). This fluorochrome quickly permeates membranes and intercalates in the DNA, allowing the investigator to gate out unstained events. Hoechst 33342 does not interfere with other fluorochromes, and its excitation/emission ranges (UV/blue) fall outside of the ranges of the most common sperm stains. This is also its main disadvantage, because of the need for a cytometer with a UV or violet laser, which may be expensive.

8. Spermatozoa from some species may require incubation at low temperatures (e.g., fishes that spawn in cold waters). Lower temperatures may require longer incubation times.

9. Note that these denominations might vary among machines, and they can even refer to different photodetectors. This is an example of a standard configuration: FL1, 530/28; FL2, 585/42, FL3, 650LP (long pass).

10. Compensation may be required in order to reduce the fluorescence spillover, especially from green fluorochromes, affecting orange and red signals.

11. Lower flow pressure helps in the formation of a laminar flow and allows better event discrimination. If needed (e.g., with highly diluted samples), it may be possible to read samples at medium or high flow pressures. Nevertheless, repeatability or the ability to compare samples could be negatively affected.

12. The manufacturer recommends incubating first with SYBR-14 and adding PI afterwards. We have not noted any difference by using the more practical and efficient approach of preparing a working solution, and incubating with both fluorochromes simultaneously.

13. Very often, a transitional population appears between the unstained (viable) and PI positive spermatozoa (non viable with unstained acrosome). It is important that this population not be included with the viable spermatozoa (see Fig. 3b).

14. Since PNA binds to the external acrosomal membrane, spermatozoa with highly degraded acrosomes might display very low PNA-FITC labeling. We have observed this event, especially in dead spermatozoa showing a loose acrosomal cap, although the occurrence seems to be rare.

15. Tubes containing mitochondrial uncouplers, such as carbonyl cyanide 3-chlorophenylhydrazone (CCCP) can be used as negative controls.

16. JC-1 staining seems to be sensitive to different staining conditions (sperm diluents, temperatures, etc.); hence, time and temperature of incubation may need to be adapted.

Acknowledgments

This work was supported in part by grants LE019A10-2 (Junta de Castilla y León), AGL2010-15758 (MICINN) and Fundación Ramón Areces. The authors were supported by the Ramón y Cajal program (MICINN). The authors thank María Mata-Campuzano for producing several of the microphotographs and cytograms.

References

1. Garrido N et al (2002) Flow cytometry in human reproductive biology. Gynecol Endocrinol 16:505–521
2. Martinez-Pastor F et al (2010) Probes and techniques for sperm evaluation by flow cytometry. Reprod Domest Anim 45(Suppl 2):67–78
3. Pena FJ (2007) Detecting subtle changes in sperm membranes in veterinary andrology. Asian J Androl 9:731–737
4. Petrunkina AM, Harrison RA (2011) Cytometric solutions in veterinary andrology: developments, advantages, and limitations. Cytometry A 79:338–348
5. Hossain MS et al (2011) Flow cytometry for the assessment of animal sperm integrity and functionality: state of the art. Asian J Androl 13:406–419
6. Garner DL et al (1994) Dual DNA staining assessment of bovine sperm viability using SYBR-14 and propidium iodide. J Androl 15:620–629
7. Grundler W et al (2004) Quantification of temporary and permanent subpopulations of bull sperm by an optimized SYBR-14/propidium iodide assay. Cytometry A 60:63–72
8. Nagy S et al (2003) A triple-stain flow cytometric method to assess plasma- and acrosome-membrane integrity of cryopreserved bovine sperm immediately after thawing in presence of egg-yolk particles. Biol Reprod 68:1828–1835
9. Bronson RA et al (1999) Progesterone promotes the acrosome reaction in capacitated human spermatozoa as judged by flow cytometry and CD46 staining. Mol Hum Reprod 5:507–512
10. Maxwell WM et al (1996) Viability and membrane integrity of spermatozoa after dilution and flow cytometric sorting in the presence or absence of seminal plasma. Reprod Fertil Dev 8:1165–1178
11. Blottner S et al (1998) Flow cytometric determination of the acrosomal status of bull and stallion spermatozoa after marking with FITC-conjugated PNA (peanut agglutinin). Tierarztl Umsch 53:442–447
12. Castro-Gonzalez D et al (2010) The acidic probe LysoSensor is not useful for acrosome evaluation of cryopreserved ram spermatozoa. Reprod Domest Anim 45:363–367
13. Cheng FP et al (1996) Use of peanut agglutinin to assess the acrosomal status and the zona pellucida-induced acrosome reaction in stallion spermatozoa. J Androl 17:674–682
14. Pena FJ et al (2009) Mitochondria in mammalian sperm physiology and pathology: a review. Reprod Domest Anim 44:345–349
15. Martinez-Pastor F et al (2004) Use of chromatin stability assay, mitochondrial stain JC-1, and fluorometric assessment of plasma membrane to evaluate frozen-thawed ram semen. Anim Reprod Sci 84:121–133
16. Guthrie HD, Welch GR (2008) Determination of high mitochondrial membrane potential in spermatozoa loaded with the mitochondrial probe 5,5′,6,6′-tetrachloro-1,1′,3,3′-tetraethylbenzimidazolyl-carbocyanine iodide (JC-1) by using fluorescence-activated flow cytometry. Methods Mol Biol 477:89–97
17. Ortega-Ferrusola C et al (2009) Apoptotic markers can be used to forecast the freezeability of stallion spermatozoa. Anim Reprod Sci 114:393–403
18. Martinez-Pastor F et al (2008) Mitochondrial activity and forward scatter vary in necrotic, apoptotic and membrane-intact spermatozoan subpopulations. Reprod Fertil Dev 20:547–556
19. Kumaresan A et al (2009) Preservation of boar semen at 18 degrees C induces lipid peroxidation and apoptosis like changes in spermatozoa. Anim Reprod Sci 110:162–171
20. Smiley ST et al (1991) Intracellular heterogeneity in mitochondrial membrane potentials revealed by a J-aggregate-forming lipophilic cation JC-1. Proc Natl Acad Sci USA 88:3671–3675
21. Shapiro HM (2003) Practical flow cytometry, 4th ed. Wiley-Liss, Hoboken, NJ

22. Lepe-Zuniga JL et al (1987) Toxicity of light-exposed Hepes media. J Immunol Methods 103:145
23. Petrunkina AM, Harrison RA (2010) Systematic misestimation of cell subpopulations by flow cytometry: a mathematical analysis. Theriogenology 73:839–847
24. Hallap T et al (2006) Usefulness of a triple fluorochrome combination Merocyanine 540/Yo-Pro 1/Hoechst 33342 in assessing membrane stability of viable frozen-thawed spermatozoa from Estonian Holstein AI bulls. Theriogenology 65:1122–1136

Chapter 17

Human Y Chromosome Microdeletion Analysis by PCR Multiplex Protocols Identifying only Clinically Relevant *AZF* Microdeletions

Peter H. Vogt and Ulrike Bender

Abstract

PCR multiplex assays are the method of choice for quickly revealing genomic microdeletions in the large repetitive genomic sequence blocks on the long arm of the human Y chromosome. They harbor the Azoospermia Factor (*AZF*) genes, which cause male infertility when functionally disrupted. These protein encoding Y genes are expressed exclusively or predominantly during male germ cell development, i.e., at different phases of human spermatogenesis. They are located in three distinct genomic sequence regions designated AZFa, AZFb, and AZFc, respectively. Complete deletion of an AZF region, also called "classical" AZF microdeletion, is always associated with male infertility and a distinct testicular pathology. Partial AZF deletions including single *AZF* Y genes can cause the same testicular pathology as the corresponding complete deletion (e.g., *DDX3Y* gene deletions in AZFa), or might not be associated with male infertility at all (e.g., some *BPY2*, *CDY1*, *DAZ* gene deletions in AZFc). We therefore propose that a PCR multiplex assay aimed to reduce only those AZF microdeletions causing a specific testicular pathology—thus relevant for clinical applications. It only includes Sequence Tagged Site (STS) deletion markers inside the exon structures of the Y genes known to be expressed in male germ cells and located in the three AZF regions. They were integrated in a robust standard protocol for four PCR multiplex mixtures which also include the basic principles of quality control according to the strict guidelines of the European Molecular Genetics Quality Network (EMQN: http://www.emqn.org). In case all Y genes of one AZF region are deleted the molecular extension of this AZF microdeletion is diagnosed to be yes or no comparable to that of the "classical" AZF microdeletion by an additional PCR multiplex assay analyzing the putative AZF breakpoint borderlines.

Key words: AZFa, AZFb, AZFc classical microdeletions, PCR multiplex assays, AZF DNA polymorphisms and border lines, *AZF* gene deletions

1. Introduction

Y chromosomal microdeletions in men with a normal karyotype (46,XY) were one of the first detected polymorphic sequence variations in the human genome: They were identified first with sophisticated molecular methods like Southern blotting, then by

PCR deletion and restriction analyses of small sequence tagged sites (STSs); and because present only on the Y chromosome of men in distinct human populations they soon became most useful to answer questions of human origin and migration (1–6). Curiously, in the same genomic sequence region of the Y long arm (Yq11) candidate genes of the so-called Azoospermia Factor (AZF) were proposed 35 years ago (7). Any AZF deletion will cause absence of sperm in the ejaculate (azoospermia), thus severe male infertility.

Consequently, to screen—by Southern blotting or PCR assays—for only those microdeletions in Yq11 proposed to contain one of the functionally relevant *AZF* Y genes it is mandatory to analyze whether the Y deletion observed in an infertile patient is a "de novo" DNA deletion, i.e., restricted to his Y chromosome. A not, polymorphic familial STS sequence variant should also be present on the Y chromosome of the patient's father or fertile brother (8, 9). Thus DNA fragment and STS deletions identified in the putative AZF region of an infertile man needs to be distinguished whether being a nonpolymorphic or familial polymorphic event in the pedigree analyzed (10–13).

Twenty years now after description of the first "de novo" *AZF* microdeletions (14) this prerequisite to identify only the clinically relevant Y chromosomal microdeletions seems to be largely forgotten. Moreover, there is a strong belief in the clinic that diagnosing *AZF* microdeletions needs only to concentrate on identification of the three now coined "classical" AZF microdeletions, AZFa, AZFb, and AZFc, respectively (15), known to be associated with a distinct testicular pathology (see Note 1). Accordingly two robust PCR multiplex assays were developed including the best practice guidelines of the European Molecular Genetics Quality Network (EMQN; http://www.emqn.org). They contain a conveniently small set of 6 STSs which are assumed to be nonpolymorphic in all human populations; thus should be safe markers for detection of the three clinically relevant AZF microdeletions worldwide (16).

However, in the meanwhile one of the AZFa STS markers (sY84) was found to be polymorphic (17) and the second AZFa STS marker was found to have multiple primer sites on the human Y chromosome (UniSTS database: DXYF134S1). Additionally, multiple polymorphic large genomic deletions were found, especially in the AZFc region due to its unstable locus specific repetitive amplicon structure (18–20). The molecular extensions of many of these partial AZF microdeletions were smaller than that of the first observed "classical" AZF microdeletions. They were also present as polymorphic length variants on the Y chromosome of fertile men from different populations (21–24).

The revised EMQN guidelines in 2004 discussed this complexity of partial AZF microdeletions and therefore advised again to prove whether the molecular extension of an AZF deletion identified by the basic 6 STS primer set is comparable or not to that of the "classical" AZF microdeletion. And again each AZF STS deletion should be also proofed to be a "de novo" DNA deletion restricted to the Y chromosome of the patient (25).

However, these remarks did not influence significantly the clinical routine practice, and the molecular extension of potentially only partial AZF microdeletions were usually not distinguished from those of the "classical" AZF microdeletions in a second PCR multiplex run as recommended by the revised EMQN guidelines (25). It has also not become a clinical practice to confirm the diagnosis of each AZF microdeletion identified as being a "de novo" microdeletion event.

A systematic screening program for the presence of Y genes in Yq11 (26) and then later knowledge of the first complete genomic Y sequence (27) then determined that many Y genes encoding proteins and probably functioning during human spermatogenesis are located in Yq11. At least 14 of them were mapped to the three "classical" AZF microdeletion intervals (26–28). Their expression in germ cells qualify them as *AZF* candidate genes of similar functional importance unless one finds their deletion on the Y chromosome of a fertile male in a distinct human population (22, 29).

Consequently, we updated the current *AZF* deletion diagnostics for our infertility clinic by establishing PCR multiplex assays with similar standard controls as described by the EMQN best practice guidelines (25) but with STS marker sets located not outside but inside the exons of all *AZF-Y* genes.

In our lab this resulted in the establishment of five novel PCR multiplex assays able, not only to detect each of the "classical" AZFa, b, c microdeletions by analyzing the AZFa, AZFb, or AZFc borderlines, but also additionally the deletions of each functionally important *AZF* candidate gene within them. In this way, we score only for AZF (gene) microdeletions with a clear impact on the patient's fertility.

2. Materials

2.1. Genomic DNA and PCR Reagents

1. Genomic DNA from *idiopathic* severe oligozoospermic and azoospermic men (see Note 2).
2. STS primer sets for four PCR multiplex assays (A, B, C, D) listed in Table 1 (see Notes 3 and 4). Unless stated otherwise, each STS *AZF* gene primer mix given in Tables 3 and 4 contains the amount of primers for 20 PCR multiplex reactions.
3. STS primer sets listed in Table 2 (see Note 5).

Table 1
STS markers in PCR multiplex mixtures for the different AZF-Y genes and Y genes controls given with lengths of PCR products and positions in current Y-reference sequence

PCR multiplex mixture	STS marker in AZF gene exon structure	Forward primer sequence	Reverse primer sequence	PCR amplification product: bp	STS position (nts) on Y-reference sequence: NC_000024.9,GRch37.p2	Y-gene	Y chromosome region
	DAZ-ex3/ex4	TAGGTTTCAGTGTTTGGATTCCG	GGAAGCTGCTTTGGTAGATAC	1368	25.314.850-25.316.218	DAZ1	AZFc
				1368	25.325.690-25.327.058	DAZ1	AZFc
				1368	25.336.538-25.337.906	DAZ1	AZFc
				1368	25.374.298-25.372.930	DAZ2	AZFc
				1368	26.950.921-26.952.289	DAZ3	AZFc
				1368	26.988.685-26.987.317	DAZ4	AZFc
				1368	26.999.533-26.998.165	DAZ4	AZFb
	RPS4Y2-in3/ex4	AGTATACCATTTGAAGCCCC	GCTGGTTATCTTCCCAGT	867	22.929.976-22.930.843	RPSAY2	AZFb
	ZFY-ex8	ACCRCTGTACTGACTGTGATTACAC	GCACYTCTTTGGTATCYGAGAAAGT	494	2.846.983-2.847.477	ZFY	Yp11.2
	DDX3Y-in2/ex3	GGTAATTCAAGATGTTTAATGGTC	CTCCTAAAGTTACTTTCTTCC	409	15.020.986-15.021.395	DDX3Y	AZFa
	SMCY-ex6	CCAATATCAAGGTCTAGATG	GGACCTGTGTAATTTAGAAAG	314	21.901.662-21.901.348	SMCY	AZFb
	TPITEY-ex10/in10	ACAATAGAGACTCATCTGAGTAG	AGACTCGATAGCGGCATTGA	273	28.702.910-28.703.183	TPITEY	Yq11.23
	UTY-ex28	AAAGTATTACGGAAGTTTGCTG	CACTGACTATATGTTTCTACTG	137	15.360.717-15.360.580	UTY	Yq11.21
	PRY-ex1/ex2	GCCATGTAATTTCCAGCACAC	CAGACATCTTGCAGTGTTTCA C	805	24.242.127-24.241.322-	PRY2	AZFb
				805	24.637.377-24.636.571	PRY1	AZFb
	GOLGA2LY-ex14	TTGGCCTGTGCTTCTAGGGTT	ACAGGAGGTGCTGTCACA	530	26.355.197-26.355.727	GOLGA 2LY1	AZFc
				530	27.607.239-27.606.709	GOLGA 2LY2	AZFc
	SRY-ex1 (sY14)	GCTGGTGCTCCATTCTTGAG	GAATATTCCCGCTCTCCGGA	469	2.655.107-2.655.576	SRY	Yp11.2
	DAZ-ex2/ex3 (sY254)	GGGTGTTACCAGAAGGCAAA	GAACCGTATCTACCAAAGCAGC	379	25.316.572-25.316.193	DAZ1	AZF c
				379	25.327.412-25.327.033	DAZ1	AZFc
				379	25.338.260-25.337.881	DAZ1	AZFc
				379	25.372.576-25.372.955	DAZ2	AZFc
				379	26.952.643-26.952.264	DAZ3	AZFc
				379	26.986.963-26.987.342	DAZ4	AZFc
				379	26.997.811-26.998.190	DAZ4	AZFc
	RBMY1-ex2	GAAGTTAGTGCCCCTTTGCTCTG	CTTCTGATATGGGACCATGTTTCCC	291	23.675.054-23.675.345	RBMY1B	AZFb
				291	23.698.595-23.698.886	RBMY1A1	AZFb
				291	24.038.843-24.038.552	RBMY1D	AZFb
				291	24.062.384-24.062.093	RBMY1E	AZFb

STS exon marker	Primer 1	Primer 2	Size	Position	Gene	Region
USP9Y-in2/in3 (sY620)	GGCTGATATATGCTGGTACTTCA	TTCACAGTACTCAAAACAACACAGTC	291	24,327,299-24,327,008	RBMT1F	AZFb
			291	24,551,412-24,551,703	RBMT1J	AZFb
			248	14,821,264-14,821,512	USP9Y	AZFa
RBMT1-ex11/ex12	ATGCACTTCAGAGATACGG	CCTCTCTCCACAAAACCAACA	852	23,663,979-23,663,127	RBMT1H	AZFb
			852	23,687,519-23,686,667	RBMT1B	AZFb
			852	23,711,059-23,710,207	RBMT1A1	AZFb
			852	24,026,376-24,027,228	RBMT1D	AZFb
			852	24,049,918-24,050,770	RBMT1E	AZFb
			852	24,314,862-24,315,694	RBMT1F	AZFb
			852	24,563,875-24,563,023	RBMT1J	AZFb
SMCY-ex12	GTACTGATCTGCAACTTGA	GGACCTCTTGTGACCAGGA	745	21,882,588-21,883,333	SMCY	AZFb
DDX3Y-inT/in1	GGGAGTTAGTGGTAGATTGTC	TCG GTAGGCTTACTTTGACTC	473	15,016,442-15,016,915	DDX3Y	AZFa
CDY1-ex1/in1	GGGCGAAAGCTGACAGCAA	GCTATGAAGATGGATTAGGGCA	367	26,192,494-26,192,127	CDY1	AZFc
			367	27,769,931-27,770,298	CDY1	AZFc
HSFY-in1/ex2	GTTGGTTCCATGCTCCATACAAAGTC	CAGGGAGAGAGCCTTTTACC	261	20,711,867-20,712,128	HSFY1	AZFb
			261	20,932,311-20,932,050	HSFY2	AZFb
UTY-ex3	TCGTGGGACTTGGAAATTTGTTG	CTTTCCTGACTGTCAGGCTAAC	155	15,591,573-15,591,418	UTY	Yq11.21
USP9Y-ex9/in11	CAGGGTTGGCTAGTGGATCTC	CTGCCATTCGCTCAGCTGTC	941	14,847,543-14,848,484	USP9Y	AZFa
BPY2-in6/ex6	GTCCATAGTGGGCTTGCTGA	GCTTGTTCCACATATGGCACA	846	25,143,591-25,144,437	BPY2.1	AZFc
			846	26,777,333-26,778,179	BPY2.2	AZFc
			846	27,185,069-27,184,223	BPY2.3	AZFc
ZFY-ex8	CGAATTCATACCGCGGAGAAGCCATACC	AAAGCTTGTAGACACACTGTTAGG	734	2,847,318-2,848,052	ZFY	Yp11.2
VCY-ex1/ex2	CTCCGGCCAAGGCCAAGGAGA	ATGGGCGCCCCTTACTCAGTGG	521	16,098,292-16,097,771	VCY1	Yq11.21
			521	16,168,198-16,168,719	VCY2	Yq11.21
DAZ-ex6ex2	CTGCTTAGCATTCTTCTATTATCTGT	ACGCAACTCTCACTGCTACATAGG	473	25,319,936-25,319,463	DAZ1	AZFc
			473	25,330,781-25,330,308	DAZ1	AZFc
			473	26,994,442-26,994,915	DAZ4	AZFc
EIF1AY-ex7	GGTTCTACAGTTGGGATTTTGGC	CCAAAAGACTGTAAACAGATTAG GTGAAGTA	343	22,754,273-22,754,616	EIF1AY	AZFb
TSPY-exons	CCTTTCATCCCAACCTTTATTTCCA	GCAGTCATGTTCAGCCAAACAGC	273	9,172,425-9,172,152	TSPY4	Yp11.1
			273	9,192,755-9,192,482	TSPY8	Yp11.1
			273	9,213,035-9,212,762	TSPY7P	Yp11.1
			273	9,233,381-9,233,108	TSPY3	Yp11.1
			273	9,301,918-9,301,645	TSPY1	Yp11.1
			273	9,322,227-9,321,954	TSPY9	Yp11.1
			273	9,342,514-9,342,241	TSP 6P	Yp11.1
			273	9,362,841-9,362,568	TSPY10	Yp11.1
			273	9,383,031-9,382,758	TSPY15P	Yp11.1

STS exon marker used for analysis of *BPY2*, *CDY*, *DAZ*, *TSPY* gene deletion is located on multiple copies of these repetitive gene families. It is therefore only shown to be deleted when all gene copies are absent

Table 2
PCR multiplex assays with STS markers for AZF deletion borderlines to proof "classical" extension of identified AZFa, AZFb, AZFc microdeletion

PCR multiplex assay	STS markers for AZF deletion borderlines		Forward primer sequence	Reverse primer sequence	PCR amplification product: bp	STS position (nts) in Y-reference sequence: NC_000024.9,GRch37.p2
AZFa extension	AZFa-prox1		CTTAAATGTTGACTCTTCACC	GCCTTGTGTAGAATAAGCAGTCA	126	14.432.192–14.432.317
	AZFa-prox2		GGTTCCTGAACAGGGACT	GGCAGCAGAAGGGCCTCTC	219	14.442.232–14.442.451
	AZFa-dist1		GGCTTCTAGTAGTATGGTC	TTGTCCTTCAATGCAGATG	389	15.222.567–15.222.956
	AZFa-dist2		GTTCCCCATTCATTATACTGTTAGC	GCACTCCAGAAAGATAATACATC	270	15.234.363–15.234.633
AZFb extension	AZFb-prox1/sY1264		GGCTTAAACTTGGGAGGGTG	GCACTTCAAACCGAGGCTTA	789	19.568.146–19.567.357
	AZFb-prox2/sY1227		GCAAAGTTCTTGCACTGTGTTT	TTGTGGCCTGTGTGTAGAA	317	20.063.027–20.063.344
	AZFb-dist1/sY1291		TAA AAG GCA GAA CTG CCA GG	GGG AGA AAA GTT CTG CAA CG	526	25.505.070–25.505.596
	AZFb-dist2/sY639		GGGCGAAAGCTGACAGCAA	GCTATGAAGATGGATTAGGGCA	367	26.192.494–26.192.127
AZFc extension	AZFc-prox1/sY1197		TCATTTGTGTCCTTCTCTTGGA	CTAAGCCAGGAACTTGCCAC	452	24.523.618–24.524.070
	AZFc-prox2/sY1192		ACTACCATTTCTGGAAGCCG	CTCCCTTGGTTCATGCCATT	254	24.873.032–24.872.778
	AZFc-dist1/sY1054		ACCTAAGGGAACCCAGGAGA	CGA CAC TTT TGGGAA GTT TCA	339	25.849.059–25.848.720
					339	28.113.309–28.113.648
	AZFc-dist2/sY1125		GTGGGGGTTTCACATTATGG	GGTCACAGACTCACATTTAAGCA	282	25.620.032–25.620.314
					282	28.342.873–28.342.591

STS markers for AZFa extension analysis are also published in ref. (30). STS markers for AZFb and AZFc extension analyses are also listed with their "sY" number (see: refs. 31, 39)

4. Cotton-filtered pipet tips.
5. PCR water (see Note 6).
6. 10× PCR buffer: 200 mM Tris–HCl (pH 8.4) and 500 mM KCl (Invitrogen, Life technologies Cooperation).
7. Recombinant *Taq* DNA polymerase (5 U/μl; Invitrogen).
8. MgCl$_2$ stock solution (50 mM).
9. Dimethylsulfoxide (DMSO).
10. 50× stock solution (25 mM) for the four Deoxynucleotide triphosphates (dNTPs; dATP, dCTP, dGTP, dTTP).

2.2. Agarose Gel Electrophoresis

1. 50× Tris-acetate-EDTA (TAE) stock solution: Disolve 484 g Tris and 37.2 g EDTA Na$_2$-salt in 1.5 L prewarmed MilliQ water. Cool to room temperature, then add 114.2 ml acetic acid (conc. 99.9%), adjust pH to 8.3 with NaOH and bring to 2 L end-volume with MilliQ water.
2. 1× TAE running buffer (10 mM Tris, 50 mM Acetic acid, 10 mM EDTA) pH 8.3: Dilute 50× TAE 1:50 in MilliQ water.
3. 10× Orange G loading buffer: Mix 16.7 mg Orange G in 5 ml MilliQ water, then add 1 ml of 1 M Tris–HCl pH 7.5 and 3.3 ml of 87% glycerol and bring up to 10 ml end-volume.
4. 2% agarose gel: Carefully melt 2 g LE-agarose in 100 ml 1× TAE buffer in a microwave for 3–5 min (adjust volume according to gel cassette volume). Pour the solution in gel mold and insert a comb. The comb should be large enough to accomodate about 20 μl volume. Allow the agarose gel to completely solidify (about 1 h).
5. 100 bp ladder-plus length marker (Fermentas).
6. Ethidium bromide (EtBr) stock solution: (10 ng/ml) in water.
7. Ethidium bromide (EtBr) working solution: Dilute 25 μl of the stock solution (10 ng/ml) in 500 ml TAE buffer for a 0.005% final concentration.
8. GelDoc2000 (BIORAD) or similar for visualizing gel.

3. Methods

3.1. PCR Multiplex Reactions

1. On ice, prepare the four *AZF-Y gene* PCR-multiplex master mixes A, B, C, and D (see Table 3), in parallel using the volumes below. The volumes given are per sample to be analyzed (see Note 7). In addition to test samples, DNA from a fertile man proven to be without any AZF deletions on the Y chromosome, as well as genomic DNA from a normal female and a water

Table 3
Primer mixes for four PCR multiplex assays (A, B, C, D) analyzing *AZF-Y* gene deletions according to EMQN best practice guidelines

Primer mix A (end-volume 100 μl)	STS marker for *AZF-* or *control*-gene	PCR product length	Forward & reverse primer stocks	μl added from each primer stock
Add 80 μl water	*DAZ-ex3/ex4*	1,368 bp	50 pmol	Each 1 μl
	RPS4Y2-in3/ex4	867 bp	100 pmol/μl	Each 1 μl
	ZFX/ZFY-ex8	494 bp	50 pmol/μl	Each 1 μl
	DDX3Y-int2/ex3	409 bp	100 pmol/μl	Each 1 μl
	SMCY-ex6	314 bp	100 pmol/μl	Each 3 μl
	TPTEY-ex10/in10	273 bp	50 pmol/μl	Each 1 μl
	UTY-ex28	137 bp	50 pmol/μl	Each 2 μl

Primer mix B (end-volume 100 μl)	STS marker for *AZF-* or *control*-gene	PCR product length	Forward & reverse primer stocks	μl added from each primer stock
Add 80 μl water	*PRY-ex1/ex2*	800 bp	100 pmol/μl	Each 1 μl
	GOLGA2Y-ex14	531 bp	100 pmol/μl	Each 3 μl
	SRY-ex1	472 bp	50 pmol/μl	Each 1 μl
	DAZ-ex2/ex3	400 bp	10 pmol/μl	Each 3 μl
	RBMY1-ex2	291 bp	50 pmol/μl	Each 1 μl
	USP9Y-ex28	249 bp	50 pmol/μl	Each 1 μl

Primer mix C (end-volume 100 μl)	STS marker for *AZF-* or *control*-gene	PCR product length	Forward & reverse primer stocks	μl added from each primer stock
Add 58 μl water	*RBMY1-ex11/ex12*	852 bp	10 pmol	Each 2 μl
	SMCY-ex12	745 bp	100 pmol/μl	Each 10 μl
	DDX3Y-inT/in1	473 bp	100 pmol/μl	Each 5 μl
	CDY1-ex1/in1p7	367 bp	50 pmol/μl	Each 1 μl
	HSFY-in1/ex2	261 bp	50 pmol/μl	Each 1 μl
	UTY-ex3	155 bp	50 pmol/μl	Each 2 μl

Primer mix D (end-volume 100 μl)	STS marker for *AZF-* or *control*-gene	PCR product length	Forward & reverse primer stocks	μl added from each primer stock
Add 58 μl water	*USP9Y-ex9/ex11*	941 bp	100 pmol	Each 4 μl
	BPY2-in6/ex6	846 bp	50 pmol/μl	Each 1 μl
	ZFY-ex8	734 bp	50 pmol/μl	Each 1 μl
	VCY-ex1/ex2	521 bp	100 pmol/μl	Each 10 μl
	DAZ-ex6/ex2	473 bp	50 pmol/μl	Each 1 μl
	EIF1AY-ex7	343 bp	10 pmol/μl	Each 3 μl
	TSPY-exons	273 bp	10 pmol/μl	Each 3 μl

control (i.e., a PCR multiplex reaction mix performed with the same cycling conditions but not including any DNA template) should be run in parallel for all PCR reactions.

Master mixes	A/μl	B/μl	C/μl	D/μl
Water	10.35	8.6	10.35	9
10× PCR buffer	3.75	5	3.75	5
50 mM MgCl$_2$	2.5	2.5	2.5	2.5
DMSO	1.5	2	1.5	1.5
25 mM dNTPs	0.5	0.5	0.5	0.5
Primer mix	5	5	5	5
Always add enzyme as last component Taq DNA polymerase	0.4	0.4	0.4	0.5

2. Place 24 μl of each master mix cocktail into separate thin-walled 0.2 ml PCR-strips or plates.
3. Add 1 μl of genomic DNA to each well for a final volume of 25 μl in all PCR multiplex reactions.
4. Maintain all reaction mixes on ice until loading plates in the thermal cycler.
5. Perform PCR reactions under the following cycling conditions (see Note 8).

	A/°C	B/°C	C/°C	D/°C
Initial melt (3 min)	94	94	94	94
Denature, anneal, and extension (35 cycles, 1 min per step)	94 59 65	94 60 65	94 60 65	94 63 72
Final extension temp. (3 min)	65	65	65	72
Total time for PCR assay	2:10	2:10	2:10	2:10

6. For samples in which a deletion in one of the three AZF microdeletion intervals is identified set up (see Fig. 1) an additional PCR multiplex mixture with four genomic STS markers located at the borderlines of the "classical" AZF microdeletion deletion interval (see Table 2) to confirm that the molecular extension of the deleted AZF region identified with the first four PCR multiplex mixtures is yes or no comparable to that of the "classical" AZFa, AZFb, AZFc microdeletion (see Fig. 2 and Note 9). The master mixes for these three PCR multiplex assays are composed as follows:

PCR multiplex amplification products of genomic Y DNA of infertile men with classical AZFa,b,c microdeletions

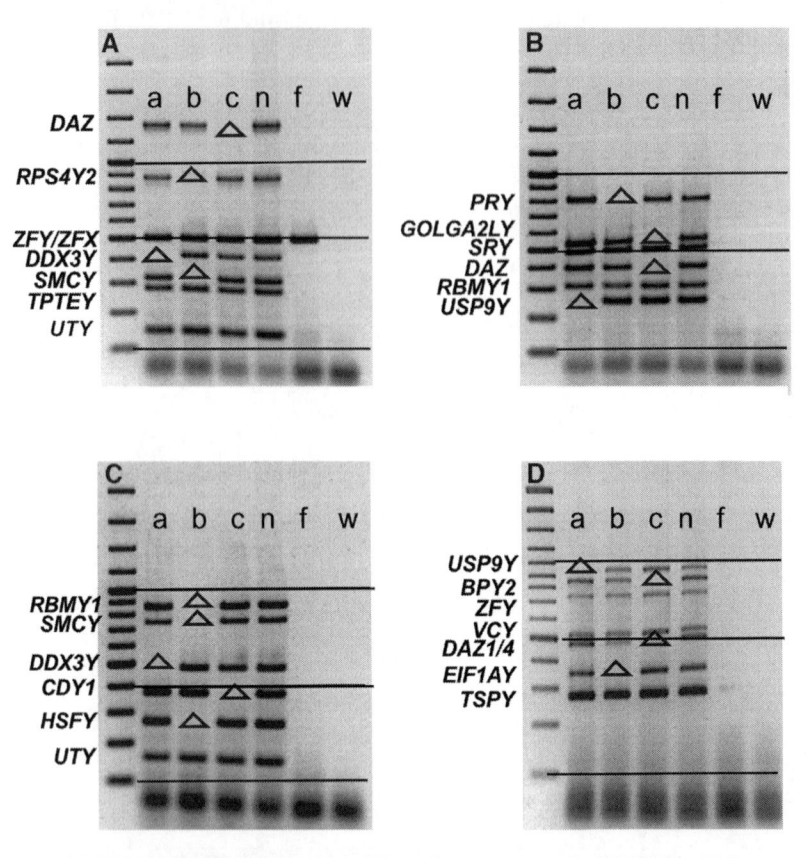

Fig. 1. Presentation of *AZF-Y* gene deletion patterns in EtBr-stained agarose gels with PCR multiplex assays A, B, C, D in infertile men with a "classical" AZFa deletion (lane "a"), "classical" AZFb deletion (lane "b"), and "classical" AZFc deletion (lane "c"). The deleted PCR amplification products are marked by a *triangle*. Lanes "n" and "f" are control lanes to display the normal pattern of all PCR amplification products in the genomic DNA sample of a proven fertile man ("n") and a fertile woman ("f"). They confirm that all STS markers analyzed are Y specific except for the *ZFY* STS in multiplex "A" which also amplifies the homologous *ZFX* gene located on the female X chromosomes. Lane "w" is the water control which should be empty because of lack of a genomic male or female DNA template in each "w" PCR multiplex reaction: Only primer oligomers are visible as background in "w" reactions. The first lane at the left displays a 100 bp ladder-plus length marker (Fermentas); the three cross lines in each picture mark the run lengths of the 100 bp, 500 bp, and 1,000 bp fragments, respectively.

Master mixes for AZF microdeletion borderlines	AZFa μl	AZFb μl	AZFc μl
Water	13.35	11.85	10.6
10×PCR buffer	3.75	3.75	3.75
50 mM MgCl$_2$	3.5	1.25	2.5
DMSO	–	1.25	1.25
25 mM dNTPs	0.5	0.5	0.5
Primer mix	2.5	5	5
Add enzyme always as last component Taq DNA polymerase	0.4	0.4	0.4

PCR multiplex assays to proof classical extension of AZFa,b,c microdeletions

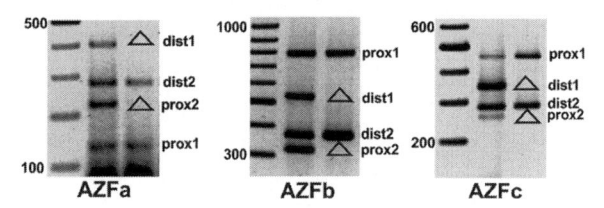

Fig. 2. Presentation of the STS deletion patterns (marked by *triangles* in the last lane at the *right*) with genomic DNA samples of infertile men suffering from the "classical" AZFa (*left* gel), AZFb (*middle* gel), and AZFc (*right* gel) deletions. For a detailed description of the STS borderline markers used see Notes 3 and 4. The first lane at the left displays a 100 bp ladder-plus length marker (Fermentas) with the range given at the left in base pairs. For comparison the pattern of PCR amplification products of a normal fertile man without any STS deletion is included in each gel (*middle* lane).

7. Perform PCR reactions under the following cycling conditions (see Note 8).

	AZFa/°C	AZFb/°C	AZFc/°C
Initial melt (3 min)	94	94	94
Denature, anneal, and extension (35 cycles, 1 min per step)	94 60 65	94 60 65	94 61 72
Final extension temp. (3 min)	65	65	72
Total time for PCR assay	2:10	2:10	2:10

3.2. Agarose Gel Electrophoresis

1. Run each PCR multiplex mixture on a 2% LE-Agarose minigel (100 ml vol).

2. After completing the PCR reactions, add 3 µl of 10× Orange G loading dye to the 25 µl reaction volume. Vortex the mixtures briefly then centrifuge to collect the mixture in the bottom of the tube.

3. Load 15 µl of each sample onto the gel, and preserve the remainder of the reaction mix on ice for a possible second gel run. In addition to the test sample(s), load the normal male control, female control, and water control reactions as well.

4. Run the gel at 120 V for about 2 h at room temperature (until the color of Orange G has nearly reached the end of the gel).

5. After electrophoresis stain agarose gels for 10–15 min with 0.005% Ethidium bromide solution on a smoothly moving shaker platform.

6. Evaluate EtBr-fluorescent patterns of PCR amplification products on an UV light transilluminator (254 nm) and photograph the gel for analysis and documentation.

4. Notes

1. The three "classical" AZF microdeletions occurring "de novo" on the Y chromosome of the infertile patient are indeed always associated with occurrence of a distinct testicular pathology. A complete Sertoli-cell-only (SCO) syndrome is observed in all patients with deletion of the complete Y sequence in the AZFa interval, i.e., only Sertoli cells, but no germ cells are visible in the tubules of their testis tissue sections. Arrest at the spermatocyte stage is observed in the testis tubules of all patients with deletion of the complete Y sequence in the AZFb interval. The populations of spermatogonia and primary spermatocytes in all tubules analyzed are in the normal range; however, no postmeiotic germ cells are identified. A variable testicular pathology is found in patients with a "classical" AZFc deletion in distal Yq11. In most tubules only Sertoli cells are identified but in some tubules germ cells of different developmental stages are clearly visible, and the occurrence of mature sperm cells albeit in low numbers has been reported from different laboratories (30–33). Hypospermatogenesis seemed therefore to be the primary result of a complete AZFc deletion. Consequently, in case one identifies a STS deletion in the AZFa or AZFb deletion interval on the Y chromosome of an infertile patient associated with a different testicular histology than that described above, one can predict, that the molecular extension of this AZF deletion likely does not correspond to that of the "classical" AZFa or AZFb microdeletions.

2. Genomic DNA samples should be only extracted from men with a diagnosis of *idiopathic* severe oligozoospermia (i.e., total sperm count was found repeatedly <5 million sperm in the patient ejaculate) or *idioapathic* azoospermia (i.e., no sperm were found repeatedly in the patient ejaculate) according to the guidelines for sperm analyses of the World Health Organisation (34). "Idiopathic" means that the karyotype in the patient's blood cells was found to be normal, 46XY, and the family history of fertility had been inconspicuous. Clinical diagnosis should have found a testis volume < 20 ml and a raised level of follicle stimulating hormone (FSH > 15 µ/L; normal range: 0.9–15 µ/L), along with normal levels of luteinizing hormone (LH: 1.5–9.3 µ/L) and testosterone (T: 2–7 ng/ml).

   The screening results for AZF microdeletions published in the scientific literature—now including several thousands of patients—indicate that strong clinical prediction values for the occurrence of an AZF microdeletion in an idiopathic infertile man cannot be given. This is especially true for partial AZF deletions and single AZF gene deletions for which a reproducible clinical phenotype has not yet been associated. A high mutation rate is expected for any AZF deletion in Yq11 because of a rather unstable sequence structure in each AZF region and especially in the repetitive ampliconic structures of AZFb and AZFc. Therefore, at the moment it may be wise to screen for single AZF Y gene deletions also in men with sperm numbers < 20 million per ejaculate.

In cases where testicular histology is known in an idiopathic azoospermic patient, incidence for the presence of an AZFa or AZFb microdeletion is higher in patients with a complete SCO syndrome (35) or meiotic arrest (36) respectively.

Therefore, genetic counselling of any couple diagnosed with idiopathic male infertility is mandatory before applying ICSI with the man's sperm sample. They should be provided with all information about the risk of producing male offspring with impaired spermatogenesis due to AZF Y gene deletions. Moreover, the patients should be informed that some AZF microdeletions diagnosed in the male partner might be premutations to the occurrence of larger Yq deletions in the male offspring potentially resulting eventually in the formation of an 45,X0 cell line associated with distinct somatic pathologies (37, 38).

3. All STS markers include at least part from an exonic sequence of the *AZF-Y* gene as indicated in Table 1. They were selected for high priming efficiency at a broad temperature range, i.e., by the robustness of their PCR-amplification process. Optimal lengths were found to be between 18 and 24 nucleotides (nts) or longer (<35 nts) depending on GC-content. All forward and reverse primer sequences are unique, i.e., without any sequence homology, especially at their 5′ and 3′ ends to any other sequence in the human genome (checked by BLAST analysis). They have a similar melting profile (within ~5°C), a purine–pyrimidine ratio of ~1 (i.e., 0.8–1.2) and no chance of primer–primer interactions due to homologous primer sequence. When possible, primers were designed with the start of 1–2 GC pairs. We designed different STS markers for the *DAZ* gene family to distinguish between complete deletion of all *DAZ* genes (A-mix, B-mix) or only the *DAZ1* and *DAZ4* gene copies (D-mix) which are distinguished by a tandem repetitive $(ex2–6)_n$ sequence block encoding the most functionally important RNA Recognition Motif (RRM).

4. Primers are set up from primer stocks with concentrations between 10 and 100 picomol/μl in sterile water and deep frozen at −20 °C for long-term storage. The individual volumes of each primer pair mixed together in the primer cocktails were determined empirically with the aim of producing a quantitatively uniform pattern of their PCR products after agarose gel

electrophoresis. Thereby composition of the STS AZF gene markers in each multiplex experiment was determined according to the EMQN best practice guidelines (16, 25) and by the necessity to display the distinct lengths of the individual PCR amplification products separately after agarose gel electrophoresis and Ethidium bromide (EtBr) staining and visualization on an UV light box. With these criteria the STS AZF gene marker patterns observed are immediately diagnostic for the presence or absence of the corresponding *AZF* gene in the patient genomic DNA sample. In our experience some STS AZF- or control-gene markers cannot be combined in any PCR multiplex assay despite their specific length differences. Such kinetic effects of the polymerase chain reaction are usually not predictable. In some cases no EtBr-stained PCR amplification products are visible, and sometimes they occur with different densities of the same STS AZF gene marker in different primer mixtures. If this occurs adjustment is possible by the use of another Taq Polymerase brand, an increase of forward and/or reverse primer concentrations or by adding more DMSO to the master mix.

According to the EMQN best practice guidelines (16, 25) the primer set for each multiplex reaction is composed of STS primers from at least one Y gene located in the AZFa, AZFb, and AZFc regions. We have added STS markers for deletion analyses of the *TSPY* gene cluster and the Y genes, *SRY*, *ZFY*, *UTY*, *VCY*, and *TPTEY* as controls to distinct PCR multiplex mixtures. They are located outside of the AZF regions and were chosen because they are also expressed exclusively or predominantly in male germ cells. The STS for *ZFY-ex8* in mix A amplifies *ZFY* and the *ZFX* homolog and therefore serves also as control in the genomic female DNA sample, which should be otherwise negative for all STS markers from the Y chromosome. The STS marker for the *TPTEY* gene located distal to the AZFc deletion interval and proximal to the heterochromatic portion of Yq (Yq12) can also indicate terminal Yq11 deletions beyond AZFc and probably also including Yq12. In such cases additional chromosome analysis should be requested.

Three additional PCR multiplex assays were established—each with four genomic STS border line markers of "classical" AZFa, or AZFb, or AZFc microdeletions. They were designated "prox1/prox2" for the proximal borderline and "dist1/dist2" for its distal borderline (see Table 4). The extension of an identified AZFa, AZFb, AZFc deletion in the first run of the basic *AZF-Y* gene deletion assay is comparable to that of the corresponding "classical" AZF microdeletion only when the STS markers "prox1" and "dist2" are present and "prox2" and "dist1" are absent. The AZFc borderline STS markers "dist1"

Table 4
Primer mixes for PCR multiplex assays proofing molecular extension of identified AZFa, AZFb, AZFc microdeletion

Primer mix AZFa breakpoints	STS marker for AZF deletion borderlines	PCR product length	Forward & reverse primer stocks	μl added from each primer stock
Add: 32 μl water to:100 μl end-volume	AZFa-prox1	126 bp	100 pmol/μl	Each 10 μl
	AZFa-prox2	219 bp	100 pmol/μl	Each 4 μl
	AZFa-dist1	389 bp	100 pmol/μl	Each 16 μl
	AZFa-dist2	270 bp	100 pmol/μl	Each 4 μl

Primer mix AZFb breakpoints	STS marker for AZF deletion borderlines	PCR product length	Forward & reverse primer stocks	μl added from each primer stock
Add: 79 μl water to:100 μl end-volume	AZFb-prox1	789 bp	100 pmol/μl	Each 3 μl
	AZFb-prox2	317 bp	100 pmol/μl	Each 2.5 μl
	AZFb-dist1	526 bp	100 pmol/μl	Each 2 μl
	AZFb-dist2	367 bp	100 pmol/μl	Each 3 μl

Primer mix AZFc breakpoints	STS marker for AZF deletion borderlines	PCR product length	Forward & reverse primer stocks	μl added from each primer stock
Add: 77 μl water to:100 μl end-volume	AZFc-prox1	452 bp	100 pmol	Each 4 μl
	AZFc-prox2	254 bp	100 pmol/μl	Each 3 μl
	AZFc-dist1	339 bp	100 pmol/μl	Each 3 μl
	AZFc-dist2	282 bp	100 pmol/μl	Each 1.5 μl

and "dist2" have two locations in the ampliconic AZFc block structure (see Table 2). These are between the "b3/yel1" and "yel2/b4" amplicons (sY1054) and between the "gr1/b3" and "b4/gr2" (sY1125) amplicons, respectively (39). They thus frame the b3 and b4 amplicons and are therefore excellent markers to distinguish between partial AZFc deletions, often resulting from recombination between b1 or b2 and b3 (40), and complete AZFc deletions which result from recombination between b2 and b4 (39, 40). Thus, only in case of a complete AZFc deletion "dist1" is absent and only the "dist2" STS marker will be amplified in the PCR multiplex reaction (see Fig. 2).

5. The STS markers used for analyzing the molecular extension of the identified AZFa, AZFb, or AZFc deletion, and proofing them to be yes or no comparable to that of the "classical" AZF microdeletion length, are listed in Table 2.

6. All solutions must be prepared with cotton-filtered pipet tips using the highest grade of purified water (e.g., Aqua ad injectabilia: Braun) and analytical grade reagents. Prepare all reagents (except DMSO) in tubes on ice. Diligently follow all waste disposal regulation when disposing of waste material.

7. It is best to start pipetting water first followed by the other ingredients as listed in the table. To minimize premature primer binding after having added the DNA template and to prevent enzyme activity before the first denaturation step, all vials should be kept on ice while pipetting the ingredients. It is recommended that work be performed under a laminar flow hood especially when cloning reactions with recombinant Y sequences are in the same lab.

8. These PCR multiplex assays can run on many cyclers. We use the Biometra T-1-96 and T-gradient-96 thermal cyclers which measure the PCR reaction temperature with 6 Peltier-elements spread across a 96 well silver block. Optimal cycling conditions for each PCR multiplex experiment in other PCR machines may be different depending on the overall temperature profile of the particular PCR machine, so conditions may need to be adapted empirically to achieve comparable kinetic conditions for each experiment. As a starting point, an initial annealing temperature of 54 °C works well for most STS primers with a length of 20 nts or more. If nonspecific PCR products form, the annealing temperature can be increased until the reaction only amplifies the fragment length of the STS markers listed in Tables 1 and 2. Ensure that the plastic caps are closed firmly for each PCR vial and the heated lid of the PCR machine is in good contact to all PCR vials. It is sometimes helpful to increase the extension time after the last cycling reaction up to 5–10 min if the amount of the PCR amplification product is too low.

9. Figure 2 illustrates the amplicons generated in a normal fertile man (middle lane) compared with men with the classical AZFa, b, and c microdeletions. Two of the Y-specific genomic STS markers are located at the proximal ("prox1" and "prox2") and two at the distal ("dist1" and "dist") borderline of the "classical" AZF microdeletion (39, 40). The extension of a "classical" AZF deletion region is indicated when the most proximal ("prox1") and the most distal ("dist2") STS borderline marker is present and both internal STS borderline markers ("prox2" and "dist1") are absent (see Fig. 2). If in a clinical setting one is interested only in screening the Y chromosome of the idiopathic infertile man for the presence or absence of a

"classical" AZFa, AZFb, or AZFc deletion (e.g., if one only wants to estimate the probability of retrieving sperm by testicular sperm extraction (TESE)), one can omit the four basic PCR multiplex *AZF-Y* gene deletion assays and directly perform the three PCR multiplex AZFa, b, c molecular extension assays. In contrast to the two PCR multiplex assays of the EAA/EMQN guidelines (16, 25), which recommend the use of only two internal STS markers to qualify each AZF deletion identified as "classical" AZF microdeletion, the STS markers used here (see Table 2) are located around the critical breakpoint regions of the three "classical" AZF microdeletion and are therefore more suitable (40).

Acknowledgments

We are indebted to the head of our department, Prof. Dr. Thomas Strowitzki, who financially supported the establishment and running of all of the PCR multiplex assays. We also thank our former colleagues Octavian Henegariu (now at Yale University School of Medicine; HTI, Immunobiology department), Angela Edelmann and Peter Hirschmann (now at Biopharm GmbH, Heidelberg), Karin Huellen (University of Heidelberg; Institute of Human Genetics), and Alexandra Schadwinkel who have contributed to earlier setups of the PCR multiplex assays for the detailed analysis of AZF (gene) microdeletions.

References

1. Ngo KY et al (1986) A DNA probe detecting multiple haplotypes of the human Y chromosome. Am J Hum Genet 38:407–418
2. Torroni A et al (1990) Y chromosome DNA polymorphisms in human populations: differences between Caucasoids and Africans detected by 49a and 49f probes. Ann Hum Genet 54:287–296
3. Jobling MA (1994) A survey of long-range DNA polymorphisms on the human Y chromosome. Hum Mol Genet 3:107–114
4. Seielstad MT et al (1994) Construction of human Y-chromosomal haplotypes using a new polymorphic A to G transition. Hum Mol Genet 3:2159–2161
5. Santos FR et al (1995) PCR haplotypes for the human Y chromosome based on alphoid satellite DNA variants and heteroduplex analysis. Gene 165:191–198
6. Jobling MA et al (1996) Recurrent duplication and deletion polymorphisms on the long arm of the Y chromosome in normal males. Hum Mol Genet 5:1767–1775
7. Tiepolo L, Zuffardi O (1976) Localization of factors controlling spermatogenesis in the nonfluorescent portion of the human Y chromosome long arm. Hum Genet 34:119–124
8. Vogt P et al (1992) Microdeletions in interval 6 of the Y chromosome of males with idiopathic sterility point to disruption of AZF, a human spermatogenesis gene. Hum Genet 89:491–496
9. Ma K et al (1992) Towards the molecular localisation of the AZF locus: mapping of microdeletions in azoospermic men within 14 subintervals of interval 6 of the human Y chromosome. Hum Mol Genet 1:29–33
10. Johnson MD et al (1989) Molecular scanning of Yq11 (interval 6) in men with Sertoli-cell-only syndrome. Am J Obstet Gynecol 161:1732–1737
11. Nagafuchi S et al (1993) A minute deletion of the Y chromosome in men with azoospermia. J Urol 150:1155–1157

12. Kobayashi K et al (1994) PCR analysis of the Y chromosome long arm in azoospermic patients: evidence for a second locus required for spermatogenesis. Hum Mol Genet 3:1965–1967
13. Pryor JL et al (1997) Microdeletions in the Y chromosome of infertile men. N Engl J Med 336:534–539
14. Vogt P et al (1991) Towards the molecular localization of AZF, a male fertility gene on the human Y chromosome by comparative mapping of microdeletions in the Y chromosome of men with idiopathic infertility. Am J Hum Genet 49:258
15. Vogt PH et al (1996) Human Y chromosome azoospermia factors (AZF) mapped to different subregions in Yq11. Hum Mol Genet 5:933–943
16. Simoni M et al (1999) Laboratory guidelines for molecular diagnosis of Y-chromosomal microdeletions. Int J Androl 22:292–299
17. Rodovalho RG et al (2008) Tracking microdeletions of the AZF region in a patrilineal line of infertile men. Genet Mol Res 7:614–622
18. Repping S et al (2003) Polymorphism for a 1.6-Mb deletion of the human Y chromosome persists through balance between recurrent mutation and haploid selection. Nat Genet 35:247–251
19. Vogt PH (2005) AZF deletions and Y chromosomal haplogroups: history and update based on sequence. Hum Reprod Update 11:319–336
20. Repping S et al (2006) High mutation rates have driven extensive structural polymorphism among human Y chromosomes. Nat Genet 38:463–467
21. Quintana-Murci L et al (2001) The relationship between Y chromosome DBA haplotypes and Y chromosome deletions leading to male infertility. Hum Genet 108:55–58
22. Fernandes S et al (2004) A large AZFc deletion removes DAZ3/DAZ4 and nearby genes from men in Y haplogroup N. Am J Hum Genet 74:180–187
23. Krausz C et al (2011) The Y chromosome-linked copy number variations and male fertility. J Endocrinol Invest 34:376–382
24. Stouffs K et al (2011) What about gr/gr deletions and male infertility? Systematic review and meta-analysis. Hum Reprod Update 17:197–209
25. Simoni M et al (2004) EAA/EMQN best practice guidelines for molecular diagnosis of y-chromosomal microdeletions. State of the art 2004. Int J Androl 27:240–249
26. Lahn BT, Page DC (1997) Functional coherence of the human Y chromosome. Science 278:675–680
27. Skaletsky H et al (2003) The male-specific region of the human Y chromosome is a mosaic of discrete sequence classes. Nature 423:825–837
28. Vogt PH (2005) Azoospermia factor (AZF) in Yq11: towards a molecular understanding of its function for human male fertility and spermatogenesis. Reprod Biomed Online 10:81–93
29. Nakahori Y et al (1994) A locus of the candidate gene family for azoospermia factor (YRRM2) is polymorphic with a null allele in Japanese males. Hum Mol Genet 3:1709
30. Girardi SK et al (1997) Submicroscopic deletions in the Y chromosome of infertile men. Hum Reprod 12:1635–1641
31. Stuppia L et al (1996) Widening of a Y-chromosome interval-6 deletion transmitted from a father to his infertile son accounts for an oligozoospermia critical region distal to the RBM1 and DAZ genes. Am J Hum Genet 59:1393–1395
32. Najmabadi H et al (1996) Substantial prevalence of microdeletions of the Y-chromosome in infertile men with idiopathic azoospermia and oligozoospermia detected using a sequence-tagged site-based mapping strategy. J Clin Endocrinol Metab 81:1347–1352
33. Foresta C et al (1997) Y-chromosome deletions in idiopathic severe testiculopathies. J Clin Endocrinol Metab 82:1075–1080
34. [Laboratory manual of the WHO for the examination of human semen and sperm-cervical mucus interaction]. Ann Ist Super Sanita 37:I–XII, 1–123 (2001)
35. Kamp C et al (2001) High deletion frequency of the complete AZFa sequence in men with Sertoli-cell-only syndrome. Mol Hum Reprod 7:987–994
36. Mirfakhraie R et al (2010) High prevalence of AZFb microdeletion in Iranian patients with idiopathic non-obstructive azoospermia. Indian J Med Res 132:265–270
37. Siffroi JP et al (2000) Sex chromosome mosaicism in males carrying Y chromosome long arm deletions. Hum Reprod 15:2559–2562
38. Patsalis PC et al (2002) Effects of transmission of Y chromosome AZFc deletions. Lancet 360:1222–1224
39. Kuroda-Kawaguchi T et al (2001) The AZFc region of the Y chromosome features massive palindromes and uniform recurrent deletions in infertile men. Nat Genet 29:279–286
40. Vogt PH (2004) Genomic heterogeneity and instability of the AZF locus on the human Y chromosome. Mol Cell Endocrinol 224:1–9

Part V

Sperm Preparation and Selection Techniques

Chapter 18

Sperm Cryopreservation Methods

Tiffany Justice and Greg Christensen

Abstract

There are multiple clinical situations where cryopreservation of sperm is required including sperm banking prior to chemotherapy or radiation treatment for cancer, donor sperm for couples without a male partner, and various etiologies of male factor infertility. Bunge and Sherman presented the first pregnancy resulting from the thawing of cryopreserved sperm in 1953 (Bunge RJ, Nature 172:767–768, 1953). Since that time, cryopreservation techniques have continued to improve. Here, we describe protocols for the cryopreservation and thawing of semen and testicular tissues using a simple liquid nitrogen vapor technique.

Key words: Sperm, Cryopreservation, Sperm-banking, Liquid nitrogen

1. Introduction

The field of sperm cryopreservation has undergone much change since its inception in 1776 when Lazaro Spallanzani, an Italian physiologist and priest, first noted a decrease in sperm motility when semen was placed in snow (2). It was not until 1953 when Bunge and Sherman described the first pregnancy from cryopreserved sperm. In their seminal article, they outlined a protocol for using a 10% glycerol solution and dry ice to preserve sperm, a technique which produced an average spermatozoa survival rate of 67% (1).

There are many clinical scenarios requiring the cryopreservation of sperm. Isolated male factor infertility affects approximately 20% of all infertile couples, with combined factors affecting another 20–40% (3). Cryopreservation of sperm discovered intraoperatively may obviate the need for donor sperm or further surgery in men with azoospermia or severe oligozoospermia. Better treatments for cancer in adolescents and adults have improved survival rates well into the reproductive years, necessitating the cryobanking of gametes prior to treatment or the use of cryopreserved donor sperm in those for whom it was not an option. In addition, cryopreserved donor sperm is being used for a variety of situations including infertile or subfertile males, lesbian couples and single women without a male partner.

Due to their small volume and relatively large surface area, sperm cells are ideal for cryopreservation. Currently, the most widely used techniques for sperm cryopreservation employ an egg yolk/glycerol preservation media with a manual or automated stepwise freeze in liquid nitrogen vapor. The use of cryoprotectants in conventional freezing media provides several benefits including provision of an exogenous energy source to prevent the exhaustion of intracellular sperm phospholipids, maintenance of osmotic pressure and pH, and the bactericidal benefit of antibiotics (4).

As mentioned, sperm can be frozen either manually or with a controlled rate freezer, and multiple variations exist on cooling rates, techniques and the equipment used. In recent publications, vitrification of small sperm volumes in a sucrose solution has also been described with good outcomes (5, 6). Regardless of the method used, sperm cells in semen or other fluids, and sperm in testicular tissue freeze and thaw well with acceptable survival rates. Due to this good survival, this chapter gives protocols for cryopreserving sperm using the simplest technique: manually in liquid nitrogen vapors.

2. Materials

2.1. Semen Collection and Analysis

1. Sterile specimen container: Each new lot of containers should be tested for sperm toxicity prior to use.
2. Microscope: Should include a minimum of a 20× and 40× objectives.
3. Cell counting chamber: Makler, Microcell or other.
4. Pipettor, disposable pipettes, and pipette tips.
5. Personal protective equipment: Gloves, eye protection.
6. 37°C Heat block or incubator to maintain temperature of sperm sample and counting chambers.

2.2. Freezing Components (Items 8–10 Needed Only for Cryopreserving Testicular Tissue)

1. Sperm freezing media: Test yolk buffer with glycerol, (available commercially from several vendors).
2. Cryovials: sterile, plastic, 1.0 ml, should have a screw-top lid such as those made by Nunc.
3. A cryomarker or cryolabel printer such as a Brady labeler for identifying vials.
4. Aluminum canes for holding cryovials.

5. Liquid nitrogen and liquid nitrogen storage dewar.
6. Serologic pipettes: sterile, plastic, disposable, for measuring and aliquoting semen and freeze media.
7. Cryobasket: Used to hold vials of semen in the liquid nitrogen vapors, above the liquid phase.
8. Hepes buffered Human Tubal Fluid (HTF) supplemented with 3% Human Serum Albumin (HSA). A Hepes buffered "sperm wash" media can also be used.
9. 100 mm petri dishes.
10. 1 cc insulin syringes.

2.3. Thawing

1. 37°C water bath.
2. Kimwipes.

3. Methods

3.1. Cryopreservation: Semen

1. As soon as the sample has been collected, place the specimen container with the semen sample on a 37°C warming block and allow it to liquefy, 20–60 min.
2. Mix sample thoroughly be gently swirling the container for approximately 10 s.
3. Perform a semen analysis on the sample including a minimum of the volume, count, and motility. After completing the analysis, the semen sample may be kept at room temperature.
4. Using a cryomarker or cryolabeler, label cryovials with the patient's first and last names, an additional unique identifier such as a medical record number or accession number, and the date. It may also be helpful to include the content of the vial (semen, fine needle aspiration [FNA] etc.), or the facility freezing the sperm on the label.
5. Draw up a quantity of room-temperature freezing media in a serological pipette equal to the measured semen volume. Slowly add the freezing media to the container with the semen while swirling it, to facilitate thorough mixing.
6. Remove the caps from the labeled cryovials, add 1 ml of the semen–freezing media mixture to each vial and recap the vials.
7. Position a cryobasket in a liquid nitrogen storage tank so that the bottom of the basket is suspended in the vapor, approximately 1–2 cm above the liquid phase.
8. Place the cryovials in the basket and keep them suspended in the vapor for 20 min. When the 20 min has elapsed, lower the bottom of the basket into the liquid nitrogen until all vials are submerged for a minimum of 1 min.

9. Working quickly, remove the vials from the nitrogen, place them on prelabeled storage canes (or in storage boxes, depending on the tank), and return them to their permanent storage locations in the liquid nitrogen tank.

3.2. Cryopreservation: Testicular Tissue

1. Testicular tissue, collected by fine needle aspiration, open biopsy or other techniques should be placed in HEPES buffer media immediately after collection. If the tissue is collected at a remote site, it should be maintained between room temperature and 37°C during transportation to the lab.
2. Place four to five, 200 µl drops of 37°C HTF in a 100 mm petri dish.
3. Working under a dissecting microscope, transfer the seminiferous tubules through the drops of HTF to rinse off excess blood. A p200 pipettor with a sterile tip or a syringe with a blunt needle works well to pick up and move pieces of tissue from drop to drop.
4. Place the rinsed tissue in a clean drop of media. Using two, 1 cc insulin syringes that have had the needles aseptically bent to approximately 90° angles, divide the tissue into portions equal to the number of vials desired for cryopreservation. The contents of a small portion of the tubules can be expressed (see below) and evaluated for the presence of sperm and motility.
5. Using a rack to hold the cryovials, transfer a portion of tissue and 0.3 ml of HTF into each prelabeled cryovial.
6. Slowly add an additional 0.3 ml of room temperature cryomedia to each vial and recap.
7. Position a cryobasket in a liquid nitrogen storage tank so that the bottom of the basket is suspended in the vapor, approximately 1–2 cm above the liquid phase.
8. Place the cryovials in the basket and keep them suspended in the vapor for 20 min. When the 20 min have elapsed, lower the bottom of the basket into the liquid nitrogen until all vials are submerged for a minimum of 1 min.
9. Working quickly, remove the vials from the nitrogen, place them on prelabeled storage canes (or in storage boxes, depending on the tank), and return them to their permanent storage locations in the liquid nitrogen tank.

3.3. Thawing of Cryopreserved Semen

1. Remove cryovial(s) from the cane and hold the vial at room temperature for 1 min.
2. Place the vial in a 37°C water bath until completely thawed and warmed, approximately 3–4 min.
3. Remove the vial from the water bath, and using a Kimwipe or other absorbent paper, remove all moisture from the outside of the cryovial.

4. Gently invert the vial two to three times to mix the contents before uncapping the vial and proceeding with the post thaw analysis and sperm preparation.

3.4. Thawing of Cryopreserved Testicular Tissue

1. Follow steps 1–3 above in semen thawing protocol
2. Prepare a 100 mm petri dish with five, 200 µl drops of HTF media.
3. Uncap vial and using a p200 pipettor or blunt needle and syringe, remove the cryopreserved tissue from the vial and transfer it to the first drop of HTF.
4. Rinse the thawed tissue through the remaining drops to remove the cryoprotectants.
5. In the final drop, using two, 1 cc insulin syringes that have had the needles aseptically bent to approximately 90° angles express the contents of the tubules into the drop. This is done by pinning the tissue down with one syringe needle and dragging the other needle across the top of the tubules with a scraping motion repeatedly until all tubules have been expressed. A few microliters of the media containing expressed tubules can be placed in a small media drop under oil and evaluated for the presence of sperm and motility.
6. Using the pipettor, draw up the media containing the sperm for processing. Depending on concentration and preference, it may be placed on a gradient column, processed with a basic wash or put directly under an oil overlay for collection of sperm for micromanipulation.

4. Notes

1. "Cryobaskets" to place vials of sperm in during the freezing process can be devised in different ways. One strategy is to fashion a simple wire mesh basket, that fits easily down into the neck of a liquid nitrogen dewar, from rabbit or chicken fencing material (available at home improvement stores). The fencing material is made of a wire mesh with openings of approximately 0.5 cm. The basket can be suspended at the desired height above the liquid nitrogen vapors and held at that position by passing a long, stiff piece of wire through the mesh openings at the top of the tank. Another alternative is to suspend a liquid nitrogen tank canister at the desired height by tying a piece of string to the canister handle and attaching the opposite end to an anchor on the wall above the tank.
2. In cases where the semen volume is very low, <0.5 ml, it may be beneficial to increase the volume by suspending the semen

in a few milliliters of HTF. This can help improve the recovery of the sperm in the container and provide a volume adequate to divide into several vials. If this is done, a volume of cryoprotectant should be added equal to the total volume of semen and HTF. Given the minimal sperm requirements necessary with advanced reproductive technologies, such vials would still be acceptable for use with ICSI.

3. When freezing samples with a high concentration of motile sperm, it may be useful to freeze vials with a volume that contains a desired concentration of motile sperm, rather than just a set volume of 1 ml per vial.

4. When twisting the caps onto vials do not over tighten, as this can blow out the seal and compromise the sample. Tighten the cap so that it is just seated.

5. Vials just removed from liquid nitrogen have the potential to release pressure suddenly, either spontaneously or when the cap is loosened. Thus, it is especially important to use proper eye protection and other appropriate protective equipment.

6. If a testicular sample is to be collected "onsite" such as in a procedure room adjacent to the lab, syringes containing aspirations can be walked directly into the lab for evaluation as they are collected. Biopsied samples could also be placed immediately into a small dish of media and taken directly into the lab for processing. For samples collected "offsite" from the lab, it is necessary to transport the tissue using a tube with an airtight, leak-proof cap. Microcentrifuge tubes work well for FNA samples, where each small aspiration can be placed in its own tube. A 15 ml centrifuge tube with 2–3 ml of HTF works well for transporting larger pieces of biopsied tissue.

7. In some scenarios it may be desirable to refreeze a portion of a sample that has already been thawed. Studies show that while the overall percentage of motile sperm decreases with each freeze–thaw cycle, some sperm are capable of surviving several cycles and retain their ability to fertilize with ICSI (7, 8).

8. A recent study suggests that thawing cryopreserved semen at 40°C may be beneficial in increasing recovery of motile sperm (9).

9. Cryopreserved sperm are not as robust as freshly ejaculated samples. If a sample is meant to be used for insemination it should be manipulated as minimally as possible, typically with a single wash of 7–8 min to remove the seminal fluids, cryoprotectants and to pellet the sperm, which can then be resuspended in fresh media.

10. Sperm motility in samples from testicular biopsies is usually very low, though sperm viability can be much higher than the percentage of motile sperm observed. A subtle "twitch" may

be the only motility observed for many sperm in these samples. To help improve motility, samples can be thawed and prepared the afternoon before they are needed for ART and incubated overnight at 37°C. Another option to improve motility in thawed testicular samples for better sperm selection is adding a solution of pentoxyphylline (10 mg/ml in HTF) to the drop containing the sperm. Improved motility should be observed within a few minutes.

11. A typical sperm freezing protocol for controlled rate freezing would be dropping the temperature from room temperature to 4°C at a rate of 0.5°C/min followed by a drop to −80°C at a rate of 10°C/min.

References

1. Bunge RJ (1953) Fertilizing capacity of frozen human spermatozoa. Nature 172:767–768
2. Anger JT et al (2003) Cryopreservation of sperm: indications, methods and results. J Urol 170:1079–1084
3. Thonneau P et al (1991) Incidence and main causes of infertility in a resident population (1,850,000) of three French regions (1988–1989). Hum Reprod 6:811–816
4. Khorram O et al (2001) Reproductive technologies for male infertility. J Clin Endocrinol Metab 86:2373–2379
5. Isachenko V et al (2012) Vitrification of human ICSI/IVF spermatozoa without cryoprotectants: new capillary technology. J Androl 33(3):462–468
6. Isachenko V et al (2004) Cryoprotectant-free cryopreservation of human spermatozoa by vitrification and freezing in vapor: effect on motility. DNA integrity, and fertilization ability. Biol Reprod 71:1167–1173
7. Rofeim O et al (2001) Effects of serial thaw-refreeze cycles on human sperm motility and viability. Fertil Steril 75:1242–1243
8. Bandularatne E, Bongso A (2002) Evaluation of human sperm function after repeated freezing and thawing. J Androl 23:242–249
9. Calamera JC et al (2010) Effect of thawing temperature on the motility recovery of cryopreserved human spermatozoa. Fertil Steril 93:789–794

Chapter 19

Density Gradient Separation of Sperm for Artificial Insemination

David Mortimer and Sharon T. Mortimer

Abstract

Human spermatozoa for clinical procedures such as IUI or IVF, or for diagnostic or research studies of sperm fertilizing ability, must be separated from the seminal plasma environment not only as soon as possible after ejaculation but also as efficiently as possible, minimizing seminal plasma and bacterial carryover. Furthermore, in addition to technical simplicity and robustness, a sperm preparation method needs to select not just the more motile and morphologically normal spermatozoa but also those spermatozoa with reduced DNA damage. Currently the most effective and efficient technique for this is density gradient centrifugation, which has been extensively validated through research and clinical application. An optimized protocol based on silane-coated colloidal silica products is provided.

Key words: Spermatozoa, Density gradient, Sperm washing, Fertilizing ability

1. Introduction

Capacitation, defined as the process by which Eutherian spermatozoa acquire the ability to undergo the acrosome reaction and fertilize oocytes, is essential for fertilization both in vivo and in vitro, and is blocked by decapacitation factor(s) present in the seminal plasma. In vivo, spermatozoa escape from the seminal plasma by swimming into the cervical mucus within a few minutes of ejaculation into the vagina, but in vitro, spermatozoa must be washed free of seminal plasma to allow capacitation to commence. Prolonged exposure to seminal plasma adversely affects sperm function, in terms of their ability to penetrate cervical mucus, undergo the acrosome reaction in vitro and to fertilize oocytes (1, 2). Even relatively short exposure to seminal plasma, e.g., for 30–90 min after ejaculation, can permanently diminish their fertilizing capacity in vitro (3), and contamination of prepared sperm populations with as little as 0.01% (v/v) seminal plasma can decrease their fertilizing capacity (4). Moreover, spermatozoa must be protected from oxidative damage caused by reactive oxygen species (ROS) (5–8).

Consequently, spermatozoa for clinical procedures such as intrauterine insemination (IUI), donor insemination (DI), in vitro fertilization (IVF), or intra-cytoplasmic sperm injection (ICSI), or for diagnostic or research studies of sperm fertilizing ability, must be separated from the seminal plasma environment not only as soon as possible after ejaculation but also as efficiently as possible, after which the spermatozoa must then be suspended in a culture medium capable of supporting capacitation.

Numerous methods for washing human spermatozoa have been described, but the most widely-used and efficient methods currently employed are based on discontinuous density gradient centrifugation (DGC), continuous gradient methods being too cumbersome for routine clinical application (2, 9–11). DGC separates spermatozoa based upon their density (i.e., mass per unit volume, or specific gravity), and results in their distribution throughout the gradient according to the locations that match their density, i.e., at their isopycnic points. Mature, morphologically normal human spermatozoa have a density >1.12 g/mL, while immature and many abnormal spermatozoa have densities between 1.06 and 1.09 g/mL (12).

DGC methods became popular in the early 1980s using the commercial product Percoll®, based on colloidal silica that had been coated with polyvinylpyrrolidone (Pharmacia Biotech, Uppsala, Sweden). Because the specific gravity of 80% (v/v) Percoll colloid is about 1.10 g/mL, only the most mature spermatozoa could penetrate the bottom gradient layer, and so Percoll dominated clinical human sperm preparation until its manufacturer withdrew it from clinical use in 1996 (2). Numerous commercial DGC products are now available for use with human spermatozoa, but care must be taken to ensure that the correct density lower layer is used, since products differ in their base colloid content (see Note 1).

The protocol provided below has been optimized over 15 years of use, and takes into account a variety of physical and practical factors that affect the yield of human spermatozoa preparation using colloidal silica-based DGC, including the selection of spermatozoa with reduced DNA damage and minimizing bacterial carryover (2, 9–11, 13–15).

2. Materials

2.1. Density Gradient Colloid

Two-step discontinuous gradients will perform well with the vast majority of human ejaculates with 1.5 mL or 2.0 mL upper and lower layers of 40% and 80% Percoll-equivalent colloid, respectively (see Note 1). Although gradients must be prepared on the day of use, bulk 40% and 80% layer media can be prepared under aseptic conditions and stored at +4°C for a few days (several weeks if sterilized using 0.22 μm Millipore Millex-GV filters); see Subheading 2.5 for preparing the layers.

2.2. Sperm Medium

This can be any bicarbonate-buffered culture medium designed to sustain human sperm metabolism and capacitation in vitro. It must be equilibrated to 37°C prior to use, and then used under an appropriate CO_2-enriched atmosphere (e.g., 6% CO_2-in-air for most such media containing 25 mEq/L bicarbonate ions when used at 37°C close to sea level) to maintain its pH.

2.3. Sperm Buffer

This is any HEPES-buffered medium designed to sustain sperm metabolism in vitro under an air atmosphere; it will not support sperm capacitation. An equivalent medium buffered using MOPS is also acceptable. Sperm Buffer is used whenever spermatozoa need to be maintained outside a CO_2 incubator (e.g., during sperm washing procedures), or when they are to be prevented from capacitating in vitro. Ideally a Sperm Buffer should be matched in its general formulation to the Sperm Medium when the two are used in combination.

2.4. Tubes

1. All disposables that come into contact with spermatozoa, e.g., centrifuge tubes, pipettes, and syringes, should be either pretested for sperm toxicity or else obtained from a trusted manufacturer (11, 16).
2. Centrifuge tubes: Must be made of polystyrene (since polypropylene tubes can be spermotoxic) and have conical bottoms to maximize recovery of the pellets, e.g., Falcon #2095 15-mL tubes or Falcon #2074 "Blue Max" 50-mL tubes (e.g., for use with urine from men with retrograde ejaculation). See Note 2.
3. Sample tubes: Also must be polystyrene, but usually with round bottoms, e.g., Falcon #2003 or #2058 7-mL culture tube.

2.5. Preparing Gradient Layers

1. Upper layer = 40% (v/v) colloid: Mix 20 mL of 100% Percoll-equivalent stock colloid (e.g., PureSperm, Nidacon International AB, Göteborg, Sweden; see Note 1) and 30 mL of Sperm Buffer.
2. Lower layer = 80% (v/v) colloid: Mix 40 mL of the stock colloid with 10 mL of Sperm Buffer.

Note: For efficiency, prepare 15-mL polystyrene conical Falcon tubes (#2095) with the upper layer and Falcon #2003 tubes with 2× volume of the lower layer. Store tubes at +4°C and equilibrate in a 37°C incubator before use; do not add the lower layer until just before use.

3. Methods

3.1. Centrifuge and Calibration

1. Must have a swing-out rotor to create the pellets at the bottom of the tubes, not smeared across the bottom and lower side of the tube as would be created using a fixed-angle rotor. Since human semen must be considered potentially infectious, buckets should have aerosol containment covers.

2. Centrifugation speeds must be expressed as g_{max} values, calculated for what the spermatozoa will experience at the bottom of the centrifuge tube (not the bottom of the centrifuge bucket). Speeds in *rpm* values vary greatly with the rotational radius of the centrifuge rotor, and hence do not provide a standardized method. The centrifuge must be calibrated. The formula describing the relationship between rotation speed, rotational radius, and g-force is:

$$g = 0.0000112 \times r \times N^2$$

where: g is the desired relative centrifugal force, r is the radius of the rotor to the bottom of the tube (in cm), N is the rotation speed in rpm.

3.2. Specimens

1. Semen (fresh):
   (a) Liquefied ejaculates ideally within 30 min of ejaculation (3).
   (b) With "dirty" specimens containing high numbers of other cells or heavy contamination with particulate debris, the "rafts" formed at the interfaces between the layers might be too dense and block the gradient, drastically reducing the yield. There are several simple options to deal with this problem:
   - Load smaller volumes of semen onto the gradients and/or use more gradients.
   - Use longer columns of colloid, e.g., 3-mL layers.
   - When loading the semen onto the gradient mix it gently with the upper quarter of the upper layer.
   - Prepare a three-step gradient using layers of 40%, 60%, and 80% colloid.
     Note: While using a less dense lower colloid layer (e.g., 72%) will increase the total yield of spermatozoa, this is only achieved by recovering less dense, and hence less good quality/less mature spermatozoa.
   (c) With highly viscous semen samples it can be extremely difficult to obtain good yields of motile spermatozoa. To reduce the viscosity, add an equal volume of Sperm Buffer and mix gently using a (sterile) Pasteur pipette; if dispersion

is not achieved within 2 min incubate at 37°C for 10 min and then mix further. The diluted sample is loaded onto the gradients as usual.

For men known to have consistently high viscosity, have them collect into chymotrypsin-coated MARQ™ Liquefaction Cups (Embryotech Laboratories, Wilmington, MA, USA).

2. Cryopreserved semen:

   (a) Because of the high osmolarity of cryoprotectant media (CPMs), cryopreserved spermatozoa will swell greatly upon entering the 40% colloid layer, thereby decreasing their specific gravity and making them too buoyant to pass through the density gradient; this swelling can also impair sperm function or survival. Cryopreserved semen must be diluted slowly with a relatively large volume of an "isotonic" medium, adding it drop-wise with constant gentle mixing over a period of at least 10 min, prior to loading onto the gradients, e.g., a 10× volume of Sperm Buffer for traditional glycerol–egg–yolk–citrate CPM, and a 5× volume for Nidacon's *Sperm CryoProtec* products.

   (b) Alternatively, but sub-optimally, if the total volume exceeds 8 mL the diluted semen can be centrifuged at $500 \times g$ for 10 min (this is safe because the cells that generate free radicals during centrifugation do not survive the freezing and thawing process). Remove the supernatant and resuspend the pellet using 1–2 mL of Sperm Buffer. Load this sperm suspension onto the gradients as usual.

3. Retrograde ejaculation urine:

   1. If essential (since pelleting the spermatozoa can cause iatrogenic damage), concentrate the specimen by centrifugation in one or more large conical tubes at $500 \times g$ for 10 min and resuspend the pellet(s) into a small volume of Sperm Buffer.

   2. Layer over density gradients and process as for a semen sample.

3.3. Loading and Running the Gradients

1. Using a (sterile) Pasteur pipette place the upper layers (each 1.5 mL or 2.0 mL of 40% v/v colloid) into two 15-mL conical tubes.

2. Using a (sterile) glass Pasteur pipette, carefully add the same volume lower layers (80% v/v colloid) beneath the upper layers. A clear interface should be visible between the two layers. A long-form (9 in.) Pasteur pipette can make this step easier. Volume control is easily achieved by attaching a 3-mL syringe to the Pasteur pipette via a short length of silicone tubing, instead of using a rubber bulb.

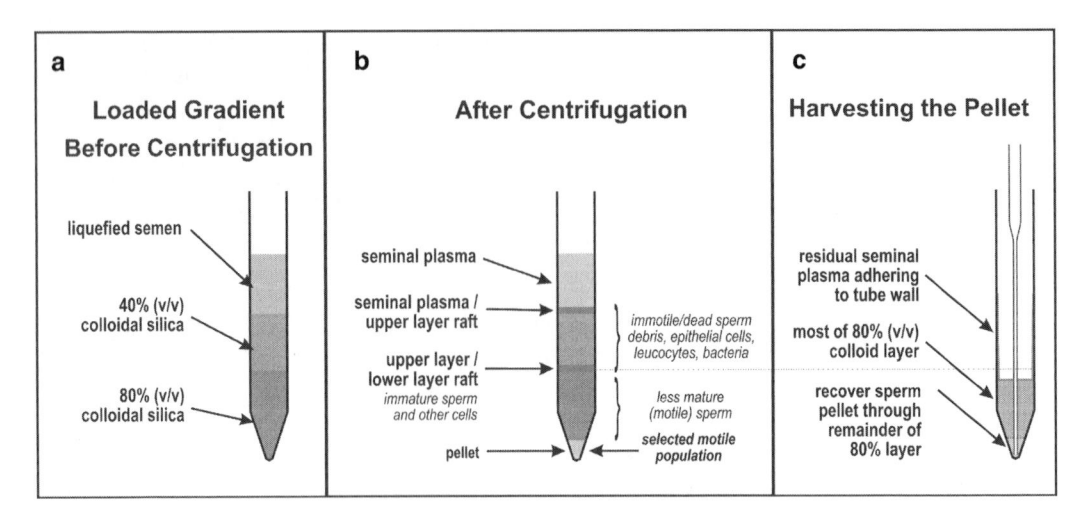

Fig. 1. Two-layer density gradient centrifugation of human spermatozoa.

Note: Alternatively, especially if one has problems achieving a sharp interface between the two layers using the underlayering method, place the lower layers in the tubes and then overlay the upper layers on top of them.

3. Overlay liquefied semen onto the gradient(s), maximum volume equals the volume of the upper layer (see Fig. 1a).

4. Centrifuge at $300 \times g$ for 20 min. See Note 3.

Notes: Do not change the centrifugation speed to try and increase the yield, it will only result in recovering either poorer quality spermatozoa or a dirty preparation.

To maximize yield and guard against problems that might arise as a result of careless handling of the gradients or sampling after centrifugation, gradients should always be run as pairs.

3.4. Harvesting the Gradients and Washing the Pellets

1. Use a (sterile) Pasteur pipette to carefully remove the seminal plasma, upper interface "raft", upper (40%) colloid layer, and the lower interface "raft" (see Fig. 1b); leave most of the lower (80%) colloid layer in place. Discard the aspirated material.

2. Using another clean (sterile) long-form Pasteur pipette remove the soft pellet by direct aspiration (maximum 0.5 mL) from the bottom of the tube beneath the lower (80%) colloid layer (see Fig. 1c). Blow a single air bubble from the tip of the pipette as it passes through the meniscus of the lower layer to minimize contamination of the pipette aperture with residual seminal plasma/raft material.

3. Transfer the pellet(s) to a clean 15-mL conical tube and resuspend in either Sperm Buffer if processing for IUI, DI or ICSI, or Sperm Medium if processing for IVF or sperm function

testing (see the relevant protocols for further information). Use at least 6 mL of medium, up to 10 mL for large pellets.

4. Centrifuge at $500 \times g$ for 10 min.

5. Use a (sterile) Pasteur pipette to aspirate the supernatant and resuspend the pellet in 1–3 mL of fresh Sperm Buffer or Sperm Medium as appropriate (see step 3 of Subheading 3.4). For very small pellets volumes as low as 0.2 mL can be used. Again, a long-form (9 in.) Pasteur pipette can make this step easier. See Note 4.

3.5. The Final Preparation

1. Transfer the sperm suspension to a small culture tube (Falcon #2003).

2. Assess the concentration and motility of the washed sperm preparation, e.g., using a Makler chamber; if available, computer-aided sperm analysis (CASA) can be used.

   Notes: The presence of even 10% immotile or dead spermatozoa in the final preparation is not a problem; there is no need to perform any further preparation (e.g., swim-up) as it will have no benefit and could compromise sperm function or survival.

   A washed sperm preparation in culture medium might contain a high proportion of hyperactivated spermatozoa (although very vigorous these spermatozoa demonstrate little progression). Report the motility as the percentages of progressive and hyperactivated cells, as well as the total motility.

3. For studies on sperm function, proceed as per the requisite protocol.

4. For IUI or DI: leave the sample at ambient temperature in a styrofoam box on the bench until it is collected for insemination.

5. For IVF: equilibrate the tube, loose-capped, in a CO_2 incubator at 37°C for 30 min.

6. For ICSI: tight-cap the tube and place it in a 37°C incubator.

3.6. Processing Multiple Specimens

In clinical settings, risk management purists insist that only a single semen specimen should be processed by a scientist at any one time, always using a separate centrifuge and workstation for each specimen. However, in many busy clinical laboratories this is often not just impracticable but impossible, requiring unrealistic numbers of staff and centrifuges. Intelligent process analysis reveals that proper observance of a carefully designed process (in conjunction with a formal risk assessment) can permit the "safe" processing of more than one sample at once provided that the following are ensured (19):

(a) Each specimen is allocated to a separate tube rack.

(b) All disposables are labeled using at least two unique identifiers (one of which must be a number, since the

(c) Keep all the materials being used for each specimen together in a defined workspace, such as a plastic box or tray.

(d) Never label racks and trays, only tubes, so as to ensure proper specimen re-identification at every step.

(e) Pipettes should be organized so they cannot be confused between specimens being processed in parallel, e.g., place in a second rack in the workspace tray, and ideally labeled.

(f) Only actually work on one specimen at any point in time, i.e., only have one rack or tray in a scientist's active work area at once, so tubes from more than one sample are never open at the same time.

(g) Multiple samples can be centrifuged together so long as the protocol requires verification of each tube's identity before it is opened.

(h) Each re-identification step should be documented and, ideally, witnessed.

4. Notes

1. Colloidal silica-based gradients.
The physical characteristics of the new silane-coated colloidal silica products are equivalent to Percoll and their clinical utility is at least as good, if not better (2). However, whereas the Percoll colloidal was in a weak inorganic aqueous buffer, requiring it to used in conjunction with a 10× buffer to make "isotonic" 90%(v/v) colloid, the silanized silica-based products are prepared in an isotonic culture medium, ready-to-use. This has created a crucial difference between modern DGC products that can adversely affect their performance unless used carefully, as in the erroneous method in the WHO laboratory manual (11, 17, 18). While PureSperm is formulated with a colloid content equivalent to 100% stock Percoll colloid as supplied by the manufacturer, ISolate (Irvine Scientific, Santa Ana, CA, USA) is produced as an equivalent to "isotonic" 90%(v/v) Percoll colloid—hence comparing lower layers of 90% ISolate (i.e., 81% Percoll colloid-equivalent, or "PCE") and 90% PureSperm (=90% PCE) will result in a lower yield from the PureSperm gradients. Similarly, an 80% ISolate lower layer (i.e., 72% PCE) will give an even higher yield, but only by allowing less good spermatozoa into the pellet; ones with a

higher ROS-generating capacity (6). Proper use of silanized silica DGC products requires that manufacturers state the actual colloid concentration, so that a lower layer of 1.10 g/mL can be employed with human spermatozoa.

2. Tube and gradient factors.
   (a) A more discrete pellet is obtained using a conical tube in a swing-out rotor, facilitating recovery of the entire pellet when harvesting the gradient.
   (b) Larger cross-sectional tube area, i.e., the area of the interface between the layers, reduces the rate of formation of the "rafts" that block the gradient, allowing more spermatozoa to reach their isopycnic point: hence higher yields with larger diameter tubes.
   (c) Longer column layers (i.e., larger volume layers) also generally increase the yield.
   (d) There is no general benefit to using more than two-layer gradients. For extremely dirty specimens a third intermediate layer of 60% colloid can reduce the formation of the interface "rafts", but routine use of three-layer gradients unnecessarily increases technical complexity and both materials and labour costs.

3. Centrifugation time and speed.
   From early empirical work with Percoll DGC, spinning the gradients at $300 \times g_{max}$ has been known to provide optimum human sperm recovery. Various studies during the past 15–20 years have shown no systematic benefit to changing either the centrifugation speed or time for this step (15): slower speeds (i.e., lower g-force) reduce the yield but do not improve sperm quality; faster speeds do not increase the yield; and by 20 min all the spermatozoa have reached their isopycnic points.

4. Harvesting technique
   Although some methods describe recovering the pellet by passing a glass Pasteur pipette straight through the entire gradient, they are based on studies conducted mostly on research donors or normal fertile men—whose semen samples were reasonably "clean". With poor quality patient specimens the interface "rafts" can be dragged down with the pipette tip, contaminating the pellet. The technique described in this protocol precludes this problem by avoiding the following *incorrect* techniques:
   (a) Removing all the bottom gradient layer to reveal the pellet for recovery will introduce at least a mild risk of contamination due to the residual seminal plasma and raft materials slowly running down the inside wall of the tube and collecting on the meniscus of the remaining 80% layer.

(b) Recovering the pellet using the same pipette as was used to remove the upper layers (moderate contamination).

(c) Resuspending the pellet in the gradient tube (certain high level contamination).

Readers should be aware that such incorrect techniques were used in several published studies, whose results were likely compromised, even to the extent of their having no practical relevance.

References

1. Yanagimachi R (1994) Mammalian fertilization. In: Knobil E, Neill JD (eds) The physiology of reproduction, 2nd edn. Raven, New York, pp 189–317
2. Mortimer D (2000) Sperm preparation methods. J Androl 21:357–366
3. Rogers BJ, Perreault SD, Bentwood BJ, McCarville C, Hale RW, Soderdahl DW (1983) Variability in the human-hamster in vitro assay for fertility evaluation. Fertil Steril 39:204–211
4. Kanwar KC, Yanagimachi R, Lopata A (1979) Effects of human seminal plasma on fertilizing capacity of human spermatozoa. Fertil Steril 31:321–327
5. Aitken RJ, Clarkson JS (1987) Cellular basis of defective sperm function and its association with the genesis of reactive oxygen species by human spermatozoa. J Reprod Fertil 81:459–469
6. Aitken RJ, Clarkson JS (1988) Significance of reactive oxygen species and antioxidants in defining the efficacy of sperm preparation techniques. J Androl 9:367–376
7. Mortimer D (1991) Sperm preparation techniques and iatrogenic failures of in-vitro fertilization. Hum Reprod 6:173–176
8. Aitken RJ, De Iuliis GN (2010) On the possible origins of DNA damage in human spermatozoa. Mol Hum Reprod 16:3–13
9. Mortimer D (1994) Sperm recovery techniques to maximize fertilizing capacity. Reprod Fertil Dev 6:25–31
10. Mortimer D (1994) Practical laboratory andrology. Oxford University Press, New York
11. Björndahl L, Mortimer D, Barratt CLR, Castilla JA, Menkveld R, Kvist U, Alvarez JG, Haugen TB (2010) A practical guide to basic laboratory andrology. Cambridge University Press, Cambridge
12. Oshio S, Kaneko S, Iizuka R, Mohri H (1987) Effects of gradient centrifugation on human sperm. Arch Androl 17:85–93
13. Sakkas D, Manicardi GC, Tomlinson M, Mandrioli M, Bizzaro D, Bianchi PG, Bianchi U (2000) The use of two density gradient centrifugation techniques and the swim-up method to separate spermatozoa with chromatin and nuclear DNA anomalies. Hum Reprod 15:1112–1116
14. Tomlinson MJ, Moffatt O, Manicardi GC, Bizzaro D, Afnan M, Sakkas D (2001) Interrelationships between seminal parameters and sperm nuclear DNA damage before and after density gradient centrifugation: implications for assisted conception. Hum Reprod 16:2160–2165
15. Nicholson CM, Abramsson L, Holm SE, Bjurulf E (2000) Bacterial contamination and sperm recovery after semen preparation by density gradient centrifugation using silane-coated silica particles at different g forces. Hum Reprod 15:662–666
16. de Ziegler D, Cedars MI, Hamilton F, Moreno T, Meldrum DR (1987) Factors influencing maintenance of sperm motility during in vitro processing. Fertil Steril 48:816–820
17. World Health Organization (2010) WHO laboratory manual for the examination and processing of human semen, 5th edn. World Health Organization, Geneva
18. Barratt CLR, Björndahl L, Menkveld R, Mortimer D (2011) The ESHRE Special Interest Group for andrology basic semen analysis course: a continued focus on accuracy, quality, efficiency and clinical relevance. Hum Reprod 26:3207–3212
19. Mortimer D, Mortimer ST (2005) Quality and risk management in the IVF laboratory. Cambridge University Press, Cambridge

Chapter 20

Recovery, Isolation, Identification, and Preparation of Spermatozoa from Human Testis

Charles H. Muller and Erin R. Pagel

Abstract

In some cases, human spermatozoa to be used for in vitro fertilization are processed from testicular or epididymal biopsies collected in the clinic or operating room. An appropriately equipped Andrology or Embryology Laboratory is required. Sterility must be maintained at all stages from collection and transport to identification and processing to insemination or cryopreservation. The technologist must be able to properly process and identify spermatozoa from aspirates, seminiferous tubules or pieces of testicular tissue. Recovery of undamaged spermatozoa from tubules or tissue requires mincing, squeezing, or vortexing the tissue, usually without the need of enzymatic digestion. A motility stimulant such as Pentoxifylline is commonly used to calculate the number of functionally competent spermatozoa. After recovery, spermatozoa may be used immediately for IVF-ICSI, incubated overnight prior to IVF-ICSI, or cryopreserved for future use. Methods for identifying, purifying, and determining the number and motility of spermatozoa during these processes are presented.

Key words: Spermatozoa, Testis, TESA, TESE, Identification, Recovery, Processing, Seminiferous tubule, Wedge biopsy, Sperm aspiration, Sperm extraction

1. Introduction

Since the advent of Intracytoplasmic Sperm Injection (ICSI)—in vitro fertilization (IVF), the possibility of using spermatozoa from the testes of men whose semen does not contain sufficient numbers of viable sperm has brought a major change to infertility treatment. Several well-described methods (1–5) are available to clinicians for obtaining samples from the testis, epididymis, and vas deferens. Andrology or embryology laboratories are faced with different challenges to process and prepare spermatozoa in a way that preserves their functional integrity and sample sterility. A variety of laboratory methods are available to deal with these samples. For some, the embryologist performing

ICSI must work laboriously on a sample containing significant cellular debris and few spermatozoa. The goal of the methods presented below is to procure as many functional spermatozoa with as little debris as possible, and to quantify the numbers of spermatozoa and their motility or viability.

It is essential that the tissue be transferred to a sterile container at the point of collection, transported to the laboratory as quickly as possible under controlled conditions of temperature, light and air, and processed in an environment that precludes contamination and infection. In general, three types of samples are presented to the laboratory from the clinic or operating room: aspirates, tubules, and tissues. Each may be dealt with in different, optimized ways. Aspirates are suspensions of isolated cells usually obtained with fine needles from the vas deferens, epididymis, ductuli efferentes, or rete testis, and occasionally from the testis. Since the cells are free, these samples may be washed either by simple centrifugation or even by density gradient if the number and motility of the spermatozoa are sufficient. Techniques for "mini-gradients" were popularized by Ord (6). The second type of sample is seminiferous tubules. These may appear as intertwined masses, isolated pieces of tubule, or small lengths of tubule within an intact piece of testis tissue. They are often obtained by large-bore (16–18 G) needle aspiration in a TESA (testicular sperm aspiration) procedure. Under a dissecting microscope, these tubules are dissected free of interstitial tissue and blood clots and transferred to clean medium in another dish. There, they may be minced with sharp instruments such as hypodermic needles, scissors, scalpels or razor blades (7). The pieces and supernatant medium are incubated and mixed, for example by vortexing, to loosen the spermatozoa from the tissue fragments. The larger fragments are allowed to settle, and the supernatant containing many of the spermatozoa is collected and processed further. Alternatively, the isolated tubules may be squeezed to obtain the tubule contents (3). The method detailed here uses specially polished jeweler's forceps. These masses of contents may be picked up using a pipettor and transferred to a centrifuge tube for simple centrifugal washing. The squeezing method usually provides a much "cleaner" sample than the mincing method. Third, and most common, the laboratory is presented with fragments or large pieces of testicular tissue from micro TESE (microscopic testicular sperm extraction) or TESE procedures. There are three alternatives for preparing sperm from these tissues. Individual seminiferous tubules may be dissected out and prepared as above; the tissue may be minced with sharp instruments as above; or the tissue may be digested (8) using collagenase to open the tubules and free the testicular spermatozoa (this chapter does not address this method, as the authors have not found it to be necessary). A combination of methods is generally advised, depending on the difficulty of finding spermatozoa (9).

Next, the isolated sperm must be identified, counted, and assessed in some way for their functional competence. Contaminating cells, nuclei and debris should be documented to determine the efficiency of the method, and so the embryologist who must select spermatozoa from the preparation will know what to expect. The sperm suspension is centrifuged if necessary to obtain a sufficient concentration for sampling purposes without compromising a large percentage of the sample. Finally, the specimen is used for IVF ICSI, incubated overnight to improve motility and as a matter of convenience, or cryopreserved for future use.

2. Materials

Transport samples from clinic or operating room to lab in an enclosure such as a portable incubator unit. All laboratory work must be performed either in a laminar flow hood (Biological Safety Cabinet) or in a HEPA-filtered clean room. Technologists must wear personal protective gear, including at a minimum, gloves, coat, mask and hair net or cap. Wipe down hood surface or table-top, pipettors, and microscope with an appropriate disinfectant before and after use. Use sterile disposable plastic-ware. Sterilize sets of 1–6 pipette tips in sterilizer bags. Sterilize jeweler's forceps and glass Petri dishes (if desired) in individual bags. Thoroughly clean forceps and dishes with an appropriate detergent after use; rinse completely with sterile deionized water, dry, and repackage for steam or gas sterilization.

1. Centrifuge tubes, 5 mL or 15 mL, plastic, sterile.
2. Petri dishes, 35 × 10 mm or larger, plastic or glass, sterile (see Note 1).
3. Serological pipettes, disposable glass, sterile, individually packaged, 1 mL, 2 mL, 5 mL, or 10 mL sizes; or sterile Pasteur pipettes. Pipette-Aid or manual pipette controlling device required.
4. Pipettors, 2–20 µL, and 10–100 µL or 20–200 µL sizes, with appropriate sterile disposable tips (in sterilizer bags for one-time use).
5. Sterile nonabsorbent pads (Kendall Telfa Nonadherent Dressing Pads) or similar (see Note 2).
6. IVF-grade culture medium such as sterile HEPES-buffered HTF, supplemented with at least 0.3% albumin or 10% serum substitute. HTF without HEPES, supplemented with protein, is suggested for overnight incubation in a CO_2 incubator. Keep medium at room temperature for processing. These media formulations may be purchased premade in sterile liquid form.

7. Pentoxifylline, 2×, made at 2 mg/mL IVF culture medium with protein: weigh 8 mg Pentoxifylline and add to 4 mL IVF culture medium with HEPES (without HEPES if incubation under CO_2 is performed). Allow powder to dissolve, and gently vortex to mix thoroughly. Filter through a 0.22 μm filter. Add protein (at least 0.3%, w/v, or 10% liquid serum substitute). For use, add 1 part of filtered 2× Pentoxifylline solution to 1 part sperm suspension, gently mix.
8. Counting chambers such as hemacytometers, Makler chambers, or fixed-coverslip slides used with counting grid.
9. No. 5 Jeweler's forceps, with polished tips, sterile (optional). A Dremel drill mounted on a drill stand, with fine polishing wheel and fine red polishing compound is required to produce these tools (see Note 3).
10. Stereo dissecting microscope with adjustable magnification and transmitted light (incident light if former not available).
11. Compound (preferably phase contrast) microscope with 10× or 20×, and 40× objectives.
12. Vortexer, adjustable speed. Set to low or medium speed.
13. Centrifuge capable of centrifuging chosen centrifuge tubes at 200–300×g. Spin at room temperature.
14. Incubator, air or 5% CO_2, set to 34–37 °C (see Note 4).
15. Analytical balance and weighing paper (optional).

3. Methods

Examine the obtained specimens while still in the tube with transport medium (see Note 5). The entire specimen may be placed into a Petri dish to examine closely under the stereomicroscope. Rinse the transport tube once with a small amount of medium to transfer most of the spermatozoa to the Petri dish. Using this observation, information from the clinician who extracted the sample, and any clinical instructions, proceed with one or more of the following methods.

3.1. Aspirated Spermatozoa in Suspension

1. Examine the tube or dish to determine if any small pieces of tissue are present. If present, they should be removed to another dish for examination. Alternatively, the tube can be gently mixed, and small pieces allowed to settle by gravity for no more than 5 min. Then, remove most of the supernatant for further processing. If the pieces are from the testis, they may be minced in a dish to obtain more spermatozoa; process as in Subheading 3.3. If the aspiration was from vas deferens, epididymis, ductuli efferentes or rete testis, there is no need to reserve or process the pieces of tissue.

2. If the supernatant medium is a relatively large volume, or the sperm concentration is too low, centrifuge the tube at 200–300 $\times g$ for 7–10 min. After centrifugation, determine if a pellet is visible. If not, consider dividing the supernatant into two or more tubes and centrifuging again to pellet all the sperm. Combine pellets (if necessary) and resuspend to a minimal volume (as little as 100 µL), until the suspension is diluted to a point at which a sperm concentration can be determined (see Note 6). Use sterile pipettes and pipette tips for these steps. Record volume.

3. Examine a wet-mount of the suspension by placing 5 µL onto a glass slide using a sterile pipette tip. Carefully cover with a 22 × 22 mm #1 coverslip, and place on the stage of the compound microscope. Examine 20 or more fields under the 40× objective (see Note 7). If there are fewer than ten sperm per ten high-power fields (400×), consider concentrating the sample further; sometimes this will not be possible. Record the % motility and motility patterns of the sperm (see Note 8). Record the apparent morphology of the spermatozoa, especially noting head shape, presence of additional cytoplasm, and structure of the midpiece and tail. Record the presence of cell debris, epithelial nuclei and red blood cells. If there are sufficient numbers of sperm, consider making a smear of 5–7 µL for staining and permanent record (see Note 9).

4. If the suspension is relatively free of extraneous cells and debris, consider using it without further purification. If the sperm are nonmotile (typical of testicular sperm) or poorly motile, do not attempt purification, since most of the sperm may be lost. If most of the sperm are progressively motile (typical of some epididymal sperm aspirations) and there are numerous nuclei, other cells or debris, consider purifying the sperm using a mini-gradient (6, 10) before proceeding to the next step (see Note 10).

5. Determine the total spermatozoa in the sample. Both the volume of the suspension and the sperm concentration will be needed for this determination. If necessary, remeasure the entire suspension volume with either a pipettor using a sterile tip (see Note 11), or a sterile serological pipette. Replace all the suspension into the tube.

6. Mix the suspension well (see Note 12), and remove 10 µL for counting (use less if there is 50 µL or less total volume). Dilute the 10 µL portion to a known volume if needed, using sterile water to kill the sperm to allow counting. Load hemacytometer or other counting chamber, and when possible count at least 200 sperm in a known number of grids. Calculate sperm concentration according to the instructions for that chamber. Alternatively, if the suspension is fairly free of debris, load an appropriate amount of suspension (usually 5–7 µL) into a

fixed-coverslip slide and analyze in a computerized sperm motility analyzer to obtain concentration, total count and motility data.

7. Multiply the sperm concentration (millions per mL) by the volume (mL) to determine the total number (in millions) of sperm in the entire sample. Record this value. Multiply the total number of sperm by the decimal equivalent of the % motile to determine the total number of motile sperm in the specimen. Record this value.

8. If the % motility of the sperm is low or few sperm are progressively motile, according to your laboratory's standards, consider a trial with the motility stimulant Pentoxifylline (see Subheading 3.2, step 6), or perform a live-dead analysis (see Note 13). Neither technique may be required if there are at least a few thousand total motile sperm in the suspension.

9. Spermatozoa may be used immediately, incubated overnight (see Note 4), or cryopreserved (see Note 14) for IVF-ICSI.

3.2. Extraction of Sperm from Isolated Seminiferous Tubules

Tubules aspirated from the testis with a large-bore needle (TESA) present the opportunity for selection of areas of high yield, quantification, and relatively clean sperm isolation. The method may also be used for micro TESE-acquired portions of seminiferous tubules. Short segments of tubules may also be isolated from wedge biopsies (TESE) with additional work under the dissecting microscope, freeing them from the interstitium.

1. Transfer the tubules to a sterile plastic or glass Petri dish, and observe using a stereo dissecting microscope. Using sterile tools (e.g., hypodermic needles mounted on small syringes, fine scissors, or polished jeweler's forceps (see Note 3)), separate tubules from one another by stretching and removing the interstitial tissue. Remove blood clots. Agitating the medium or the tubules will help free interstitial cells and red blood cells. If necessary, move the tubules to sequential Petri dishes with fresh medium during this process.

2. If desired, sperm recovery may be normalized to tissue weight or tubule length, or both. When the tubules (or a subset of them) are sufficiently cleaned, they may be weighed if desired (see Note 15). The length of selected tubules or the entire set may also be measured (see Note 16). In either case, the total number of sperm recovered from the measured portion may be divided by the weight or length to acquire data on sperm per mg or sperm per cm.

3. Examine the tubules under light and magnification conditions that allow the contents to be differentiated. Unlike rodents, the human spermatogenic stages take up very small wedges of tubule, and are closely applied one to another. However, it is

20 Recovery, Isolation, Identification, and Preparation of Spermatozoa...

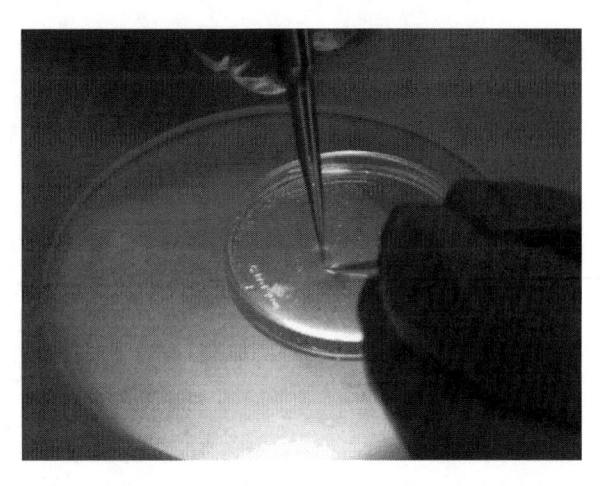

Fig. 1. Close view of dissecting microscope stage during the tubule squeezing stage of sperm recovery. Combined tubule contents are at the far left of the dish (*white mass*).

Fig. 2. Close-up of a single seminiferous tubule (between the forceps tips) being squeezed.

possible to distinguish areas likely to contain few sperm (more translucent, thinner tubules) from those more likely to contain more spermatozoa (translucent, brownish-grayish, refractile points, thicker). Choose areas to collect and measure their length if desired. Move selected tubules to another dish with fresh medium containing protein.

4. To extract the cells from the separated segments, grasp the end of the tubule firmly with polished forceps and with a second polished forceps closed over the tubule at a right angle, move the second forceps from the grasped end to the open end using slight pressure. Too much pressure and the tubule will snap; not enough, and the cells won't move. The action is similar to working the last of the toothpaste out of the tube (Figs. 1 and 2).

Fig. 3. Aspiration of tubule contents with a pipettor.

The cells will push out the open end, sometimes keeping the tubule shape, other times rolling up into balls. If there appear to be cells still in the tubules, the squeezing motion can be repeated. Remove the empty tubule from the dish and discard. Repeat with each selected segment.

5. As the dish fills with "wooly masses" of cells they can be transferred to a centrifuge tube using a micro-pipettor or Pasteur transfer pipette (Fig. 3). Once all the segments have been squeezed, the remaining media from the dish can be added to the centrifuge tube. Concentrate the cell suspension into a smaller volume (usually 100 µL, or more, depending on pellet size) for wet mount and counting chamber analysis as in Subheading 3.1, steps 2, 3 and Subheading 3.1, steps 5–8.

6. If sperm are present, but they exhibit little (<5%) or no "twitching" motility, consider treating with Pentoxifylline to stimulate motility. Dilute all or a portion of the sperm sample 1:1 with 2 mg/mL Pentoxifylline and incubate 15 min at 37 °C. Reexamine a wet-mount. Record the stimulated motility. Spermatozoa must be washed to remove Pentoxifylline before being used for IVF-ICSI, but may be left in during cryopreservation.

7. Spermatozoa may be used immediately, incubated overnight (see Note 4), or cryopreserved (see Note 14) for IVF-ICSI.

3.3. Extraction of Spermatozoa from Testicular Tissue by Mincing

Isolated tubules, wedge biopsies of testis (TESE), and tubule segments collected by micro TESE may be minced in the operating room or laboratory to quickly determine if spermatozoa are present, and to efficiently collect testicular spermatozoa for IVF-ICSI. The finer the mincing, the more spermatozoa are likely to be freed.

However, the finely minced tubular and interstitial tissues will be more difficult to separate from the spermatozoa, yielding a preparation with more nuclei and debris than a coarsely minced preparation, or one prepared by squeezing the tubules.

1. Testicular tissue may be weighed before proceeding (see Note 15).

2. Place the tissue into a sterile Petri dish containing about 100–500 μL IVF medium (with Pentoxifylline (10) if desired; (see Note 17)). Hold the tissue with sterile forceps, needle, or other tool, and mince with fine scissors, razor or scalpel blade, or other sharp tool (some use the edge of a glass slide). Continue mincing until the tissue is a slurry of fine to moderately coarse pieces.

3. Transfer all the contents of the dish to a 5 or 15 mL centrifuge tube, using a Pasteur pipette, pipettor and forceps as needed. The dish may be rinsed with an additional 1–2 mL of medium, which is added to the same tube. Cap and vortex the tube for 30 s to 1 min to free spermatozoa from the surrounding tissues (11). A moderate setting suffices to provide vigorous mixing; avoid extremely high settings. Flush any tissue fragments adhering to the sides of the tube down into the suspension.

4. Incubate the suspension in an incubator for 30–60 min (10) (see Note 4). This step is optional if fast processing is required. Vortex the tube again for 30–60 s.

5. Remove as many pieces of tissue as possible, using long sterile forceps or a Pasteur or transfer pipette. Tissue pieces may be pulled up the sides of the tube and left there. Other pieces will settle to the bottom.

6. Aspirate the tissue-free portion of the suspension and place into a clean tube.

7. Concentrate the suspension by centrifuging at $200–300 \times g$ for 7–10 min. Remove the supernatant and resuspend the pellet in 100–250 μL IVF medium with protein, depending on pellet size. Medium may contain Pentoxifylline at this step, and during cryopreservation, but spermatozoa must be washed to remove Pentoxifylline before IVF.

8. Determine the concentration, motility, and volume as in Subheading 3.1, steps 2, 3 and Subheading 3.1, steps 5–8. Calculate the total number of sperm and total motile sperm. Note the presence of debris, nuclei, and other contaminants. Note the appearance of sperm and motility characteristics.

9. Adjust the volume as needed, and process for IVF-ICSI, incubate overnight prior to IVF-ICSI (see Note 4), or cryopreserve using standard methods (see Note 14).

4. Notes

1. Reusable glass Petri dishes provide a more firm support for mincing with a blade. Be sure to thoroughly clean and sterilize the top and bottom between uses. Larger diameter dishes are best for mincing methods, allowing more room to use a blade. We routinely use 35 × 10 mm plastic disposable dishes for all other procedures, including squeezing tubules. The lid of a 35 mm dish works best for squeezing.

2. Sterile pads are useful for providing a clean resting place for tools. If working in an open area, a second pad may be placed upside down over tools to help protect them.

3. Jeweler's forceps have fine tips, but there are burrs and rough spots that typically "hook" tissues. The tips of these forceps may be polished to prevent this. First, use a small rubber band to hold the tips of the forceps just closed, without forcing them. Do not further squeeze the forceps. Set up a Dremel or similar mini-drill fitted with a fine round polishing wheel on a stable drill stand in a vertical position. Place the apparatus in an area that is not critical for fumes or debris. Wear protective eyewear. Turn on the drill at a medium speed and load the polishing wheel with fine (red) polishing compound. Using medium-to-high speed, polish the outer portions of the tips (about 1–2 cm) by carefully bringing each side of the forceps to the polishing wheel. Be careful to polish in a direction leading toward the tips, and to not jam the tips into the wheel. The polishing compound will remove rough areas and will also grind down the tips to finer points if desired. Remove any polishing compound that gets between the tips using lens paper or Kimwipe before proceeding. It is critical that the tips remain in close contact. When the outer surfaces are sufficiently polished or shaped, the inner flat surface of each tip may be very slightly polished if desired. This step requires practice so that the tips continue to meet each other correctly. The forceps must be widely opened while one tip at a time is placed in very light contact with the polishing wheel. Test the forceps using paper, hair, or other thin objects to ensure that they function correctly. Wash thoroughly and carefully with an appropriate detergent, rinse well in deionized water, and package for sterilization. We typically have four or five pairs of forceps in use.

4. Incubation conditions vary with preference and choice of medium. Although sperm develop at several degrees below body temperature, once isolated they do well at 37 °C. If the medium has HEPES buffering, no CO_2 is needed. However, for overnight incubation, the medium should be buffered with bicarbonate and incubated under 5% CO_2. Typically, testicular

sperm motility naturally improves after overnight incubation, from mostly nonmotile or slight twitching to many twitching and even a few with progressive motility.

5. Transport medium, and all media used in room air, should be HEPES-buffered to maintain a slightly alkaline pH. Transport medium should be IVF culture medium (e.g., HTF), and need not have protein added for this step.

6. A fast and convenient method for estimating that the sperm concentration is within range for obtaining a concentration measurement, add medium to the pellet gradually with mixing until it is transparent enough to see color differences (i.e., a colored sticky note or a black line) through the tube. Typically, we post a note on the back wall of the Biological Safety Cabinet for this purpose.

7. A wet mount is best observed using phase contrast optics. Spermatozoa are easily seen and differentiated from other cells. For testicular preparations, Sertoli nuclei may be common, and are identified by a prominent nucleolus. In epididymal preparations, epididymal epithelial cell nuclei and cell fragments are common. If the aspiration was in the caput epididymis, it is not unusual to find cuboidal cells with beating apical cilia from the ductuli efferentes. The presence and relative abundance of these nuclei and cells should be noted.

8. Sperm motility varies along the reproductive tract. Epididymal sperm often exhibit progressive motility once placed into IVF medium, although they may have residual cytoplasm on their midpiece. Testicular sperm typically have no motility or a few may twitch when first isolated.

9. A smear of the preparation can be a permanent record of the presence of cell types, nuclei and debris. Make a smear as usual for semen. Be sure not to "run over" the drop while pulling the liquid forward on the slide with a second slide held at about a 30° angle. Dry, fix, and stain with any usual stain for semen or blood. Coverslip over Permount or similar mounting medium for a permanent slide (see Figs. 4 and 5).

10. Use density gradient medium (e.g., PureSperm, Isolate, etc.) and diluent approved for human use, following standard protocols. A mini-gradient is a variation of the standard protocol in which a smaller, narrower tube (such as a 5 mL culture tube) is used, with proportionately lower volumes of gradient mixtures. Sperm recovery may also be improved by using a single 60% or 70% gradient mixture instead of the typical 40%/80% mixtures. Red blood cells will pellet below the spermatozoa. We do not recommend lysing RBCs.

11. Pipette tips may be packaged in sterile boxes, but they lose their sterility once the box is opened. Instead, package 1–6 tips

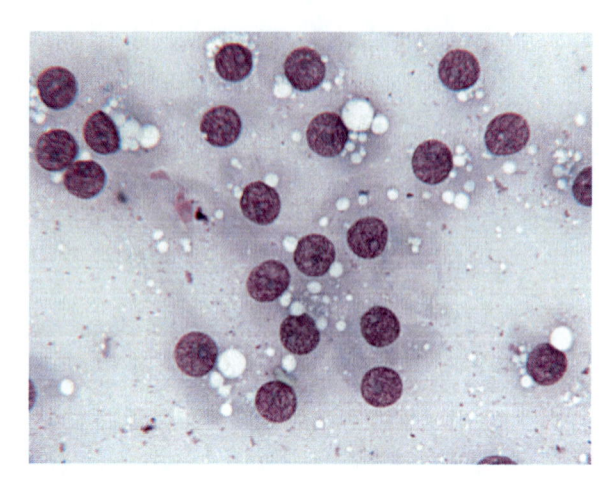

Fig. 4. Stained slide of a testicular cell smear. Only Sertoli cells (note nucleoli) are present. 40× oil immersion objective.

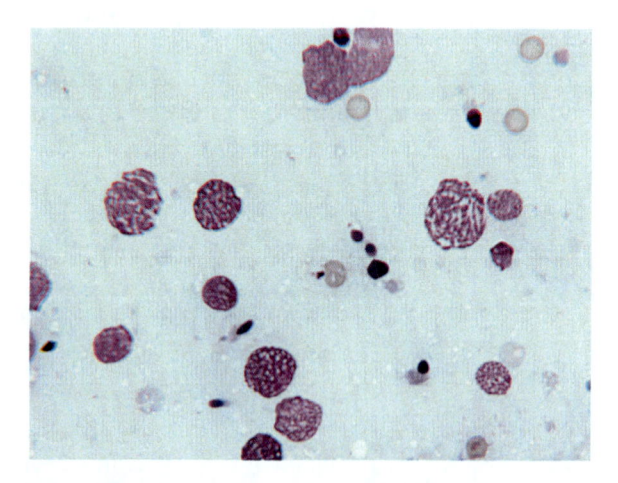

Fig. 5. Stained slide of testicular cell smear. Testicular sperm (small condensed nuclei with lighter-staining acrosomes), primary spermatocyte nuclei (large, with long condensed chromosomes), sertoli cell nuclei, and red blood cells (round cells without nuclei) are present. 40× oil immersion objective.

(depending on expected need for one procedure) into sterilizer bags, and autoclave or gas sterilize (too high a temperature can melt some tips). We typically have 20 or more such sterile packages on hand, color labeled for each size of pipettor.

12. Good mixing of the suspension is critical to obtain accurate counts. A vortexer set on the lowest speed that efficiently swirls the medium, for about 30 s is sufficient. Immediately take a sample from the middle of the suspension. Repeat vortexing and take a second sample to load the second side or chamber. Use the average of the two counts if they are within 10% of each other; otherwise repeat.

13. Viability tests are the only way to determine if nonmotile sperm are alive, as defined by an active exclusion of extracellular components. Dead sperm should not be used for IVF-ICSI. Typical viability tests use Eosin-Nigrosin, Trypan Blue, Hoechst 33258, or Hypo-osmotic swelling.

14. Testicular and epididymal spermatozoa may be cryopreserved using variations of common sperm freezing methods.

15. There are alternatives for weighing small amounts of tissue. Use an analytical balance capable of measuring in the mg range. A simple method is to cover the Petri dish being used and weigh the entire contents. Remove the tubules to a fresh dish, reweigh the same dish and cover to determine the net "wet weight" of the tubules. Second, the tissue may be briefly placed on a sterile pad to remove most liquid, then placed on a tared or preweighed sterile pad (cut to fit the pan) to obtain a wet weight. The weight obtained by either method will include some medium, but the tissue must not be allowed to dry before processing.

16. We find that a convenient way to measure tubule length is to place two dots one cm apart on the glass stage above the mirror of our Wild stereomicroscope. Then, tubules can be quickly moved one cm at a time to accumulate measurements. It is important to straighten the tubules using forceps, but not stretch them during measurement.

17. A theoretical advantage of using Pentoxifylline at this step is that testicular spermatozoa will usually start flagellating, increasing the chances that they will break free of the surrounding tissue, or even remain in the suspension supernatant longer to assist purification from tissue debris. A known advantage of including Pentoxifylline at this or the next step is that a measurement of the enhanced % motility can be made, often obviating a viability test. Alternatively, one may divide the sample into one tube with and the other without Pentoxifylline to determine if it has an effect, or if the IVF lab wishes to use untreated sperm.

References

1. Schlegel PN (2009) Nonobstructive azoospermia: a revolutionary surgical approach and results. Semin Reprod Med 27:165–170
2. Schlegel PN, Li PS (1998) Microdissection TESE: sperm retrieval in non-obstructive azoospermia. Hum Reprod Update 4:439
3. Steele EK et al (2000) Testicular sperm extraction by Trucut needle and milking of seminiferous tubules: a technique with high yield and patient acceptability. Fertil Steril 74:380–383
4. Okada H et al (2002) Conventional versus microdissection testicular sperm extraction for nonobstructive azoospermia. J Urol 168:1063–1067
5. Khadra AA et al (2003) Efficiency of percutaneous testicular sperm aspiration as a mode of sperm collection for intracytoplasmic sperm injection in nonobstructive azoospermia. J Urol 169:603–605
6. Ord T et al (1992) The role of the laboratory in the handling of epididymal sperm for assisted

reproductive technologies. Fertil Steril 57: 1103–1106
7. Kamal A et al (2004) Selection of individual testicular tubules from biopsied testicular tissue with a stereomicroscope improves sperm retrieval rate. J Androl 25:123–127
8. Crabbe E et al (1998) Enzymatic digestion of testicular tissue may rescue the intracytoplasmic sperm injection cycle in some patients with non-obstructive azoospermia. Hum Reprod 13:2791–2796
9. Aydos K et al (2005) Enzymatic digestion plus mechanical searching improves testicular sperm retrieval in non-obstructive azoospermia cases. Eur J Obstet Gynecol Reprod Biol 120:80–86
10. Hammitt DG et al (2002) Development of a new and efficient laboratory method for processing testicular sperm. J Assist Reprod Genet 19:335–342
11. Tokuyama O et al (2003) Using a vortex mixer for testicular sperm collection. J Reprod Med 48:865–868

Chapter 21

Enhancement of Sperm Motility Using Pentoxifylline and Platelet-Activating Factor

Shilo L. Archer and William E. Roudebush

Abstract

Enhancement of sperm motility can effectively improve assisted reproductive technique outcomes. Here we describe two (pentoxifylline and platelet-activating factor) popular sperm motility enhancers and their respective methods.

Key words: Sperm, Motility, Pentoxifylline, Platelet-activating factor, PAF

1. Introduction

Spermatozoa prior to any assisted reproductive technique (e.g., intrauterine insemination; IUI) must be first processed (e.g., density/gradient separation, wash, or swim-up). This processing of spermatozoa insures that the most normal motile population is used to help facilitate conception. For IUI procedures, semen washing is important as it removes prostaglandins, debris and white blood cells while also reducing the number of nonmotile and morphologically abnormal sperm cells. Additionally, removal of seminal plasma also benefits the sperm by enhancing capacitating conditions resulting in the enhancement of sperm motility. A number of additives that improve sperm motility have been investigated, including but not limited to the following: cAMP, human follicular fluid, xanthine, caffeine, pentoxifylline, and platelet activating factor (PAF). This article focuses only on pentoxifylline and PAF, primarily due to the author's experience (PAF), and because pentoxifylline is routinely used in assisted reproductive technology (ART) laboratories.

Pentoxifylline is a Food & Drug Administration (FDA), approved drug that is used for treating peripheral arterial disease by improving blood flow in patients with circulation problems to reduce aching, cramping, and tiredness in the hands and feet. Pentoxifylline works by decreasing the viscosity of blood, allowing it to flow more easily, especially in the small blood vessels of the hands and feet (1). Pentoxifylline is routinely used clinically in ART as a chemical sperm stimulant (2–5). Pentoxifylline is a phosphodiesterase inhibitor that promotes the accumulation of intracellular adenosine triphosphate and intracellular calcium, resulting in the enhancement of sperm motility (2, 3). While pentoxifylline increases sperm motility by intracellular calcium, the drug is considered toxic thus its use is controversial (6, 7).

PAF is a unique and novel signaling phospholipid that has pleiotropic biologic properties in addition to platelet activation (8). PAF influences ovulation, fertilization, preimplantation embryo development, implantation, and parturition (9). It is present in human spermatozoa, and its endogenous content has a significant and positive relationship with motility and pregnancy rate (10). Research has demonstrated PAF to be a nontoxic treatment to increase sperm motility and intrauterine insemination pregnancy rates (11). Although the exact mechanism for PAF action remains unclear, its importance for normal reproductive function is significant. Additionally, endogenous PAF may serve as a biomarker for normal sperm function (12). One distinctive advantage in using PAF is that it is a natural substance produced by sperm and is nontoxic unlike some other motility stimulators such as pentoxifylline (13).

2. Materials

2.1. Pentoxifylline Materials

1. Sperm separation medium: e.g., Spermcare (45, 90%; InVitroCare, Inc., Frederick, MD).
2. Sperm wash medium (InVitroCare, Inc., Frederick, MD).
3. 5 mL syringe.
4. BD Precisionglide® syringe needles L1½″, size 21 g needle.
5. Centrifuge.
6. Falcon 15 mL polystyrene centrifuge tubes.
7. Pentoxifylline: P1784, Sigma-Aldrich.

2.2. Platelet-Activating Factor Materials

1. Sperm separation medium: e.g., Spermcare (45, 90%; InVitroCare, Inc., Frederick, MD).
2. Sperm wash medium: InVitroCare, Inc., Frederick, MD.
3. 5 mL syringe.
4. BD Precisionglide® syringe needles L1½″, size 21 g needle.

5. Centrifuge.
6. Falcon 15 mL polystyrene centrifuge tubes.
7. Platelet-activating factor (0.5 µg; Fertility Technology Resources, Marietta, GA).
8. Kimble borosilicate glass culture tubes (75 × 12).

3. Methods

3.1. Pentoxifylline (1.76 mM) Method

1. Bring all components of the system and samples to room temperature or to 37°C.
2. Transfer 1 mL of sperm separation medium upper layer (e.g., 45% silica density solution in HTF) to a sterile centrifuge tube.
3. Using a 3 cm³ syringe with a 1½″, 21 g needle, place 1 mL of sperm separation medium lower layer (e.g., 90% silica density solution in HTF) under the upper layer (see Note 1). Gently place up to 2 mL of liquefied semen onto the upper layer using a transfer pipette or syringe.
4. Repeat for additional tubes if semen volume is >2 mL.
5. Centrifuge for 15–20 min at 350–400 × g. When this centrifugation is complete you may not be able to visibly see a pellet. If this is the case, it is essential to continue the procedure with a second centrifugation of 5 min.
6. Remove supernatant down to the 1.0 mL mark above the pellet.
7. Prepare a stock solution of pentoxifylline (1 mg/mL) in sperm wash medium.
8. Add 1 mL of stock pentoxifylline, to the resuspended sperm pellet. This will provide a final pentoxifylline concentration of 1.76 mM.
9. Incubate for 15 min at 37°C.
10. Centrifuge for 8–10 min at 300 × g (see Note 2).
11. Remove supernatant down to the pellet and add 4 mL of sperm wash medium.
12. Centrifuge for 8–10 min at 300 × g (see Note 2).
13. Remove supernatant and replace with a suitable volume of appropriate medium.

3.2. PAF (10^{-7} M) Method (See Note 3)

1. Bring all components of the system and samples to room temperature or to 37°C.

2. Transfer 1 mL of sperm separation medium upper layer (e.g., 45% silica density solution in SWM) to a sterile centrifuge tube.
3. Using a 3 cm^3 syringe with a 1½″, 21 g needle, place 1 mL of sperm separation medium lower layer (e.g., 90% silica density solution in SWM) under the upper layer (see Note 1).
4. Gently place up to 2 mL of liquefied semen onto the upper layer using a transfer pipette or syringe.
5. Repeat for additional tubes if semen volume is >2 mL.
6. Centrifuge for 15–20 min at 350–400×g. When this centrifugation is complete you may not be able to visibly see a pellet. If this is the case, it is essential to continue the procedure with a second centrifugation of 5 min.
7. Remove supernatant down to the 1.0 mL mark above the pellet.
8. Add 10 mL of sperm wash medium to the PAF (0.5 µg; Fertility Technology Resources, Marietta, GA) vial and vortex vigorously for 1 min just prior to use (see Note 3).
9. Add 3 mL of PAF in sperm wash medium and resuspend the sperm pellet. This will provide a final PAF concentration of 10^{-7} M.
10. Incubate for 15 min at 37°C.
11. Centrifuge for 8–10 min at 300×g (see Note 4).
12. Remove supernatant down to the pellet and add 4 mL of sperm wash medium.
13. Centrifuge for 8–10 min at 300×g (see Note 4).
14. Remove supernatant and replace with a suitable volume of appropriate medium.

4. Notes

1. Take care that the two layers are distinctly separated. This is done by placing the tip of the needle on the bottom of the test tube and slowly dispensing the lower layer. Most two-layer gradients are stable for up to 2 h.
2. Higher sperm concentrations will require the maximum 10 min centrifugation to ensure a complete and thorough sperm wash.
3. For optimal results, PAF is best administered to washed sperm, and with a protein source, it should be handled using borosilicate glass coated with silicon.
4. Higher sperm concentration will require the maximum 10 min centrifugation to ensure a complete and thorough sperm wash.

References

1. Ward A, Clissold SP (1987) Pentoxifylline. A review of its pharmacodynamic and pharmacokinetic properties, and its therapeutic efficacy. Drugs 34:50–97
2. Sikka SC, Hellstrom WJ (1991) The application of pentoxifylline in the stimulation of sperm motion in men undergoing electroejaculation. J Androl 12:165–170
3. Tesarik J et al (1992) Effect of pentoxifylline on sperm movement characteristics in normozoospermic and asthenozoospermic specimens. Hum Reprod 7:1257–1263
4. Lewis SE et al (1993) Effects of pentoxifylline on human sperm motility in normospermic individuals using computer-assisted analysis. Fertil Steril 59:418–423
5. Pang SC et al (1993) Effects of pentoxifylline on sperm motility and hyperactivation in normozoospermic and normokinetic semen. Fertil Steril 60:336–343
6. Centola G et al (1995) Differential responses of human sperm to varying concentrations of pentoxyfylinne with demonstration of toxicity. J Androl 16:136–142
7. Patrizio P et al (2000) Effect of pentoxifylline on the intrinsic swimming forces of human sperm assessed by optical tweezers. J Androl 21:753–756
8. Roundebush W (2007) Seminal platelet-activating factor. Semin Thromb Hemost 33:69–74
9. Roundebush W, Purnell E (2000) Platelet-activating factor content in human spermatozoa: predicting pregnancy outcome. Fertil Steril 74:257–260
10. Harper MJ (1989) Platelet-activating factor: a paracrine factor in preimplantation stages of reproduction? Biol Reprod 40:907–913
11. Roudebush WE et al (2004) Platelet-activating factor significantly enhances intrauterine insemination pregnancy rates in non-male factor infertility. Fertil Steril 82:52–56
12. Levine AS et al (2002) A review of the effect of platelet-activating factor on male reproduction and sperm function. J Androl 23:471–476
13. Henkel RR, Schill WB (2003) Sperm preparation for ART. Reprod Biol Endocrinol 1:108

Chapter 22

Intracytoplasmic Morphology-Selected Sperm Injection

Luke Simon, Aaron Wilcox, and Douglas T. Carrell

Abstract

Approximately 40% of sterility in couples can be attributed to male subfertility and intracytoplasmic sperm injection (ICSI) has become a powerful tool in assisted reproduction to overcome male infertility. Intracytoplasmic morphologically selected sperm injection (IMSI) is an advanced and sophisticated method of ICSI, where prior to sperm injection the morphology of the sperm is evaluated under high magnification. In addition, the IMSI procedure involves a few minor modifications in sperm preparation which are not carried out during the conventional ICSI procedure, such as the use of MSOME criteria, the requirement for a glass-bottomed dish for selection, prolonged sperm manipulation following separation from the seminal fluid, and sperm storage prior to microinjection. These variations are discussed in this chapter.

Key words: High magnification, Intracytoplasmic morphologically selected sperm injection, Male infertility, Motile sperm organelle morphology examination, Sperm morphology, Sperm selection

1. Introduction

Sperm morphology evaluation plays an important role in the diagnosis of male fertility (1) and has been demonstrated to influence in vitro pregnancy outcomes (2, 3). Increase in sperm with morphological abnormalities in the ejaculate has been associated with high incidence of aneuploidy, abnormal protamination, impaired chromatin integrity, and immature sperm (4–6). The introduction of intracytoplasmic sperm injection (ICSI) in 1992 involved the selection of morphologically normal sperm and has became a common laboratory procedure allowing the embryologist to select morphologically normal and motile sperm under a magnification of 200× or 400× for injection into the oocyte. Such sperm selection has been reported to improve assisted treatment outcomes (7, 8).

A decade later in 2002, an Israeli team of male infertility researchers led by Professor Bartoov developed a new high magnification method of motile sperm organelle morphology examination (MSOME) (9), which amplifies the image of individual sperm up to 6,600 times through high powered microscopy coupled with digital magnification. This higher magnification makes it possible to identify sperm whose nuclei have an abnormal shape or contents and which are less likely to produce a healthy embryo (10). In 2003, the MSOME criteria were first applied to the selection of motile sperm to be used for ICSI, termed intracytoplasmic morphologically selected sperm injection (IMSI) (11). Sperm selection based on MSOME criteria, for the use in assisted reproduction yielded significantly better results in terms of pregnancy, implantation and miscarriage rates (12).

The IMSI technique is reported to select the best sperm for microinjection enabling the selection of motile sperm with fine nuclear morphology and without any head vacuoles (13). Sperm with nuclear vacuoles have been reported to have poor mitochondrial function, increased chromatin abnormalities including DNA damage and a higher incidence of sperm aneuploidy (14, 15). The use of sperm selected by IMSI method in assisted treatment is reported to improve embryo quality, the percentage of embryos developed into blastocyst, implantation rate, clinical pregnancy rate and a has been reported to reduce miscarriage rates (12, 16–23). A recent meta-analysis of randomized control trials comparing ICSI and IMSI cycles demonstrated a statistically significant improvement in implantation and pregnancy rates and a statistically significant reduction in miscarriage rates after sperm selected through IMSI rather than conventional ICSI procedures (24).

2. Materials

2.1. Consumables

1. Centrifuge tubes, 15 ml.
2. Falcon 14 ml culture tube.
3. Glass slide, plain.
4. Syringe, 3 ml sterile.
5. Blunt Luer-lock needle 18 GA, 1.5 in.
6. Commercial medium (sperm wash medium from Cooper Surgical).
7. Isolate® (Irvine Scientific).
8. Pipette tips (10, 100, 200 and 1,000 µl).
9. Glass-bottomed dish (MatTek Corporation).
10. Holding pipette (Humagen).
11. ICSI needle (Humagen).

2.2. Solutions

1. Washed Mineral Oil (AmerisourceBergen).
2. Quinn's Advantage w/ HEPES flushing medium (Cooper Medical).
3. Serum protein substitute (SPS; Cooper Medical).
4. Polyvinylpyrrolidone (PVP) solution (MEDICULT).
5. Microscopy immersion oil.

2.3. Equipment

1. Nikon Eclipse Ti-U inverted microscope.
2. Nikon DSQi1Mc camera.
3. Narishige micromanipulators.
4. Micropipette.
5. Speed adjustable centrifuge with swinging rotor.
6. Microscope—Olympus BH2.
7. Incubator (37°C).
8. Warming plate (37°C).
9. Makler® Chamber.

3. Methods

3.1. Sperm Preparation by Density Gradient Centrifugation

3.1.1. Solutions

1. Prepare 90% working Isolate®: Add 9.0 ml isotonic Isolate® to a sterile 15 ml centrifuge tube. Add 1.0 ml commercial medium to Isolate®. Cap the tube and mix by inverting. Store at 2–6°C, expires in 2 weeks.
2. Prepare 35% working Isolate®: Add 3.5 ml isotonic Isolate® to a sterile 15 ml centrifuge tube. Add 6.5 ml commercial medium to Isolate®. Cap the tube and mix by inverting. Store at 2–6°C, expires in 2 weeks.
3. Aspirate 1.5 ml of prepared 90% Isolate® using a fresh syringe and blunt needle. Dispense into a centrifuge tube that has been labelled with the patient name and accession number. Draw a line with a marker on the tube at the level of the meniscus.
4. Aspirate 1.5 ml of prepared 35% Isolate® using a syringe and blunt needle. Gently dispense into the centrifuge tube on top of the 90% layer. Draw a line with a marker on the tube at the level of the meniscus, cap and warm on the 37°C dry block for 15–30 min before use.
5. After the semen sample has liquefied, mix by swirling the collection container then aspirate the sample using a syringe and blunt needle. Carefully transfer up to 3 ml of sample on top of the gradient column in the centrifuge tube (see Notes 1–4).
6. Centrifuge for 15 min at $300 \times g$.

7. After centrifugation remove the sperm layer, down to the mark on the tube with the syringe and blunt needle. Then remove the 35% layer to the mark in the tube with a separate syringe and blunt needle.

8. Add 3–5 ml of commercial medium to the centrifuge tube with the 90% Isolate® layer. Cap and gently invert 3–5 times to mix. Centrifuge for 10 min at $300 \times g$.

9. Decant the supernatant from the centrifuge tube. Aspirate 0.5–0.6 ml of fresh commercial medium into the tube and resuspend the sperm pellet by repeatedly aspirating and expelling the sample with the syringe until homogenous (see Note 5).

10. In oligozoospermic patients, the sperm preparation can be changed to a sperm wash procedure (see Note 3).

11. Bring the final concentration of motile sperm to 4×10^6/ml by adding commercial medium.

12. Incubate the prepared sperm at 37°C until use in IMSI.

3.2. Sperm Preparation for MSOME Retrieval (Fig. 1)

1. Place a drop of PVP just to the right of center of the glass-bottomed dish.
2. Place a vertical line of PVP in the center of the glass-bottomed dish.
3. Place six, 20 µl drops of flushing media with 3% SPS above the PVP.

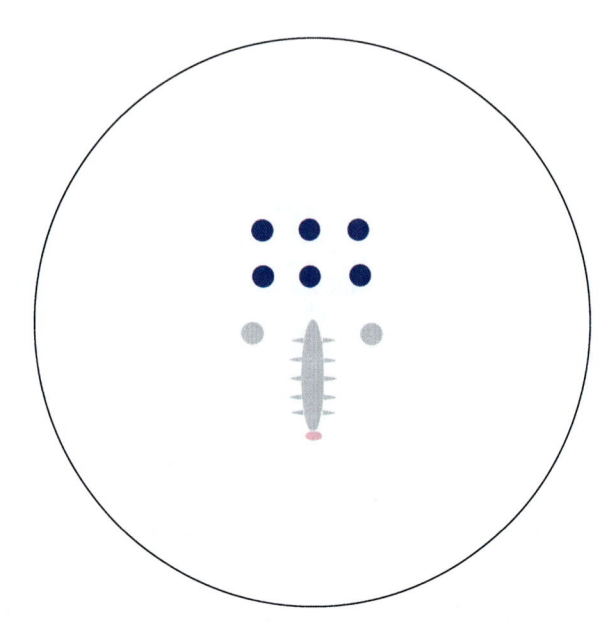

Fig. 1. A schematic overview of a petri dish prepared for IMSI procedure. *Pink*: droplet of unselected sperm, retrieved via density gradient centrifugation. *Grey*: streak and recipient droplets containing PVP solution for sperm selection and retrieval. Blue: droplets with retrieved oocyte for sperm injection.

4. Cover the entire dish with washed mineral oil to prevent evaporation and place it on a microscopic stage over the top of an objective lens previously covered by a droplet of immersion oil.

5. Place a 3 µl drop of the prepared sperm sample, containing approximately 12,000 sperm at the bottom end of the PVP line (see Notes 6–8).

6. Place the dish in the incubator and allow the sperm to swim up the PVP line for at least 5 min.

7. View the motile sperm, suspended in the observation droplet with lower magnification first and then at high magnification using the Nikon Eclipse inverted microscope (see Notes 9–12).

3.3. MSOME Criteria for Sperm Selection for IMSI

The MSOME criteria for the selection of morphologically normal sperm is according to the arbitrary descriptive approach reported by Bartoov et al. (25) comparing the ultrastructure of sperm from fertile and infertile males.

According to this criterion, the morphological normalcy of a sperm is defined by the presence of a normal nucleus, acrosome, postacrosomal lamina, mitochondria, neck, tail, and no cytoplasmic droplets around the head. In each sperm, three different parameters are evaluated: head shape and dimensions in both axes, head vacuoles and head base normality.

1. A normal shaped nucleus should be smooth, symmetric, and oval with a regular outline.

2. The average length and width limits for a normal head should be approximately 4.75 and 3.28 µm, respectively (see Note 13).

3. Any extrusion or invagination of the nuclear mass is defined as a regional nuclear shape malformation, and sperm with such features should be excluded.

4. The nuclear chromatin mass should contain no more than one vacuole, which occupies less than 4% of the nuclear area (see Figs. 2 and 3).

5. A normal head base is U-shaped.

6. The acrosome and postacrosomal lamina is considered abnormal if they are absent, partial, or vesiculated.

7. The mitochondria is considered abnormal if it is absent, partial or disorganized.

8. An abaxial neck, with the presence of disorders or cytoplasmic droplets, is considered abnormal.

9. The presence of a coiled, broken, short, or double tail is considered as abnormal.

Fig. 2. Normal sperm with one vacuole containing less than 4% of the sperm head.

Fig. 3. Abnormal sperm morphology using IMSI. Note that multiple vacuoles are present.

10. Mathematical scoring of sperm is used in some clinics to identify morphologically normal sperm (see Note 14).

3.4. Sperm Retrieval After MSOME

1. Turn on the Nikon DSQi1Mc camera and set up the ICSI tool.
2. It is also necessary to prime micropipettes with medium before use so that the selected sperm and later on the retrieved oocytes never come into contact with air or oil.
3. Prime the ICSI needle with PVP prior to sperm selection.
4. Place the retrieved, cumulus-free, MII oocytes were into drops of flushing medium with 3% SPS.
5. Focus the microscope on the interface of the PVP line to visualize the sperm.
6. Using the ICSI needle, select the sperm according to the criteria discussed in Subheading 3.3 while observing at a magnification exceeding 6,000×.

7. MSOME selected sperm are retrieved from the observation droplet and placed into a recipient selection droplet, containing 4 μl of PVP medium in the same dish.

8. The microdroplet containing selected sperm is subsequently observed using the monitor and the Nikon DSQi1Mc camera to verify that the sperm meet the MSOME criteria.

3.5. Microinjection

1. The recipient droplet should contain the sperm, morphologically selected for ICSI (Fig. 1).

2. Perform the microinjection procedure according to standard procedures at a magnification of 400×.

3. Following microinjection immediately transfer each oocyte to a culture dish, incubated in a 30 μl HEPES buffered medium droplet and covered with 0.5 ml of mineral oil at 37°C in an atmosphere of 5% CO_2 and 6% O_2.

4. Notes

1. Record volume of the semen sample, motility and concentration. Note any abnormal parameters such as agglutination, amorphous cells, bacteria, red blood cells, and/or altered viscosity.

2. If the semen sample is 3 ml or less, the semen can be directly layered onto the gradient column. Volumes greater than 3 ml need to be either washed or divided into smaller volumes and layered on separate columns.

3. Washing of semen: Add 5–12 ml commercial medium to achieve a ratio of at least two parts medium to one part semen in a 15 ml centrifuge tube. Cap the tube and invert 3–5 times to mix. Centrifuge for 10 min at 300×g. Decant the supernatant from the centrifuge tube. Aspirate 0.8–0.9 ml of fresh commercial medium into the syringe and gently resuspend the pellet by repeatedly aspirating and expelling the sample with the syringe until homogenous.

4. Add the gradient and semen sample slowly taking care not to disturb the interface formed between the layers.

5. Perform the motility evaluation and determine the concentration. Record the results. Make a thin smear on a labelled slide for a morphology assessment.

6. The temperature of the sperm sample and the PVP concentration should be coordinated with the intensity of the sperm motility. With poor sperm motility, the sample temperature

should be raised to 37°C, on the other hand, with high sperm motility, the suspension temperature can be lowered to 20°C.

7. To reduce PVP toxicity, which has to be used to slow down highly motile sperm from disappearing from the monitor screen, the concentration of the PVP is adjusted to a minimum (range 0–8%). With poor sperm motility, no PVP is added to the culture medium, while 6% human serum albumin is added to the recipient droplet. However, with normal motility samples PVP should be added to a final concentration of 8%.

8. In order to gain control of the moving sperm cells and facilitate their morphological evaluation, a series of small bays extruding from the rim of the droplets can be created to block the heads of the motile sperm (Fig. 1).

9. To observe every single sperm at high magnification, an inverted microscope equipped with Nomarski differential interference contrast optics, an Uplan Apo oil/1.35 objective lens, and a 0.55 NA condenser lens should be used.

10. This system should be coupled with an image capturing high definition color video camera and a high definition color video monitor for morphological assessment.

11. The resulting magnification is based on four parameters: (1) objective magnification × 100; (2) magnification selector × 1.5; (3) video coupler magnification × 0.99; (4) calculated video magnification (CCD × monitor diagonal dimension) × 44.45. Therefore, total magnification = 100 × 1.5 × 0.99 × 44.45 = × 6,600.

12. In order to estimate the morphological state of the sperm nucleus, one has to follow the motile sperm cell by moving the microscopic stage in the x, y and z directions for about 20 s.

13. Computer assisted sperm selection is used in some clinics where a fixed, celluloid form of a sperm nucleus representing the normal criteria is superimposed on the examined cell; any sperm cell with a normal nuclear shape has to be excluded from selection if it varies in length or width by 2 SDs from the normal mean values.

14. The head is scored according to Knez et al. (20) as follows: 1 point for normality in one or two axes, and 0 points if both axes are abnormal. Sperm with no vacuoles in the head or a single small diameter vacuole are scored with 1 point, and sperm with larger vacuoles score 0 points. Sperm with a normal shape of the sperm head base is scored with 1 point and abnormal with 0 points. High-quality sperm with a calculated score of 4–6 should be used for injection based on the following formula for scoring: Score of sperm = (2 × head score) + (3 × vacuole score) + (base score).

Acknowledgment

This work was supported by University of Utah, and Andrology and IVF laboratories.

References

1. Kruger TF et al (1987) New method of evaluating sperm morphology with predictive value for human in vitro fertilization. Urology 30:248–251
2. Kruger TF et al (1986) Sperm morphologic features as a prognostic factor in in vitro fertilization. Fertil Steril 46:1118–1123
3. Donnelly ET et al (1998) In vitro fertilization and pregnancy rates: the influence of sperm motility and morphology on IVF outcome. Fertil Steril 70:305–314
4. Huszar G, Vigue L (1993) Incomplete development of human spermatozoa is associated with increased creatine phosphokinase concentration and abnormal head morphology. Mol Reprod Dev 34:292–298
5. Kahraman S et al (2006) Preliminary FISH studies on spermatozoa and embryos in patients with variable degrees of teratozoospermia and a history of poor prognosis. Reprod Biomed Online 12:752–761
6. Faure AK et al (2007) Predictive factors for an increased risk of sperm aneuploidies in oligo-astheno-teratozoospermic males. Int J Androl 30:153–162
7. Miller JE, Smith TT (2001) The effect of intracytoplasmic sperm injection and semen parameters on blastocyst development in vitro. Hum Reprod 16:918–924
8. De Vos A et al (2003) Influence of individual sperm morphology on fertilization, embryo morphology, and pregnancy outcome of intracytoplasmic sperm injection. Fertil Steril 79:42–48
9. Bartoov B et al (2002) Real-time fine morphology of motile human sperm cells is associated with IVF-ICSI outcome. J Androl 23:1–8
10. Antinori M et al (2008) Intracytoplasmic morphologically selected sperm injection: a prospective randomized trial. Reprod Biomed Online 16:835–841
11. Bartoov B et al (2003) Pregnancy rates are higher with intracytoplasmic morphologically selected sperm injection than with conventional intracytoplasmic injection. Fertil Steril 80:1413–1419
12. Hazout A et al (2006) High-magnification ICSI overcomes paternal effect resistant to conventional ICSI. Reprod Biomed Online 12:19–25
13. Berkovitz A et al (2005) The morphological normalcy of the sperm nucleus and pregnancy rate of intracytoplasmic injection with morphologically selected sperm. Hum Reprod 20:185–190
14. Franco JG Jr et al (2008) Significance of large nuclear vacuoles in human spermatozoa: implications for ICSI. Reprod Biomed Online 17:42–45
15. Garolla A et al (2008) High-power microscopy for selecting spermatozoa for ICSI by physiological status. Reprod Biomed Online 17:610–616
16. Cohen-Bacrie P et al (2007) Indications for IMSI. J Gynecol Obstet Biol Reprod (Paris) 36(Suppl 3):S105–108
17. Vanderzwalmen P et al (2008) Blastocyst development after sperm selection at high magnification is associated with size and number of nuclear vacuoles. Reprod Biomed Online 17:617–627
18. Cassuto NG et al (2009) A new real-time morphology classification for human spermatozoa: a link for fertilization and improved embryo quality. Fertil Steril 92:1616–1625
19. Balaban B et al (2011) Clinical outcome of intracytoplasmic injection of spermatozoa morphologically selected under high magnification: a prospective randomized study. Reprod Biomed Online 22:472–476
20. Knez K et al (2011) The IMSI procedure improves poor embryo development in the same infertile couples with poor semen quality: a comparative prospective randomized study. Reprod Biol Endocrinol 9:123
21. Oliveira JB et al (2011) Pregnancy outcomes in women with repeated implantation failures after intracytoplasmic morphologically selected sperm injection (IMSI). Reprod Biol Endocrinol 9:99
22. Sermondade N et al (2011) Successful childbirth after intracytoplasmic morphologically

selected sperm injection without assisted oocyte activation in a patient with globozoospermia. Hum Reprod 26:2944–2949

23. Wilding M et al (2011) Intracytoplasmic injection of morphologically selected spermatozoa (IMSI) improves outcome after assisted reproduction by deselecting physiologically poor quality spermatozoa. J Assist Reprod Genet 28:253–262

24. Setti AS et al (2010) Intracytoplasmic sperm injection outcome versus intracytoplasmic morphologically selected sperm injection outcome: a meta-analysis. Reprod Biomed Online 21:450–455

25. Bartoov B et al (1994) Improved diagnosis of male fertility potential via a combination of quantitative ultramorphology and routine semen analyses. Hum Reprod 9:2069–2075

Chapter 23

Sperm Selection for ICSI Using Annexin V

Sonja Grunewald and Uwe Paasch

Abstract

Annexin-V magnetic-activated cell sorting (MACS) is a new tool to optimize sperm selection in assisted reproduction. The technique is based on the binding of superparamagnetic Annexin-microbeads to externalized phosphatidylserine at the outer leaflet of the plasma membrane of sperm with activated apoptosis signaling or membrane damage. The combination of Annexin-V MACS and density gradient centrifugation was demonstrated to enhance clinical pregnancy rates in ICSI cycles. This chapter focuses on the practical details of Annexin-V MACS.

Key words: Annexin V, MACS, ICSI, Sperm, Apoptosis

1. Introduction

Intracytoplasmic sperm injection (ICSI) into oocytes is to date the gold standard for treatment of severe male factor infertility, and indications for this procedure are expanding (1). However, the current success rates of the procedure remain suboptimal. What might be the limiting factors?

Diagnosis of male factor infertility is based largely on the spermiogram parameters of viability, motility, and morphology. Moreover, the same parameters are used for the selection of the injected sperm (2). These criteria are clearly inadequate to detect abnormalities at the molecular level, which may impact fertilization. Multiple studies have established that spermatozoa in patients diagnosed with male infertility display a higher incidence of apoptotic features (3–7). Because spermatozoa are transcriptionally inactive cells and the DNA is densely packed, the apoptosis process is different from that in somatic cells (8, 9). Nevertheless, key features of activated apoptosis signalling like disruption of the mitochondrial transmembrane potential, activation of caspase-3, externalization of phosphatidylserine, and abnormalities such as sperm DNA fragmentation are present in human sperm and have been directly linked to failure of fertilization during assisted reproduction (10–12). Therefore, sperm preparation protocols based on the selection of nonapoptotic cells might be a useful option to improve ICSI success rates.

Apoptotic cells can be depleted from a cell suspension by selection based on the externalization of phosphatidylserine (EPS) from the inner to the outer leaflet of the plasmamembrane. EPS marks the beginning of the terminal phase of apoptosis.

The covalent binding of Annexin V, a phospholipid binding protein that has high affinity for phosphatidylserine and lacks the ability to pass through an intact sperm membrane, can be used to label sperm with EPS (13). Subsequently, Annexin V-conjugated superparamagnetic microbeads can be used effectively to separate nonapoptotic spermatozoa from those with deteriorated plasma membranes based on the externalization of phosphatidylserine using magnetic-activated cell separation (Annexin V MACS). The separation of sperm yields two fractions: EPS-negative (vital, nonapoptotic sperm with intact membranes) and EPS-positive (14, 15). In the EPS-positive sperm subpopulation apoptotic cells with externalized phosphatidylserine are found, but the Annexin V-conjugated microbeads also label dead cells and cells with exposed inner leaflet of the plasma membrane—e.g., acrosome-reacted sperm. Repeated investigations did not reveal any annexin V microbeads in the EPS-negative sperm fraction (16). The separation process itself does not affect the sperm integrity or viability (14, 16, 17); however, one additional centrifugation step might reduce motility particularly in altered sperm from subfertile patients (18).

Combination of Annexin V-MACS with density gradient centrifugation allows superior selection of nonapoptotic spermatozoa compared to Annexin V MACS or density gradient centrifugation alone (18–20). Thus, it is most feasible to integrate the Annexin V MACS into standard sperm preparation protocols of IVF labs.

2. Materials

All buffers and solutions required for Annexin V separation are commercially available in one kit (Death cell removal kit, Miltenyi Biotec, Bergisch Gladbach, Germany). Although the first clinical studies do not report any side effects (20, 21), the kit is currently designated for research use only. It will be available soon as a GMP conforming product (MACS® GMP Annexin V Kit, Miltenyi Biotec, Bergisch Gladbach, Germany).

1. Standard IVF Lab centrifuge
2. Magnetic cell separator MiniMACS™ (for a single separation) or OctoMACS™ (for up to eight separations at the same time).
3. MS MACS Columns.
4. Components of the MACS® Dead Cell Removal Kit
   - 10 mL Annexin V Microbeads.
   - 25 mL MACS 20× Binding Buffer stock solution (The 20×Binding Buffer has to be diluted 20-fold with sterile, double distilled water prior to use.).

Per semen sample (containing up to ~50 million sperm) 100 µL Annexin V Microbeads and one MS column are necessary to separate EPS negative and positive sperm. The volume of the Binding buffer varies depending on the volume of the semen sample (see Note 1).

3. Methods

For optimal sperm selection in ICSI it is highly recommended that the Annexin V MACS separation be integrated with the standard sperm preparation procedure and combined with density gradient centrifugation. Annexin V MACS can be performed before or after the density gradient centrifugation (see Note 2).

1. Dilute the sperm sample 1:2 in Binding buffer.
2. Centrifuge the sperm suspension for 4 min at $300 \times g$.
3. Remove supernatant.
4. Resuspend the sperm pellet in 400 µL Annexin V Binding Buffer.
5. Add 100 µL Annexin V conjugated superparamagnetic microbeads, mix gently, and incubate for 15 min at room temperature (19–25°C).
6. Preparation of the magnetic separation should be started a few minutes before the end of incubation time:
   (a) Place the MS column in the magnet and rinse it with 1.0 mL Annexin V binding buffer to keep the column moist.
   (b) Put a new suitable vial below the column, in which the EPS negative sperm fraction will be collected.
7. Load the sperm/microbead suspension onto the rinsed MS column. Wait until the suspension has passed through completely. Labelled, EPS-positive sperm are retained in the column by

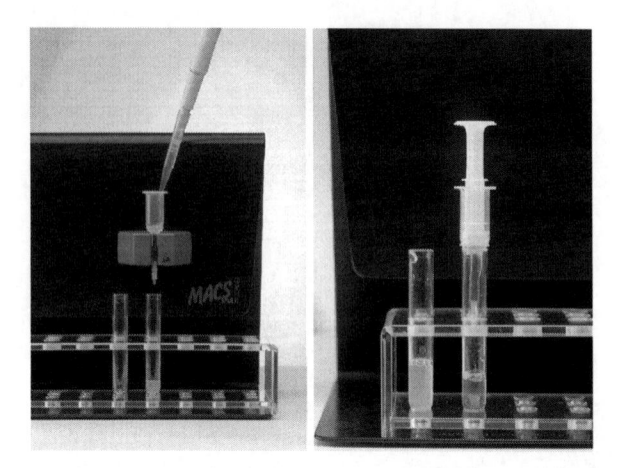

Fig. 1. Annexin V MACS separation. *Left*: Generation of EPS-negative sperm by depleting EPS-positive sperm with Annexin V microbeads in the MS column in the magnetic field. *Right*: After collection of EPS-negative sperm, EPS-positive sperm might be obtained for research or diagnostic purposes by removing the MS column from the magnetic field and rinsing it with 1 mL buffer with pressure.

the magnetic field, while nonlabelled, EPS-negative sperm flow through and are collected in the vial placed below the column (see Fig. 1).

8. Optional, the MS column can be rinsed with 500 µL Annexin V binding buffer to elute all EPS-negative sperm (see Notes 3 and 4).

9. Remove the MS column with the retained EPS-positive sperm from the magnetic field. Add 1.0 mL Annexin V-binding buffer onto the column and rinse with pressure (see Note 5 and 6 and Fig. 1).

4. Notes

1. The protocol should work for semen samples up to 50—(maximum 100) million sperm in total. For more sperm in one sample it is recommended that the sample be split and the separation process performed twice.

2. As mentioned in the introduction, many studies have demonstrated the superiority of combining Annexin-V MACS with density gradient centrifugation compared to Annexin V MACS and density gradient centrifugation alone (19, 22, 23). Annexin V MACS can be performed before or after density gradient centrifugation to prepare human sperm for ICSI (18, 20). The main advantage of performing it before density gradient

centrifugation is to save one centrifugation step. The Annexin V binding buffer is a HEPES buffer which needs to be removed before ICSI—this step would be performed by density gradient centrifugation.

3. Rinsing with 500 μL Annexin V binding buffer after the sperm suspension has passed the column can be omitted if the sperm concentration is very low to avoid further sperm dilution before ICSI.

4. Sometimes the MS column can be blocked by the sperm-microbead suspension. This happens particularly in semen samples with gelatinous masses, or if too many sperm are used per separation column. In case the column is blocked the sperm microbead suspension may be retrieved, and the separation can be reinitiated with a fresh column. The risk for a blockage of the column appears to be reduced if density gradient centrifugation is performed before the Annexin V MACS (see Note 2).

5. If needed the EPS-positive sperm fraction can be collected, e.g., for research or diagnostic purposes. EPS-positive sperm are not suitable for ICSI.

6. Cryopreservation and thawing increases the percentages of sperm with activated apoptosis signalling including externalization of phosphatidylserine (7, 24, 25). It is also beneficial to include the Annexin V MACS procedure in cryopreservation-thawing protocols of sperm designated for ICSI. Annexin V MACS can be performed before cryopreservation. Sperm cryotolerance is increased in EPS-negative compared to EPS-positive sperm (26, 27). Moreover, sperm with activation of apoptosis signalling and severe membrane damage due to the cryopreservation and thawing procedure can be depleted when the Annexin V MACS is performed after thawing (14, 16, 24, 25, 28).

References

1. Jain T, Gupta RS (2007) Trends in the use of intracytoplasmic sperm injection in the United States. N Engl J Med 357:251–257
2. Henkel RR, Schill WB (2003) Sperm preparation for ART. Reprod Biol Endocrinol 1:108
3. Sakkas D et al (1999) Abnormal sperm parameters in humans are indicative of an abortive apoptotic mechanism linked to the Fas-mediated pathway. Exp Cell Res 251:350–355
4. Gandini L et al (2000) Study of apoptotic DNA fragmentation in human spermatozoa. Hum Reprod 15:830–839
5. Taylor SL et al (2004) Somatic cell apoptosis markers and pathways in human ejaculated sperm: potential utility as indicators of sperm quality. Mol Hum Reprod 10:825–834
6. Barroso G et al (2006) Mitochondrial membrane potential integrity and plasma membrane translocation of phosphatidylserine as early apoptotic markers: a comparison of two different sperm subpopulations. Fertil Steril 85:149–154
7. Grunewald S et al (2005) Sperm caspases become more activated in infertility patients

than in healthy donors during cryopreservation. Arch Androl 51:449–460
8. Grunewald S et al (2005) Mature human spermatozoa do not transcribe novel RNA. Andrologia 37:69–71
9. Aitken RJ, Koppers AJ (2011) Apoptosis and DNA damage in human spermatozoa. Asian J Androl 13:36–42
10. Henkel R et al (2004) Influence of deoxyribonucleic acid damage on fertilization and pregnancy. Fertil Steril 81:965–972
11. Grunewald S et al (2008) Relationship between sperm apoptosis signalling and oocyte penetration capacity. Int J Androl 31:325–330
12. Borini A et al (2006) Sperm DNA fragmentation: paternal effect on early post-implantation embryo development in ART. Hum Reprod 21:2876–2881
13. van Heerde WL et al (1995) The complexity of the phospholipid binding protein Annexin V. Thromb Haemost 73:172–179
14. Grunewald S et al (2001) Enrichment of non-apoptotic human spermatozoa after cryopreservation by immunomagnetic cell sorting. Cell Tissue Bank 2:127–133
15. Glander HJ et al (2002) Deterioration of spermatozoal plasma membrane is associated with an increase of sperm lyso-phosphatidylcholines. Andrologia 34:360–366
16. Paasch U et al (2003) Deterioration of plasma membrane is associated with activated caspases in human spermatozoa. J Androl 24:246–252
17. Said TM et al (2008) Utility of magnetic cell separation as a molecular sperm preparation technique. J Androl 29:134–142
18. Grunewald S et al (2009) Increased sperm chromatin decondensation in selected non-apoptotic spermatozoa of patients with male infertility. Fertil Steril 92:572–577
19. Said TM et al (2005) Advantage of combining magnetic cell separation with sperm preparation techniques. Reprod Biomed Online 10:740–746
20. Dirican EK et al (2008) Clinical outcome of magnetic activated cell sorting of non-apoptotic spermatozoa before density gradient centrifugation for assisted reproduction. J Assist Reprod Genet 25:375–381
21. Rawe VY et al (2010) Healthy baby born after reduction of sperm DNA fragmentation using cell sorting before ICSI. Reprod Biomed Online 20:320–323
22. Said T et al (2006) Selection of nonapoptotic spermatozoa as a new tool for enhancing assisted reproduction outcomes: an in vitro model. Biol Reprod 74:530–537
23. de Vantery Arrighi C et al (2009) Removal of spermatozoa with externalized phosphatidylserine from sperm preparation in human assisted medical procreation: effects on viability, motility and mitochondrial membrane potential. Reprod Biol Endocrinol 7:1
24. Paasch U et al (2004) Activation pattern of caspases in human spermatozoa. Fertil Steril 81(Suppl 1):802–809
25. Said TM et al (2010) Implication of apoptosis in sperm cryoinjury. Reprod Biomed Online 21:456–462
26. Said TM et al (2005) Effects of magnetic-activated cell sorting on sperm motility and cryosurvival rates. Fertil Steril 83:1442–1446
27. Grunewald S et al (2006) Magnetic-activated cell sorting before cryopreservation preserves mitochondrial integrity in human spermatozoa. Cell Tissue Bank 7:99–104
28. Paasch U et al (2004) Cryopreservation and thawing is associated with varying extent of activation of apoptotic machinery in subsets of ejaculated human spermatozoa. Biol Reprod 71:1828–1837

Chapter 24

Sperm Selection for ICSI Using the Hyaluronic Acid Binding Assay

Mohammad Hossein Nasr-Esfahani and Tavalaee Marziyeh

Abstract

Sperm selection is an important part of the ICSI procedure. Routinely, sperm selection for ICSI is based solely on sperm morphology and motility. These latter parameters may not be sufficient to select sperm with intact chromatin. Therefore, sperm selected based on sperm functional characteristics may result in the most appropriate sperm for the ICSI procedure. The methodology explained below describes the selection of sperm based on the ability of sperm to bind solid-state hyaluronic acid as its receptor, present on mature sperm with intact chromatin.

Key words: Sperm selection, Intra cytoplasmic sperm injection, Hyaluronic acid, Motility

1. Introduction

Following the first pregnancy achieved by intracytoplasmic injection of a single spermatozoon into an oocyte (ICSI) in 1992, this procedure has been considered as a breakthrough in fertility treatment (1). Despite the fact that this procedure opens doors to parenthood for hopeless couples with male factor infertility, its success is limited. In addition, in some of these cases, there is a price to pay for the effectiveness of the intra cytoplasmic sperm injection (ICSI) procedure including the encouragement of vertical transmission of male subfertility to the next generation, increases in miscarriage and birth defect rates, and potentially reduced health and well-being of the offspring (2, 3). A novel sperm selection technique based on the ability of sperm to bind hyaluronic acid was recently developed in an effort to minimize some of the shortcomings of the ICSI procedure, especially in terms of chromatin and DNA integrity, which have a profound impact on fertilization, embryo and fetal development and may affect the health and the well-being of offspring (4–10).

Hyaluronan or hyaluronic acid was discovered by Karl Meyer in the 1930s, as a polymer of disaccharides composed of D-glucuronic acid and D-N-acetylglucosamine, linked via alternating β-1,4 and β-1,3 glycosidic bonds (11). Hyaluronan is secreted by many tissues including cumulus cells, which are embedded in a thick matrix of polymerized hyaluronic acid. In the reproductive tract, two main receptors have been recognized for HA, the receptor for HA-mediated motility (RHAMM) and PH-20. PH-20 is present on the plasma membrane and on the inner acrosomal membrane of mammalian sperm heads (11). Several functions have been envisaged for PH-20 including hyaluronidase activity, intracellular signaling, and zona pellucida adhesion and penetration. It is believed that its interaction with HA in the cumulus oophorus primes the sperm for interaction with zona pellucida proteins. Therefore, sperm heads carrying PH-20 can bind to hyaluronan (12). The formation of zona binding sites and HA binding sites takes place during the late stages of spermatogenesis concomitant with cytoplasmic extrusion and nuclear histone–protamine replacement. Therefore, only mature sperm can bind to HA (13). Thus, taking this property along with the sperm tail movement, it is possible to select or separate sperm that have the ability to bind HA. Benefits of sperm selected based on HA are well covered by the famous work of Hauzer and colleagues in the literature, and presently there are two commercial products available for the selection of sperm based on HA binding potential (14, 15). The section below describes a procedure for the assessment of in-vitro sperm HA binding. Although these methods can be implemented for assessment of sperm HA binding potential and sperm selection, the commercial products are recommended for clinical applications.

2. Materials

1. PureSperm® 40 (Nidacon International AB; Sweden).
2. PureSperm® 80 (Nidacon International AB; Sweden).
3. PureSperm® wash (Nidacon International AB; Sweden).
4. Hyaluronic acid (Juvederm 30, LEA Derm; Paris, France).
5. Mineral oil (Vitrolife; Sweden).
6. G-MOPS™ medium (Vitrolife; Sweden).
7. ICSI™ (PVP with recombinant human albumin; Vitrolife; Sweden).
8. Micromanipulators.
9. Inverted microscope.

3. Methods

3.1. Preparation of Sperm Sample for ICSI Technique

1. Transfer liquefied semen to a conical tube.
2. Add 2 ml of PureSperm® wash to a conical tube and mix gently (see Notes 1 and 2).
3. Spin the conical tube into the centrifugation for 5 min at 300×*g* (see Note 3).
4. Remove supernatant and resuspend the sperm pellet in a suitable volume of PureSperm® wash (1 ml) and mix well.
5. Calculate sperm concentration using Makler counting chamber.

3.2. Density Gradient Centrifugation

1. Add 2 ml of 80% PureSperm® to a new conical tube using a sterile pipette (lower layer).
2. Add 2 ml of 40% PureSperm® gently on top of 80% PureSperm® using a sterile pipette (upper layer) (see Note 4).
3. Place the washed semen sample on top of the upper layer.
4. Centrifuge for 20 min at 300×*g*.
5. Carefully aspirate the supernatant using a sterile pipette.
6. Transfer the sperm pellet to a new tube with a new pipette.
7. Resuspend the pellet in 5 ml G-MOPS™, and centrifuge at 300×*g* for 5 min.
8. Repeat step 7 with a second wash.
9. Aspirate G-MOPS™ supernatant.
10. Add G-MOPS™ to sperm pellet, and adjust sperm concentration (see Note 1) using Maker counting chamber to $0.2–1 \times 10^6$ cells in 1 ml (see Note 5).

3.3. Preparation of HA-Coated Dishes for (ICSI) Technique

1. Dilute hyaluronic acid with sterile water (1:40) under sterile conditions (see Note 5).
2. Place 10 or 50 µl drops in a 50 mm Falcon dish.
3. Allow to dry in sterile conditions (under a laminar hood).
4. Replace the lid and store the dishes at 4°C.

3.4. HA-ICSI Procedure

1. Prepare the ICSI dish (or other alternative) as routinely prepared, using ICSI™ or other PVP alternatives, G-MOPS™ and mineral oil, and cover the coated HA spots with 15 or 75 µl of G-MOPS™ or other suitable diluents, according to the initial size of HA drop (Fig. 1).
2. Wash HA drops twice with G-MOPS™ or other suitable sperm diluents to remove excessive or uncoated HA (see Note 6).

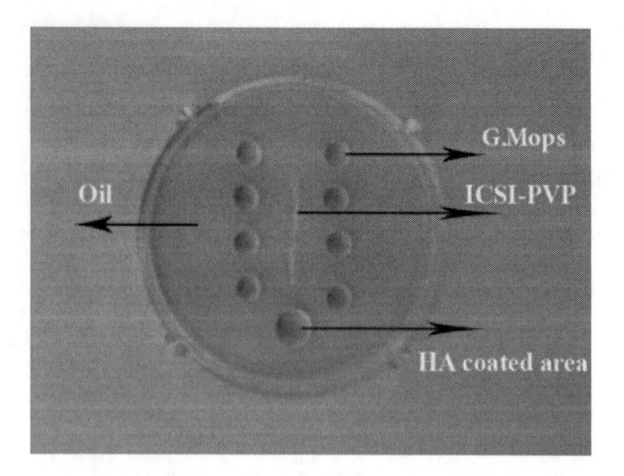

Fig. 1. Hyaluronic acid coated dish prepared for ICSI.

3. Remove the G-MOPS™ medium from the HA coated dish and replace with 50 µL of G-MOPS™ or other suitable sperm diluents containing around 10,000–50,000 processed sperm.
4. Transfer denuded oocytes into G-MOPS™ drops in the ICSI dish.
5. Select several HA-bound sperm with normal morphology with the aid of an ICSI needle. HA-bound sperm show vigorous tail movement without forward motility, while HA-unbound sperm present progressive movement (see Notes 7 and 8).
6. Transfer the selected sperm into the ICSI™ drop in the center of the ICSI dish, immobilizing or inactivating the tail of the sperm.
7. Wash the sperm in different regions of the ICSI™ drop.
8. Select the best sperm based on morphology.
9. Load a single sperm into the ICSI needle for injection into the oocytes according to your routine injection protocol (see Notes 9–11).

4. Notes

1. If the sample is highly viscous, add a higher volume of PureSperm® wash and mix gently.
2. Prepare two PureSperm® gradients for each semen sample, when volumes of sample are greater than 3 ml.
3. After centrifugation, if you do not see any pellet, remove all fluid except the lowest 0.5 ml.

4. Try not to interrupt the two layers and to preserve a sharp interface between them.

5. Juvederm 30 composition: cross-linked hyaluronic acid (24 µg) in phosphate buffer pH 7.2 (1 ml).

6. When adding or washing the HA drop, try to avoid contact of the micropipette with the center of the spot by adding or removing medium from the edge of the drop.

7. When possible, avoid selecting loosely bound sperm or sperm which is temporarily bound to HA and then show progressive movement.

8. Avoid selecting immotile sperm without tail movement as HA-bound sperm.

9. During selection and aspiration of HA-bound sperm into the ICSI needle, use care not to pierce and tear the coated HA with the ICSI needle.

10. In most cases HA-bound sperm can be found within minutes, when sperm is added to the HA-drop; however, in some cases more time is required to select a suitable number of sperm.

11. Due to uneven hydration of HA or duration of the washing procedure, areas of coated HA might be removed. Try to avoid selecting sperm from these areas. The HA-bound sperm can be easily distinguished by its tail movement and nonprogressive motility.

Acknowledgments

The authors express their gratitude to the Royan Institute for its financial support, as well as to the staff of Isfahan Fertility and Infertility Center for their kind collaboration.

References

1. Palermo G et al (1992) Pregnancies after intracytoplasmic injection of single spermatozoon into an oocyte. Lancet 340:17–18
2. Hansen M et al (2008) Admission to hospital of singleton children born following assisted reproductive technology (ART). Hum Reprod 23:1297–1305
3. Aitken RJ, Koppers AJ (2011) Apoptosis and DNA damage in human spermatozoa. Asian J Androl 13:36–42
4. Razavi S et al (2003) Effect of human sperm chromatin anomalies on fertilization outcome post-ICSI. Andrologia 35:238–243
5. Nasr-Esfahani MH et al (2008) Evaluation of sperm selection procedure based on hyaluronic acid binding ability on ICSI outcome. J Assist Reprod Genet 25:197–203
6. Fleming SD et al (2008) Prospective controlled trial of an electrophoretic method of sperm preparation for assisted reproduction: comparison with density gradient centrifugation. Hum Reprod 23:2646–2651
7. Deemeh MR et al (2007) Evaluation of protamime deficiency and DNA fragmentation in two globozoospermia patients undergoing ICSI. Iran J Fertil Steril 1:85–88

8. Tavalaee M et al (2009) Influence of sperm chromatin anomalies on assisted reproductive technology outcome. Fertil Steril 91:1119–1126
9. Khajavi NA et al (2009) Can Zeta sperm selection method recover sperm with higher DNA integrity to compare to density gradient centrifugation? Iran J Reprod Med 7:73–77
10. Deemeh MR et al (2010) The first report of successfully pregnancy after ICSI with combined DGC/Zeta sperm selection procedure in a couple with eleven repeated fail IVF/ICSI cycles. Iran J Fertil Steril 4:41–43
11. Garg HG, Hales CA (2004) Chemistry and biology of hyaluronan, Hardbound, Elsevier Ltd.
12. Li MW et al (2002) Importance of glycosylation and disulfide bonds in hyaluronidase activity of macaque sperm surface PH-20. J Androl 23:211–219
13. Park CY et al (2005) Increase of ICSI efficiency with hyaluronic acid binding sperm for low aneuploidy frequency in pig. Theriogenology 64:1158–1169
14. Prinosilova P et al (2009) Selectivity of hyaluronic acid binding for spermatozoa with normal Tygerberg strict morphology. Reprod Biomed Online 18:177–183
15. Yogev L et al (2010) Assessing the predictive value of hyaluronan binding ability for the freezability potential of human sperm. Fertil Steril 93:154–158

Chapter 25

Sperm Selection Based on Electrostatic Charge

Luke Simon, Shao-qin Ge, and Douglas T. Carrell

Abstract

Charge is a fundamental property of all forms of matter that exhibit attraction or repulsion in the presence of another charged particle. This electrokinetic property occurs when the particles exhibiting a net negative or positive charge are subjected to an external electric field that exerts an electrostatic force between them. Sperm surface membranes exhibit varying levels of electrostatic potential that are proportional to the levels of sialic acid residue acquired on the cell surface during maturation. Electrostatic charge-based sperm separation is a recently developed technique that uses an electric field to isolate mature sperm with reduced levels of DNA fragmentation. Two methods for the separation of sperm based on electrostatic charge, the Zeta method and a commercially available electrophoretic method using the SpermSep Cell Sorter 10, are discussed in this chapter including a detailed protocol for sperm separation based on the Zeta method.

Key words: DNA fragmentation, Electrostatic charge, Electrophoretic sperm separation, Mature sperm, Sperm plasmalemma, Zeta potential

1. Introduction

1.1. Electrostatic Properties of Sperm

The sperm membrane plays a major role in sperm maturation and is important for cellular interactions, such as capacitation, cell-to-cell recognition, and sperm–egg interaction during fertilization. The membrane is covered by a negatively charged 20–60 nm thick coating and the glycocalyx to facilitate the interaction with its extracellular environment. Mature sperm possess an electric charge of −16 to −20 mV (1). The highly negatively charged glycocalyx on the sperm plasma membrane helps to prevent self-agglutination and nonspecific binding with the genital tract epithelium during its transport and storage (2).

Studies on sperm surface membrane reveal a surprisingly highly organized topography (3, 4) not found in most somatic cells. Electron microscopy observations of mature sperm have revealed the presence of an electro-dense layer containing branch-like electro-dense strands (5) indicating a strong negative membrane potential. In a normal mature sperm, the membrane glycocalyx is rich in sialic acid (6). High levels of sialic acid residues on the sperm membrane increase its net negative charge, which may play a role in capacitation and the formation of binding bridges between sperm membrane proteins and the oocyte (5).

In mammalian species, CD52 is a bipolar glycopeptide of epididymal origin that forms a major component of the sperm glycocalyx (2). CD52 is a highly sialated glycosylphosphatidylinositol anchored protein that is acquired during epididymal transit and is located on the sperm plasmalemma (7). During sperm maturation, the lipid-anchored CD52 antigen is transferred to the sperm's membrane resulting in its net negative charge (8). Giuliani et al. (7) reported that the expression of CD52 is higher in the sperm of fertile men compared with a sub-fertile group. The same group also showed that CD52 expression is associated with capacitation and normal sperm morphology.

The presence of high concentrations of sialic acid residue in the sperm's membrane may be indicative of normal spermatogenesis and maturation status of sperm. If this is true, mature sperm would be expected to have a higher net negative charge compared to abnormal and immature sperm. Recently, Aitken's research group demonstrated that sperm selection based on high net negative charge results in isolation of sperm that are mature, viable, motile, morphologically normal, and relatively free of DNA damage (9, 10). The difference in charge exhibited by the sperm plasmalemma is exploited by two methods of sperm separation: a simpler version known as Zeta test (11, 12) and a more sophisticated model electrophoretic sperm separation (13, 14). These two methods are presented in this chapter.

1.2. Sperm Selection Using Zeta Test

The negative electrical charge of the sperm's membrane, known as Zeta potential or electrokinetic potential, causes the sperm to adhere to surfaces (ICSI needle/plate) in a protein-free medium. This can be avoided by using serum-supplemented medium for the process of priming. Zeta potential was first identified by Ishijima et al. (1), and is defined as the electrostatic potential between the sperm membrane and the surrounding medium. Thus sperm selection based on its surface charges was established as the Zeta method. The Zeta potential is showed to be higher in mature sperm with intact chromatin and low DNA damage (11). Chan et al. (11) were the first to develop the Zeta test to select sperm according to the sperm's electrical potential. They demonstrated the method's efficiency to recover morphologically normal and mature sperm with low levels of histone retention. The Zeta potential is known to decrease following capacitation (15). The presence of neuraminidase in the uterus and follicular fluid hydrolyzes the extracellular terminus of sialic acid, resulting in reduced surface charge on the sperm membrane (16, 17).

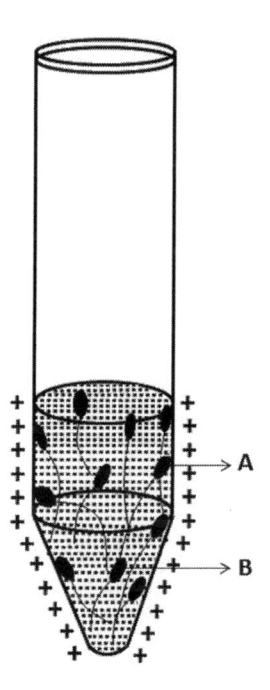

Fig. 1. Diagram of sperm selection using Zeta test. (a) Negatively charged mature sperm will adhere to the positively charged tube surface. (b) Non-mature sperm that are not adhered to the tube surface will be discarded.

The Zeta method is inexpensive, easy to perform, and does not require any complex equipment or an electrophoresis unit (see Fig. 1). It has been demonstrated that the highest quality sperm in an ejaculate display the most electronegativity (2, 7). A recent study reported that sperm selected based on Zeta potential are more mature when assessed for markers such as protamine content, ability to resist DNA fragmentation, apoptotic markers such as terminal deoxynucleotidyl transferase-mediated deoxyuridinetriphosphate nick-end labeling (TUNEL), or acridine orange (18). This method of sperm selection isolates sperm with significantly improved morphology, hyperactivation, DNA integrity, and maturity, compared with control samples processed by density gradient centrifugation (DGC); however, the process of sperm binding to the surface charge of the container reduces its motility (11, 12, 18–21). Another advantage of the Zeta method is that it can be performed effectively on cryopreserved sperm (12). In addition, sperm selected with the Zeta method have higher fertilization, implantation, and pregnancy rates (18, 22). Despite its advantages, the low sperm recovery rate is a limitation for oligozoospermic patients. In addition, this method of sperm selection may not be suitable for testicular or caput epididymal sperm (11) as sperm from these sources lack sufficient net electrical charge on the sperm membrane (23).

1.3. Sperm Preparation by Electrophoresis

Researchers at the University of Newcastle in New South Wales, Australia, developed a new technique of sperm selection known as electrophoresis sperm separation based on the idea that negatively charged sperm are mature and have normal chromatin stability (13).

Such negatively charged sperm when suspended in an electrophoretic buffer are attracted to the positive electrode. The electrophoretic unit uses this property to isolate negatively charged sperm in a large quantity (24). In this method, the electrophoretic unit applies an electric potential through the chambers to separate sperm across a separation membrane. The 5-µm pore size of the polycarbonate separation membrane facilitates the movement of morphologically normal sperm, while larger cells such as immature germ cells and leukocytes are restricted. In addition, sperm with low or no negative charge have less electrophoretic mobility and do not manage to reach the collection chamber through the separation membrane during a short electrophoresis time period of 5 min (see Fig. 2).

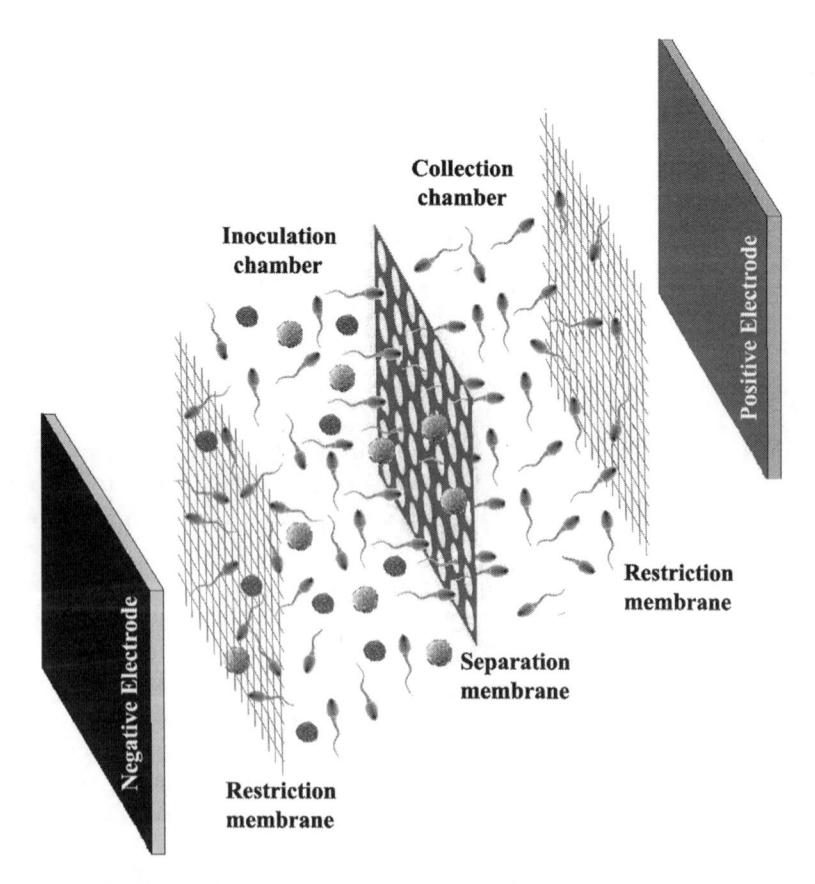

Fig. 2. Schematic diagram showing the apparatus for the electrophoretic mobility sperm sorter.

As soon as seminal plasma is removed by sperm preparation methods (swim-up, swim-down, or DGC), the sperm become vulnerable to free radical attack (25, 26). Therefore, rapid isolation of viable sperm from semen is important to prevent oxidative damage. In addition, semen is a heterogeneous mixture of various cell types (precursor germ cells, leukocyte subtypes, viable and nonviable sperm) and debris, which has to be removed during the process of sperm preparation.

The electrophoretic approach of sperm separation is rapid, free from contaminant cells, and able to isolate normal sperm with high percentage of morphologically normal and motile sperm with intact DNA (13). In the asymmetric model of the electrophoresis unit SpermSep Cell Sorter 10, about 2 mL of semen is loaded in the inoculation chamber, and 400 μL of sperm is obtained from the collection chamber after a 5-min equilibration step and then subjected to electrophoresis for 5 min. Conveniently, the obtained 400 μL of sperm can be directly used for intrauterine insemination procedures or for the purposes of in vitro fertilization and intracytoplasmic sperm injection if the appropriate medium is used within the collection chamber.

The 5 μm polycarbonate membrane that separates the inoculation and collection chambers has an active area of 20×15 mm for sperm passage and is bounded by polyacrylamide restriction membranes that prevent cross-contamination between the semen and electrophoresis buffer while permitting free transit of electrolytes. The electrophoresis buffer contains 10 mM Hepes, 30 mM NaCl, and 0.2 mM sucrose, having an osmolarity of 310 mOsm kg^{-1} and a pH of 7.4 (14). Overheating of the instrument during electrophoresis is prevented by maintaining the buffer at 25°C and circulating the excess buffer stored in the reservoir around the instrument using a pump. Electrophoresis is achieved by applying a constant current of 75 mA at a variable voltage of 18–21 V for a period of 5 min.

The only drawback of the electrophoretic system is the laborious procedure of cleaning the instrument when compared with other sperm preparation methods. The components of the separation cartridge have to be autoclaved to ensure sterility. After each sperm separation the electrophoresis buffer in the system is removed and the entire unit is rinsed by sterile distilled water. At the end of each day, the sterile distilled water is replaced by a cleaning buffer (0.1 M NaOH) and circulated in the electrophoresis unit for 30 s using the buffer pump. The cleaning buffer is left in the system overnight, and the following day it is removed and the system is thoroughly rinsed out with at least three washes of sterile distilled water (24).

The complete validation of the electrophoresis unit was done by the research group led by Prof. John Aitken, University of

Newcastle, Australia. All experiments were performed using the SpermSep Cell Sorter 10 system and sperm samples were obtained from normozoospermic men. Adequate recovery of sperm is an important factor for any sperm preparation method, as in cases of oligozoospermic men the initial concentration is too low. In this electrophoretic system, when a mean sample concentration of $52 \pm 5.2 \times 10^6$ mL^{-1} was loaded, after the 5-min equilibration period only 3.2% ($1.7 \pm 0.6 \times 10^6$ mL^{-1}) of sperm was recovered in the collection chamber as a consequence of its motility. However, after electrophoresis for 30 s the recovery of sperm increases to 6.8% ($3.55 \pm 0.42 \times 10^6$ mL^{-1}) reaching a maximum recovery of 42.9% ($22.31 \pm 5.85 \times 10^6$ mL^{-1}) after 15 min (13).

For assisted reproduction, sperm motility and viability are the key factors for successful treatment. Particularly, sperm progressive motility has been reported to influence in vitro fertilization rates (27), and thus it is considered a useful parameter to determine the type of ART with which to proceed. Using the electrophoretic system, the percentage of motile and viable sperm following isolation was consistent with the values recorded for the original ejaculates (13). Furthermore, the duration of electrophoresis did not affect the percentage of motile or viable sperm, and the speed of motile sperm measured by the kinematic analysis using computer-assisted semen analysis was likewise not detrimentally affected after electrophoresis.

An ideal sperm preparation method should select all physiologically normal sperm from the ejaculate (28). While this is not feasible with current technologies, any improvement in the percentage of genetically normal sperm is advantageous. A number of studies have demonstrated that abnormal sperm of infertile men exhibit poor chromatin compaction and damaged DNA (29, 30). Sperm selection by electrophoresis has been shown to significantly increase the percentage of morphologically normal sperm (13). Likewise, the sperm deformity index, another expression for sperm morphology and a known predictor of fertilization in vitro (31), was significantly lower than 0.93 (a threshold value for sperm deformity index to determine in vitro fertilization efficiency) following electrophoretic separation (32)). Sperm deformity index is also known to be positively associated with oxidative DNA damage (33). In accordance with the hypothesis, the sperm DNA damage was significantly reduced in the sperm obtained after sperm separation by electrophoresis (13), and this reduction was observed at all time points up to 10 min of electrophoresis.

While electrophoretic separation increases the percentage of morphologically normal sperm and decreases levels of DNA damage it does not result in improved sperm motility. One possible explanation for such an effect is that the process of electrophoresis may disrupt motility in a subset of the sperm population by interfering with the regulation of ion flux across the sperm plasma membrane (13).

In 2007, the first successful clinical pregnancy following ICSI was reported using sperm separated by electrophoresis (14). However, there have been no large prospective controlled trials to evaluate the suitability of electrophoretically separated sperm in clinical practice. In 2008, Fleming et al. reported the first prospective controlled clinical trial, involving 28 couples, of which 17 were undergoing IVF and 11 were undergoing ICSI. This was a split-sample cohort study design where sperm was prepared by DGC and electrophoresis (24). There was no significant difference in fertilization rates (63.6 vs. 62.4%), embryo cleavage rates (88.5 vs. 99.0%), or the percentage of top-quality embryos (26.1 vs. 27.4%) obtained following insemination using sperm preparation by DGC and electrophoresis, respectively (24). Similarly, there were no significant differences in fertilization rate and embryo quality in either the IVF or ICSI groups of patients. Although the study size was insufficient to demonstrate significant benefit, the study reported two ongoing pregnancies after transfer of 13 DGC-derived embryos compared to five ongoing pregnancies after transfer of 23 embryos derived from electrophoretically separated sperm. This report provides the proof of principle that electrophoretically separated sperm can be used in assisted reproduction and may be beneficial in selecting improved quality sperm and improving ART success.

Electrophoretic sperm selection has also been evaluated on cryopreserved semen and testicular biopsy samples. Following thawing, samples were subjected to vitality, motility, morphology, and DNA damage analysis (14). After 5 min of electrophoresis, this method generated 27% sperm recovery from cryopreserved semen and 28.4% sperm recovery from testicular biopsies (14). In both cryopreserved semen and testicular biopsies, the recovered subpopulations showed improved vitality and morphologically normal sperm along with a significant reduction in the levels of DNA damage.

In principle this electrophoretic sperm separation procedure has great potential as an extremely versatile and cost-effective method to prepare sperm. The sperm isolated by this method is reported to have good recovery rate, an improvement in morphologically normal sperm, and a reduction in sperm with DNA damage. Additional studies will be required to confirm the effectiveness of this electrophoretic method in the management of male infertility and as a technique for improvement of ART success.

2. Materials

1. 15 mL polystyrene plastic centrifuge tube (Fisher Scientific, Dallas, TX, USA).
2. HEPES–HTF medium (InVitroCare Inc., San Diego, CA, USA).
3. Serum (Irvine Scientific, Santa Ana, CA, USA).
4. Latex gloves.

3. Methods

All steps are performed at room temperature unless otherwise specified.

1. It is essential to use a new centrifuge tube (15 mL) for this method, as the electrostatic charge is higher on new tubes (see Notes 1–3).
2. Pipet 0.1 mL of washed sperm into the tube and dilute with 5 mL of serum-free HEPES–HTF medium.
3. Place the tube inside a latex glove up to the cap and hold the cap at all times (see Note 4).
4. Rotate the sperm sample two or three turns in a clockwise direction.
5. Incubate the tube at room temperature (23°C) for 1 min to allow adherence of the charged sperm to the wall of the centrifuge tube.
6. After 1 min, slowly invert the tube to drain out all non-adhering sperm and other contaminant cells.
7. Centrifuge the tube at $300 \times g$ for 5 min (see Note 5).
8. Place the tube upside down on a tissue paper to blot off the excess liquid at the mouth of the tube.
9. Pipet 0.2 mL of 3% serum supplemented with HEPES–HTF medium (0.2 mL) into the tube allowing the medium to trickle down the side of the wall. This process helps to neutralize the charge on the wall of the tube and detach the adhering sperm.
10. Re-pipet the collected medium at the bottom of the tube to rinse the wall of the same tube several times to increase the concentration of recovered sperm (see Note 6).

4. Notes

1. The zeta method of sperm selection should be carried out immediately after semen liquefaction as sperm cells become less negatively charged with the onset of capacitation.
2. The electrostatic charge of a new centrifuge tube can be verified using an electrostatic volt meter. To ensure that the tube has adequate positive charge, the volt meter should read 2–4 kV per square inch.
3. Glass centrifuge tubes tend to permit more sperm adherence when compared with polystyrene tubes. However, glass tubes should be rinsed and soaked prior use to reduce contaminants.
4. Handling the centrifuge tubes is very important throughout the procedure. Hold each tube by the cap, and avoid grounding the tube.
5. Centrifugation of the tube will not alter the net electrostatic charge of the tube.
6. The use of culture medium with a higher percentage of serum for discharging the tube might improve recovery of detached sperm in cases with low sperm concentrations.

Acknowledgment

This work was supported by the University of Utah Andrology and IVF laboratories.

References

1. Ishijima SA et al (1991) Zeta potential of human X- and Y-bearing sperm. Int J Androl 14:340–347
2. Kirchhoff C, Schroter S (2001) New insights into the origin, structure and role of CD52: a major component of the mammalian sperm glycocalyx. Cells Tissues Organs 168:93–104
3. Holt WV (1984) Membrane heterogeneity in the mammalian spermatozoon. Int Rev Cytol 87:159–194
4. Naz RK et al (1984) Monoclonal antibody to a human germ cell membrane glycoprotein that inhibits fertilization. Science 225:342–344
5. Calzada L et al (1994) Presence and chemical composition of glycoproteic layer on human spermatozoa. Arch Androl 33:87–92
6. Kallajoki M et al (1986) Surface glycoproteins of human spermatozoa. J Cell Sci 82:11–22
7. Giuliani V et al (2004) Expression of gp20, a human sperm antigen of epididymal origin, is reduced in spermatozoa from subfertile men. Mol Reprod Dev 69:235–240
8. Schroter S et al (1999) Male-specific modification of human CD52. J Biol Chem 274:29862–29873
9. Aitken RJ et al (2011) Electrophoretic sperm isolation: optimization of electrophoresis conditions and impact on oxidative stress. Hum Reprod 26:1955–1964
10. Ainsworth CJ et al (2011) The electrophoretic separation of spermatozoa: an analysis of genotype, surface carbohydrate composition and

potential for capacitation. Int J Androl 34 (5 Pt 2):e422–e434
11. Chan PJ et al (2006) A simple zeta method for sperm selection based on membrane charge. Fertil Steril 85:481–486
12. Kam TL et al (2007) Retention of membrane charge attributes by cryopreserved-thawed sperm and zeta selection. J Assist Reprod Genet 24:429–434
13. Ainsworth C et al (2005) Development of a novel electrophoretic system for the isolation of human spermatozoa. Hum Reprod 20:2261–2270
14. Ainsworth C et al (2007) First recorded pregnancy and normal birth after ICSI using electrophoretically isolated spermatozoa. Hum Reprod 22:197–200
15. Della Giovampaola C et al (2001) Surface of human sperm bears three differently charged CD52 forms, two of which remain stably bound to sperm after capacitation. Mol Reprod Dev 60:89–96
16. Hartree EF, Srivastava PN (1965) Chemical composition of the acrosomes of Ram spermatozoa. J Reprod Fertil 9:47–60
17. Srivastava PN, Farooqui AA (1980) Studies on neuraminidase activity of the rabbit endometrium. Biol Reprod 22:858–863
18. Kheirollahi-Kouhestani M et al (2009) Selection of sperm based on combined density gradient and Zeta method may improve ICSI outcome. Hum Reprod 24:2409–2416
19. Nasr-Esfahani MH et al (2008) The comparison of efficiency of density gradient centrifugation and Zeta methods in separatio of mature sperm with normal chromatin structure. Yakhteh Med 11:168–175
20. Khajavi N et al (2009) Can Zeta sperm selection method recover sperm with higher DNA integrity compared to density gradient centrifugation? Iranian J Reprod Med 7:73–77
21. Razavi SH et al (2010) Evaluation of zeta and HA-binding methods for selection of spermatozoa with normal morphology, protamine content and DNA integrity. Andrologia 42:13–19
22. Deemeh MR et al (2010) The first report of successfully pregnancy after ICSI with combined DGC/Zeta sperm selection procedure in a couple with eleven repeated fail IVF/ICSI cycles. Int J Fertil Steril 4:41–43
23. Stoffel MH et al (2002) Density and distribution of anionic sites on boar ejaculated and epididymal spermatozoa. Histochem Cell Biol 117:441–445
24. Fleming SD et al (2008) Prospective controlled trial of an electrophoretic method of sperm preparation for assisted reproduction: comparison with density gradient centrifugation. Hum Reprod 23:2646–2651
25. Aitken RJ, Clarkson JS (1987) Cellular basis of defective sperm function and its association with the genesis of reactive oxygen species by human spermatozoa. J Reprod Fertil 81:459–469
26. Aitken RJ, Clarkson JS (1988) Significance of reactive oxygen species and antioxidants in defining the efficacy of sperm preparation techniques. J Androl 9:367–376
27. Turner TT et al (2006) Sonic hedgehog pathway inhibition alters epididymal function as assessed by the development of sperm motility. J Androl 27:225–232
28. Hammadeh ME et al (2001) Comparison of sperm preparation methods: effect on chromatin and morphology recovery rates and their consequences on the clinical outcome after in vitro fertilization embryo transfer. Int J Androl 24:360–368
29. Kosower NS et al (1992) Thiol-disulfide status and acridine orange fluorescence of mammalian sperm nuclei. J Androl 13:342–348
30. Manochantr S et al (2011) Relationship between chromatin condensation, DNA integrity and quality of ejaculated spermatozoa from infertile men. Andrologia 44(3):187–199
31. Aziz DM (2006) Assessment of bovine sperm viability by MTT reduction assay. Anim Reprod Sci 92:1–8
32. Panidis D et al (1998) The sperm deformity and the sperm multiple anomalies indexes in patients who underwent unilateral orchectomy and preventive radiotherapy. Eur J Obstet Gynecol Reprod Biol 80:247–250
33. Said TM et al (2005) Novel association between sperm deformity index and oxidative stress-induced DNA damage in infertile male patients. Asian J Androl 7:121–126

Chapter 26

Sex-Sorting Sperm Using Flow Cytometry/Cell Sorting

Duane L. Garner, K. Michael Evans, and George E. Seidel

Abstract

The sex of mammalian offspring can be predetermined by flow sorting relatively pure living populations of X- and Y-chromosome-bearing sperm. This method is based on precise staining of the DNA of sperm with the nucleic acid-specific fluorophore, Hoechst 33342, to differentiate between the subpopulations of X- and Y-sperm. The fluorescently stained sperm are then sex-sorted using a specialized high speed sorter, MoFlo® SX XDP, and collected into biologically supportive media prior to reconcentration and cryopreservation in numbers adequate for use with artificial insemination for some species or for in vitro fertilization. Sperm sorting can provide subpopulations of X- or Y-bearing bovine sperm at rates in the 8,000 sperm/s range while maintaining; a purity of 90% such that it has been applied to cattle on a commercial basis. The sex of offspring has been predetermined in a wide variety of mammalian species including cattle, swine, horses, sheep, goats, dogs, cats, deer, elk, dolphins, water buffalo as well as in humans using flow cytometric sorting of X- and Y-sperm.

Key words: Sperm sorting, Flow cytometry, Predetermination of sex, X- and Y-sperm, Mammals, Hoechst 33342, DNA content

1. Introduction

Mammalian sperm can be effectively separated into relatively pure living populations of X- and Y-chromosome-bearing sperm using flow cytometry/cell sorting. In addition to describing procedures, this review provides the background on how conceptual advances originating from genetics, applied gamete biology, biophysics, and computer science were integrated in development of a commercial method to predetermine the sex of mammalian offspring.

1.1. Biological Basis Examination of bovine chromosomes indicated that a total chromosome length difference of 4.2% existed between karyotypes of bulls and cows due to X-and Y-chromosome differences (1). This finding suggested a potential size difference between a sperm carrying the X-chromosome compared to one carrying a

Y-chromosome. It also had been reported that a potential difference in dry mass exists between sperm heads of X- and Y-chromosome-bearing sperm (2). Thus, a difference in DNA content between sperm bearing the X-chromosome and those bearing the Y-chromosome existed. However, measurement of the DNA content of individual sperm proved to be difficult due to the asymmetry of mammalian sperm head shape (3, 4).

1.2. Precise Measurement of Sperm DNA Content

The initial indication that differences in DNA content of mammalian sperm could be determined utilized coaxial flow cytometry and revealed two fluorescent populations of demembranated nuclei of X- and Y-human sperm (5, 6). This was accomplished using the first fluorescence-based commercial analytical flow cytometer developed by Wolfgang Göhde in 1968–1969 and marketed as the Phywe ICP 11(7) (German patent application DE1815352). The DNA content of asymmetric cells also could be measured precisely in an orthogonal flow cytometer if the cells were hydrodynamically orientated so that fluorescence could be accurately measured from the flat surface of the nucleus (8). Pinkel et al. (9) demonstrated that such an orienting system could readily resolve two fluorescent populations of sperm stained with a nucleic acid specific dye 4′-6-diamidino-2-phenylindole (DAPI) to reflect DNA content differences. The presumed sex-chromosomal origin of these fluorescence differences was supported by measurements of sperm from mice possessing the Cattanach translocation, where a piece of chromosome 7 had been translocated to the X-chromosome (10). Flow cytometric examination of the DNA content of sperm from Cattanach mice showed a 5.1% difference in the two sperm populations compared to the 3.3% for sperm from normal mice. Quantitative differences in the DNA content of the X- and Y-chromosome-bearing sperm of several heterogametic mammals including cattle (*Bos taurus* and *Bos indicus*), sheep (*Ovis aries*), swine (*Sus scrofa*), and rabbits (*Oryctolagus cuniculus*) were demonstrated using the homogametic Z sperm of roosters (*Gallus domesticus*) as a standard (11). The above described results strongly suggested that the two peaks demonstrated in mammalian sperm represent the X- and Y-chromosome-bearing sperm. This flow cytometric approach identified two populations of sperm with characteristic suggesting that they were the X- and Y-chromosome-bearing populations, but this coaxial analytical system could not physically sort cells.

1.3. Developments in Flow Cytometry

The development of flow cytometry began in the 1960s when Louis Kamentsky and associates first measured ultraviolet absorption in cancer cells using a fluidic system (12, 13). Subsequent work at the Los Alamos National Laboratory (8) and Stanford University (14) resulted in development of a flow sorting system called a fluorescence-activated cell sorter (FACS) that used fluorescence rather than absorption as a measurement criterion.

The first actual sorting of mammalian sperm into separate populations was achieved with sperm nuclei of the vole, *Microtus oregoni* (15). The DNA content difference of the sex determining Y- and O-sperm from this rather unique species is 9% because half of the sperm contain a Y-chromosome, while the other half of the sperm have no sex chromosome. The X-chromosome is reconstituted at fertilization in this species (15). Although not very efficient, sort purities of 82–95% were achieved with nuclei of the "O" sperm and 72–83% for the Y fraction. Thus, the ability to separate the nuclei of sperm of mammals into two populations relative to DNA content was established using flow cytometric sorting.

1.4. DNA Content Measurement in Living Sperm

The demonstrated DNA content difference between the two populations of mammalian sperm is large enough that the possibility of sorting the sperm into separate living populations was undertaken. Sperm sorting technology was established at the USDA Beltsville Agricultural Research Center (BARC) by purchase and modification of a Coulter EPICS V® Flow Cytometer/Cell Sorter (Beckman Coulter, Hialeah, FL and Fullerton, CA, USA) with the intent of separating the X- and Y-sperm of domestic livestock. At that time the membrane impermeant fluorescent staining system that was used to measure sperm for DNA content required removal of the cell membranes, thereby killing sperm (11, 15). Initial efforts demonstrated that dead X- and Y-sperm populations could be produced with purities greater than 85–90% for each population (16). It wasn't until Johnson et al. (17) altered the staining process that living sperm could be sorted according to their DNA content. The membrane permeant bisbenzimidazole fluorescent dye, Hoechst 33342, readily differentiated between the two populations of living sperm according to their DNA content (18). This initial advancement in staining with Hoechst 33342 was followed by development of the ability to sort living sperm into populations of the purportedly viable X- and Y-sperm populations at 85–90% purity (see Fig. 1) (17, 18). The efficacy of this sperm sorting technology was tested using surgical insemination of the two sorted sperm populations into the oviducts of rabbits (17). In rabbits, the resultant offspring from inseminations with the purported X-sperm populations were 94% female, while those females inseminated with the purported Y-sperm population produced 81% males (17). Similar data were obtained when this approach was applied to the domestic swine (19). This successful production of offspring with significantly skewed sex ratios in rabbits and swine provided the basis of a patent (US Patent #692958, 04/26/1991). The USDA then licensed the technology to Animal Biotechnology, Cambridge, Ltd, which was later to become Mastercalf Ltd, Cambridge, UK, for field testing. In this first field testing of the sorting technology, a Facstar Plus® (BD Biosciences, San Jose, CA, USA) was modified to sort living bovine sperm. This initial project utilized in vitro

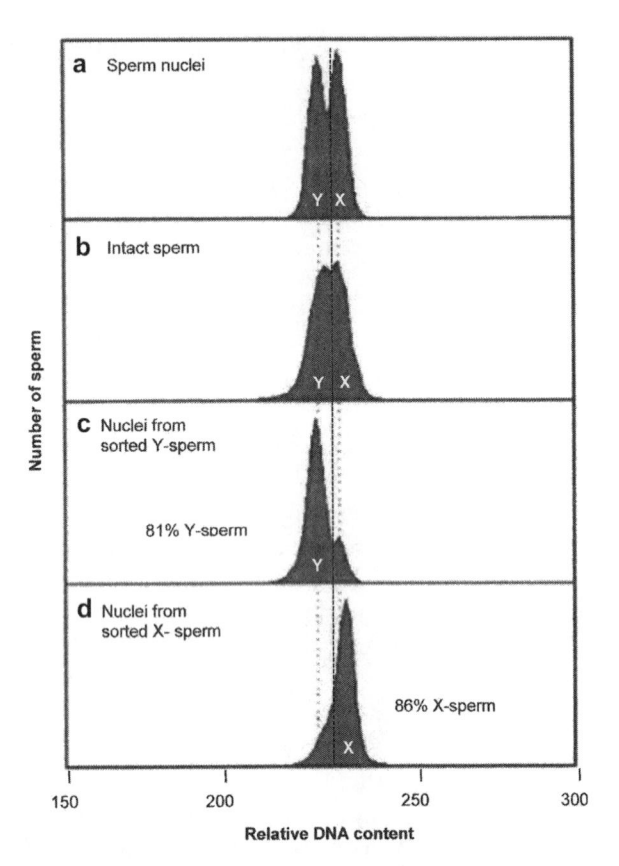

Fig. 1. Illustration showing a comparison of the relative DNA content of Hoechst 33342-stained rabbit sperm nuclei, living X- and Y-sperm, nuclei recovered from Y-sorted sperm and X-sorted sperm. (a) Example illustration of the 3.0% DNA content difference between X- and Y-sperm nuclei; (b) Example of the relative DNA content difference of intact living sperm stained with Hoechst 33342; (c) Representation of the frequency distribution of DNA content from reanalysis of rabbit sperm nuclei prepared from an aliquot of Y-sorted living sperm; and (d) Representation of the frequency distribution of DNA content from reanalysis of rabbit sperm nuclei prepared from an aliquot of X-sorted living sperm (Adapted from Johnson et al., 1989 (17)).

fertilization (IVF) with sex-sorted bovine Y-sperm in an effort to produce male embryos. The resultant embryos were then cryopreserved and later transferred to recipients on 12 Scottish farms (20, 21). From 106 twin transfers of the bovine embryos that were fertilized by Y-sorted sperm to recipient cows, 37 male and 4 female calves were born. Thus, 90% were male as predicted from the Y-sperm DNA content observed during the sex-sorting process. The sorting process at that time, about 600 sperm/s, was not fast enough to produce adequate numbers of sperm for the use of sex-sorted sperm with artificial insemination (AI).

The membrane impermeant DNA-specific dye, propidium iodide (PI), was added during the staining process to identify dead and injured sperm (22). The PI quenches the Hoechst 33342 stain, thereby providing a means of selectively eliminating these damaged cells from the sorted product by selective gating. This staining process was further improved by substituting the food dye, FD&C #40 (Warner Jenkinson, St. Louis, MO, USA) for the PI, which is a potential mutagen (23, 24).

Development of a high speed sorter at Lawrence Livermore National Laboratory (25) and significant improvements in data acquisition speed (26) increased the speed at which sperm could be sorted. This high speed sorting system was commercialized by Cytomation, Inc. (Now part of Beckman Coulter, Inc, Miami, FL, USA) as the MoFlo®. This sperm sorting system was further enhanced by development of a novel nozzle-tip that oriented approximately 70% of the sperm in the laser beam as they passed through the sorter compared to 30% orientation capable with the original sperm sorter (27, 28). An original MoFlo instrument was installed at the BARC and modified for sperm sorting by installing a cell orienting nozzle-tip, making it capable of measuring and handling analysis rates close to 15,000 sperm/s (Rens et al., US Patent 5,985,216). The instrument, which had been modified specifically for sperm sorting, is called a MoFlo® SX.

2. Flow Cytometric Sorting of X- and Y-Sperm

2.1. Sperm Sample Preparation

Semen is collected and prepared for sorting using carefully cleansed equipment and sterilized media. The semen sample is examined for volume, and antibiotics in 20 µL/mL of semen [Tylosin (100 µg/mL), Gentamicin (500 µg/mL), and Linco-Spectin (300/600 µg/mL; final concentrations)] are added to the raw semen prior to determinations of sperm concentration and motility. The raw semen is then stored up to 7 h with aliquots diluted approximately hourly to a sperm concentration of 200×10^6/mL with a modified Tyrode's balanced salt solution (staining TALP) consisting of 95.0 mM NaCl, 3.0 mM KCl, 0.3 mM NaH_2PO_4, 10.0 mM $NaHCO_3$, 0.4 mM $MgCl \cdot 6H_2O$, 2.0 mM pyruvic acid, 5.0 mM glucose, 25 mM Na lactate, 40.0 mM HEPES, 3 mg/mL BSA, and 30 µg/mL gentamycin sulfate (24).

2.2. Sperm Staining

An aliquot of the bull semen which had been diluted to 200×10^6 sperm/mL in staining TALP is then stained with 50.4 µM Hoechst 33342 (9.0 µL Hoechst 33342/mL, from a 5 mg/mL distilled H_2O stock solution of Hoechst 33342) and placed in a 34°C water bath for incubation for 45 min. After removal from the water bath, 100 µL/mL of 20% egg yolk in a TALP is added, which makes the final egg yolk concentration, 2%. The stained sperm samples are then filtered using 50 µm sterile nylon mesh filters (CellTrics®, Partec, Münster, Germany) to remove clumped sperm, media aggregates and seminal debris. Identification of nonviable sperm is accomplished by adding 2 μL/mL of a 1% FD&C #40 food coloring in TALP (23) to quench Hoechst staining of the DNA in membrane-damaged sperm.

2.3. Sorting Equipment

After stoichiometrically staining the sperm with Hoechst 33342, they are pumped in a stream in front of a UV laser beam (wavelength ~355 nm) to excite the Hoechst 33342-stained DNA to differentiate between the X- and Y-chromosome-bearing sperm. Most stained sperm are oriented as they pass in front of the laser so that the flat surface of the sperm nucleus can be used to precisely measure the DNA content (see Fig. 2). This orientation of the sperm nucleus is accomplished using the hydrodynamic forces of

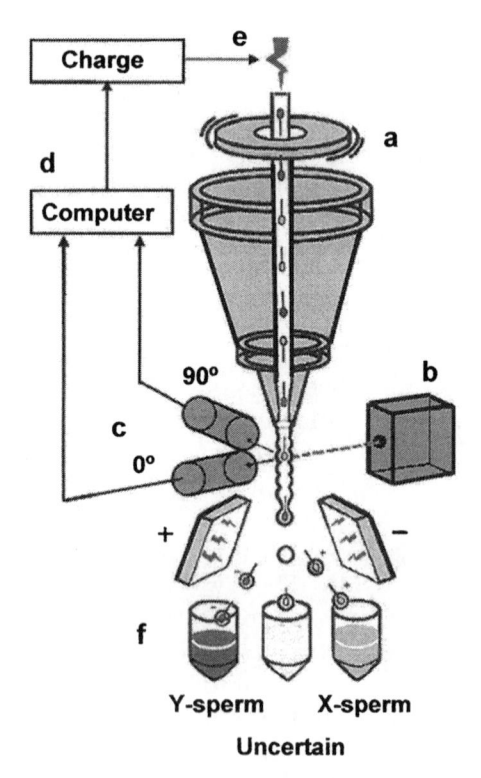

Fig. 2. Flow cytometer/sperm sorter system showing how Hoechst 33342-stained sperm are pumped through the sorting system including, (a) the piezoelectric crystal vibrator (~70,000/s) that causes the stream to form droplets as it exits the system; (b) the pulsed UV laser that illuminates the sperm as they flow through the system in the stream; (c) the two fluorescence photodetectors that capture the fluorescence emissions at 0° and 90°; (d) the fluorescence signal is quantified and categorized as X, Y, or uncertain; (e) positive, negative, or no charge is applied to the droplets as they emerge from the stream; (f) as the charged droplets pass between two oppositely charged plates they are deflected into either the X- or Y-catch tubes according to their DNA content while droplets without sperm, those with compromised cell membranes, or those with sperm of uncertain DNA content pass directly into the unsorted container. Adapted from Garner, 2001 (46).

the pressurized sheath fluid that surrounds the sample stream as it is pumped into the nozzle of the flow sorter. The sheath fluid used for this purpose is a *TRIS*-based buffer consisting of *TRIS* (hydroxymethyl) aminomethane (*tris*, 197.0 mM), citric acid monohydrate (55.4 mM), and fructose (4.75 mM) (24). The emitted fluorescence (~460 nm) of the oriented sperm is digitally quantified by dual orthogonal photodetectors which are situated at 0° and 90° to the laser beam to optimize capture of the fluorescence and to determine if a sperm is properly oriented for precise measurement of the DNA content using the 0° detector. This system can provide sorted subpopulations of X- or Y-bearing sperm at rate of 8,000 sperm/s at a purity of 90%. As the sperm exit the orienting nozzle, the stream is broken into individual droplets by a piezoelectric crystal vibrator. Either a negative or positive charge is applied to droplets containing sperm relative to their DNA content. The charged droplets then pass in front of two charged plates, one positive and one negative, such that they are deflected into one of three collection tubes, X-bearing sperm, Y-bearing sperm, and one for undetermined DNA content, membrane-damaged sperm with quenched fluorescence or no sperm (see Fig. 2).

The current state-of-the-art sperm sorter is a MoFlo® SX XDP, which uses digital technology to overcome problems associated with the coincidence when two sperm fall within the processing time window of the sorting electronics while exiting the nozzle at nearly the same time. Previous versions of the MoFlo® SX instrument were not able to differentiate between two closely positioned sperm due to processing speeds of the electronics and overlap of their fluorescence emissions. The configuration of a dual headed MoFlo® SX XDP includes two sort chambers supported by one laser (see Fig. 3).

2.4. Instrument Configuration

The excitation and fluorescence collection scheme for sorting sperm according to their DNA content requires that the stained sperm nucleus be oriented with its flat surface perpendicular to the 0° detector (see Fig. 4a). The 90° detector is utilized simultaneously to determine if that particular sperm is properly oriented. When the edge, rather than the flat surface, of the sperm head is oriented towards a detector this produces an approximate 2:1 relative fluorescence compared to that which is measured by the flat surface (see Fig. 4b). This is due to a lensing effect of the asymmetric sperm head. The dead sperm along with those with damaged membranes are identified by the uptake of the red food dye, which quenches the emission of the Hoechst 33342, thereby resulting in less fluorescence emission (see Fig. 4b).

2.5. Setting Sort Gates

The sperm populations identified by Hoechst 33342 staining, X- and Y- and dead sperm, are gated to achieve the desired purity of the population to be sorted and to identify damaged and dead

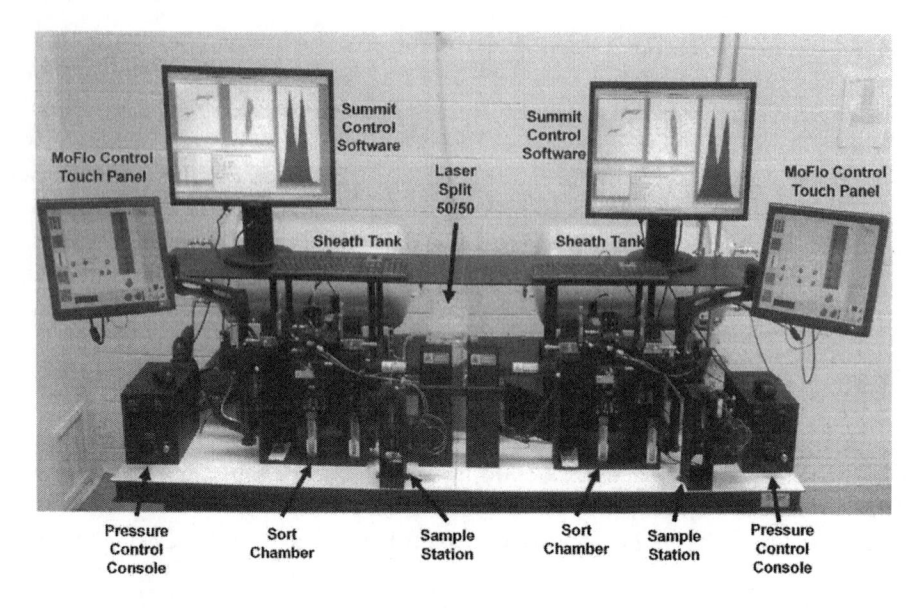

Fig. 3. The configuration of a dual-headed flow cytometer/sperm sorter, the MoFlo® SX XDP (Beckman Coulter, Inc., Miami, FL, USA), is shown illustrating the use of one pulsed laser with two sorting chambers. The various components; pulsed NdYag: Quazi-CW Mode Tripled laser, pressurized sheath tanks, sample stations, sort chambers, pressure control consoles, MoFlo® control touch panels, and the Summit® control software monitors are shown.

sperm (see Fig. 5). Once the desired population has been identified through bivariate analyses, it can be secondarily gated and magnified using the 0° detector to enhance the purity and recovery of the sorted population (see Fig. 5).

2.6. Sperm Sorting Procedure

Once the sort gates are set, the selected populations are sorted into a 15 or 50 mL conical centrifuge tube containing approximately 4 mL of 4% egg yolk in a TALP medium. The actual volume of EYC medium in this "catch tube" may vary based on the number of sperm to be sorted. Approximately 8,000 sperm/s of one sex can be sorted with the MoFlo®SX. Thus, approximately 28×10^6 sperm can be sorted during a 60 min sorting session using this equipment. The sort speeds are somewhat slower if both X- and Y-sperm populations are sorted. The catch tubes should be swirled gently every few minutes, after about 500,000 sperm have been sorted, to mix the sorted sperm with the egg yolk-containing medium in the catch tube to minimize oxidative damage to the sperm membranes.

2.7. Reconcentration of Sperm

The sorting process results in considerable dilution of the sperm, which necessitates centrifugation of the sperm to achieve a sperm concentration suitable for subsequent packaging and utilization in AI or IVF. The sperm that had been sorted into the medium in the catch tube are then slowly cooled to 5°C to prevent cold shock by placing the catch tube containing the prepared sperm in a 600 mL

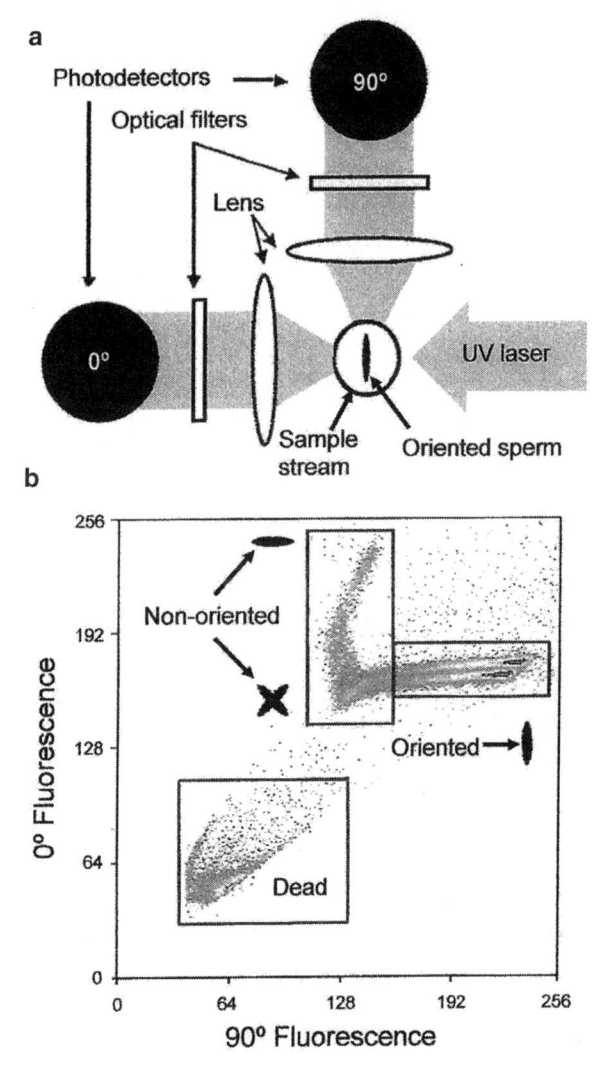

Fig. 4. Excitation and detection scheme for the flow cytometric sexing of sperm showing sperm orientation. (a) Configuration of laser and detectors to provide accurate measurement of DNA content of living sperm. (b) Bivariate histogram produced by flow cytometry showing the fluorescence measurement of live-oriented sperm with the edge toward the 90° detector (Region R1) and non-oriented sperm where the edge is toward the 0° detector or somewhere in between (R2) along with the sperm where the red food dye quenches the Hoechst 33342-stained sperm revealing the dead or damaged sperm (R3). The orientations of the sperm heads are provided adjacent to each population (below R1, for oriented sperm) and left of R2 for incorrectly oriented sperm heads Adapted from Sharpe and Evans 2009 (48).

beaker containing room temperature (~18–20°C) water prior to placing the beaker in a 5°C cold room for approximately 1.5 h. Twelve percent glycerol is added in two equal portions 15 min apart. The cooled sperm sample is then centrifuged at 5°C at $850 \times g$ for 20 min to concentrate the sperm. The supernatant is removed carefully by pouring off, leaving a 400–500 µL sperm

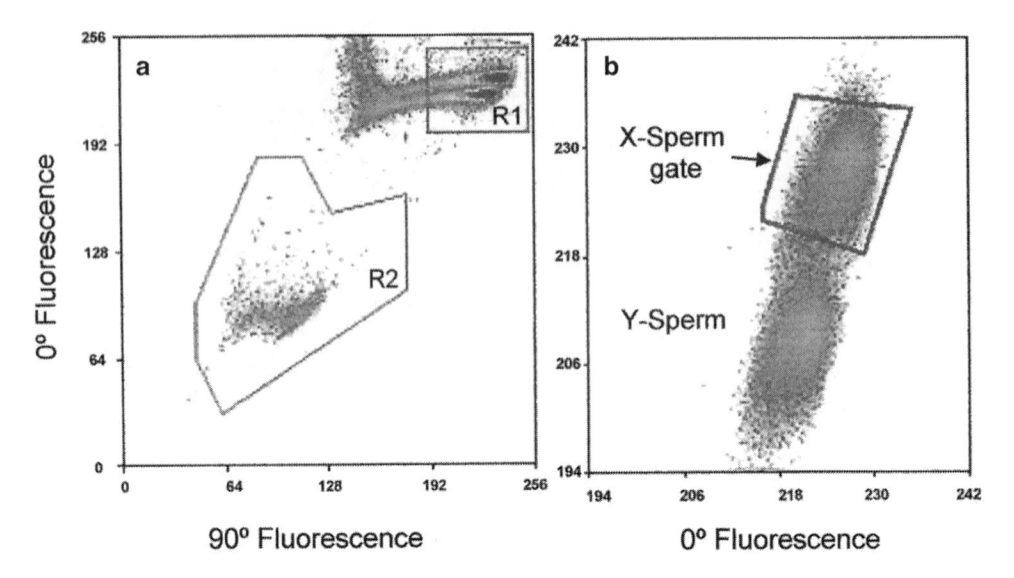

Fig. 5. An example of how the various sperm populations are gated for sorting bovine X-sperm. (a) Bivariate plot of 90° (x-axis) and 0° (y-axis) of the fluorescence emitted from Hoechst 3342-stained bovine sperm where properly oriented sperm are displayed in R1 and dead or damaged sperm are displayed in R2 (uptake of red food dye), (b) Bivariate plot of the sperm populations, X- and Y-sperm showing the gate (*box*) set for X-sperm sorting. Adapted from Sharpe and Evans 2009 (48).

pellet containing >90% of the sorted sperm. Additional Tris extender containing 20% egg yolk and 6% glycerol is added, and the sperm pellet is resuspended carefully to minimize introduction of air into the sample. This results in a final concentration of approximately 20% egg yolk and a sperm concentration of approximately 9.8×10^6/mL. The cooled sperm are examined for actual sperm concentration and for the percentage of progressively motile sperm. If the sex-sorted sperm sample exhibits a minimum of 70% progressive motility, they are then packaged at a concentration of 10×10^6 sperm/mL which results in a dose of 2.1×10^6 sperm per straw in a volume of 210 μL. The sperm are packaged into the 0.25 mL French straws and then cryopreserved.

2.8. Cryopreservation of Flow-Sorted Sperm

Cryopreservation of sexed sperm has been performed by procedures in routine use for unsexed sperm (24); however, 0.25-mL plastic straws are used instead of 0.5-mL straws, the more common packaging container for cryopreserving bovine sperm in North America. The typical sperm concentration is 10^7 sperm/mL, and the actual volume of liquid is about 0.21 mL due to the seals at each end of the straw.

The most commonly used medium for cryopreservation of sexed bovine sperm is a 20% egg yolk-Tris-based extender with 6% glycerol (v/v) as explained above and described by Schenk et al. (24). Important principles not to be violated include slow cooling over about 90 min from ambient temperature to ~5°C to prevent cold shock, and then holding the sperm at ~5°C for 3–6 h allowing them to adjust to this low temperature before freezing; the 18 h holding at 5°C sometimes used for unsexed sperm is too long (24).

The actual freezing procedure can be done in a variety of ways as long as cooling rates are carefully controlled. The simplest and most common approach is to place straws horizontally in racks positioned a few centimeters above a pool of liquid nitrogen (which constitutes static liquid nitrogen vapor mixed with a bit of air) for 20 min or more, and then plunging the straws into liquid nitrogen (−196°C). Straws then usually are packaged into plastic goblets clipped to metal canes and stored indefinitely in liquid nitrogen tanks in the liquid or vapor phases.

Thawing of sexed sperm also is done by conventional procedures, placing the straws into a 35–37°C water bath for at least 30 s within <3 s of removing straws from the liquid nitrogen container. Straws are then ready for use for AI, IVF, or quality control purposes. The latter is essential for each batch frozen, and typically consists of evaluation of post-thaw progressive sperm motility and cell membrane integrity. Accuracy of sexing can be evaluated by resort analysis (23) continually during sorting, in a batch prior to cryopreservation, or post-thaw after processing to remove most of the egg yolk. A few percent of batches of sexed bull sperm must be discarded due to poor sperm viability post-thaw, as well as occasional batches that do not meet accuracy standards, typically at least 87% of the desired sex.

Cryopreservation procedures have been described for bovine sperm, which constitute >90% of sperm sexed commercially; species-specific variations are used as appropriate, e.g., for sheep (29) or men (30).

3. Advances in Flow Sorting of Sperm

Improvements have been made in the detector circuitry of sperm sorters, which allow the fluorescence detectors to operate at voltages lower than typically used in flow sorting. The lower detector voltage provides an improved signal to noise ratio by making the detectors nonlinear in their response, thereby creating more distance between the means of X- and Y-sperm populations (Evans et al., US Patent 7,371,517 B2). Another important advancement was development of beam shaping lenses to minimize the time that a sperm is in the laser beam path. The laser path is shaped from a round beam path to a thin elliptical, ribbon-like beam using special lenses such that the time that a sperm is irradiated at 355 nm is reduced, thus minimizing the exposure of the sperm to potential UV irradiation damage.

3.1. Damage to Sperm During the Sorting Process

Sex-sorting mammalian sperm using flow cytometry requires several steps that have the capability of damaging the gametes. Initial approaches using sperm motility evaluations to determine which steps in the process such as the mechanical aspects of sorting and centrifugation, as well as the combination of dye and laser exposure either singularly or in combination, had the potential to damage sperm were not definitive. However, flow cytometric assessment of the proportion of viable sperm at each step in the process revealed that the most damaging aspect of the process was the mechanical stress of sorting (31, 32). This problem was partially overcome by reducing the sheath pressure of the sorter from 50 to 40 psi (32).

3.2. Pulsed Laser

Significant improvements were made with the implementation of pulsed lasers compared to the original water-cooled continuous wave (CW) versions. These older CW lasers operate with a continuous beam as contrasted to the pulsed version, which pulses several hundred times as a sperm traverses the path of the beam. The Pulsed NdYag: Quazi-CW Mode Tripled laser, which emits at 355 nm, has proven to be a reliable, safe, and effective illumination system for sorting sperm. Pulsed lasers can operate for as long as 25,000 h with minimal interaction before any maintenance requirements, making the operating cost 1/10th that of the earlier water-cooled lasers. Furthermore, a pulsed laser beam can be readily split such that two or more sperm sorters can be operated simultaneously from one laser. Operation of the detector system under this pulsation not only provides lower background radiation, but also improves resolution and life span of the laser. Pulsed lasers are air cooled thereby eliminating the need for the expensive and cumbersome water cooling systems that were necessary with the CW lasers that were previously used with the sperm sorting system (33).

3.3. Digital Output

Recent research efforts have focused on innovations to increase production speeds through improved hardware, optics, and electronics. The original MoFlo® SX instrument was an analog-based system that is limited by coincidence problems such as when two sperm flow through the instrument in close proximity. This coincidence leads to discarding both sperm because the instrument cannot differentiate between the two events when two sperm are in the laser beam at the same time. A newer version of the sperm sorter, the MoFlo® SX XDP (see Fig. 3), which utilizes digital technology to overcome this coincidence problem, was recently developed through collaboration of Sexing Technologies (Navasota, TX, USA) and Beckman Coulter instruments (Beckman Coulter, Hialeah, FL and Fullerton, CA, USA). The processing electronics of the digital-based MoFlo® SX XDP can differentiate between two closely positioned sperm such that sorting yields are greatly improved. Thus, technological improvements in the current sorters provide the ability to sort up to 10,000 X- or Y-sperm/s while achieving 90% or greater purities. The speed of sorting also is dependent on the quality of the sperm sample that is to be sorted and type of product being produced. There are now two sexed semen products available to the market 75–86% product and 87–98% product.

3.4. Profile Examples of Flow-Sorted Sperm from Different Species

Among mammalian species, both the shape of the sperm head and the X–Y sperm DNA content difference contribute to the effectiveness of utilizing DNA content to sex sort sperm (34). Although the univariate profiles of the X- and Y-sperm populations of six of the domesticated species that have been successfully sorted for pre-determination of sex are similar (see Fig. 6), their DNA content differences vary somewhat; 3.8% for cattle (*Bos taurus*), 4.1% for horses (*Equus caballus*), 3.6% for swine (*Sus scrofa*), 4.2% for sheep (*Ovis aries*), 4.2% for domestic cats (*Felis catus*), and 4.4% for White-Tailed Deer (*Odocoileus virginianus*) (34–36).

3.5. Purity of Flow-Sorted Sperm Samples

A small analytical flow cytometer, the STS Purity Analyzer, was developed by Sexing Technologies to determine the purity of sorted samples, thereby eliminating the need to use the high demand sperm sorters for such quality control analyses (33, 37). This instrument assists operators in maintaining and optimizing the purity of sorted sperm by providing continual monitoring the X- or Y-sperm populations during the actual sorting process. Use of this analytical

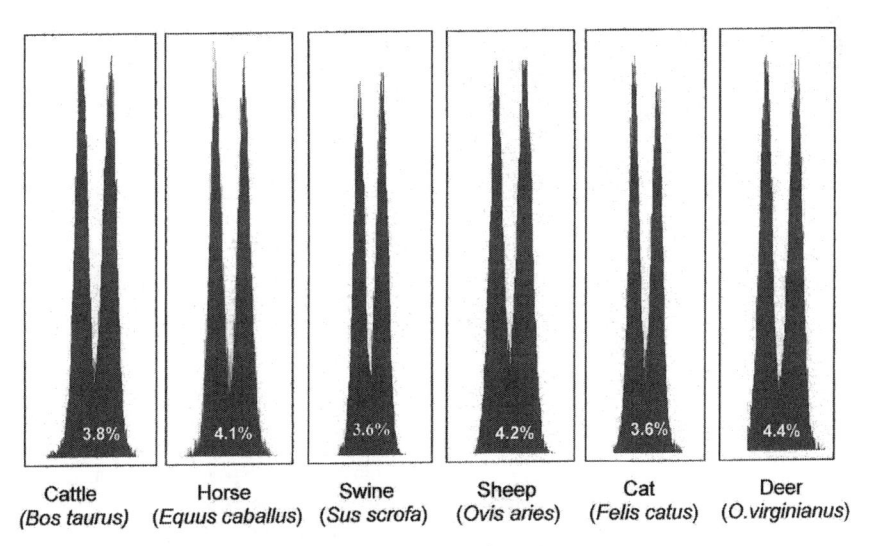

Fig. 6. Examples of univariate fluorescence profiles of Hoechst 33342-stained X- and Y-sperm populations from cattle (*Bos taurus*), horses (*Equus caballus*), swine (*Sus scrofa*), sheep (*Ovis aries*), cats (*Felis catus*), and White Tail deer (*Odocoileus virginianus*) showing the relative DNA content differences the two sperm populations for each species.

instrument as a routine monitoring effort considerably decreased production batch losses due to the failing sort purity (33, 37). Efficiency of producing sexed sperm is enhanced by setting the sort purity to approximately 90% rather than trying to achieve maximum purity of sexed sperm (see Fig. 7).

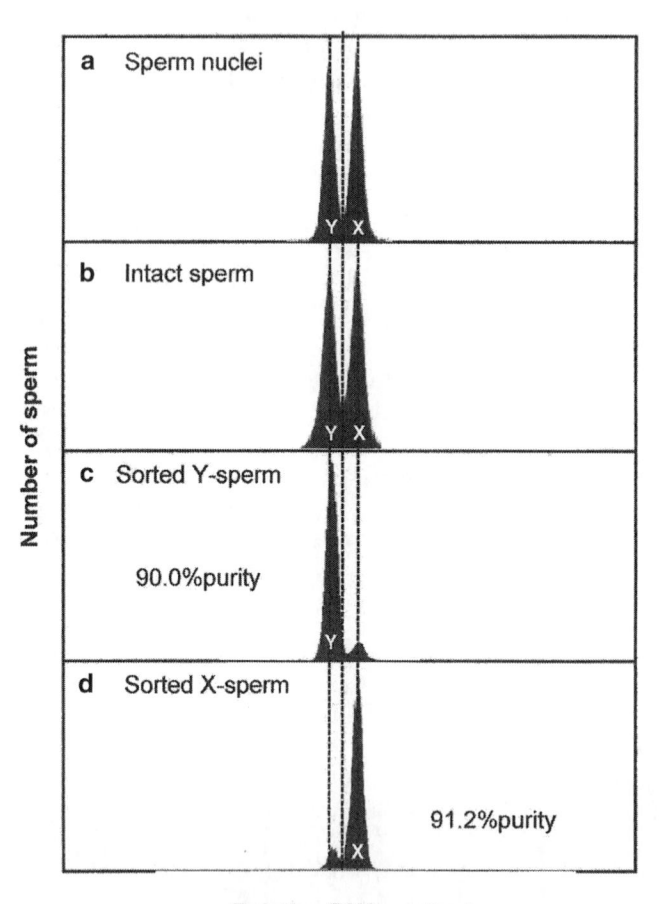

Fig. 7. Illustration of purity analyses of sex-sorted bovine sperm showing a comparison of the relative DNA content of Hoechst 33342-stained bovine sperm nuclei, living X- and Y-sperm, the Y-sorted sperm population, and sperm recovered from the X-sorted population. (a) Example illustration of the 3.8% DNA content difference between X- and Y-sperm nuclei; (b) Example of the relative DNA content difference of intact living bovine sperm stained with Hoechst 33342; (c) Representation of the frequency distribution of DNA content from reanalysis of living sex-sorted bovine Y-sperm prepared by thawing the sperm sample and removing the egg yolk-based medium by aspiration following centrifugation and restaining the sperm with Hoechst 33342; (d) Representation of the frequency distribution of DNA content from reanalysis of living sex-sorted bovine X-sperm prepared by thawing the sperm sample and removing the egg yolk-based medium by aspiration following centrifugation and restaining the sperm with Hoechst 33342. This particular example, which was sorted at relatively high sort rates, shows purities of 90.0% Y-sperm and 91.2% X-sperm for these sorted populations of living sperm.

3.6. Fertility of Sex-Sorted Mammalian Sperm

Mammalian sperm normally are available in the huge numbers required for applications such as AI. However, the process of sex-sorting sperm constitutes a bottleneck, limiting numbers of sorted sperm available for the assisted reproduction techniques required to apply sex-sorting sperm, such as IVF and AI.

The first evidence of the fertility of sex-sorted sperm was demonstrated by surgical insemination of small numbers of sperm into the reproductive tract of the rabbit, close to the normal in vivo site of fertilization (17). However, the first attempts to apply this technology were via IVF because many fewer sperm are needed for this approach than for AI (20, 21); IVF also is used routinely for human applications (30) and has been used to a limited extent with cattle. However, higher concentrations of sex-sorted sperm are needed to achieve equivalent fertilization rates in vitro than for unsorted sperm, and there are huge male-to-male differences (38, 39) with unacceptably low fertilization rates for some bulls.

Artificial insemination is a much more practical way to apply sexed semen than IVF, but requires many more sperm. Despite huge advances in flow cytometry/cell sorting, it still is impractical and uneconomical to provide $\geq 10^7$ sexed bovine sperm per insemination dose, which would be the typical dose for unsexed sperm. The compromise has been to use $\sim 2 \times 10^6$ sexed sperm per insemination dose for cattle. For horses and pigs, with typical unsexed insemination doses >100 and $1,000 \times 10^6$ sperm, respectively, it has thus far remained impractical to apply sexed semen commercially.

With good animal management, pregnancy rates with 2×10^6 sexed sperm per dose in cattle have usually been 70–90% of the standard $\geq 10 \times 10^6$ unsexed sperm per dose. This low dose product has been favorably received commercially, despite the lower fertility. Attempts have been made to determine if fertility is compromised mainly due to the lower sperm numbers per dose or damage to sperm during sorting. Frijters et al. (40) concluded that about one-third of the lower fertility was due to fewer sperm. However, with rare exceptions (41), it has not been possible to compensate for the reduced fertility with sexed sperm by increasing the sperm dose (42–46). Research is continuing to address this problem.

3.7. Commercial Development

The flow cytometric sperm sexing technology described above has been utilized commercially in cattle (*Bos taurus and Bos indicus*) for nearly 10 years resulting in more than 10 million live births of calves of a predetermined sex. Sperm from a variety of other mammalian species including horses (*Equus caballus*), boars (*Sus scrofa*), rams (*Ovis aries*), dogs (*Canis lupis familiaris*), cats (*Felis catus*), elk (*Cervus canadiensis*), water buffalo (*Bubalus bubalis*), and White-Tail deer (*Odocoileus virginianus*) have been successfully sex sorted and inseminated resulting in offspring of a predetermined sex (33–36, 45, 47). The application of this sex-sorting technology to humans by Genetics and IVF Institute, Fairfax, VA has resulted in more than 1,000 babies (30).

Acknowledgments

The authors are grateful to Rick Lenz for his suggestions and to Justine O'Brien and David Del Olmo Llanos for providing fluorescence profile data.

References

1. Moruzzi JF (1979) Selecting a mammalian species for the separation of X- and Y-chromosome-bearing spermatozoa. J Reprod Fertil 57:319–323
2. Sumner AT, Robinson JA (1976) A difference in dry mass between the heads of X- and Y-bearing human spermatozoa. J Reprod Fertil 48:9–15
3. Gledhill BL et al (1982) Identifying X- and Y-chromosome-bearing sperm by DNA content: retrospective perspective and prospective opinions. In: Prospects for sexing mammalian sperm. Colorado Associated University Press, Boulder, CO, pp 177–191
4. Van Dilla MA et al (1977) Measurement of mammalian sperm deoxyribonucleic acid by flow cytometry. Problems and approaches. J Histochem Cytochem 25:763–773
5. Meistrich ML et al (1979) Resolution of x and y spermatids by pulse cytophotometry. Nature 274:821–823
6. Otto FJ et al (1979) Flow cytometry of human spermatozoa. Histochemistry 61:249–254
7. Gohde W (1973) Automation of cytofluorometry by use of the impulse microphotometer. In: Thaer AA, Sernetz M (eds) Fluorescence techniques in cell biology. Springer, Berlin, pp 79–88
8. Fulwyler MJ (1977) Hydrodynamic orientation of cells. J Histochem Cytochem 25:781–783
9. Pinkel D et al (1982) High resolution DNA content measurements of mammalian sperm. Cytometry 3:1–9
10. Cattanach BM (1961) A chemically-induced variegated-type position effect in the mouse. Z Vererbungsl 92:165–182
11. Garner DL et al (1983) Quantification of the X- and Y-chromosome-bearing sperm of domestic animals by flow cytometry. Biol Reprod 28:312–321
12. Kamentsky LA et al (1963) Ultraviolet absorption in epidermoid cancer cells. Science 142:1580–1583
13. Kamentsky LA et al (1965) Spectrophotometer: new instrument for ultrarapid cell analysis. Science 150:630–631
14. Herzenberg LA et al (1976) Fluorescence-activated cell sorting. Sci Am 234:108–117
15. Pinkel D et al (1982) Sex preselection in mammals? Separation of sperm bearing Y and "O" chromosomes in the vole *Microtus oregoni*. Science 218:904–906
16. Johnson LA, Pinkel D (1986) Modification of a laser-based flow cytometer for high-resolution DNA analysis of mammalian spermatozoa. Cytometry 7:268–273
17. Johnson LA et al (1989) Sex preselection in rabbits: live births from X and Y sperm separated by DNA and cell sorting. Biol Reprod 41:199–203
18. Johnson LA et al (1987) Flow sorting of X and Y chromosome-bearing spermatozoa into two populations. Gamete Res 16:1–9
19. Johnson LA (1991) Sex preselection in swine: altered sex ratios in offspring following surgical insemination of flow sorted X- and Y-bearing sperm. Reprod Dom Anim 309–314
20. Cran DG et al (1993) Production of bovine calves following separation of X- and Y-chromosome bearing sperm and in vitro fertilisation. Vet Rec 132:40–41
21. Cran DG et al (1995) Sex preselection in cattle: a field trial. Vet Rec 136:495–496
22. Johnson LA et al (1994) Improved flow sorting resolution of X- & Y-chromosome bearing viable sperm separation using dual staining and dead cell sorting. Cytometry Suppl 7: Abstr 476D
23. Johnson LA, Welch GR (1999) Sex preselection: high-speed flow cytometric sorting of X and Y sperm for maximum efficiency. Theriogenology 52:1323–1341
24. Schenk JL et al (1999) Cryopreservation of flow-sorted bovine spermatozoa. Theriogenology 52:1375–1391
25. Peters D et al (1985) The LLNL high-speed sorter: design features, operational characteristics, and biological utility. Cytometry 6:290–301
26. van den Engh G, Stokdijk W (1989) Parallel processing data acquisition system for multilaser flow cytometry and cell sorting. Cytometry 10:282–293

27. Rens W et al (1998) A novel nozzle for more efficient sperm orientation to improve sorting efficiency of X and Y chromosome-bearing sperm. Cytometry 33:476–481
28. Rens W et al (1999) Improved flow cytometric sorting of X- and Y-chromosome bearing sperm: substantial increase in yield of sexed semen. Mol Reprod Dev 50–56
29. de Graaf SP et al (2007) Birth of offspring of pre-determined sex after artificial insemination of frozen-thawed, sex-sorted and re-frozen-thawed ram spermatozoa. Theriogenology 67:391–398
30. Karabinus DS (2009) Flow cytometric sorting of human sperm: MicroSort clinical trial update. Theriogenology 71:74–79
31. Garner DL, Suh TK (2002) Effect of Hoechst 33342 staining and laser illumination of viability of sex-sorted bovine sperm. Theriogenology 746 (Abstr)
32. Suh TK et al (2005) High pressure flow cytometric sorting damages sperm. Theriogenology 64:1035–1048
33. Evans KM (2010) Interpretation of sex-sorting process and new developments. In: Proceedings of the 23rd technical conference on artificial insemination & reproduction, National Association of Animal Breeders, pp 93–98
34. Garner DL (2006) Flow cytometric sexing of mammalian sperm. Theriogenology 65:943–957
35. Garner DL, Seidel GE Jr (2008) History of commercializing sexed semen for cattle. Theriogenology 69:886–895
36. O'Brien JK et al (2009) Application of sperm sorting and associated reproductive technology for wildlife management and conservation. Theriogenology 71:98–107
37. M. E. Kjelland et al (2011) DNA fragmentation kinetics and postthaw motility of flow cytometric-sorted white-tailed deer sperm. J Anim Sci 89:2–12
38. Xu J et al (2009) Optimizing IVF with sexed sperm in cattle. Theriogenology 71:39–47
39. Barcelo-Fimbres M et al (2011) In vitro fertilization using non-sexed and sexed bovine sperm: sperm concentration, sorter pressure, and bull effects. Reprod Domest Anim 46:495–502
40. Frijters AC et al (2009) What affects fertility of sexed bull semen more, low sperm dosage or the sorting process? Theriogenology 71:64–67
41. Schenk JL et al (2009) Pregnancy rates in heifers and cows with cryopreserved sexed sperm: effects of sperm numbers per inseminate, sorting pressure and sperm storage before sorting. Theriogenology 71:717–728
42. Seidel GE Jr, Schenk JL (2008) Pregnancy rates in cattle with cryopreserved sexed sperm: effects of sperm numbers per inseminate and site of sperm deposition. Anim Reprod Sci 105:129–138
43. DeJarnette JM et al (2008) Effect of sex-sorted sperm dosage on conception rates in Holstein heifers and lactating cows. J Dairy Sci 91:1778–1785
44. DeJarnette JM et al (2010) Effects of 2.1 and $3.5 \times 10(6)$ sex-sorted sperm dosages on conception rates of Holstein cows and heifers. J Dairy Sci 93:4079–4085
45. Dejarnette JM et al (2011) Effects of sex-sorting and sperm dosage on conception rates of Holstein heifers: is comparable fertility of sex-sorted and conventional semen plausible? J Dairy Sci 94:3477–3483
46. Garner DL (2001) Sex-Sorting mammalian sperm: concept to application in animals. J Androl 22:519–526
47. DeYoung RW et al (2004) Do *Odocoileus virginianus* males produce Y-chromosome-based ejaculates? Implications for adaptive sex ration theories. J Mammal 85:768–773
48. Sharpe JC, Evans KM (2009) Advances in flow cytometry for sperm sexing. Theriogenology 71:4–10

Part VI

Staging and Histology Techniques

Chapter 27

Assessment of Spermatogenesis Through Staging of Seminiferous Tubules

Marvin L. Meistrich and Rex A. Hess

Abstract

Male germ cells in all mammals are arranged within the seminiferous epithelium of the testicular tubules in a set of well-defined cell associations called stages. The cellular associations found in these stages and characteristics of the cells used to identify the stages have been well described. Here we present a binary decision key roadmap for identifying stages and present several examples of how staging tubules can be used to better assess the developmental profile of gene expression during spermatogenesis and defects in spermatogenesis arising in pathological conditions resulting from genetic mutations in mice. In particular, when one or more cells of a cellular association cannot be clearly identified or are missing, the cell types that should be present can be precisely identified by knowledge of the approximate or exact stage of the tubule cross section.

Key words: Stages, Seminiferous epithelium cycle, Spermatids, Kinetics, Acrosome

1. Introduction

Spermatogenesis is a continuous process by which stem spermatogonia transform within the seminiferous tubules through the differentiated spermatogonial, spermatocyte, and round and elongated spermatid stages to become spermatozoa. Because the differentiating cells move only towards the lumen and because new clones of cells enter the differentiation process at regular intervals that are about one-fourth of the time for stem cells to become sperm, about four generations of germ cells can be observed in any given testicular tubule cross section. Also the kinetics of germ cell differentiation is precisely regulated so that the same stages of spermatogonia, spermatocytes, round spermatids, and late spermatids are always found in association. Finally, entry of cells into the differentiation process is spatially synchronized over some distance in many species so that an entire tubule cross section should most often have cells at the same stage.

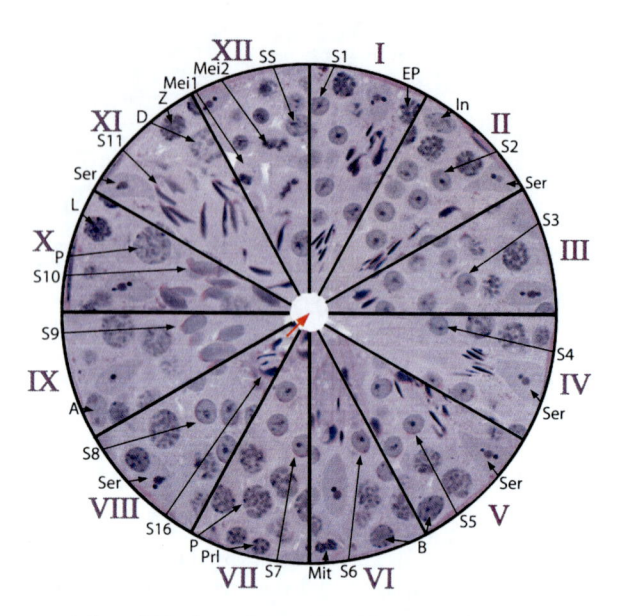

Fig. 1. Representation of tubule sections at the 12 stages of the mouse seminiferous epithelium with nuclei of characteristic cell types indicated. Images were from tissue fixed by vascular perfusion with 3% glutaraldehyde, embedded in glycol methacrylate, and stained with PAS/hematoxylin. The circular arrangement emphasizes the continuous nature and lumenal movement of the cells as they progress and shows the relative positions of the cells. The stage is given the *Roman numeral* corresponding to the *Arabic numeral* of the step of the younger generation of spermatids observed. *Ser* Sertoli cell, *A* type A spermatogonium, *In* intermediate spermatogonium, *B* type B spermatogonium, *Prl* preleptotene spermatocyte, *L* leptotene spermatocyte, *Z* zygotene spermatocyte, *EP* early pachytene spermatocyte, *P* pachytene spermatocytes, *D* diakinesis, *Mei1–2* first and second meiotic divisions, *SS* secondary spermatocyte, *S1–8* round spermatids, *S9–11* elongating spermatids, *S16* late spermatids, *red arrow* spermiation.

This presence of repeated cell associations enabled Leblond and Clermont (1) to identify 14 stages in the rat and Oakberg (2) to describe 12 stages in the mouse (Fig. 1). Additional guides for staging in the mouse and rat have subsequently been published (3, 4). Here we focus on the application of staging to studies of the mouse. Currently the mouse is the most widely used species for generation of gene mutations and many show pathology in spermatogenesis. The knowledge of these stages during histological assessment of spermatogenesis adds power to the analysis and avoids errors that occur when an observer is unaware of these stages.

Since the kinetics of the process is precisely regulated, the durations of each stage and the time to pass through the entire sequence of stages are known; this time to complete the 12 stages, known as the duration of the cycle of the seminiferous epithelium, is 8.6 days in the mouse (2, 5). Knowledge of the time intervals required for a cell in one stage to differentiate into a more mature cell is very useful when assessing spermatogenesis at a given time after a toxicant exposure or a stage-specific developmental change due to expression (or lack thereof) of a gene. Although it is possible to manually calculate these times, the calculations are greatly facilitated, and the relationship between the cell stages is more easily visualized, using the STAGES software (6) (see Note 1).

2. Materials

The materials required for Bouin's fluid or paraformaldehyde fixation of tissue, paraffin embedding, and periodic acid Schiff (PAS)–hematoxylin staining have been outlined on page 265 in a recent methods paper in this series, Methods in Molecular Biology, vol. 558, 2009 (7). Reagents needed for modified Davidson's fluid are a 37–40% solution of formaldehyde, ethanol, glacial acetic acid, and deionized water.

3. Methods

3.1. Sample Preparation and Staining

Likewise the methods for immersion fixation in Bouin's, paraffin embedding, and PAS–hematoxylin staining have been also outlined on page 266 of Methods in Molecular Biology, vol. 558, 2009 (7). Although immersion in Bouin's fixative is excellent for general use, comparisons of different routes of fixation, fixatives, and embedding media have been previously discussed (8, 9). Modified Davidson's fluid provides equally good preservation of morphology and also provides better accessibility of epitopes for immunohistochemical reactions (10). Modified Davidson's fluid is prepared by adding 0.3 volumes of a 37–40% solution of formaldehyde, 0.15 volumes of ethanol, and 0.05 volumes of glacial acetic acid to 0.5 volumes of water.

When optimum preservation of morphological structures is required, fixation by vascular perfusion is superior to immersion and may be worth the extra effort involved (11). Paraffin embedding is adequate for most studies, but embedding in plastics such as glycol methacrylate does provide superior morphology, although immunohistochemistry is not usually possible (8). When morphological identification of the different subtypes of type A spermatogonia is desired, perfusion fixation with glutaraldehyde and embedding in an epoxy, such as Araldite, is required (11).

3.2. Recognition of the Stages of the Cycle of the Seminiferous Epithelium

For most practical purposes, it is highly recommended that PAS–hematoxylin staining of sections of paraffin-embedded, Bouin's or modified Davidson's fluid-fixed, testicular tissues be used. The methods for distinguishing stages in such material have been well outlined on pages 266–270 of Methods in Molecular Biology, vol.

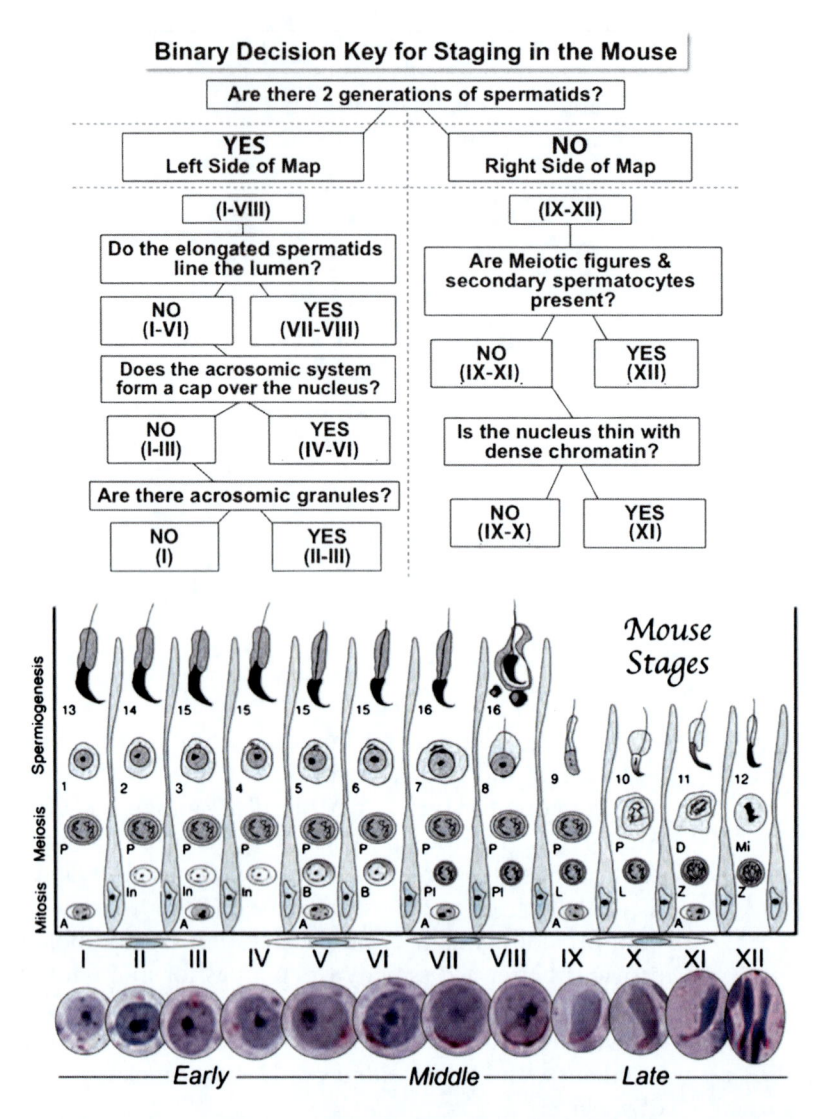

Fig. 2. Binary decision key for rapid identification of specific stages of spermatogenesis, using staging maps, as previously published (3) and illustrated in the lower portion of the figure. The first question of whether two generations (round and elongated) or a single generation (elongating or early elongated) of spermatids are present within the same seminiferous tubular cross section enables identification of the tubule as being either in the early/middle or the late set of stages. Adapted with permission from Dr. Robert E. Braun and ref. 19.

558, 2009 (7). Here we provide a roadmap for the staging process in the form of a binary decision key, simplified from one that was previously formulated for the rat (4) involving a series of yes/no answers (Fig. 2). The first two questions are used to narrow the number of stages to four groups out of the 12 described for the mouse. An observer should then use these four groups to perform a more detailed examination to identify individual stages.

With PAS–hematoxylin-stained material, the development of the acrosome cap or the morphology of the younger generation of spermatids should be used to determine the exact stage (Fig. 1). Steps 1–7 spermatids are defined by the development of the acrosomic granule and spreading of the acrosome over the nuclear surface. Step 8 is primarily distinguished by the release of step 16 elongated spermatids through spermiation and the movement of the nucleus from a central position in the cell so that the acrosome approaches the surface of the cytoplasm and sometimes points towards the basal membrane of the tubule. Steps 9 through 12 are defined by the shape and chromatin condensation of the spermatid nucleus as well described by Oakberg (2) and also presented by Meistrich (12).

Often the material may not be ideal for identification of all 12 stages of the cycle of the seminiferous epithelium. This occurs in the following instances: (a) when cryosections are used, (b) when only hematoxylin and not PAS staining can be observed, (c) when performing immunofluorescence with only a DNA-binding fluorochrome, and (d) when cell generations may be missing or abnormal because germ cell development has not yet been achieved (young mice) or is blocked at some stage because of activation of a mutant gene, or (e) because a toxicant has eliminated a subset of sensitive cells. In such cases the first two questions on the binary decision key are useful to narrow down the possible stages to one of the four groups.

Other cell types in the cell associations can also be used to localize the tubule to a limited set of stages. These include the morphology of differentiating spermatogonia, increasing number of differentiating spermatogonia or preleptotene spermatocytes during stages I–VIII, transitions of preleptotene spermatocytes (stage VIII) to leptotene (stages IX–X), major changes in the size and position of the pachytene spermatocytes (stages I–III vs.VI–XI), and formation of the residual body of the late spermatids (stages VII and VIII). Many of these principles are included in the description of staging from hematoxylin staining and when cell types are missing as described on pages 270–275 and Table 2 of Methods in Molecular Biology, vol. 558, 2009 (7). It should be noted that the use of the morphology of the differentiated spermatogonia and preleptotene spermatocytes to distinguish stages I–VIII requires excellent quality preparations and a high level of skill.

4. Examples of Application

4.1. Loss of Entire Layers of Germ Cells in the Etv-5 Knockout Mouse

The seminiferous epithelium is capable of losing entire layers of germ cells and yet remaining otherwise intact. In such cases, recognition of an abnormality could be overlooked at low magnification and without an understanding of stages it would be possible to misread the histopathological changes. The transcription factor

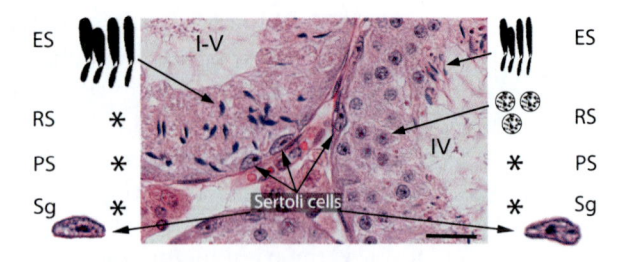

Fig. 3. *Etv5–/–* mutant mouse testis at 6 weeks of age showing two seminiferous tubules (stages I–V and IV) with diagrams of germ cell layers that should be present and missing layers indicated (*asterisk*). *Sg* spermatogonia, *PS* pachytene spermatocytes, *RS* round spermatids, *ES* elongating spermatids. Bar, 20 μm.

Ets-variant gene 5 (*Etv5*) is essential for maintaining self-renewal of spermatogonia in the adult seminiferous epithelium, and although the deletion of *Etv5* (*Etv5–/–*) allows the progression of a first wave of spermatogenesis, it results in the eventual loss of subsequent layers of germ cells (13). This model provides an example of "maturation depletion" of the germinal epithelium, whereby at one point there will be normal spermatogenesis, but at subsequent time points, seminiferous tubules will lack complete layers of specific germ cells (Fig. 3). Using staging recognition patterns, it was possible to determine the specific germ cell types that are missing and, with the STAGES program, show that in these tubules there was no wave of spermatogenesis initiated in these tubules for 29 days prior to tissue harvest, which would indicate the loss of the stem cells by 13 days of age.

4.2. Failure to Form Acrosomes in the Fads2 Knockout Mouse

Docosahexaenoic acid (DHA) is an omega-3 fatty acid that is synthesized by a Sertoli cell enzyme encoded by the *Fads2* gene. *Fads2–/–* mice are infertile due to the formation of abnormal spermatids, sloughing of germ cells, and failure to form acrosomes (14). This animal model is an example of why it is important to use proper fixation and PAS staining for the evaluation of testicular histopathology. In Fig. 4, the seminiferous epithelium of the *Fads2–/–* testis shows normal cellularity, although the heads of the late spermatids appear abnormal. However, careful examination of the round spermatids revealed that the acrosomic cap, which is prominent at stage VIII in the wild-type testis, is missing in the knockout mice. Although round spermatids without acrosomic caps are usually found in stages I–III, this tubule is shown to be in stage VIII by the number and morphology of the preleptotene spermatocytes, large size of the pachytene spermatocytes, translocation of the round spermatid nucleus to the surface of the cytoplasm, and alignment of the late spermatids at the lumen. Thus in the absence of acrosome formation, it was necessary to use these other stage characteristics to localize the onset of this specific lesion.

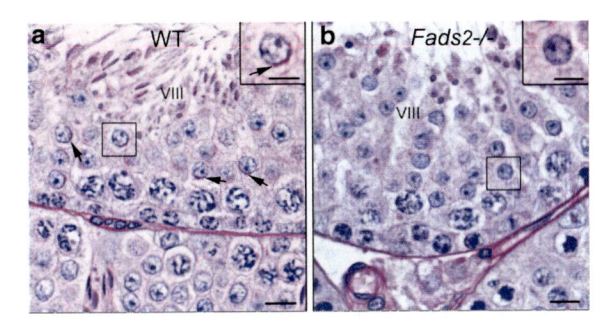

Fig. 4. (a) Wild-type (WT) testis showing stage VIII with step 8 round spermatids capped with prominent acrosomes that are PAS+ (*arrows*). (b) In the *Fads2−/−* testis, spermatids show a complete lack of acrosome formation. Without acrosomic information, recognition of the tubules as being in stage VIII depended upon other features, such as movement of the round spermatid nucleus to the edge of the cytoplasm. Bars in main panels, 10 μm. Bars in the insets, 5 μm.

4.3. Loss of Late Spermatids in Tnp1, Tnp2 Double Knockout Mouse

Mice lacking the genes for transition nuclear proteins (*TNPs*) 1 and 2, the proteins that replace the histones and are subsequently replaced by the protamines, are sterile and have extremely low epididymal sperm counts (15). The *TNPs* are normally first synthesized in elongating spermatids (steps 10 and 11). Histological examination indicated that in *Tnp1,Tnp2* null mice the previous stages of spermatogenesis were normal, but late stages of spermiogenesis were affected. Since it is not possible to reliably distinguish step 13–16 spermatids, these cells were staged by their specific associations with spermatids of the next generation (steps 1–8) in stages I–VIII. The number of step 1–13 spermatids in the *Tnp1,Tnp2*-null mice was identical to controls although some of the step 13 spermatids had reduced chromatin condensation (Fig. 5). There was a progressive decline in spermatid number beginning at step 14. Furthermore the presence of a few late spermatids in stage IX–XII tubules was noted, showing that there was failure of release of the abnormal spermatids accounting for the low epididymal sperm counts. Thus staging allowed the determination of specific steps of spermatid development that were affected as a result of the absence of the *TNPs*.

4.4. Differential Expression of Spermatogonial Markers

It is known that the protein *PLZF* (official designation *ZBTB16*) is expressed in the stem spermatogonia and early progeny (16) whereas *SOHLH1* is expressed in differentiating spermatogonia (17). To determine more precisely in which spermatogonial subtypes the proteins were present, it was necessary to visualize both proteins simultaneously and the cells present in the tubules using 2-color immunofluorescence with DAPI staining for DNA. With DAPI staining it was not possible to identify all stages, so tubules were staged in three groups, I–VI, VII–VIII, and XI–XII, by the presence, morphology, and position of the nuclei of the older

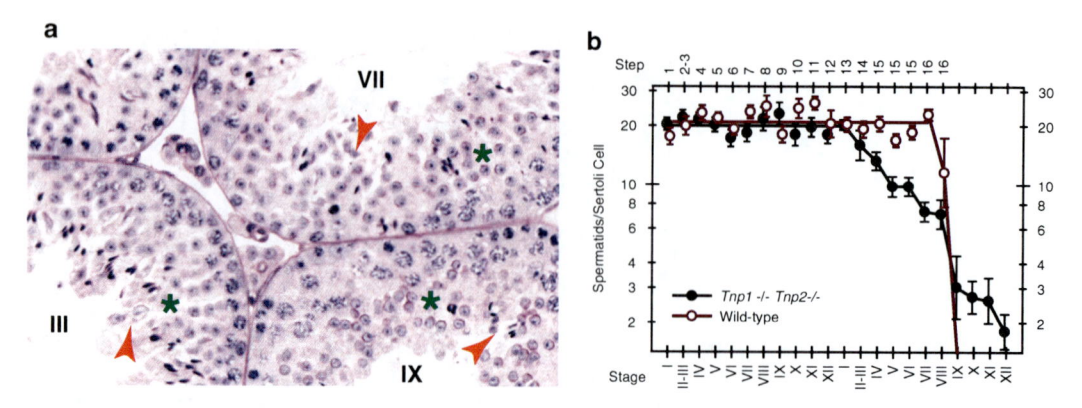

Fig. 5. (a) Staged seminiferous tubules in *Tnp1,Tnp2*-null mouse testes. (*Asterisk*) indicates younger generation of spermatids used to determine the stage. *Arrowheads* indicate abnormalities in older generation of spermatids: Stage III—some step 14 spermatids are decondensed; Stage VII—markedly reduced number of step 16 spermatids; Stage IX—failure of release of mature spermatids. (b) Decline in spermatid number beginning at step 14 in *Tnp1,Tnp2*-null mouse testes determined from analysis of staged tubules. Modified with permission from ref. 15.

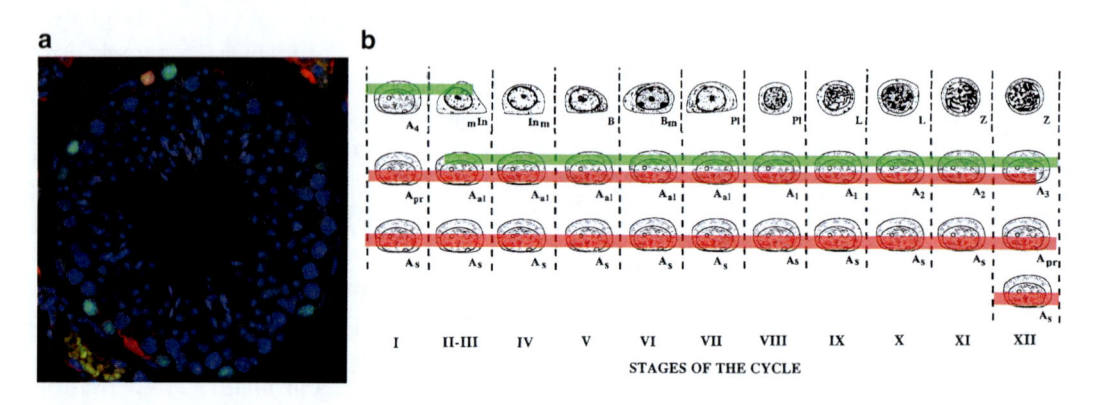

Fig. 6. (a) Immunofluorescence of a mouse seminiferous tubule that is in Stages I–VI of the cycle using antibodies to *PLZF* (*red*) and *SOHLH1* (*green*); a cell positive for both proteins appears *yellow*. DAPI was used to stain DNA (*blue*) and used to identify the stage group by the presence of round spermatids and elongated spermatids that are not aligned at the lumen. (b) Spermatogonial and early spermatocyte types at the 12 stages of the cycle, showing those containing *PLZF* (*red bar*) and those containing *SOHLH1* (*green bar*).

generation of spermatids. The number of spermatogonia stained with an antibody to *PLZF* (Calbiochem), stained with an antibody to *SOHLH1* (Abcam), and double stained (Fig. 6a) was counted in each tubule and averaged for each group of stages. The expected number of different types of spermatogonia in each stage was calculated from the number of spermatogonia in specific stages (18) and the duration of the stages. Then different assumptions were made as to the stages and cell types at which *SOHLH1* is first detected, *PLZF* becomes undetectable, and then when *SOHLH1* becomes undetectable. The appearance of *SOHLH1* in A_{al} spermatogonia at stage II, loss of *PLZF* in A_3 spermatogonia at stage XII, and loss of *SOHLH1* in intermediate spermatogonia in stage

II (Fig. 6b) gave the best fit to the data. Thus, knowledge and application of the stages were essential to the localization of expression of proteins in spermatogonia, which cannot be simultaneously identified by morphology.

5. Note

1. STAGES: Interactive software on spermatogenesis (1998) is available from Vanguard Productions, Inc., 1113W. John St., Champaign, IL 61821, or e-mail: stagescycle@me.com.

References

1. Leblond CP, Clermont Y (1952) Definition of the stages of the cycle of the seminiferous epithelium in the rat. Ann N Y Acad Sci 55:548–573
2. Oakberg EF (1956) A description of spermiogenesis in the mouse and its use in analysis of the cycle of the seminiferous epithelium and germ cell renewal. Am J Anat 99:391–413
3. Russell LD et al (1990) Staging for laboratory species. In: Russell LD, Ettlin RA, Hikim APS, Clegg ED (eds) Histological and histopathological evaluation of the testis. Cache River Press, Clearwater, FL, pp 62–194
4. Hess RA (1990) Quantitative and qualitative characteristics of the stages and transitions in the cycle of the rat seminiferous epithelium: light microscopic observations of perfusion-fixed and plastic embedded testes. Biol Reprod 43:525–542
5. Clermont Y, Trott M (1969) Duration of the cycle of the seminiferous epithelium in the mouse and hamster determined by means of ³H-thymidine. Fertil Steril 20:805–817
6. Hess RA, Chen P (1992) Computer tracking of germ cells in the cycle of the seminiferous epithelium and prediction of changes in cycle duration in animals commonly used in reproductive biology and toxicology. J Androl 13:185–190
7. Ahmed EA, de Rooij DG (2009) Staging of mouse seminiferous tubule cross-sections. Methods Mol Biol 558:263–277
8. Hess RA, Moore BJ (1993) Histological methods for the evaluation of the testis. In: Chapin RE, Heindel JJ (eds) Methods in toxicology, vol 3A. Academic, San Diego, CA, pp 52–85
9. Creasy DM (2002) Histopathology of the male reproductive system I: techniques. Curr Protoc Toxicol Unit16 13
10. Latendresse JR et al (2002) Fixation of testes and eyes using a modified Davidson's fluid: comparison with Bouin's fluid and conventional Davidson's fluid. Toxicol Pathol 30:524–533
11. Chiarini-Garcia H, Meistrich ML (2008) High resolution light microscopic characterization of spermatogonia. Methods Mol Biol 450:95–107
12. Meistrich ML (1993) Nuclear morphogenesis during spermiogenesis. In: Kretser DM (ed) Molecular biology of the male reproductive system. Academic, San Diego, pp 67–97
13. Chen C et al (2005) ERM is required for transcriptional control of the spermatogonial stem cell niche. Nature 436:1030–1034
14. Roqueta-Rivera M et al (2011) Deficiency in the omega-3 fatty acid pathway results in failure of acrosome biogenesis in mice. Biol Reprod 85:721–732
15. Zhao M et al (2004) Transition nuclear proteins are required for normal chromatin condensation and functional sperm development. Genesis 38:200–213
16. Buaas FW et al (2004) Plzf is required in adult male germ cells for stem cell self-renewal. Nat Genet 36:647–652
17. Ballow D et al (2006) Sohlh1 is essential for spermatogonial differentiation. Dev Biol 294:161–167
18. Huckins C, Oakberg EF (1978) Morphological and quantitative analysis of spermatogonia in mouse testes using whole mounted seminiferous tubules. I. The normal testes. Anat Rec 192:519–528
19. Hess RA, Renato de Franca L (2008) Spermatogenesis and cycle of the seminiferous epithelium. Adv Exp Med Biol 636:1–15

Chapter 28

Immunohistochemical Approaches for the Study of Spermatogenesis

Cathryn A. Hogarth and Michael D. Griswold

Abstract

Immunohistochemistry is an important technique that uses specific antibodies to determine the cellular localization of proteins/antigens in highly complex organs and tissues. While most immunohistochemistry experiments target protein epitopes, nonprotein antigens including BrdU may also be detected. Briefly, tissues are fixed, processed, sectioned, and then probed by a primary antibody while preserving the integrity of the tissue and cellular morphology. There are various methods available for visualization of the bound primary antibody that involve a reporter molecule which can be detected using light or fluorescent microscopy. Here we describe a basic immunohistochemistry protocol for identifying protein localization in testis sections using protein-specific antibodies.

Key words: Immunohistochemistry, Testis, Antigen retrieval, Antibody, Antigen, Biotin, Fluorophores

1. Introduction

The localization of proteins to single cells within thin sections of tissue, known as immunohistochemistry, was first described in 1941 (1), and this technique has been used extensively to study protein localization patterns in almost every organ. The basic principle behind immunohistochemistry is the recognition of antigens within sections of tissue by protein-specific antibodies and the use of secondary and tertiary molecules to detect the antigen/antibody complex (Fig. 1). These colored reagents are then visualized, photographed, and analyzed using either light or fluorescent microscopy. The technique is made more complex by the vast array of choices now available for different detection methods, multiple options for tissue fixation and preparation, and the need for

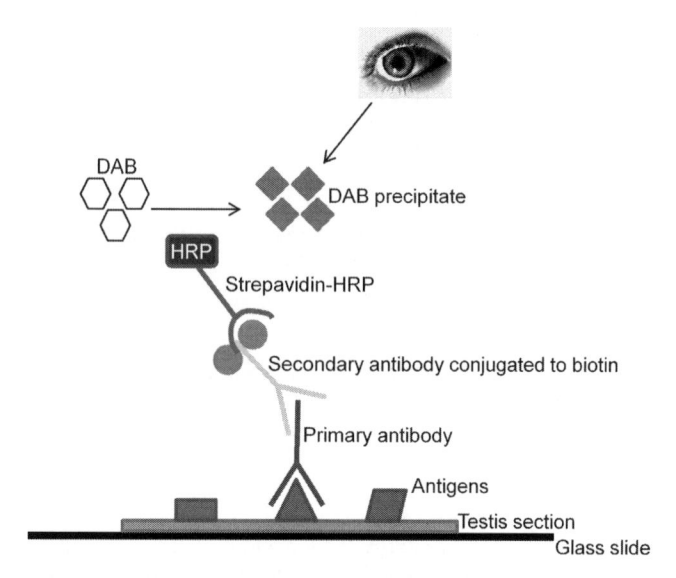

Fig. 1. The basic principle of immunohistochemistry. Immunohistochemistry is the detection of proteins within thin sections of tissue using antibody/antigen complexes. A 4 μm section of testis tissue is fused to a glass slide. This piece of tissue contains proteins which present antigens that may be bound by a primary antibody raised against a particular antigen (*triangle*). In order to visualize the binding of this primary antibody to its antigen, a secondary antibody conjugated to biotin is added to the slide. These molecules are then bound by streptavidin conjugated to horseradish peroxidase (HRP). The enzymatic activity of the HRP is then activated by the addition of the HRP substrate, 3′ Diaminobenzidine (DAB) which, when catabolized, produces a brown precipitate that can be visualized with a light microscope.

stringent controls to determine the sensitivity and specificity for each study (2, 3).

With regard to the study of spermatogenesis, an immunohistochemistry experiment requires fixed testicular tissue that has been embedded in paraffin wax so that thin tissue sections can be cut and mounted on a glass slide. These slides are then subjected to several different treatments to prepare the tissue for incubation with an antibody: (1) tissue deparaffinization using organic solvents and a graded ethanol series, (2) retrieval of antigens masked by fixation procedures using a combination of heat, pH, and/or enzymatic digestion (3, 4), (3) quenching of endogenous enzyme activity if using an enzymatic detection method, and (4) blocking to prevent nonspecific antibody binding. Following the incubation of slides with primary antibody, the detection of antibody/antigen complexes can be performed in various ways, all usually involving the visualization of a secondary antibody conjugated to a fluorophore or an enzyme which can cleave a substrate to produce a detectable product. The end result is testis tissue sections within which cells containing the target protein/antigen can be viewed using light or fluorescence microscopy. Here, we describe a complete immunohistochemical procedure for detecting proteins in sections of testis using STRA8 as the target protein of interest.

2. Materials

Prepare all solutions with deionized water (dH$_2$O). There is no requirement for the solutions to be sterile; however, autoclaving is recommended if solutions are to be stored for long periods of time (>1 month). Prepare and store all reagents at room temperature unless otherwise indicated.

2.1. Fixative Preparation

1. Bouin's fixative: (71% saturated picric acid (21 g/L), 24% formaldehyde (use 37% stock, no methanol), 5% glacial acetic acid). To a 1-L graduated cylinder, add 750 ml saturated picric acid, 250 mL 37% formaldehyde, and 50 mL glacial acetic acid (see Note 1). Cover cylinder with laboratory parafilm and mix by gentle inversion. Transfer Bouin's fixative to a glass bottle for long-term storage.
2. Four percent paraformaldehyde (PFA): Transfer 1 g PFA to a 50 mL tube containing 20 mL dH$_2$O. Add two drops of 1 N NaOH, and incubate at 60°C until PFA has dissolved (see Note 2). Place tube on ice to cool to room temperature, then add 2.5 mL 10× phosphate-buffered saline (PBS) (see Subheading 2.2 for recipe). Adjust pH to 7.4 and add dH$_2$O to a final volume of 25 mL. Four percent PFA is stable at 4°C for up to 1 week.

2.2. Immunohistochemistry

1. Xylene.
2. Graded ethanol series: 100% ethanol, 95% ethanol, 70% ethanol.
3. Antigen retrieval (see Note 3): 0.01 M Citrate Buffer, pH 6.0. To make a 0.1 M (10×) solution, dissolve 117.64 g sodium citrate dihydrate in dH$_2$O, adjust pH to 6.0, then adjust final volume to 4 L with dH$_2$O (see Note 4).
4. Three percent hydrogen peroxide: Dilute 30% hydrogen peroxide stock 1/10 with dH$_2$O (see Note 5).
5. PBS (10×): 1.37 M NaCl, 27 mM KCl, 100 mM Na$_2$HPO$_4$, 24 mM KH$_2$PO$_4$, pH 7.4 (see Note 6).
6. PB: PBS (1×) containing 0.1% bovine serum albumin (BSA). Weigh 0.1 g BSA and transfer to a glass beaker containing 100 mL PBS. Stir to dissolve BSA. Store at 4°C for up to 1 month. For long-term storage, store in 10 mL aliquots at −20°C.
7. Blocking solution: Normal goat serum diluted in PB to 5% (see Note 7). Store at 4°C for 2 weeks.
8. Hematoxylin solution: Prepare a 33% Harris Hematoxylin (Sigma-Aldrich, St. Louis, MO, USA) solution by diluting stock Hematoxylin in dH$_2$O.

9. DPX mounting medium (BDH, Radnor, PA, USA).
10. Superfrost® plus glass slides.
11. Paraffin wax.
12. Embedding cassettes.
13. Embedding oven.
14. Microtome.
15. Slide warmer.
16. Immunohistochemistry slide containers.
17. Immunohistochemistry slide racks.
18. Ice bath.
19. PAP pen (Invitrogen).
20. Glass coverslips (22 mm × 50 mm).
21. Humid chamber: Lay paper towels in a plastic container, moisten paper towels with dH$_2$O, and cover towels with laboratory film.
22. Aluminum foil.
23. 30 G needles.

2.3. Antibodies and Conjugates

1. Rabbit polyclonal STRA8 antibody immune serum (see Note 8).
2. Rabbit polyclonal STRA8 antibody pre-immune serum (see Note 9).
3. Goat anti-rabbit biotinylated secondary antibody (Histostain® bulk kit, Invitrogen) (see Note 10). Store at 4°C.
4. Ready-to-use streptavidin–horseradish peroxidase (HRP) (Vector Laboratories, Burlingame, CA, USA) (see Note 11). Store at 4°C.
5. Liquid 3,3′-Diaminobenzidine (DAB)-Plus substrate kit (Invitrogen): Add 50 µL of reagents #1, #2, and #3 to 1 mL of dH$_2$O. Solution is light sensitive, so cover tube with aluminum foil. Prepare solution fresh before each use.

3. Methods

All procedures are to be performed at room temperature unless otherwise specified.

3.1. Testis Fixation, Embedding, and Sectioning

1. Collect testes from adult male mice. Puncture the testes 10–15 times using a 30 G needle and remove the tunica (see Note 12) before immersing in Bouin's fixative for 5–8 h (see Note 13).

2. Using a scalpel, cut the testes in a transverse fashion and wash testis halves in a graded series of ethanol (see Note 14): wash three times for 1 h each in 70% ethanol, wash one time for 8 h in 80% ethanol, wash one time for 8 h in 95% ethanol, and wash one time for 4 h in 100% ethanol (see Note 15).

3. Dehydrate and embed testis halves in paraffin wax: wash two times for 10 min each in xylene followed by two 1-h immersions in paraffin wax. Embed testis halves cut-side down in paraffin wax. Leave tissue to cure in wax overnight at room temperature before sectioning.

4. Using a microtome, cut 4 μm sections, transfer to Superfrost® Plus glass slides, and seal by incubating slides on a warmer overnight at 37°C (see Note 16).

3.2. Immunohistochemistry

1. Places slides in immunohistochemistry slide containers.

2. Deparaffinize and rehydrate tissue sections using xylene and a graded series of ethanol washes: wash 2 times for 10 min each in xylene (see Note 17), two times for 5 min each in 100% ethanol, one wash for 5 min in 95% ethanol, and one wash for 5 min in 70% ethanol. Rinse using running tap water for 5 min to remove any remaining ethanol and then finish with a 5-min dH$_2$O wash.

3. Antigen retrieval: Place slides in a plastic immunohistochemistry slide rack in a microwaveable container (see Note 18). Immerse slides in 0.01 M citrate buffer (pH 6), microwave on HIGH power until a heavy, rolling boil is achieved, and then continue to microwave on HIGH power for an additional 5 min (see Notes 3 and 19). Place container in an ice bath and allow the liquid to cool for approximately 20 min.

4. Wash briefly with dH$_2$O and using a PAP pen, circle each tissue section (see Note 20).

5. To quench endogenous peroxidase activity, pipet 100 μL of 3% hydrogen peroxide onto each section and lay slides in the humid chamber for 5 min (see Note 21).

6. Wash slides briefly in dH$_2$O.

7. Wash slides two times for 5 min each in PBS.

8. To block tissue sections, pipet 100 μL of blocking solution (see Note 22) onto each section and incubate slides in the covered humid chamber for at least 20 min (see Note 23).

9. Prepare an appropriate amount of STRA8 immune serum or STRA8 pre-immune serum (negative control) in blocking solution (see Note 24).

10. Drain blocking solution by tapping the edge of the slides on a paper towel and pipet 100 μL of prepared STRA8 antibody

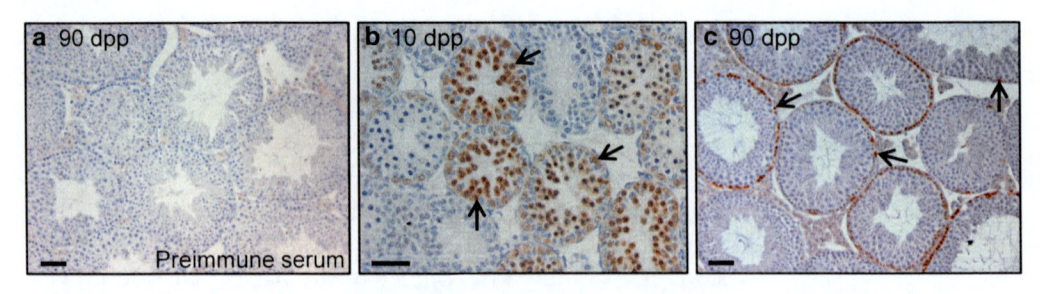

Fig. 2. STRA8 localization in the 10 and 90 dpp testis. (**a–c**) Immunohistochemistry detecting STRA8 protein in 10 and 90 dpp mouse testis sections. A section of adult mouse testis incubated in pre-immune serum serves as the negative control and is shown in Panel (**a**). Scale bars represent 50 µm. *Black arrows*: STRA8-positive germ cells. These sections were incubated in DAB substrate for 60 s.

solution (or negative control solution) onto sections. Incubate slides in a well-sealed humid chamber overnight (see Note 25).

11. Wash slides three times for 5 min each in PBS.
12. Prepare an appropriate amount of secondary antibody (see Note 26).
13. Pipet 100 µL of the goat anti-rabbit biotinylated secondary antibody onto each section and incubate slides in the humid chamber for 1 h (see Note 27).
14. Wash slides three times for 5 min each in PBS (see Note 28).
15. Pipet 100 µL of the ready-to-use streptavidin–HRP onto each section and incubate slides in the humid chamber for 30 min.
16. Wash slides three times for 5 min each in PBS (see Note 29).
17. Prepare liquid DAB substrate.
18. Pipet liquid DAB substrate onto each section and through a microscope, watch for cells to develop a brown precipitate. Place slide in dH$_2$O when the desired amount of signal has been reached (Fig. 2). Record the development time for each sample (see Note 30).
19. Wash slides briefly in dH$_2$O.
20. Counterstain tissue sections with a drop of Hematoxylin solution for 10 s and terminate staining by placing slide in dH$_2$O (see Note 31).
21. Dehydrate tissue by placing slides in running tap water for 3 min followed by one wash for 3 min in 70% ethanol, one wash for 5 min in 95% ethanol, two washes for 5 min each in 100% ethanol, and finally, two washes for 5 min each in xylene.
22. Mount slides under glass coverslips using DPX. Allow the slides to dry for approximately 2 h before visualizing under a light microscope.

4. Notes

1. Saturated picric acid and formaldehyde are both hazardous chemicals. Carefully add each component of Bouin's solution to a graduated cylinder in an approved chemical hood.
2. In order for the PFA to go into solution, heat and a slightly basic pH are required.
3. Antigen retrieval is performed in order to expose epitopes within the tissue that may have been masked by the fixative. There are several different solutions that can be used for the antigen retrieval process. The most common solutions, their components, and incubation procedures are given in Table 1. Citrate buffer is the most common antigen retrieval solution; however, it has been reported that the immunodetection of nuclear antigens can be improved by using a 1% SDS solution antigen retrieval method (5).
4. In our laboratory, we make a 10× stock of citrate buffer for long-term storage at room temperature and make fresh 1× buffer before each use. The 1× buffer should be adjusted to pH 6.0 before use.
5. We prepare fresh 3% hydrogen peroxide before each use.
6. Tris buffered saline (TBS) can be substituted for PBS.
7. For optimal blocking performance, the serum used in the blocking solution should be derived from the animal that the secondary antibody was raised in. For example, the STRA8 antibody we use is a rabbit polyclonal. The secondary antibody we use to detect our rabbit polyclonal antibody is a goat anti-rabbit antibody. Therefore we use normal goat serum in our blocking solution. Kits such as the Histostain® Bulk kit (Invitrogen, Carlsbad, CA, USA) contain ready-to-use blocking

Table 1
Common antigen retrieval solutions and procedures for their use

Antigen retrieval	Solution	Incubation procedure
Citrate	0.01 M Sodium citrate, dihydrate, pH 6	Achieve a heavy, rolling boil and then maintain for 5 min, e.g., Microwave HIGH 7.5 min
Glycine	50 mM Glycine, pH 3.5	Achieve a rolling boil and then maintain a slow boil for 7 min, e.g., Microwave HIGH for 3 min, LOW for 7 min
HCl	1 M Hydrochloric acid	Maintain a slow boil for 10 min, e.g., Microwave MED-LOW for 10 min
SDS	1% SDS in dH$_2$O	Leave at room temperature for 10 min

and secondary antibody solutions for primary antibodies raised in specific animals. If nonspecific signal occurs in your study, the amount of serum in the blocking solution can be increased to 10%.

8. We produced our own STRA8 antibody by injecting rabbits with full-length recombinant STRA8 protein. We collected serum pre- and post-injection. Antibodies can also be generated by synthesizing small peptides specific to antigenic regions of your protein of interest and injecting these peptides into animals. However, commercial antibodies to most proteins are available.

9. The most stringent negative control for any antibody used in an immunohistochemical study is to incubate a section of tissue in serum collected from an animal before it is injected with the protein/peptide you wish to raise an antibody against (pre-immune serum). The pre-immune serum should not bind to your tissue; therefore, all signal that you detect after incubating tissue with the immune serum is from the antibody raised against the protein/peptide antigen the animal was injected with and is not a result of the binding of an endogenous serum protein that can be recognized by the secondary antibody. When purchasing commercial antibodies, rarely is the pre-immune serum also available. However, most companies will sell the peptide that was used to generate the antibody. If so, this peptide can be incubated with primary antibody prior to the immunohistochemical study taking place. If the antibody is specific, this peptide will block the antibody from binding to the section and act as a negative control. This "peptide blocking" also confirms that the antibody is specific to the target protein/peptide. If neither the pre-immune serum nor the peptide is available, then the most common negative control performed is to incubate a tissue section in secondary antibody only.

10. There are many choices regarding commercial secondary antibodies and the reporter molecules with which they are conjugated. In general, you will need to use a secondary antibody which is (A) appropriate for the species your primary antibody was generated in and (B) conjugated to either a fluorophore (for example the AlexaFluor® range, Invitrogen) or to an enzyme (for example HRP). You can also choose to amplify your signal by adding a tertiary level of detection to your study. This involves a secondary antibody usually conjugated to biotin, which can be detected with a streptavidin reagent conjugated to either a fluorophore or an enzyme. We routinely use the tertiary method when detecting antigen/antibody complexes using an enzyme and light microscopy and use the directly labeled AlexaFlour® secondary antibodies for detection via fluorescence microscopy.

11. Another common choice for enzyme detection is to use secondary antibodies or streptavidin molecules conjugated to alkaline phosphatase which can be detected using BCIP®/NBT (Sigma-Aldrich) as the substrate.

12. The testis is covered by a tough fibrous capsule known as the tunica (6). For optimal conservation of testis morphology using immersion fixation, the tunica should be removed by pulling it out and away from the rest of the testis tissue using two pairs of forceps, much like peeling the skin away from a grape. Puncturing the tissue with a 30 G needle also aids by allowing the fixative to diffuse quickly through the tissue. For testes collected from mice aged 30 days post partum (dpp) and older, both detunication and puncturing should be performed. For testes collected from animals younger than 30 dpp, only puncturing is necessary. Puncturing is not required when fixing fetal gonads. Perfusion fixation can also be performed if the resulting cellular morphology from immersion fixation is not satisfactory.

13. Tissue fixation times are dependent on the type of fixative and size of the tissue. Table 2 describes the fixation times for testis tissue collected from mice of different ages for both Bouin's solution and 4% PFA. Over-fixing the tissue can lead to excessive nonspecific binding or make the antigen of interest very difficult to detect within the tissue sections.

14. When performing 4% PFA fixation, the tissue should be washed three times in PBS for 10 min per wash, and then washed in a 1:1 solution of PBS:70% ethanol for 30 min before the graded series of ethanol washes are performed. These wash steps may not be necessary if you plan to have a histology Core laboratory perform your tissue processing and embedding for you.

Table 2
Fixation times for testis tissue collected from mice of different ages

Age	Bouin's solution[a]	4% PFA[b]
Fetal gonads	30 min	1 h
0–10 dpp	3 h	4 h
10–30 dpp	6 h	8 h
30 dpp and older	8 h	8 h to overnight

[a]All Bouin's fixations are performed at room temperature on a rotating wheel so that the tissue can move through the liquid
[b]All PFA fixations are performed at 4°C on a rotating wheel so that the tissue can move through the liquid.

15. The time for the 70, 80, and 95% ethanol washes is extremely flexible. These washes can be performed for a minimum of 1 h to indefinitely. Bouin's fixed tissue can be stored indefinitely in 70% ethanol. The 100% ethanol wash, however, should not exceed 4 h as the tissue will dry out and become difficult to section.

16. We routinely put three sections of tissue onto one glass slide to allow a negative control and two different primary antibody concentrations to be tested on the same block of tissue on the same slide.

17. Histolene can be used in place of xylene if a less potent organic solvent is required, e.g., when dehydrating X-gal-stained tissue if the X-gal stain is to be preserved. If detecting antibody/antigen complexes in Bouin's fixed tissue using fluorescent detection reagents, extending the xylene washes to two washes for 20 min each can aid in reducing the amount of background auto-fluorescence that can be present in Bouin's fixed tissue.

18. Specialized immunohistochemistry slide racks and containers can be purchased from several different commercial vendors (e.g., Electron Micrscopy Sciences). Slide racks are worth purchasing. For the slide incubation steps, rectangular plastic containers purchased from any general retailer will suffice and are much cheaper than the specialized containers.

19. Antigen retrieval is only necessary if the antigen is masked by the fixation procedure. Each antibody can be tested in an immunohistochemistry experiment without antigen retrieval, especially if nonspecific binding is an issue when antigen retrieval is performed. If it is difficult to detect antibody binding even after a heat/buffer-based antigen retrieval is performed, an enzymatic antigen retrieval, e.g., trypsinization, can be performed either in place of, or in addition to, the heated buffer. Trypsinizing sections may facilitate the breakdown of cross-links formed during the fixation process, thereby exposing epitopes for better antibody recognition (7).

20. The PAP circles allow tissue sections on the same slide to be incubated in different solutions.

21. The quenching of endogenous enzyme activity step is dependent on the selected method of detection. It is performed to eliminate the possibility of any endogenous enzymes being able to catalyze the substrate reaction used to detect antigen/antibody recognition, thereby inducing background signal (8). This step is not required if detecting the antigen/antibody complexes using fluorescence.

22. The pipetting volume of 100 µL is given as a guide only. The liquid only has to cover the tissue section. Small pieces of tissue could require as little as 20 µL, with larger pieces requiring up to 200 µL.

23. The times given for the blocking step or any of the antibody incubation or detection steps (e.g., streptavidin–HRP) serve as a guide only. These can be adjusted depending on the needs of the researcher.

24. The appropriate concentration of primary antibody will need to be determined empirically for each new antibody tested. As a guide, start with a final concentration of 2–5 μg/mL of primary antibody (usually a 1/100 dilution of a commercial antibody in blocking solution) and adjust this concentration based on staining results. With respect to the volume of primary antibody to prepare, this will depend on the number of sections to be stained and on the volume of liquid required to cover the sections during the incubation steps (see Note 22). Prepare as little diluted primary antibody as required for your experiment.

25. The incubation temperature and time will also need to be empirically determined for each new antibody. The common incubation temperatures range from 4 to 37°C and the incubation times range from 1 h to overnight. As a guideline, begin with a room-temperature, overnight incubation and adjust these conditions accordingly based on results. For example, if background staining is a problem, decrease the temperature and/or time frame of the primary antibody incubation step. If no signal is present, increase the temperature and/or time frame. In addition, the humid chamber should be tightly sealed during an overnight incubation to avoid the slides from drying out. Drying of slides will cause excessive background staining of the tissue.

26. The optimal concentration of secondary antibody will depend on each new primary antibody tested and on the type of secondary antibody used. As a guideline, biotinylated commercial secondary antibodies are often used at a 1/500 dilution whereas fluorescent antibodies, such as the Alexa Fluor (Invitrogen), should be used at a 1/1000 dilution. The amount of secondary antibody to prepare will be dependent on the number of sections in your experiment and the size of these sections (see Note 22). Some secondary antibodies come ready-to-use (e.g., Histostain® Bulk Kit (Invitrogen)).

27. If you have chosen a fluorescent secondary antibody, the humid chamber, and any other container used to house the slides from this point forward, they should be covered with aluminum foil to protect the fluorescently labeled sections from the light.

28. If you have chosen a fluorescent secondary antibody method of detection, the slides are now ready for mounting under glass coverslips using a fluorescent mounting medium such as Vectashield (Vector Laboratories). There are many different commercially available fluorescent mounting media, and some include counterstains such as 4′,6-diamidino-2-phenylindole

(DAPI), which stains the nucleus of the cell. This allows the opportunity to compare the cellular location of your protein of interest with the nucleus of the cell and can also aid in identifying cell types. Leave slides to dry, protected from the light, for about 1 h before visualizing under a fluorescent microscope.

29. If you have chosen a fluorescent tertiary molecule for detection, the slides are now ready for mounting under glass coverslips using a fluorescent mounting medium.

30. The substrate development time is antibody and sample dependent. Typically, development times range between 30 s and 10 min for immunohistochemical studies. If you wish to compare controls with treated samples, optimize the substrate development time with a control sample and then incubate all other sections in substrate for the same amount of time.

31. The Harris Hematoxylin counterstain will stain cell nuclei blue and allow the easy identification of cell types within the tissue section. It also provides a nice contrasting color against the brown DAB precipitate for photographing results. There are other commercially available colored stains for histological purposes depending on your individual needs.

Acknowledgement

The authors would like to thank Chris Small and Ryan Evanoff for their critical reading of the manuscript and Ryan for supplying the STRA8 immmunohistochemistry pictures. The work was supported by NIH grant HD RO1 10808 to MDG.

References

1. Coons A, Creech H, Jones R (1941) Immunological properties of an antibody containing a fluorescent group. Proc Soc Exp Biol Med 47:200–202
2. Ramos-Vara JA (2005) Technical aspects of immunohistochemistry. Vet Pathol 42:405–426
3. Mighell AJ, Hume WJ, Robinson PA (1998) An overview of the complexities and subtleties of immunohistochemistry. Oral Dis 4:217–223
4. Yamashita S (2007) Heat-induced antigen retrieval: mechanisms and application to histochemistry. Prog Histochem Cytochem 41:141–200
5. Wilson DM III, Bianchi C (1999) Improved immunodetection of nuclear antigens after sodium dodecyl sulfate treatment of formaldehyde-fixed cells. J Histochem Cytochem 47:1095–1100
6. Russell LD et al (1990) Histological and histopathological evaluation of the testis, 1st edn. Cache River Press, St. Louis, MO
7. Daneshtalab N, Dore JJ, Smeda JS (2010) Troubleshooting tissue specificity and antibody selection: Procedures in immunohistochemical studies. J Pharmacol Toxicol Methods 61:127–135
8. Miller RT, Groothuis CL (1990) Improved avidin-biotin immunoperoxidase method for terminal deoxyribonucleotidyl transferase and immunophenotypic characterization of blood cells. Am J Clin Pathol 93:670–674

Chapter 29

Ultrastructural Analysis of Testicular Tissue and Sperm by Transmission and Scanning Electron Microscopy

Hector E. Chemes

Abstract

Transmission electron microscopy (TEM) studies have provided the basis for an in-depth understanding of the cell biology and normal functioning of the testis and male gametes and have opened the way to characterize the functional role played by specific organelles in spermatogenesis and sperm function. The development of the scanning electron microscope (SEM) extended these boundaries to the recognition of cell and organ surface features and the architectural array of cells and tissues. The merging of immunocytochemical and histochemical approaches with electron microscopy has completed a series of technical improvements that integrate structural and functional features to provide a broad understanding of cell biology in health and disease. With these advances the detailed study of the intricate structural and molecular organization as well as the chemical composition of cellular organelles is now possible. Immunocytochemistry is used to identify proteins or other components and localize them in specific cells or organelles with high specificity and sensitivity, and histochemistry can be used to understand their function (i.e., enzyme activity). When these techniques are used in conjunction with electron microscopy their resolving power is further increased to subcellular levels. In the present chapter we will describe in detail various ultrastructural techniques that are now available for basic or translational research in reproductive biology and reproductive medicine. These include TEM, ultrastructural immunocytochemistry, ultrastructural histochemistry, and SEM.

Key words: Transmission electron microscopy, Ultrastructural immunocytochemistry, Ultrastructural histochemistry, Scanning electron microscopy, Testes, Semen, Spermatozoa, Spermatogenesis

1. Introduction

The late seventeenth century witnessed the birth of fine structural studies of cells with the introduction of the first optical microscopes equipped with single biconvex lenses like the ones developed by Antoni Von Leeuwenhoek that allowed him to identify in 1677 the "animacula" (microscopic animals) in the semen of various species including humans. He produced the first scientific account of the basic structural features of spermatozoa originally published in the

Philosophical Transactions of the Royal Society. Technical improvements in the design, complexity, and resolving power of microscopes followed in the nineteenth and twentieth centuries and afforded the descriptions of spermatogenesis, Sertoli and Leydig cells by European scientists Albert Von Kölliker, Franz Leydig, Enrico Sertoli, and others, who in the mid-1800s greatly contributed to the knowledge of components and microscopic organization of mammalian testes. Further examinations by Branca and Stieve in the early 1900s of the structural details of seminiferous tubules were followed by modern accounts of the dynamics of testicular germ cells by Roosen Runge, Clermont, and others that laid the foundations of our present understanding of spermatogenesis. However, modern insight about the internal structure of testicular cells and spermatozoa awaited the breakthrough of electron microscopy. In the words of D. W. Fawcett (1), "The introduction of commercial electron microscopes in the 1950s and of microtomes capable of cutting ultrathin sections, initiated an exciting and remarkably fruitful period of exploration of biological structure at magnifications up to half a million times and resolutions that have now reached 4–5 Å" (Angstrom, a microscopic unit of length equivalent to 1×10^{-4} μm). These advances provided the basis for understanding the cell biology and physiology of the testis and male gametes and emphasized the functional role played by specific organelles in spermatogenesis and sperm function (Fig. 1). The introduction of scanning electron microscopes (SEMs) extended these boundaries to the recognition of cell and organ surface features and the architectural array of cells and tissues (Figs. 2 and 3). The merging of immunocytochemistry and histochemistry with electron microscopy has further integrated structural and functional features to provide a broad understanding of cell biology and its association with health and disease. With these advances the detailed study of cell organelles is now possible, not only in their intricate structural and molecular organization but also in their chemical composition. Immunocytochemistry identifies cell components and localizes them with high specificity and sensitivity (Fig. 4), and histochemistry can be used to understand their functions (i.e., enzyme activity, Fig. 5). When these techniques are used in conjunction with electron microscopy their resolving power is further increased to subcellular levels.

In the present chapter we will describe in detail various ultrastructural techniques that are now available for basic or translational research in reproductive biology and reproductive medicine. These include transmission electron microscopy (TEM), ultrastructural immunocytochemistry, ultrastructural histochemistry, and SEM. The content is not meant to be addressed to specialists, but to all researchers and clinicians interested in exploring the utility of basic ultrastructural studies of the testis and spermatozoa, their application to the understanding of the biological basis of abnormal cell function, and their use as a diagnostic tool in Andrology.

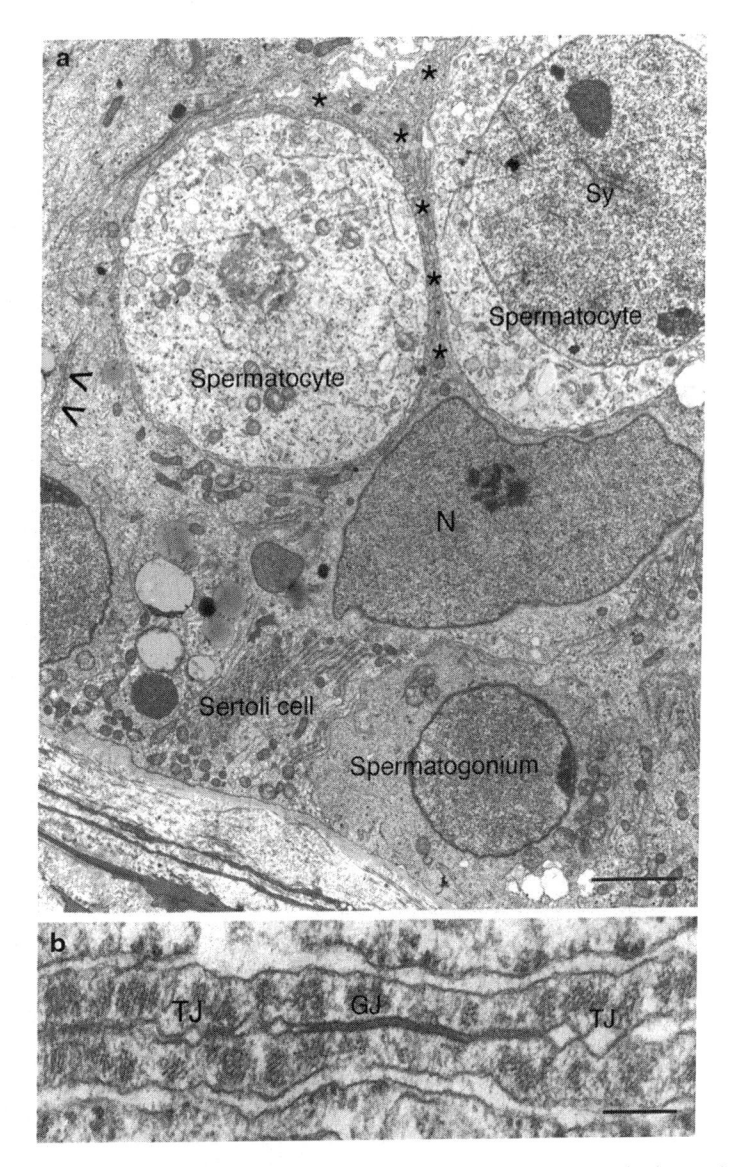

Fig. 1. Transmission electron microscopy. (**a**) Normal adult human testis. Immersion fixation, Epon–Araldite embeddment, Uranyl–lead double staining. A Sertoli cell partially surrounds two spermatogonia resting on the basal lamina and completely encircles two pachytene spermatocytes clearly separated by Sertoli cell processes (*asterisks*). Note synaptonemal complexes (Sy) in the spermatocyte to the *right*. An inter Sertoli junctional complex is seen to the *left* (*arrowheads*). There is a thin knife mark crossing obliquely over the two spermatocytes. Even though it is not desirable, this artifact does not disrupt tissue architecture. Modified from ref. 17. Bar = 1 μm. (**b**) Normal adult rat testis. Perfusion fixation, Epon–Araldite embeddment, Uranyl–lead double staining. A high magnification detail of an inter Sertoli junctional complex. There is a large gap junction (GJ) of close membrane apposition and various point tight junctions (TJ) with fusion of the outer leaflets of both cell membranes. At each side of the intercellular junction there are various filament bundles and a cistern of the rough endoplasmic reticulum. Bar = 0.2 μm.

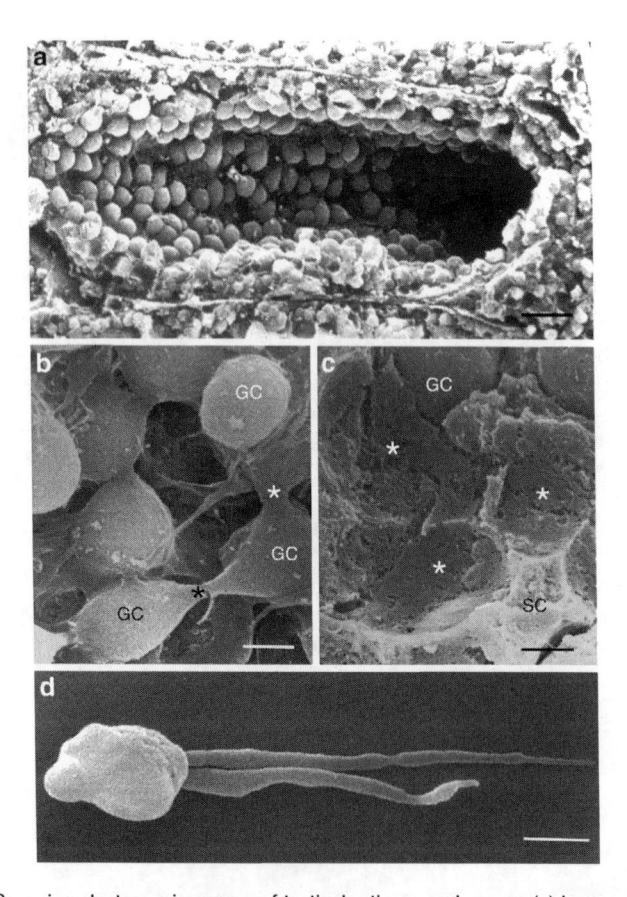

Fig. 2. Scanning electron microscopy of testicular tissue and sperm. (**a**) Low power SEM of a seminiferous tubule in an immature rat. Perfusion fixation. The surface view afforded by the scanning electron microscope clearly reveals the architectural organization of the testis and the tunnel-like appearance of a seminiferous tubule. Its inner surface is covered by spherical immature spermatids. Bar = 40 μm. (**b**) SEM view of the luminal surface of a seminiferous tubule in an immature rat. Various germ cells (GC, early spermatids) are joined by intercellular bridges (*asterisks*). (**c**) SEM of adult rat testis. A close view of the thickness of the seminiferous epithelium shows the body of a Sertoli cell (SC) and numerous SC processes around spherical germ cells (GC). The *asterisks* mark the concave on-face view of SC processes at the sites formerly occupied by three shed germ cells. Compare these Sertoli cell processes (and their relationship with spherical germ cells) whith those in Fig. 1a (as seen in thin sections). Bars = 4.5 μm (**b**), 3 μm (**c**). (**d**) SEM of a two-tailed human spermatozoon from a patient with dysplasia of the fibrous sheath (15, 18). SEM clearly reveals thick irregular tails shorter than 10 μm (normal length 60 μm) (Taken from ref. 18 with permission).

Since electron microscopy allows for the visualization of the interior of the cell and its organelles, a word of caution in relation to subcellular dimensions is appropriate for all those not currently experienced in ultrastructural studies. It is important to understand how small the microscopic field is for electron microscopic evaluation. Metal grids serve as a mechanical support for the thin sections of tissue to be studied. They are assembled by a network of parallel metal bars, crossing at right angles leaving square openings through

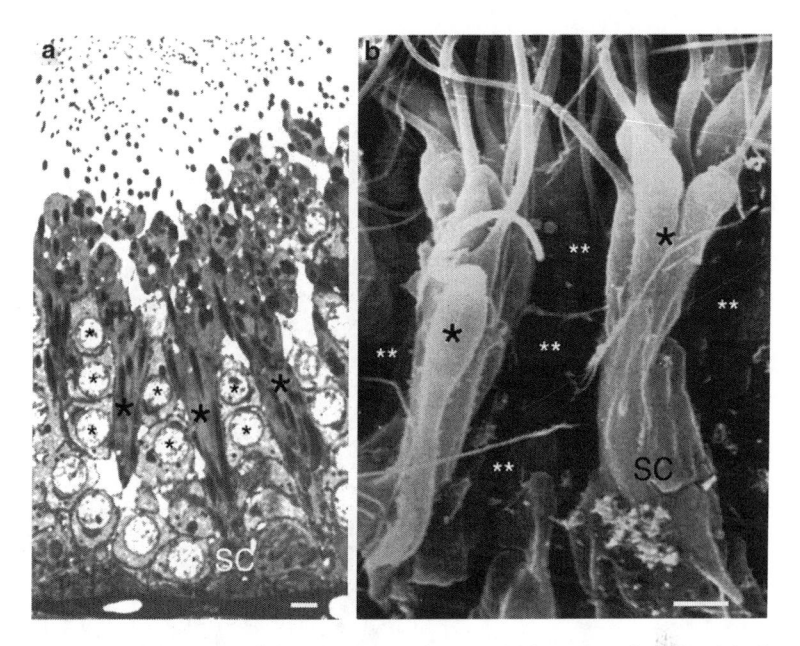

Fig. 3. (a) High-resolution light microscopy of 1 μm thick section of adult rat testis. Perfusion fixation, Epon–Araldite embeddment, toluidine blue staining. Late spermatids are arranged in conspicuous bundles that penetrate deeply into the seminiferous epithelium (*large asterisks*). Rows of early spermatids with prominent round nuclei (*small asterisks*) intercalate between bundles. (b) Scanning electron microscopy of adult rat testis. Perfusion fixation. Similar seminiferous tubule as that depicted in (a). Spermatid bundles (*large asterisks*) stand out dramatically because early round spermatids have detached from the seminiferous epithelium during SEM processing leaving large empty spaces between them (*double small asterisks*). The deepest ends of spermatid bundles are partially covered by Sertoli cell processes (SC). Panel (b) is a good example of a processing artifact (the shedding of round spermatids) that allows better visualization of remaining structures. Bars (a and b) = 5 μm.

which tissue can be observed. The space between the bars is denoted by a number followed by the word mesh. There is an enormous variety of grids, but the most commonly used are 200 or 300 mesh. The surface of a single square is 10,000 μm² in a 200 mesh grid and 2,500 μm² in a 300 mesh grid. *This means that between 100 and 400 squares should be screened to cover 1 mm² of tissue.* Therefore, tissue or cell sampling is of paramount importance when evaluating how significant a finding may be. As a general rule it is desirable to include samples from different regions of an organ. In the case of pellets from cell suspensions (including semen), this requirement is less important assuming that the suspension is homogeneous and the distribution of cells is uniform throughout. The final areas selected for ultrastructural studies should be chosen after careful light microscopic evaluation of semi-thin sections from different tissue blocks to guarantee that all important regions are represented.

The validity of an ultrastructural observation requires consistency of the findings throughout a tissue or cell type. This is

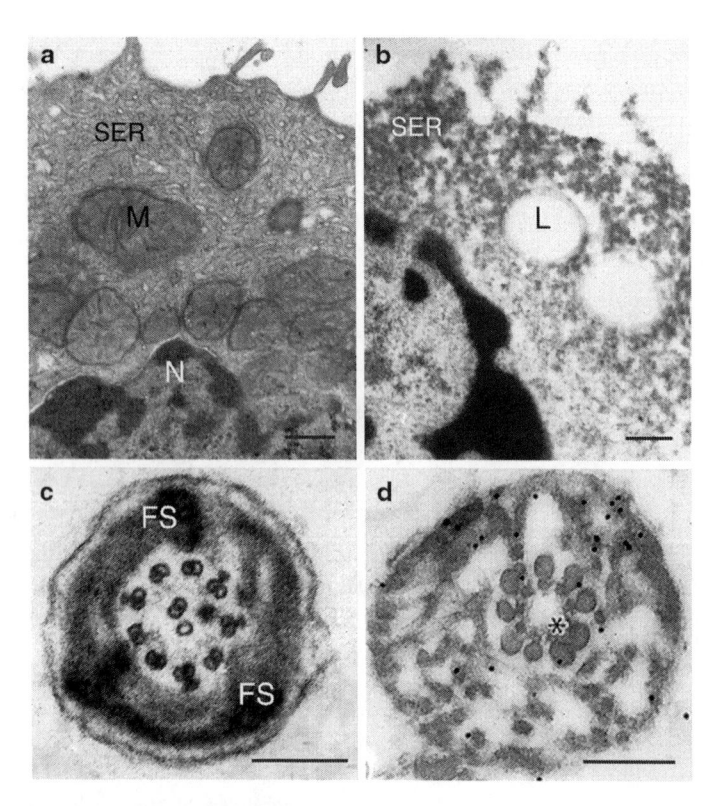

Fig. 4. Pre- and post-embedding ultrastructural immunocytochemistry. Immersion fixation of cell suspensions. (a) Regular TEM. Mature Leydig cell isolated from an adult rat. Note prominent smooth endoplasmic reticulum cisterns (SER), mitochondria (M), and nucleus (N). (b) Pre-embedding immunocytochemistry of a similar Leydig cell using a biotin streptavidin-peroxidase detection system. The cisterns of the smooth endoplasmic reticulum (SER) show dark diaminobenzidine deposits localizing 3 β-HSD, a Leydig cell steroidogenic enzyme. Two lipid droplets (L) are negative. Nuclear chromatin is stained with Uranyl acetate. (c) Regular TEM. Cross section of a human sperm flagellum from a patient with Dysplasia of the fibrous sheath (2, 18). The axoneme is surrounded by a thickened fibrous sheath (FS). Note the absence of dynein arms in the axoneme microtubular doublets. (d) Post-embedding colloidal gold immunocytochemistry of a similar flagellum. LR White embeddment. Localization of AKAP4 (the main protein component of the FS). Gold particles are deposited over the thickened FS (modified from ref. 15). The axoneme (*asterisk*) is not labeled. Bars = 1 μm (**a**), 0.25 μm (**b–d**).

particularly true if the goal is to assign biological or physiopathological meaning to a particular observation. An extraordinary finding should make sense and be consistently present to be considered an original structure or a pathological trait. An *interesting feature* present only in a limited sample from a single patient or animal would probably lack significance or be the result of processing artifacts. These comments are not meant to discourage the search for new features that constitute the basis for scientific advancement. When such results are being considered, their importance as promising findings should be analyzed in the context of current structural and physiological knowledge.

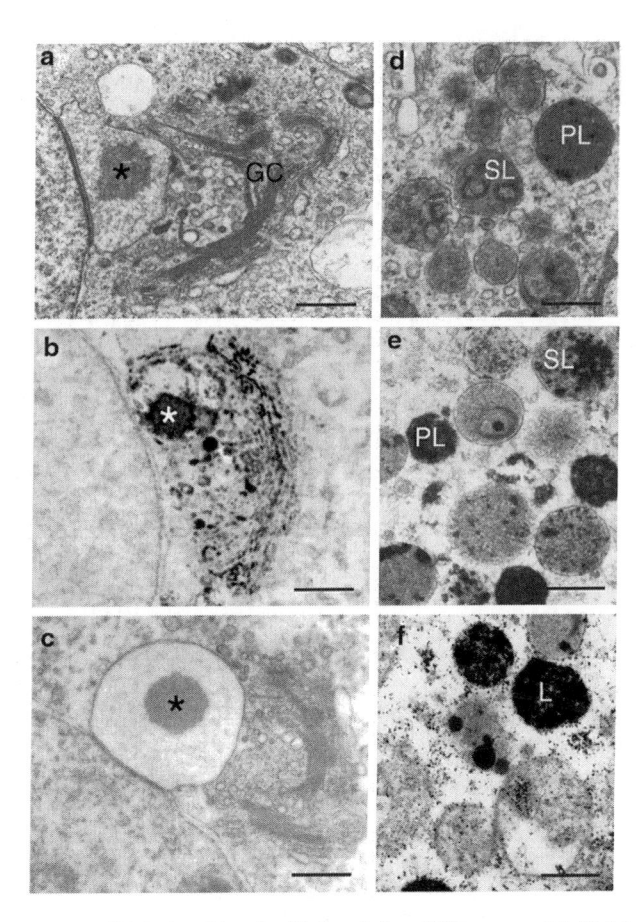

Fig. 5. Ultrastructural histochemistry of acid phosphatase (AP) in spermatids (Golgi complex and acrosome) and Sertoli cells (lysosomes). (a) Regular TEM. An ealy rat spermatid shows a well-developed Golgi complex (GC), an acrosomic vesicle and granule (*asterisk*). (b) AP histochemistry. Golgi cisterns, vesicles, and acrosomic granule depict dense lead deposits indicative of AP enzymatic activity. There is no background staining. (c) A lightly stained negative control (incubaton media contained NaF that inhibits AP activity). No lead deposits. (d) Regular TEM. Sertoli cell primary lysosomes (PL, homogeneous content) and secondary lysosomes (SL, heterogeneous content). (e) AP cytochemical localization. Dark lead deposits in both types of lysosomes, no background. (f) With longer incubation times or higher incubation temperatures AP localization in lysosomes is intense but there is excessive background. All bars 0.25 μm.

2. Materials

2.1. Fixatives/Buffers/Antibodies

1. Glutaraldehyde (8–25–50%), EM grade, sealed in ampoules under inert gas or in larger volume bottles.
2. Formaldehyde, EM grade.
3. 4% Paraformaldehyde-0.25% glutaraldehyde fixative (for immunocytochemistry): For 1,000 ml, heat 400 ml of distilled water to approximately 80°C. Add 40 g of paraformaldehyde and six drops of 1 M NaOH. Maintain solution below 80°C with

constant agitation. When paraformaldehyde is dissolved filter over ice. Add 10 ml of 25% glutaraldehyde and distilled water to complete 500 ml. Add 500 ml of 0.2 M phosphate buffer (combine 90 ml of 0.2 M dibasic Na or K phosphate with 10 ml 0.2 M monobasic Na or K phosphate).

4. Phosphate buffer, 0.1 M, pH 7.4 (used for glutaraldehyde and osmium solutions and for rinses between fixatives; see Note 1): To prepare, combine 90 ml of 0.2 M dibasic Na or K phosphate with 10 ml 0.2 M monobasic Na or K phosphate. Dilute it to 0.1 M by adding 100 ml distilled water and adjust pH to 7.4 with 0.2 M monobasic phosphate.

5. Blocking Buffer (TBS-BSA, to block nonspecific binding in immunocytochemistry): Prepare 50 mM Trizma Hydrochloride and 150 mM NaCl in distilled water. Adjust pH to 7.4 with NaOH 1 N. Add normal goat serum or BSA to 10% concentration.

6. 10 mM Na-Citrate Buffer (for microwave antigen retrieval): Correct pH to 6.0 with NaOH.

7. Osmium tetroxide comes as crystals sealed in ampoules (0.25, 0.5, and 1 g). To prepare solution, immerse ampoule in 40–60°C tap water until crystals melt. Remove the ampoule from the hot water, and rotate it until the liquid osmium solidifies again as a thin layer on the inner face of the glass ampoule. Wash well with detergent, rinse in abundant distilled water, and place the ampoule within a thick walled glass container. Close the bottle and shake vigorously until ampoule breaks. Add enough distilled water to make a 2% solution, close tightly, and leave overnight at room temperature. Keep well sealed at 4°C. To prepare, mix two parts 2% osmium with one part phosphate buffer. Caution: Osmium is very volatile and is an irritant. Avoid contact with skin and eyes, and do not inhale vapors. Always use under hood. Dispose used solutions with toxic materials.

8. Primary antibodies: according to antigens to be localized.

9. Detection kits: Super sensitive link-label immunocytochemistry detection systems are sold for light microscopy but can also be used in electron microscopy. Kits contain biotinylated secondary antibodies with streptavidin-peroxidase and peroxidase detection systems (diaminobenzidine-H_2O_2).

10. Colloidal gold conjugated secondary antibodies.

2.2. Dehydration/Embedding

1. Ethanol 50, 70, 96, and 100%. Prepare lower dilutions from 100% ethanol.

2. Propylene Oxide EM grade.

3. Absolute acetone for dehydration in SEM.

4. Eponate 12—Araldite 502 resin embedding kit (see Note 2). To prepare Eponate–Araldite mix, combine 10 parts Eponate 12 Resin, 10 parts of Araldite 502 Resin, and 30 parts of dodecenyl

succinic anhydride (DDSA). Mix well. The mixture can be moderately heated to facilitate mixing. Keep at 4°C. Before using, add benzyldimethylamine (BDMA) at 3% final concentration (90 μl BDMA for each 3 ml resin mixture).

5. LR white resin, medium grade.

2.3. Staining Solutions

1. Toluidine blue (to stain semi-thin 0.5–2 μm sections for light microscopy): Prepare a 0.1% solution of Na borate in distilled water and dissolve toluidine blue at a concentration of 1%. Filter and use.

2. Lead citrate solution (use to stain thin 80–100 nm sections for electron microscopy): Dissolve 100 mg NaOH in 25 ml of distilled water. The water should be previously boiled to eliminate CO_2 and allowed to cool to room temperature before dissolving NaOH. Add 50 mg lead citrate and dissolve well. Avoid breathing over the solution, and keep it in syringes eliminating all air (lead can combine with CO_2 and form insoluble crystals that may contaminate sections).

3. Uranyl acetate solution (to stain EM sections): Prepare a saturated solution of Uranyl acetate in distilled water. Keep well sealed at 4°C. To use, mix equal amounts of Uranyl solution and absolute acetone.

2.4. Consumables/Tools/Equipment

1. Conical centrifuge tubes (15 ml) (used to obtain sperm or cell pellets).
2. Beem conical tip capsules (0.5 ml) (used to pellet cells or sperm when the number is extremely low; see Subheading 3.1.3).
3. 10-, 20-, 200-, and 1,000-μl Pipetters.
4. Pipetter tips.
5. Flat embedding molds (used to polymerize embedding resins).
6. Filter paper.
7. Single edge stainless steel blades (to dice tissue to be fixed or to trim resin blocks before cutting).
8. Wax plates or trimming boards.
9. Glass strips (to make glass knives).
10. Glass knives.
11. Glass knife boats (to hold water for section flotation after cutting).
12. Fine wooden sticks fitted with a fine bristle or eyelash (to handle floating sections or transfer them to microscopy slides). These can be purchased or prepared in the laboratory.
13. Soft brushes.
14. Glass slides (charged or poly-L-lysine coated to assure cell or section adherence).

15. Diamond knives (light and EM).
16. Diamond knife cleaners: Polystyrene sticks or knife washer.
17. Copper grids (200–300 mesh).
18. Nickel grids (200–300 mesh).
19. Fine tweezers with straight or curved ends (to handle grids).
20. Clamping O-rings (used to secure tweezers in a closed position).
21. Humid chamber (to incubate samples for immunocytochemistry).
22. Bench-top centrifuge.
23. Timers.
24. Hot plate (used to stain semi-thin sections for light microscopy).
25. Polymerization oven.
26. Glass knife maker.
27. Ultramicrotome (to obtain thin sections for electron microscopy).
28. Microslicer or vibratome (to obtain 30–100 μm tissue slices for histochemistry or immunocytochemistry).
29. Conductive adhesives (liquids or paste containing fine particulate silver used to attach SEM samples to metal specimen mounts). Conductive, double-sided, adhesive tabs can be used for the same purpose.
30. Metal specimen mounts for SEM (see Note 3).
31. Hexamethyldisilazane or Tetramethylsilane (used for the preparation of SEM samples as an alternative to critical point drying (CPD)).

3. Methods

The methods described below outline procedures to study testicular tissue, immature testicular germ cells in suspension, and ejaculated spermatozoa in semen. Complete technical details will be provided for regular TEM, ultrastructural histochemistry and immunocytochemistry, and scanning electron microscopy (SEM). The author has extensively used all of these methods with consistently good results for basic research purposes and for diagnostic purposes in the Andrology laboratory. Ultrastructural analysis can provide excellent results, but methods should be meticulously followed. Ensure that working surfaces and tools are very clean and avoid contamination of solutions or mechanical damage to small samples. A clear identification method should be used for samples to be studied by electron microscopy. This is important because thin sections for TEM or small pieces of tissue for SEM cannot be directly identified by labels as those applied to glass slides for light microscopy.

3.1. Processing of Testicular Tissue for Regular TEM

The fixative of choice is EM grade glutaraldehyde, in concentrations ranging from 3 to 5% in phosphate buffer. Glutaraldehyde and buffers are kept at 4°C, and working solutions should be prepared immediately prior to fixation. Testicular tissue can be fixed by immersion or perfusion. Perfusion fixation is preferred because testicular structure is very delicate and can be mechanically disrupted by manipulation of fresh tissue. However, when this is not possible, as for testicular biopsies, immersion fixation can be successfully used.

1. Fixation by immersion: Cover a small testicular fragment with a few drops of cold 3–5% glutaraldehyde on a wax plate or trimming board that allows cutting with no damage to edges of razor blades or scalpels. Delicately dice the sample in small pieces not larger than 1–2 mm^3 *without using forceps to grab the tissue or scissors to cut it* (see Note 4). Transfer fragments to a glass vial containing X10 volume of fixative and maintain at 4°C for 2–3 h or overnight. Following fixation, wash the fixative from the tissue by two changes of cold phosphate buffer, 30 min each (see Note 5).

2. Perfusion fixation (see Note 6): Connect two 500 ml intravenous solution bottles (one for saline solution and the other one for glutaraldehyde solution or other fixative) with flexible tubing to a three-way stopcock. Add an additional piece of tubing and fit with a hypodermic needle at the free end (18–25 g, according to the size of the aorta). Place the bottles 120–150 cm above the level of the animal. Under deep anesthesia open the abdomen with a longitudinal incision, expose the abdominal aorta, and clear it from connective tissue and fat. The inferior vena cava and testicular arteries should be identified and not disturbed. Recline the left kidney to an anteromedial position, and open the retroperitoneal space above it with a blunt iris forceps to expose the abdominal aorta proximal to the origin of the easily identified left renal artery. Pass a silk suture around this section of the aorta with fine curved forceps and leave it loosely tied. Grasp the abdominal aorta just above its distal bifurcation with a curved Halsted forceps and elevate it slightly. Insert the hypodermic needle at the end of the perfusion tubing into the abdominal aorta in a retrograde direction. Open the stopcock to allow saline flow, and tie the suture around the aorta (to avoid perfusion of supra-diaphragmatic organs). Open the left renal vein by a small cut with fine scissors to allow clearance of perfusion fluids. When the vasculature of the testes is cleared from blood (see Note 7), shift the flow of saline to 3–5% glutaraldehyde for 10 min, then remove the testes and immerse them in fixative for 30 min. After this,

the testes are hard enough to be cut in 1–2 mm thick slices and then diced into small cubes of no more than 1–2 mm^3 that are left in fixative for 3–4 h (see Note 8). Following fixation, wash the fixative from the tissue by two changes of cold phosphate buffer, 30 min each.

3. Fixation of suspensions of immature germ cells after enzymatic digestion-separation: Fix separated germ cells by resuspending the cell pellet and allow for 10–20 min in the fixative. Centrifuge fixed suspensions at 150–350×g in a bench-top centrifuge for 10 min, resuspend the pellet in 0.5 ml buffer, and re-centrifuge in small Beem conical tip capsules (0.5 ml; see Note 9). Following fixation, wash the fixative from the pellet by two changes of cold phosphate buffer, 30 min each.

4. Fixation of semen samples for regular electron microscopy: The following technique has been used for the study of semen samples for research or diagnostic purposes (2). Samples should be processed after complete liquefaction (30–40 min after ejaculation, no more than 60 min). Dilute semen to 12–14 ml with saline solution or phosphate buffer at room temperature in a centrifuge tube (conical bottom) and mix well. If the resulting sample is too viscous or the suspension is not uniform, see Note 10. Spin the sample at 350–580×g for 10 min in a bench-top centrifuge. Once the pellet is obtained discard the supernatant by gently tilting the tube. DO NOT use aspiration with pipettes as this can remove parts of the pellet by creating fluid turbulence. Gently add 2 ml of fixative taking care not to unsettle the pellet. If the pellet is thin (up to 2 mm thick) it can be fixed in situ. When it is thicker it should be very gently dislodged from the bottom of the tube with a fine spatula without breaking it by inserting the tip of the spatula between the pellet and the tube wall and making rotating movements. The pellet should be exposed to fixative on both faces. If it is too thick (more than 3–4 mm) gently press it against tube walls to make it thinner. Fix for 3–4 h at 4°C (it can be left overnight). Following fixation, wash the fixative from the tissue by two changes of cold phosphate buffer, 30 min each. If spermatozoa are severely asthenozoospermic or completely immotile this procedure should be modified slightly (see Note 11).

5. Perform secondary fixation (post-fixation) with osmium tetroxide: For secondary fixation immerse tissue blocks or pellets in 1 ml cold 1.3% solution of OsO_4 in 0.1 M phosphate buffer (two parts 2% osmium solution and one part phosphate buffer) for 2–4 h at 4°C (can be left overnight) followed by two washes in cold phosphate buffer, 30 min each. Perform all manipulations with osmium in the hood (see Note 12).

6. Embeddment of fixed tissue in plastic resins for regular TEM (see Notes 13 and 14): Dehydrate tissue blocks using an

ascending series of ethanol (50, 70, 96%) at 4 °C. Perform three rinses of 10 min for each ethanol concentration. Perform the last 96% rinse at room temperature. To achieve complete dehydration immerse tissue blocks in 100% ethanol (three changes of 20 min each) at room temperature, then perform three rinses in propylene oxide (20 min each) at room temperature. After dehydration, infiltrate in a 1:1 mixture of propylene oxide—Eponate–Araldite mix, followed by pure, undiluted, Eponate–Araldite resin (see Note 15).

7. Section resin blocks for ultrastructural studies (see Note 16): Prepare resin blocks for sectioning by eliminating as much as possible all resin material surrounding the tissue. Trim the block cutting surface and sides using single edge razor blades. Secure the block on the ultramicrotome stage under adequate illumination. The first trimming should be parallel to its surface to reach the level of the embedded tissue that is readily visible because of the black color resulting from osmium fixation. Once the tissue is reached, carefully trim the sides of the block to a clean-cut trapezoid with its long axis parallel to the direction of sectioning and its longer base always oriented so that it is the first side to touch the knife edge. The proportion between length and width should be approximately 2:1.5 (see Note 17). For instructions regarding the use of glass and diamond knives, see Notes 18 and 19. Cut semi-thin sections first (1–2 μm thick, see below for details on sectioning), then pick them up from the water using fine sticks fitted with bristles or eyelashes and float them on a drop of water on a clean glass slide that is immediately placed on a hot plate until water dries and sections remain firmly attached to the glass. Without removing slides from the hot plate, cover the sections with a few drops of toluidine blue solution until edges of the drops start to dry and turn iridescent. Remove the slides from the hot plate and wash them in a container filled with tap water, then dry them with filter paper and a hot plate and observe unmounted sections with a light microscope to select the appropriate areas to be studied with the electron microscope. These areas should be carefully trimmed to smaller trapezoids of no more than 0.5 mm in their long axis and ultrathin sections obtained from them. Pick up appropriate sections with copper or nickel grids held with fine tweezers (see Notes 20–22).

8. Staining ultrathin sections: Once the sections are mounted on grids, allow them to dry and subsequently stain them with heavy metal solutions (containing uranium and lead), so that specific deposition of metal salts in different subcellular organelles assures scattering of electrons that will be the basis for image formation in the electron microscope. The general staining of sections for ultrastructural studies is double staining with lead citrate and Uranyl acetate (see Note 23).

3.2. Processing for Ultrastructural Immunocytochemistry

This method is designed to immuno-localize tissue antigens such as enzymes, proteins, or their receptors. Positive results indicate the presence of antigen under study but do not give any information on its functional state. Ultrastructural localization is particularly informative since it identifies particular organelles/subcellular structures where antigens are localized.

1. Fix tissue fragments and semen pellets for ultrastructural immunocytochemistry for 1 h at 4°C either in 5% formaldehyde in phosphate buffer (0.1 M, pH 7.4) or in a mixture of 4% paraformaldehyde and 0.25% glutaraldehyde in the same buffer, and rinse twice in buffer after fixation. Osmium postfixation is to be omitted.

2. Dehydrate samples in an increasing series of ethanol, infiltrate in medium grade LR White Resin, and polymerize at 60°C for 24 h (see Note 24). Mount thin sections (silver to pale golden) on 300 mesh nickel grids and dry them at room temperature (see Note 25).

3. Perform antigen retrieval (3–5) (see Note 26).

4. Wash the grids, and incubate them for 30 min at room temperature in Blocking Buffer (TBS + 10% normal goat serum) to block nonspecific binding and then float them on drops of primary antibody at appropriate dilutions, and incubate in a humid chamber overnight at 4°C.

5. Perform three washes in TBS then further incubate the grids for 1 h at 4°C with Blocking Buffer containing colloidal gold labeled secondary antibody (goat anti-rabbit IgG 10, 15, or 20 nm gold particles) at 1:25 or 1:50 dilutions, and rinse three times in TBS (see Notes 27 and 28).

3.3. Processing for Ultrastructural Histochemistry

Histochemistry is designed to detect cell components such as enzymes by their specific activity, and therefore the results are functionally informative. A suitable substrate and detection system should be used depending on the enzyme under study. The purpose of this section is to describe the general procedures when this technique is used for ultrastructural localization of enzymes or other chemical cell components. A protocol for detection of acid phosphatase activity will be described in detail (see Note 29). We have used this technique to characterize lysosomal activity, distribution, and cyclic variations in rat seminiferous tubules (6).

1. Fix with 2.5% glutaraldehyde for 60 min. The choice of fixatives will depend on their capability to preserve the activity of specific enzymes. Perform perfusion or immersion fixation as appropriate (see Note 30).

2. After fixation, rinse the tissue with buffer, and slice in a Vibratome at 50 μm thick sections.

3. Incubate sections in the medium containing Na β-glycerophosphate at pH 5.0 for acid phosphatase localization (7, 8) (see Note 31). Optimize incubation time and temperature (room temperature or 37°C) to maximize enzyme activity while avoiding background staining as much as possible (Fig. 5).

4. Observe semi-thin 1 μm sections without any further staining for a black deposit indicative of acid phosphatase activity under a light microscope.

5. Analyze ultrathin sections by electron microscopy either unstained or briefly stained with Uranyl acetate and lead citrate.

3.4. Preparing Samples for Scanning Electron Microscopy

1. Fix and dehydrate tissues: Perform primary fixation of tissue fragments or cell/sperm suspensions with glutaraldehyde and secondary fixation with osmium tetroxide (see Note 32). Sediment fixed cells or spermatozoa on positively charged or poly-L-lysine coated small glass slide fragments to ensure cells/sperm adherence to the glass. Control sedimentation under the light microscope to obtain an adequate microscopic density avoiding clumping or overlapping of cells or spermatozoa (9). Dehydrate tissue fragments or cells adhered to glass fragments in a graded series of ethanol (50, 70, and 96%, three changes, 10 min each, at 4°C) followed by absolute ethanol and acetone at room temperature (three 10 min changes for each).

2. Dry the samples: After dehydration remove fluids from tissue/cells taking care to avoid surface tension forces (see Notes 33 and 34).

3. Expose tissue surface features: When studying spermatozoa or spermatogenic cells by SEM, surface features of cell membranes are readily visible (Fig. 2d). To examine testicular tissue some procedures may be necessary to better expose surface features or intercellular relationships. This may require some dissection with rather blunt instruments to promote cell shedding and exposure of intercellular spaces (Figs. 2c and 3b). To observe the inner surface of seminiferous tubules a cut with a sharp razor blade parallel to the longitudinal axis of seminiferous tubules would be helpful (Fig. 2a, b). These manipulations can be performed in the soft tissue after osmium fixation. After drying, tissue is fragile and manipulations should be very gentle. Some degree of experimentation and trust in serendipity is necessary as well as intuitive tissue exploration in search of interesting features. Actually, some artifacts may help when studying some structures as illustrated in Fig. 3b for late spermatid bundles.

4. Mount specimens and apply metal coating: After drying, mount tissue fragments or cells attached to glass slide fragments on metal holders adapted for different SEMs. Fasten the tissue/

cells to holders with special adhesives containing metal particles or doubled sided, adhesive, conductive tapes to ensure conductivity. After adhesion, coat the samples with a thin fine-grain coat of gold-palladium to be observed with the SEM (see Note 35).

Tissue samples or slide fragments with attached cells/spermatozoa are now ready to be studied by SEM (Figs. 2 and 3).

4. Notes

1. Cacodylate Buffer and Collidine Buffer are considered excellent all-purpose buffers; however, Cacodylic acid is highly toxic and Collidine solutions are somewhat difficult to prepare. Phosphate buffers are inexpensive nontoxic and easy to prepare.
2. We prefer this mixture because it has the combined advantages of Eponate (it stains well) with those of Araldite (it cuts very well).
3. These come in various shapes including pin stubs, cylinders, mounts with clips for thin specimens, etc. for use with SEMs of different brands. There are also adaptable holders to adjust the viewing angle.
4. Care should be taken to avoid mechanical disruption that may occur by grasping or pressing the tissue with forceps or scissors while cutting it.
5. If the tissue is too soft and difficult to cut into small pieces without distortion it can be first immersed in the fixing solution for 5–10 min. This will harden it and facilitate cutting. Immersion fixation as described is particularly suited for testicular biopsies of primates (including humans) but not for rodent testes. Rodent testes have extensive peritubular lymphatic spaces (Fig. 6) and cannot be cut before fixation without considerable architectural distortions. Seminiferous tubules shrink, and lymph spaces around them widen, seriously compromising structural preservation. In cases when perfusion cannot be performed it is better to immerse the whole uncut organ in cold glutaraldehyde, and leave it for about 30 min. This will harden the superficial tissue immediately underneath the albuginea allowing it to be cut into smaller pieces with less damage. These should be immersed in X10 volume of fixative for three to four more hours. *The deepest softer parts of the testis, where the fixative did not penetrate, should not be used.* To facilitate fixative penetration a few punctures can be made in the albuginea at both poles of the organ prior to immersion.

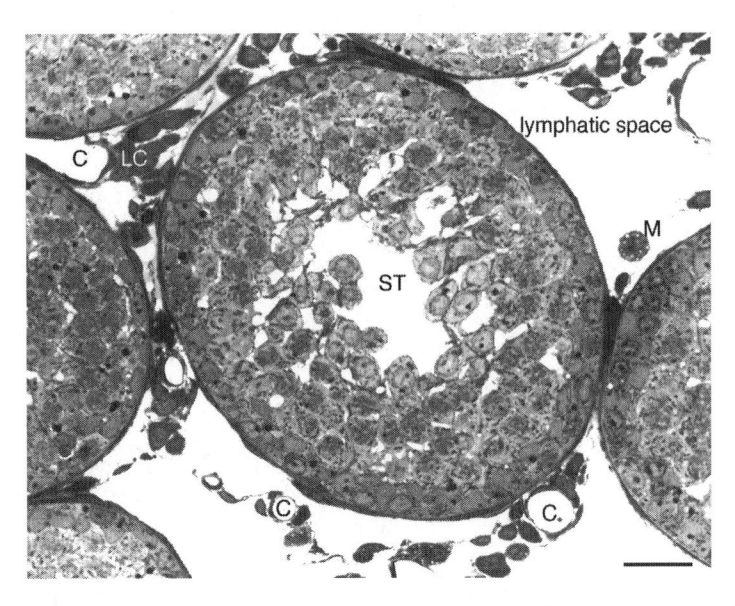

Fig. 6. High-resolution light microscopy of 1 μm thick section of a 35-day-old rat testis. Perfusion fixation, Epon–Araldite embeddment, toluidine blue staining. Perfusion has emptied erithrocytes from blood capilaries (C). The lymphatic space, Leydig cells (LC), macrophages (M), and seminiferous tubules (ST) are very well preserved. Tubule diameter and germ cell density are diminished because of FSH neutralization with anti-FSH antibodies. Modified from ref. 12. Bar = 20 μm.

6. Testes can be successfully fixed via retrograde perfusion of the abdominal aorta as reported in detail by Vitale et al. (10) from whom this protocol is taken. We have applied this method with excellent results (11, 12) (Fig. 6).

7. To visualize testicular arteries, the testes should not be moved from the scrotum. Instead, the lower abdominal wall and inguinal canal can be opened toward the scrotum and the testes exposed. To avoid drying of the albuginea, cover the testis with a piece of gauze moistened with saline solution. This can be lifted to observe the testicular vasculature while perfusing.

8. When this perfusion method cannot be used it is still possible to perfuse primate or rodent testes by delivering saline and fixative from a perfusion tubing to a fine needle inserted in the testicular artery. In the case of rat testes the needle should be inserted in the readily visible straight initial course of the testicular artery just underneath the tunica albuginea. Extreme care should be exercised to place the needle *inside* the testicular artery to avoid injecting fixative into the interstitial space.

9. These small capsules are placed at the conical bottom of regular 15 ml tubes where they can be centrifuged for 10 min. A conical pellet is obtained that can be further processed in situ (osmium

fixation, embedment, and polymerization). This in situ method is particularly useful when the number of cells is relatively low ($1–5 \times 10^6$) to avoid loosing cells during changes of solutions.

10. If the suspension is not homogeneous let it stand for a few minutes in a rack so that solid material settles in the bottom and remove by decanting the supernatant into a clean centrifuge tube to eliminate all small gelatinous particulate or irregular solid fragments. A homogeneous and fluid sample should be obtained. If solid material still remains in suspension it is advisable to briefly spin the suspension for a few seconds to settle solid material at the bottom (discard sediment). If semen is very viscous it will require more thorough mixing or the use of repeated aspirations through a Pasteur pipette or a disposable 10 ml syringe fitted with a 21 gauge needle. In these cases a higher dilution may be needed, but the pellets obtained will be more prone to disruption because most of the seminal plasma is removed and there is less cell cohesion after fixation. Following centrifugation the pellet should be compact with a very well-defined upper border. If semen is too viscous the resulting pellet will not have a clear interface with the supernatant ("fuzzy" interface). In this case discard as much supernatant as possible, resuspend the pellet in fresh buffer, and centrifuge again. During further processing of semen samples, care should be exercised not to disturb the pellet. This is accomplished by adding all solutions drop by drop onto the tube wall and discarding supernatants by slowly pouring them. Avoid use of pipettes or any other device that may create liquid turbulences and resuspend the pellet at these steps. Fixation is accomplished by slowly adding 3% glutaraldehyde in phosphate buffer down the tube wall at 4°C.

11. If spermatozoa are severely asthenozoospermic or completely immotile two changes should be introduced in the procedure to enhance staining of axonemal microtubular doublets and dynein arms (13). For fixation, glutaraldehyde should be prepared with phosphate buffer containing 2 mM Mg SO_4, and at the last dehydration step 0.1% tannic acid in absolute ethanol should be used.

12. When recently prepared, osmium solutions are lightly yellow and transparent. If kept for prolonged periods of time they may discolor and exhibit fine black particles in suspension. If discoloration is not pronounced solutions can be centrifuged and used again, but should be discarded when turned completely grayish.

13. The introduction of plastic resins has opened new horizons in ultrastructural studies since they yield tissue blocks with optimal cutting and staining properties. Among the many varieties

available, we recommend the use of embedding kits combining Eponate and Araldite. Eponate was introduced some years ago and successfully replaced Epon 812 which had been widely used before.

14. After secondary osmium fixation, tissue fragments are ready for embedment. In the case of cell or semen pellets, after osmium fixation these should be fragmented into smaller pieces, no larger than 1–2 mm^3. When cell or sperm numbers are low, pellets can be obtained in small (0.5 ml) Beem conical tip capsules and, without being removed, further processed within the capsules until resin blocks are obtained. This will ensure that all materials are embedded without losses due to unwanted fragmentation during processing.

15. Eponate–Araldite mix and Embedding protocol: Combine 10 parts Eponate 12 Resin with 10 parts Araldite 502 Resin and 30 parts DDSA. Mix well. The mixture can be moderately heated to facilitate mixing. Unless resin mix is to be used frequently, it is better to prepare small volumes that can be kept at 4°C. Before using for embedding BDMA should be added at 3% final concentration (90 µl BDMA in 3 ml resin mixture should suffice to fill six embedding cells of flat silicone rubber molds).

    After complete dehydration and propylene oxide infiltration of pellets or tissue fragments the following procedure should be performed:

    – Infiltrate for 2 h at room temperature in a mixture of equal parts of Eponate–Araldite mix (containing 3% BDMA) and propylene oxide (can be left overnight in the refrigerator).

    – Transfer to pure Eponate–Araldite mix for 2 h at room temperature (or overnight in the refrigerator).

    – Change to fresh Eponate–Araldite mix.

        At this point samples are ready for polymerization. Place them in numbered cells of flat silicone rubber embedding molds or in embedding capsules together with an identification tag and fill with fresh Eponate–Araldite mix. Place embedding molds into a polymerization oven and cure at 60°C for 24–48 h. Tags, properly identified with pencil (do not use ball pens), can be cut from white cards and placed flat on the bottom of embedding cells.

        When polymerization is carried out within small Beem conical tip capsules (0.5 ml volume, used when the number of cells is below 5×10^6) it is advisable to cure for 24 h, then peel off the capsule and continue polymerization for an additional 24 h.

16. Procedures in this section are applied to tissue embedded either in Eponate–Araldite or in LR White. After adequate resin

polymerization, blocks are ready to be cut. This is accomplished using ultramicrotomes that will give semi-thin sections for high-resolution light microscopy and ultrathin sections for electron microscopy. Various brands of automatic ultramicrotomes are available in the market. Operating procedures are beyond the scope of this chapter. Specific details on different brands and models can be found in catalogs and manuals provided with the instrument. All ultramicrotomes are equipped with stages where properly illuminated blocks can be fastened for trimming under the observation with a binocular stereomicroscope. Light comes from different sources and directions as diffuse light, backlighting, fiber optic transillumination, or "spot light." These can be used for general observation or for specific purposes such as adjusting the orientation of knives, monitoring their advance in relation with blocks, etc. The stage also provides for the location and movement of knife holders. High precision micrometers or special knobs allow the movement and rotation of resin blocks and knives in all directions so that proper orientation and contact between blocks and cutting knives can be achieved. Section thickness (from the nm to the μm range) and cutting speed are automatically controlled. Ultramicrotomes are mounted on special tables to isolate them from outside vibrations that can compromise section quality.

17. Avoid very large blocks, square or horizontal rectangular shapes because they can damage the knife edge by exerting excessive force on it and will likely vibrate during sectioning.

18. Glass and diamond knives can be used to obtain semi-thin and ultrathin sections. Glass knives can be obtained with knifemakers from specially designed glass strips provided for this purpose. Details on obtaining glass knives are beyond the scope of this chapter and should be obtained from technical books on TEM techniques or from instrument catalogs. Properly prepared glass knives can be successfully used to obtain semi-thin and ultrathin sections, but even though excellent results can be obtained, it is advisable to use diamond knives, which require less effort and provide more reproducibility. Diamond knives can also be used for semi-thin sectioning (knives are available for semi- and ultrathin sectioning, as well as for special uses such as cryoultramicrotomy). When using the ultramicrotome, sections float, as they are being cut, in boats filled with water that are located behind the knife edge (see above). There are a few practical details that should be attended to. The water level should be such as to form a slightly concave meniscus with water reaching (wetting) the knife edge. To neutralize surface tension forces that sometimes may keep the water from wetting the knife edge, a few drops of 96% ethanol should be added to the distilled water. A convex meniscus is to be avoided because

water tends to "jump" and wet the surface of approaching resin blocks. Another important detail is that the knife clearance cutting angle should ideally be set between 2 and 10 with the clearance angle adjustment knob (we set it at 4–6). This angle is different from the predetermined 45–55° angle which is set by the manufacturer during diamond knife mounting and cannot be changed. Glass knives are discarded after cutting a few blocks, but diamond knives should be meticulously cleaned after each use. There are inexpensive diamond knife cleaning rods made of strong pressed Styrofoam or pith wood that can be utilized after soaking in 96% ethanol. However, their use demands personal expertise to avoid damaging the extremely delicate diamond resulting from excessive pressure exerted on the knife edge. For operators with less experience it is advisable to use any of the Diamond Knife Cleaners available in the market that will remove dirt from the knife by rotating it in a bath with water and detergent without any physical contact with the edge.

19. To use glass knives an ad hoc container should be attached to them (knife boat) and filled with water for flotation of sections as they are being cut. Hard plastic knife boats can be bought from EM suppliers or the boats can be prepared with adhesive tape. Diamond knives are mounted in pre-designed metal holders that include small water containers for section flotation. The ad hoc boat in glass knives or the built-in container in diamond knives, when properly filled, should provide a water surface just at the level of the cutting edge.

20. It is important to use sharp diamond knives without scratches to obtain good ultrathin sections. The section thickness control should be set to 80–100 nm, but actual thickness is to be evaluated by the interference colors that sections display while they float in the water. Depending on the physical characteristics of some blocks, it may be necessary to turn the thickness knob above 100 nm to obtain sections of the appropriate thickness. The best sections for ultrastructural studies should be *silver to pale golden in color*. Dark golden sections are too thick for good results because they will not allow obtaining a good focus under the electron beam. When sections come from the diamond edge they can be somewhat compressed giving darker interference colors than the real thickness. Sections should be "extended" to eliminate compression folding. This is achieved by approaching the sections (but not touching them!) with a wood stick soaked in chloroform while they are floating (sticks should be kept in small glass vials filled with chloroform). Vapors will stretch the sections, which will then display the real interference color from which thickness can be determined.

21. The physical properties of resin blocks are critical to obtaining good ultrathin sections. Polymerization should be complete

(not less than 24–48 h at 60°C). Blocks that are too soft will produce compressed or vibrated sections. Also, the trapezoid shape obtained after trimming should be such that the base is the first point of contact with the knife edge and the longer trapezoid axis is oriented parallel to the direction of sectioning. Ideal cutting speed is around 0.6 mm/s. Good ultrathin sections display silver to pale-gold interference colors after being stretched with chloroform vapors. They will be suitable to study subcellular structures in detail and should ideally be free from longitudinal lines that result from imperfections in the knife edge. These areas of the knife should be avoided. Imperfections will inevitably appear on the edge with use and are acceptable if they do not produce coarse lines (Fig. 1). Horizontal parallel lines or frosty (not shiny) appearance of sections are usually due to vibrations or compression due to blocks which are soft, too large, or incorrectly shaped. If problems persist after blocks are adequately polymerized and shaped, they may be due to dull diamond knives that should be sent to the suppliers for re-sharpening.

22. Picking up sections from knife boats with grids and lowering them on filter paper:

    Copper or nickel grids should first be cleaned by placing them inside a glass vial filled with absolute acetone and agitated to ensure a cleaning action. They should be later transferred to a filter paper-covered Petri dish so that acetone dries before using grids to mount thin sections. Once thin sections are ready they should be mounted on copper or nickel grids. With a fine tweezer (preferably with curved ends) hold one grid, shiny side up, and keep it firmly held by sliding a rubber O-ring on the tweezer arms so that they will remain firmly closed. Approach the grid to the sections from above, and gently touch the water surface so that sections will stick in the center of the grid (on its dull side). Lift grids from the water, turn them dull side up, and release them with a sliding motion on a filter paper-covered Petri dish. The paper should be previously wet so that grids will "land" on the paper and not "jump back," driven by surface tension forces, to the small drop of water that remains between tweezer arms. An alternative to this is to gently absorb the water between the tweezer arms with a fine piece of filter paper before landing the grid on the paper. Grids now will rest on filter paper-covered Petri dishes, sections side up.

23. Procedure to stain EM sections:
    - Preparing the working surface.
    - Prepare a square piece of parafilm or wax plate under a large Petri dish and arrange drops in various horizontal rows according to indications below.

- Row 1. Line up a first horizontal row of as many water drops as grids from different blocks will be stained. Up to two grids from the same block can be accommodated in each drop, but the ideal number is 1 to keep the procedure simple.
- Row 2. Line up a second row of Uranyl–acetone drops. To prepare, mix a few drops of saturated Uranyl acetate in water with an equal number of absolute acetone drops on a corner of the working surface, recover with a Pasteur pipette, and line up drops in the row.
- Rows 3 and 4. Same number of distilled water drops as in previous rows.
- Row 5. Same number of lead citrate drops. These should be put in place just before using (lead citrate can combine with CO_2 from air or breathing and form insoluble crystals that may contaminate sections).
- Row 6. Same number of distilled water drops.
- Before starting, note that grids from different blocks do not bear marks and can be easily confused. To stain sections on grids the procedure below should be performed with maximal attention.
- Pick up grids with fine tweezers and float them with sections facing down on the first line of distilled water drops to hydrate them. It is suggested that the order of grids in drops be the same as that of aligned Petri dishes from which grids are removed.
- Transfer grids (sections facing down) to row 2 (Uranyl–acetone mix). Stain for 2 min.
- Transfer grids to distilled water in row 3.
- Pick up grids one by one with tweezers, and gently rinse them by immersion in three changes of distilled water. For this purpose three small (20 ml) glass vials filled with distilled water can be used. Rinsed grids are transferred to water drops in row 4 until all of them are rinsed. After finishing, change distilled water in vials for the next rinse (after lead citrate staining).
- Transfer grids (sections facing down) to row 5 (lead citrate drops should be set just before using). Stain for 2 min.
- Transfer grids to distilled water drops in row 6.
- Rinse well in the three vials of distilled water and place them back in the original Petri dishes. Allow grids to dry face up on filter paper.
- Grids are now ready for electron microscopy studies. They should be safely stored in special grid boxes with proper identification.

24. LR White embedding protocol: LR white is an aromatic acrylic embedding resin for on-grid ultrastructural immunocytochemistry. Osmium post-fixation is omitted because it compromises tissue antigenicity. Tissue fragments or pellets are dehydrated in ascending concentrations of ethanol (50, 70, and 96%, three 10 min washes at each concentration). It is not necessary to reach 100% ethanol. Infiltrate in a mixture of equal parts of LR White and 96% ethanol, for 60 min. Transfer to pure LR White resin and change 4–6 times, 30 min each, at room temperature. Change to fresh LR White and polymerize for 24 h in 60°C oven in molds closed without air bubbles.

25. Use of copper grids should be avoided. Ultrastructural immunocytochemical localization of selected cell components is performed on nickel grids that were first hydrated by flotation on distilled water drops.

26. To perform antigen retrieval place the grids with the sections facing up on the bottom of a covered glass Petri dish containing 10 mM Na–Citrate buffer pH 6.0 and subject them to 1–15 min of microwave irradiation at 800 W (time varies with different tissues and should be experimentally determined). This method allows for significant increases of labeling density without modifying antibody concentrations (3, 14, 15) (Fig. 4). This is due to enhanced immunoreactivity by microwave irradiation that "unmasks" tissue antigens concealed by aldehyde fixation-induced protein cross-linking. A short 1-min exposure in a family type microwave oven results in a dramatic two- or threefold increase in particle density with no detectable changes in specificity and very low levels of background labeling. It is possible that longer exposure times can lead to further increases in immunoreactivity, which is particularly important in cases of low or negative labeling. In fact, sometimes antigen retrieval can be extended to as much as 10–15 min microwave irradiation (14). When extending microwaving time, the buffer level should be controlled and restored with distilled water to neutralize changes in concentrations due to evaporation.

27. Grids can be subsequently lightly treated with Osmium Tetroxide for 5 min followed by 1:1 aqueous Uranyl acetate: absolute acetone to increase contrast, or alternatively can be left without any further staining. Negative controls are processed identically, replacing the primary antibody with similar dilutions of primary antibodies pre-adsorbed with excess antigen or omitting the first antibody.

28. If colloidal gold-coupled secondary antibodies are not available a pre-embedding protocol can be attempted utilizing a commercial immunocytochemistry peroxidase detection system. This method is particularly suited for cell suspensions that should be pelleted and washed with buffer after each step.

Fixation is similar to that previously indicated using either 5% formaldehyde or a mixture of 4% paraformaldehyde and 0.25% glutaraldehyde in phosphate buffer (see above). Cell suspensions are permeabilized with 0.5% Triton X-100 for 45 min at 4°C, endogenous peroxidase neutralized with H_2O_2 for 10 min, and nonspecific binding neutralized with blocking buffer. Cells are then incubated in the primary antibody at the appropriate dilution, washed twice with buffer, and treated with a commercial ultra sensitive link-label immunocytochemistry detection system (peroxidase) according to kit directions. Localization of peroxidase activity is achieved by treating cells with Diaminobenzidine-H_2O_2. After this step cells are post-fixed with Osmium Tetroxide for 10 min. This will produce an electron opaque deposit easily recognizable under the electron microscope (Fig. 4). Cells are subsequently pelleted and embedded in Eponate–Araldite as described above. This simple protocol gives good results when gold-coupled antibodies are not available.

29. For demonstration of other enzymes or cell components there are specific substrates or detection systems that can be found in the literature.

30. As in the case of immunocytochemistry, osmium fixation should be avoided prior to enzyme detection because it severely diminishes enzyme activity.

31. The idea of including acid phosphatase cytochemical demonstration in this chapter is to show how these methods can be applied to TEM. For complete details on different solutions, buffers, etc., please refer to Miller and Palade (16). After fixation, 50 μm sections are obtained with a vibratome or cryostat. We prefer vibratome sections (no freezing involved). All procedures should be performed at pH 5.0. After buffer rinses incubate sections in the following recently prepared and filtered mixture: dissolve 2 g sucrose in 100 ml 0.05 M Sodium Acetate Maleate pH 5.0, add the three following components stirring well after each addition: 0.12 g Lead Nitrate (Pb $(NO_3)_2$), 0.17 g Magnesium Sulfate (Mg (SO_4)), and 0.3 g Sodium-beta-glycerophosphate. Filter solution, and incubate sections for 15–60 min at room temperature or 37°C. Time and temperature should be experimentally determined to achieve good enzyme labeling with minimal background. Control sections should be incubated either without substrate or in complete medium to which 0.01 M NaF was added to inhibit acid phosphatase activity. After incubation, sections are rinsed in buffer and post-fixed in 1.3% Osmium Tetroxide for 30 min. Repeat rinse. The 50 μm thick sections should now be divided into square flakes of no more than 2 mm by side, dehydrated, and embedded in Eponate–Araldite. To obtain optimal results

place resin-infiltrated sections over a piece of aluminum foil, add a drop of resin, cover with another piece of aluminum foil, press flat with a weight on top, and polymerize as usual. When polymerized, peel foil, cut around individual sections, and using Epoxi adhesives, attach the thin flake to the cutting face of a sample-free resin block specially prepared for this purpose. Blocks are now ready to be cut, but special precautions should be taken. Vibratome sections are just below the cutting surface; therefore, no trimming is necessary, and care should be exercised not to use up all embedded material with orientation sections. It is advisable to check the first sections until tissue is reached.

32. Fixing testicular tissue fragments and cell/sperm suspension for SEM: Testicular tissue for SEM can be fixed by immersion or perfusion with 3% glutaraldehyde and post-fixed in 1.3% osmium tetroxide as for TEM. In the case of sperm or spermatogenic cell pellets 2 ml of primary fixative (3–5% glutaraldehyde in 0.1 M phosphate buffer, pH 7.4) is added to a fraction of the pellet, which is resuspended and cells fixed in suspension for 10–20 min at 4°C. Suspensions are pelleted and resuspended at each step. After two buffer washes secondary fixation is accomplished with 1.3% osmium tetroxide in phosphate buffer for 10–20 min at 4°C.

33. When fluids evaporate from soft biological material high surface tension forces are exerted as the liquid/gas interphase recedes around and through tissue or cells during drying. As a consequence, delicate surface features such as cilia, microvillae, or other cell membrane features and intercellular relationships can be seriously damaged. Special methods have been devised to avoid distortions. The more classically used is critical point drying (CPD) that gives excellent results. This is accomplished using a special CPD apparatus usually available at all SEM facilities. CO_2 is used as an intermediary fluid that, when subjected to high pressure and temperature, reaches a "critical point" where surface tension forces equal 0 and the liquid/gas interphase disappears. At this point, gaseous CO_2 is slowly released and surface-tension-free tissue drying is accomplished.

34. An alternative simplified method for drying samples for SEM studies using special compounds: The classical method used in SEM is the CPD that requires special equipment generally available at SEM facilities (described above). If this method cannot be used there are alternatives such as the use of Hexamethyldisilazane (HMDS) or Tetramethylsilane (TMS) that yield similar results to CPD. We have successfully used Peldri II, one compound of this type that is not currently available on the market and has been replaced by HMDS (we have no direct experience with this last compound).

35. This coating is achieved in special devices (Sputter Coaters) that accommodate the sample in a vacuum chamber where it is metal coated avoiding high temperatures. Sputter Coaters are normally available at SEM facilities.

Acknowledgments

Supported by Grants from CONICET (PIP 112200901 00615) and ANPCyT (PICT 2005 38229). Figures 6 and 1 are modified from refs. 12 and 17, Fig. 4d is taken from ref. 15, and Fig. 2d is taken from ref. 18 with permission (see figure legends).

The author is indebted to Prof. I. von Lawzewitsch and R. Mancini from the Histology Department, Buenos Aires University and Prof. Don W. Fawcett and all staff from the Anatomy Department, Harvard Medical School where his initial training in electron microscopy took place. Special thanks to R. Burghardt Ph.D., M. Musse M.Sc., and other colleagues with whom the author has collaborated over the years.

References

1. Fawcett DW (1975) The mammalian spermatozoon. Dev Biol 44:394–436
2. Chemes HE et al (1998) Ultrastructural pathology of the sperm flagellum: association between flagellar pathology and fertility prognosis in severely asthenozoospermic men. Hum Reprod 13:2521–2526
3. Musse M, Chemes HE (2005) Ultrastructural immunocytochemistry of Leydig cell enzyme and sperm tail protein using antigen retrieval by microwave irradiation. Treballs de la Societat Catalana de Biología 56:101–106
4. Leong AS, Sormunen RT (1998) Microwave procedures for electron microscopy and resin-embedded sections. Micron 29:397–409
5. Xiao JC et al (1996) A comparison of methods for heat-mediated antigen retrieval for immunoelectron microscopy: demonstration of cytokeratin No. 18 in normal and neoplastic hepatocytes. Biotech Histochem 71:278–285
6. Chemes H (1986) The phagocytic function of Sertoli cells: a morphological, biochemical, and endocrinological study of lysosomes and acid phosphatase localization in the rat testis. Endocrinology 119:1673–1681
7. Gomori T (1952) Microscopic Histochemistry: Principles and practicies. University of Chicago Press, Chicago
8. Polasaki Z et al (1968) Hydrolytic enzymes during spermatogenesis in rats. An electron microscopic and histochemical study. J Histochem Cytochem 16:249–262
9. Chemes HE et al (1999) Acephalic spermatozoa and abnormal development of the head-neck attachment: a human syndrome of genetic origin. Hum Reprod 14:1811–1818
10. Vitale R et al (1973) The normal development of the blood-testis barrier and the effects of clomiphene and estrogen treatment. Anat Rec 176:331–344
11. Chemes HE et al (1979) Hormonal regulation of Sertoli cell differentiation. Biol Reprod 21:251–262
12. Chemes HE et al (1979) The role of gonadotropins and testosterone on initiation of spermatogenesis in the immature rat. Biol Reprod 21:241–249
13. Wilton LJ et al (1985) Structural heterogeneity of the axonemes of respiratory cilia and sperm flagella in normal men. J Clin Invest 75:825–831
14. Alvarez Sedo, C., et al. (2012) Inmucytochemical and ultrastructural study of the acrosome formation in human globozoospermia. Human reproduction 27:1912–1921
15. Turner RM et al (2001) Molecular genetic analysis of two human sperm fibrous sheath proteins, AKAP4 and AKAP3, in men with dysplasia of the fibrous sheath. J Androl 22:302–315

16. Miller F, Palade GE (1964) Lytic activities in renal protein absorption droplets. An electron microscopical cytochemical study. J Cell Biol 23:519–552
17. Chemes HE et al (1977) Patho-physiological observations of Sertoli cells in patients with germinal aplasia or severe germ cell depletion. Ultrastructural findings and hormone levels. Biol Reprod 17:108–123
18. Chemes EH, Rawe YV (2003) Sperm pathology: a step beyond descriptive morphology. Origin, characterization and fertility potential of abnormal sperm phenotypes in infertile men. Hum Reprod Update 9:405–428

Part VII

Oxidative Stress Testing

Chapter 30

Assessment of Oxidative Stress in Sperm and Semen

Anthony H. Kashou, Rakesh Sharma, and Ashok Agarwal

Abstract

The chemiluminescence method is the most commonly employed technique as a direct measurement of reactive oxygen species (ROS) generation by spermatozoa. This assay is capable of quantifying both intracellular and extracellular ROS. Moreover, the use of various probes allows for differentiation between superoxide and hydrogen peroxide production by spermatozoa. When the total antioxidant reserves are overwhelmed by excessive production of ROS, it results in oxidative stress. Therefore correct measurement of both ROS and total antioxidant capacity (TAC) is essential in the assessment of oxidative stress in sperm and semen. This chapter describes the methodological approach for measuring seminal oxidative stress through the use of chemiluminescence assay for accurate measurement of ROS and the colorimetric assay for measurement of TAC.

Key words: Chemiluminescence, Seminal plasma, Spermatozoa, Antioxidant capacity, Oxidative stress

1. Introduction

1.1. Reactive Oxygen Species

Male infertility represents a vexing problem of great magnitude. Reactive oxygen species (ROS) affect and influence spermatozoa in their local environments. Any imbalance between ROS production and the biological system's ability to scavenge these reactive intermediates by antioxidants (both enzymatic and nonenzymatic) results in a condition known as oxidative stress (OS). Disturbances in the normal redox state can have deleterious effects on the body. OS has been implicated in several diseases, along with impaired sperm parameters. Moreover, OS levels have been shown to correlate with reduced fertility (1).

Therefore, accurate ROS measurement provides a vital tool in the initial evaluation and follow-up of infertile male patients.

Establishing reference values may help in identifying the potential causes of reduced fertility and developing strategies to reduce OS and improve sperm quality. ROS can be produced both by peroxidase-positive leukocytes and defective morphologically abnormal spermatozoa with excessive residual cytoplasm present in the midpiece area. Common ROS are superoxide anion ($O2^{\bullet-}$), hydroxyl radical ($OH^{\bullet}$), and the strong oxidizer hydrogen peroxide (H_2O_2). ROS or their oxidized products can be measured both by direct assays such as chemiluminescence, nitroblue tetrazolium test, cytochrome *c* reduction, flow cytometry, electron spin resonance, and xylenol orange-based assay. Indirect methods include measurement by myeloperoxidase test, measurement of redox potential, lipid peroxidation levels, chemokines, antioxidants, and antioxidant enzymes and measuring levels of DNA damage (2, 3).

Two probes may be used with the chemiluminescence assay: luminol and lucigenin. A luminol-mediated chemiluminescence signal in spermatozoa occurs when luminol oxidizes at the acrosomal level. Luminol reacts with a variety of ROS and allows both intracellular and extracellular ROS to be measured. Lucigenin is more specific for superoxide anions released extracellularly (2, 3).

The luminol assay is more advantageous for a number of reasons. It can measure H_2O_2, $O_2^{\bullet-}$, and $OH^{\bullet-}$ levels. However, it cannot distinguish these oxidants from one another (4). In addition, luminol is easy to use and can measure the global level of ROS under physiological conditions. Since the assay can measure both extracellular and intracellular ROS, it has a high sensitivity (4). Multiple studies have correlated high chemiluminescent signals using luminol as a probe with adverse effects on sperm function.

This chapter aims to provide a step-by-step approach in the use of chemiluminescence assay utilizing luminol as a probe for the assessment of ROS in sperm and semen. The instrument used to measure the light intensity resulting from chemiluminescence reaction is called luminometer. These can be single tube, multiple tube, or plate luminometers. Depending on the nature of the manner in which the signal input is measured they can be either the photon-counting luminometers that count individual photons or the direct current luminometers that measure electric current that is proportional to the photon flux passing through the photomultiplier tube. The results can be expressed as relative units (RLU), counted photons per minute (cpm), or millivolts/second. The method described below utilizes a multiple tube luminometer, which is a photon-counting instrument that covers a spectral range from 390 to 620 nm. ROS can be measured in either neat or unprocessed liquefied seminal ejaculate, washed sperm, or sperm prepared by swim up or by density gradient separation.

1.2. Total Antioxidant Capacity

Living organisms have developed a complex antioxidant system to counteract the effects of ROS and reduce damage. The antioxidant system of living organisms includes enzymes such as superoxide dismutase, catalase, and glutathione peroxidase; macromolecules such as albumin, ceruloplasmin, and ferritin; and an array of small molecules, including ascorbic acid, α-tocopherol, β-carotene, reduced glutathione, uric acid, and bilirubin. The sum of endogenous and food-derived antioxidants represents the total antioxidant activity of the extracellular fluid. Thus, the overall antioxidant capacity may give more relevant biological information compared to that obtained by the measurement of individual components, as it considers the cumulative effect of all antioxidants present.

A simple assay kit is available from Cayman Chemicals (antioxidant assay kit, Ann Arbor, MI). We have standardized this assay and use it in our laboratory setting. The assay relies on the ability of antioxidants in the sample to inhibit the oxidation of 2,2′-azino-di-[3-ethylbenzthiazoline sulfonate] (ABTS) to ABTS+ by met-myoglobin. Under the reaction conditions used, the antioxidants in the seminal plasma cause suppression of the absorbance at 750 nm to a degree that is proportional to their concentration. The capacity of the antioxidants present in the sample to prevent ABTS oxidation is compared with that of the standard—Trolox, a water-soluble tocopherol analogue. Results are reported as micromoles of Trolox equivalent. This assay measures the combined antioxidant activities of all its constituents including vitamins, proteins, lipids, glutathione, uric acid, etc.

1.3. Conclusion

Accurate measurement of both ROS and total antioxidant capacity (TAC) is important to determine if oxidative stress is the underlying cause of male infertility. In such conditions, it is important to reduce the ROS levels by eliminating the ROS generation by leukocytes in case of an underlying infection, and antioxidant supplementation to improve TAC may prove to be beneficial.

2. Materials

2.1. Equipment and Consumables for ROS Assay

1. Disposable polystyrene tubes with caps (15 mL).
2. Eppendorf pipets (5, 10 µL).
3. Serological pipets (1, 2, 10 mL).
4. Desktop centrifuge.
5. Disposable sperm-counting chamber.
6. Dimethyl sulfoxide (DMSO).
7. Luminol (5-amino-2,3-dihydro-1,4-phthalazinedione).
8. Polystyrene round-bottom tubes (6 mL).
9. Multiple tube luminometer
10. Dulbecco's Phosphate-Buffered Saline Solution 1× (PBS).

2.2. Reagents for ROS Assay

1. DMSO solution: Provided ready to use (see Note 1).
2. Stock Luminol (100 mM): Weigh out and add 177.09 mg of luminol to 10 mL of DMSO in a polystyrene tube (see Notes 2 and 3).
3. Working Luminol (5 mM): Mix 380 µL DMSO with 20 µL of prepared luminol stock solution in a foil-covered polystyrene tube (see Note 4).

2.3. Equipment and Materials for TAC Assay

1. Antioxidant assay kit (Cayman Chemical, Ann Arbor, MI).
2. Absorbance Microplate Reader (e.g., BioTek Instruments, Inc., Winooski, VT).
3. Pipettes (20, 200, and 100 µL).
4. Pipette tips (20, 200, and 100 µL).
5. Multichannel pipettes (eight channel, 30–300 µL).
6. Aluminum foil.
7. Microfuge tubes.
8. Deionized water.
9. Polystyrene centrifuge tubes (50 and 15 mL).
10. Round-bottom tubes (12 × 75 mm).

3. Methods

3.1. Specimen Preparation and ROS Measurement

1. Herein we describe the details of measurement of ROS by chemiluminescence assay. ROS can be measured by the chemiluminescence method using a probe called luminol. Luminol is extremely sensitive and reacts with a variety of ROS at neutral pH. It has the ability to measure both extracellular and intracellular ROS. Free radicals have a very short half-life and are continuously produced. The free radical combines with luminol to produce a light signal that is then converted to an electrical signal (photon) by the instrument called luminometer. The number of free radicals produced is measured as $RLU/s/ \times 10^6$ sperm.
2. Upon collection of semen sample, incubate at 37°C for 20 min to allow for liquefaction. Record initial characteristics, including semen volume, pH, and color, and manually assess sperm count, concentration, and motility. Thereafter, process the semen specimen for ROS measurement as described below (5) (Fig. 1).
3. The following samples/preparations can be used for ROS measurement: neat or unprocessed sample (seminal ejaculate after

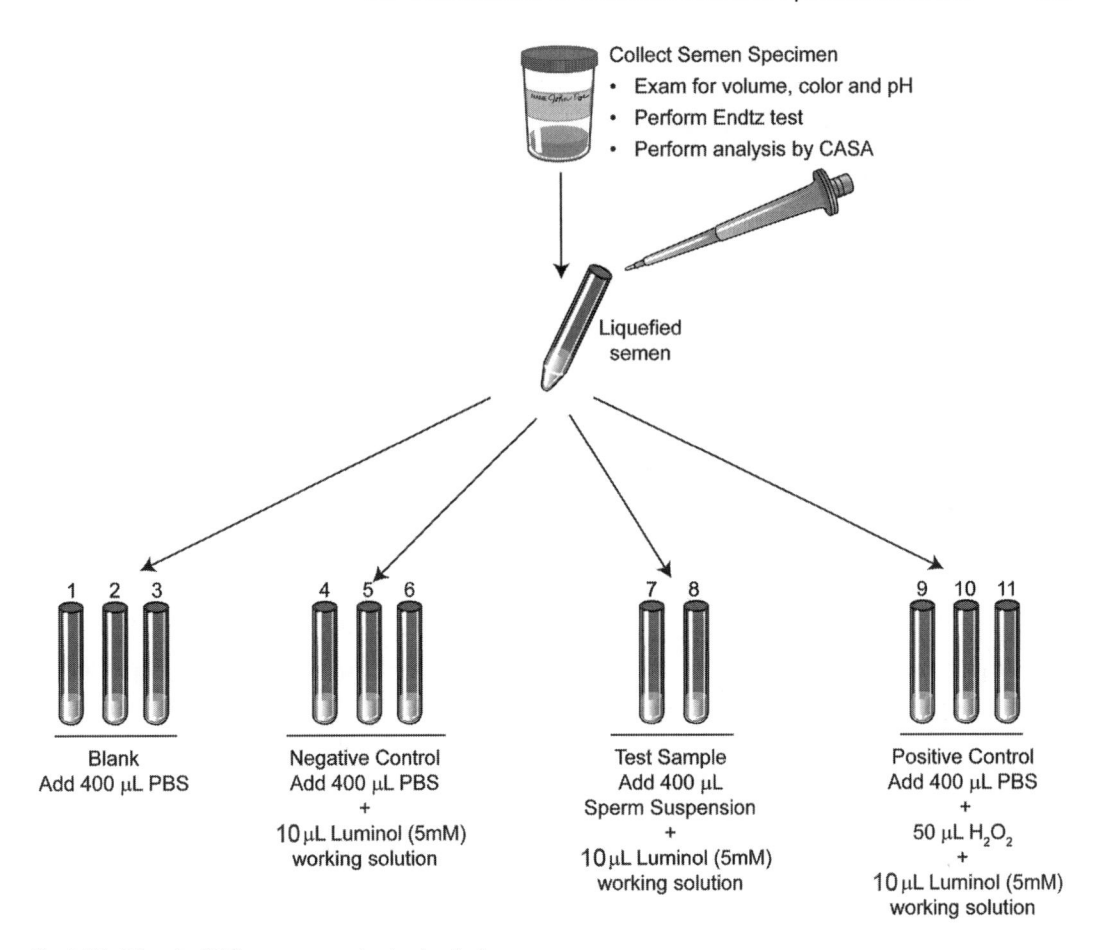

30 Assessment of Oxidative Stress in Sperm and Semen   355

Fig. 1. Workflow for ROS measurement using luminol.

3.2. ROS Measurement with Luminometer (see Fig. 2)	liquefaction) (6), washed sample (see Note 5), sample prepared by swim-up method (see Note 6), or sample prepared by density gradient centrifugation (see Note 7).
1. Turn on the luminometer and ensure that it is ready for measurements.
2. Enter information in the luminometer including the patient information, number of samples, single measuring time, and data points measured in the integrated mode. Generally, the measurement time is 15 min.
3. Label 6 mL tubes for blank, negative control, test sample and positive control and add reagents as shown in Table 1 (see Note 8). The blank, negative controls and positive controls are run in triplicate, and patient samples are typically run in duplicate, though they can be run in triplicate if ejaculate volume is sufficient.

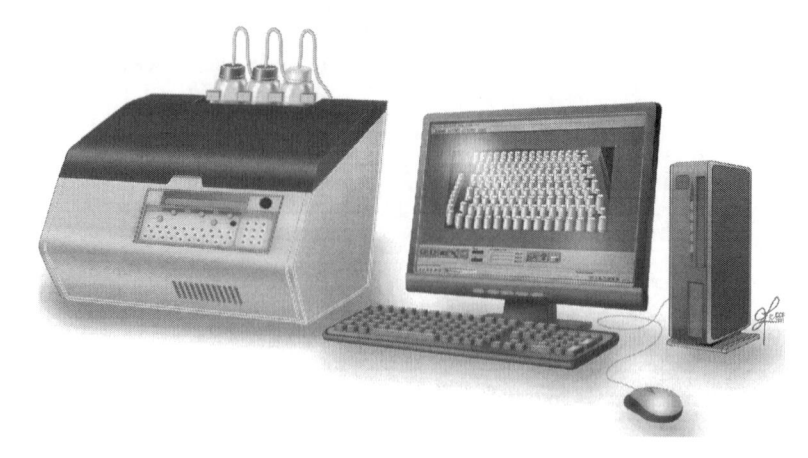

Fig. 2. Illustration of a typical luminometer setup for luminol based ROS analysis. Reprinted with permission from the Cleveland Clinic Center for Medical Art & Photography © 2011.

Table 1
Components and volumes for preparation of the various tubes for measuring ROS

No.	Labeled tube	PBS volume (µL)	Specimen volume (µL)	Luminol (5 mM)	Hydrogen peroxide (8–9 M)
1	Blank (tubes S1–S3)	400	–	–	–
2	Negative control (tubes S4–S6)	400	–	10 µL	–
4	Patient (tubes S7–S8)	–	400	10 µL	–
5	Positive control (tubes S9–S11)	–	400	10 µL	50 µL

4. Immediately after adding the probe, initiate measurements. Load the tubes and operate the luminometer according to manufacturer's recommendations.

5. After finishing the measurements, save the data for analysis.

3.3. Calculating Results for the ROS Assay

1. Calculate the "average RLU" for Negative control, Samples, and Positive control.

2. Calculate sample ROS by subtracting its average from negative control average: Sample ROS = Average "RLU mean" for sample—Average "RLU mean" for negative control.

3. Correct the sample ROS by dividing it by "Sperm concentration/mL" (see Note 9).

4. Each lab can establish its reference value for ROS by running a large number of samples from healthy and infertile subjects.

In our laboratory, we have established the reference values as follows: *Normal range*: <20 RLU/s/10⁶ sperm. *Critical values*: >20 RLU/s/10⁶ sperm.

5. Evaluate replicates to ensure that they are concordant within acceptable limits and ensure that control values are within the normal range (see Note 10).

3.4. Preparation of Assay Reagents for TAC Assay

All reagents should be equilibrated at room temperature before beginning the assay and prepared according to the manufacturer's instructions provided with the assay kit and also available at www.caymanchem.com.

1. Prepare Antioxidant assay buffer (10×) (vial # 1): Dilute 3 mL of assay buffer concentrate with 27 mL of HPLC-grade water in a 50 mL conical tube (see Note 11).

2. Prepare chromogen (vial # 2): Reconstitute the chromogen (containing ABTS) with 6 mL of water and vortex it. This volume is sufficient for 40 wells (see Note 12).

3. Prepare metmyoglobin (vial # 3): Reconstitute the lyophilized powder with 600 µL of assay buffer and vortex it. Once reconstituted, it is sufficient for 60 wells (see Note 13).

4. Prepare Trolox (vial # 4): This vial contains the standard Trolox (6-hydroxy-2,5,7,8-tetramethylchroman-2-carboxylic acid). Reconstitute the lyophilized powder in the bottle with 1 mL of water and vortex it. This is used to prepare the standard curve (see Note 14).

5. Prepare hydrogen peroxide (vial # 5): This vial contains 8.82 M solution of hydrogen peroxide. Dilute 10 µL of hydrogen peroxide reagent with 990 µL of water. Further dilute by removing 20 µL and diluting with 3.98 mL of water to give 441 µM working solution (see Note 15).

3.5. Specimen Preparation for TAC Assay

1. Bring frozen seminal plasma to room temperature and centrifuge in a microfuge at high speed for 5 min. Remove clear seminal plasma and dilute each sample 1:10 (10 µL sample + 90 µL assay buffer) in a microfuge tube (see Note 16).

2. Use a plate template to note the locations of each sample (run standard and test samples in duplicate (see Note 17)).

3.6. TAC Determination

1. Prepare the standards in seven clean tubes and mark them A–G. Add the amount of reconstituted Trolox and Assay buffer to each tube as shown in Table 2.

2. Add 10 µL of Trolox standard (tubes A–G) or sample in duplicate + 10 µL of metmyoglobin + 150 µL of chromogen per well (see Note 18).

Table 2
Preparation of the Trolox standards

Tube	Reconstituted Trolox (µL)	Assay buffer (µL)	Final concentration (mM Trolox)
A	0	1,000	0
B	30	970	0.044
C	60	940	0.088
D	90	910	0.135
E	120	880	0.18
F	150	850	0.225
G	220	780	0.330

3. Initiate the reaction by adding 40 µL of hydrogen peroxide working solution using a multichannel pipette (see Note 19).
4. Cover the plate with the plate cover and incubate on a shaker for 5 min at room temperature.
5. Remove the cover and read the absorbance at 750 nm using a plate reader.

3.7. Calculation of TAC Results

1. Calculate the average absorbance of each standard and sample.
2. Calculate the antioxidant concentration of the samples using the equation obtained from the linear regression of the standard curve by substituting the average absorbance values for each sample into the following equation (see Note 20):

$$\text{Antioxidant } (\mu M) = \frac{\text{Unknown average absorbance} - Y \text{ intercept}}{\text{Slope}} \times \text{dilution} \times 1,000$$

3. Evaluate results against a reference range (see Note 21).
4. Evaluate the ROS-TAC score for each sample (see Note 22).

4. Notes

1. Store at room temperature until the expiration date.
2. Cover the tube with aluminum foil to protect the luminol from light.
3. Store stock solution at room temperature in the dark until expiration date.

4. Prepare working luminol prior to each use.
5. Seminal plasma is removed by washing and resuspending the sample in culture media (7). Measurement is done after liquefied semen specimens are centrifuged at 300×g for 7 min and seminal plasma is removed. The sperm pellet is then washed and resuspended in 1 mL PBS. While this approach allows for the removal of seminal plasma and other dissolved components, all cellular components—round cells, while blood cells, leukocytes, and debris—are still present in the sample.
6. Sperm preparation by the swim-up method: Following liquefaction, mix specimen with sperm wash media using a sterile Pasteur pipette. Centrifuge at 330×g for 10 min. Aspirate supernatant and resuspend the pellet in 3 mL of fresh sperm wash media. Transfer the resuspended sample in equal parts to two 15 mL sterile round-bottom test tubes, and centrifuge at 330×g for 5 min. Incubate test tubes at a 45° angle in 5% CO_2 at 37°C for 1 h to allow motile sperm to swim up. Aspirate supernatant into a clean test tube and centrifuge at 330×g for 7 min. Aspirate final supernatant and resuspend sperm pellet in 0.5 mL of sperm wash media. Measure the final volume, and perform semen analysis on an aliquot of the sample (8).
7. Sperm preparation by density gradient: A double density gradient (lower density; 40–47% "upper phase" and 80–90% "lower phase") is used (8). Transfer 2.0 mL of the "lower phase" into a 15 mL conical centrifuge tube using a sterile pipette. Carefully add 2.0 mL of the "upper phase" to the top of the "lower phase." Place the liquefied semen sample (1–2 mL) on top of the "upper phase" layer and centrifuge at 330×g for 20 min. Aspirate the upper and lower layers without disturbing the pellet. Add 2–3 mL of sperm wash media, and spin at 330×g for 7 min. Remove the supernatant and resuspend the pellet in 1.0 mL of sperm wash media. Measure sperm count and motility in recovered fractions. Perform all steps at 37°C.
8. This step must be performed in subdued light.
9. Example corrected ROS calculation:

    Sperm Count = 12.6 × 10^6/mL.

    Patient average ROS = 12,161 RLU/s.

    Negative control average ROS = 8,850.3 RLU/s.

    Unadjusted ROS = 12,161 – 8,850.3 RLU/s = 3310.7 RLU/s.

    Adjusted ROS = 3,310.7/12.6.

    =262.7 RLU/s/10^6 sperm.

    Result = ROS positive.
10. In our lab, the criterion for acceptance is that control reads are <20 RLU/s/10^6 sperm. Values >20 RLU/s/10^6 sperm for

controls are rejected and the assay is rerun. The reagent lot numbers and expiration dates are recorded.

11. The reconstituted vial is stable for 6 months when stored at 4°C.
12. The reconstituted vial is stable for 24 h at 4°C.
13. The reconstituted reagent is stable for 1 month when stored at −20°C.
14. The reconstituted vial is stable for 24 h at 4°C.
15. The working solution is stable for 4 h at room temperature.
16. Label the vials for correct identification.
17. Any errors during pipetting can be highlighted on the template for any discrepancy in the final results.
18. A multichannel pipette should be used to pipette chromogen. Chromogen can be pipetted from a flat container.
19. This step should be completed as quickly as possible (within 1 min). Hydrogen peroxide can be pipetted from a flat container using a multichannel pipette.
20. In addition the information can also be derived from the worksheet provided by the manufacturer and plugging each value in the spreadsheet available at www.caymanchem.com.
21. Reference ranges established in our lab are the following: Normal value: >2,000 µM Trolox, abnormal value: <2,000 µM Trolox. Since seminal OS results from an imbalance between ROS production and antioxidant defense, measurement of both ROS and TAC is important. It is essential to have a reproducible and reliable method for ROS measurement for clinical purposes. Strict quality control must also be taken into account for valid assessment in a clinical laboratory setting.
22. While measurement of ROS levels or TAC alone cannot precisely quantify OS, it is possible to combine these two parameters into one index score called ROS-TAC score (1). The ROS and TAC values from the controls can generate a reference scale of these two variables. In order to normalize values to the same distribution, log (ROS + 1) is used in calculations. The reason for this is that ROS levels if measured as ×10^6 cpm/20 million sperm can have a very wide range including zero, where no detectable ROS is generated. Therefore it is more accurate to convert these to log values and add a positive number such that the ROS value is always positive. Both TAC and log (ROS + 1) are initially standardized to z-scores so that both will have the same variability. These standardized scores are calculated by subtracting the mean values of the controls from the mean value of the patients, and dividing this number by the standard deviation of the control population.

For details refer to our earlier study (1) where ROS values were expressed as cpm.

ROS-TAC score minimizes the variability of ROS and TAC alone. The ROS-TAC score reported by us earlier (1) was based on a group of normal healthy fertile men who had very low levels of ROS measured as $\times 10^6$ cpm/20 million sperm. It is easy to convert cpm to RLU and vice versa (10 RLU = 1 cpm).

The ROS-TAC score was found to be better than ROS or TAC alone in discriminating between fertile and infertile men. Infertile men with male factor infertility or an idiopathic diagnosis had significantly lower ROS-TAC scores than the healthy controls, and men with a male factor diagnosis who eventually were able to initiate a pregnancy had significantly higher ROS-TAC scores than those who were not able to do so. The average ROS-TAC score for fertile healthy men was 50.0 ± 10, which was significantly higher ($p \pm 0.0002$) than that of the infertile patients (35.8 ± 15). The probability of successful pregnancy was estimated at <10% for values of ROS-TAC <30, but increased as the ROS-TAC score increased. Measurement of ROS in neat semen has proved to be an accurate and reliable test for assessing the OS status (6). Assessing ROS directly in neat semen also has diagnostic and prognostic capabilities identical to that of the ROS-TAC score. This approach overcomes the limitations of earlier methods where processing of semen could generate ROS by itself, thereby accurately representing the true in vivo OS status of an individual. While ROS levels were significantly lower in neat semen than in washed spermatozoa, ROS levels in neat semen demonstrated a strong positive correlation with ROS levels in washed semen (9, 10).

References

1. Sharma RK et al (1999) The reactive oxygen species-total antioxidant capacity score is a new measure of oxidative stress to predict male infertility. Hum Reprod 14: 2801–2807
2. Aitken RJ, Buckingham D (1992) Enhanced detection of reactive oxygen species produced by human spermatozoa with 7-dimethyl aminonaphthalin-1, 2-dicarbonic acid hydrazide. Int J Androl 15:211–219
3. Aitken RJ et al (1992) Reactive oxygen species and human spermatozoa: analysis of the cellular mechanisms involved in luminol- and lucigenin-dependent chemiluminescence. J Cell Physiol 151:466–477
4. Sharma RK, Agarwal A (1996) Role of reactive oxygen species in male infertility. Urology 48:835–850
5. Kobayashi H et al (2001) Quality control of reactive oxygen species measurement by luminol-dependent chemiluminescence assay. J Androl 22:568–574
6. Allamaneni SS et al (2005) Characterization of oxidative stress status by evaluation of reactive oxygen species levels in whole semen and isolated spermatozoa. Fertil Steril 83:800–803
7. Agarwal A et al (2004) Chemiluminescence technique for measuring reactive oxygen species. Reprod Biomed Online 9:466–468
8. Allamaneni SS et al (2005) Comparative study on density gradients and swim-up preparation techniques utilizing neat and cryopreserved spermatozoa. Asian J Androl 7:86–92
9. Saleh RA, Agarwal A (2002) Oxidative stress and male infertility: from research bench to clinical practice. J Androl 23:737–752
10. Saleh RA et al (2002) Leukocytospermia is associated with increased reactive oxygen species production by human spermatozoa. Fertil Steril 78:1215–1224

Chapter 31

Improved Chemiluminescence Assay for Measuring Antioxidant Capacity of Seminal Plasma

Charles H. Muller, Tiffany K.Y. Lee, and Michalina A. Montaño

Abstract

An improved enhanced chemiluminescence antioxidant assay utilizes horseradish peroxidase conjugate and luminol to produce a cell-free oxygen radical generating system. We introduce the use of a peroxidase enzyme stabilizer to prolong the production of oxygen radicals at a steady rate. Addition of antioxidants temporarily interrupts oxygen radical generation, resulting in an inhibition curve. A linear relationship exists between the area of the inhibition curve and the molar quantity of added antioxidant used to quantify total nonenzymatic antioxidant capacity (TAC) in biological fluids including seminal plasma. We streamline the existing enhanced chemiluminescence technique by using a microtiter plate luminometer. A plate luminometer is as accurate as a tube luminometer in measuring TAC, using identical reaction volumes. As little as 1–50 µL of sample may be analyzed. A plate luminometer can detect molar Trolox equivalents as low as 12.5 µM, compared to 25 µM in tube luminometer, using identical volumes. The plate luminometer assay is made even more rapid with use of an injector.

Key words: Oxygen radicals, Antioxidant, Total antioxidant capacity, Oxidative stress, Luminometer, Chemiluminescence, Semen, Seminal plasma

1. Introduction

Oxidative stress (OS) results from a combination of the formation of oxygen radicals and insufficient scavenging mechanisms to protect against their damage (1). In the laboratory, OS is measured as an imbalance between high reactive oxygen species (ROS) production and low antioxidant capacity. OS in biological systems has been widely studied in the past decades. In the male reproductive system, seminal fluid OS is associated with male infertility (2–11) and infection of the male reproductive tract (12). It also may be related to pain symptoms in patients with chronic prostatitis/chronic pelvic pain syndromes (13–15). Oxidative stress is a cause of sperm DNA damage during cryopreservation (16).

Although high levels of ROS are harmful to many biological systems, organisms have developed primary systems to protect themselves against OS. These defense mechanisms of detoxification counterbalance the highly toxic byproducts of OS. The antioxidant defense system includes specific enzymes such as superoxide dismutase, catalase, and glutathione peroxidase, and nonspecific, nonenzymatic antioxidants such as proteins, amino acids, and vitamins A, E, and C (17).

The balance between free radical formation and antioxidant activity determines the extent of OS; therefore, it is important to accurately and reliably quantify ROS scavenging ability in biological fluids. There are a number of methods commonly used for evaluation of OS, including lipid peroxidation, DNA damage, nitric oxide production, plasma carbonyl content, antioxidant enzyme levels, and antioxidant capacity.

The primary focus of this chapter is a method of measuring "total antioxidant capacity" (TAC), or the ability of a compound or biological fluid to scavenge a free radical with any specific or nonspecific mechanism available. Different methods in determining TAC in biological fluids include enhanced chemiluminescence (18), phycoerythrin fluorescence-based assays (19), total radical trapping antioxidant parameter (TRAP) (20–22), oxygen radical absorbance capacity (ORAC) (23–25), and ferric reducing ability of plasma (FRAP) (26). A colorimetric assay is available in kit form and has successfully been used with seminal plasma (27). Of these methods, an enhanced chemiluminescence reaction involving a cell-free ROS-generating system of horseradish peroxidase (HRP) hydrogen peroxide and luminol is the most commonly used for measuring nonenzymatic TAC in seminal plasma (28). This system produces oxygen radicals at a known and steady rate, where the luminescence intensity remains almost constant for more than several minutes. The steady-state light output is temporarily interrupted when an antioxidant is added to the chemiluminescent reaction. The light emission is restored after a variable period of time once the free radical scavenging ability is depleted.

Our laboratory modified the enhanced chemiluminescence assay developed by Whitehead's laboratory (18) and subsequently used by Agarwal and Thomas' group to study seminal plasma (29). We decided to streamline the existing enhanced chemiluminescence technique by using a multiwell plate luminometer instead of a conventional tube luminometer. By analyzing our samples on plates, we lowered the cost by using lower volumes of reagent and reduced the time by running multiple samples without compromising the precision and sensitivity of the TAC measurement.

In this improved enhanced chemiluminescence assay, we introduced the use of an enzyme stabilizer to prolong the constant rate of the chemiluminescence reaction. In addition, we compared different ways of quantifying antioxidant capacity, including a method based on the "area of inhibition" (termed area of "protection" by Cutler's laboratory) (23). The results are quantified by measuring the total area under the curve during inhibition of ROS by antioxidants. This bypasses problems associated with recovery or lag-time measurements, as used by Whitehead et al. (18) and others. We find a linear relationship between the "area of inhibition" and the molar quantity of a standard antioxidant. The final result is calibrated and expressed with reference to a known amount of antioxidant (e.g., Trolox, a water-soluble vitamin E analogue). Finally, we introduce the use of a plate luminometer injector to increase the efficiency of the method.

2. Materials

1. Luminometer, multiwell plate (injector optional) or tube model. Preferably with data download capability. Luminometer tubes, cuvettes or microtiter plates (e.g., Microlite 1) designed for luminometry.

2. Computer with NIH Image or other image analysis software and spreadsheet or statistical software.

3. Hydrogen peroxide, 30%. Fresh bottle opened weekly, store refrigerated. Use protective clothing and nitrile gloves when handling.

4. Dimethyl sulfoxide (DMSO). Store in flammable storage cabinet. *Caution*: Flammable, harmful vapors, irritant, avoid skin contact. Use in a fume hood, and wear protective clothing and nitrile gloves when handling.

5. 4-Iodophenol (p-Iodophenol), 41.8 mM stock: Add 0.023 g to 2.5 mL DMSO. *Caution*: Corrosive, harmful, irritant chemical. Wear nitrile gloves and protective mask when weighing. Store in a light-protected container in the dark at room temperature for no more than 1 month (see Note 1).

6. Luminol (5-amino-2,3-dihydro-1,4-phthalazinedione), 282.2 mM stock: Add 0.125 g Luminol to 2.5 mL DMSO. *Caution*: Irritant chemical. Wear nitrile gloves and protective mask when weighing. Store in a light-protected container in the dark at room temperature for no more than 1 month.

7. Tris–HCl Buffer, 0.1 M, pH 8.0 at 25°C: Dissolve 2.65 g Trizma Base and 4.44 g Trizma HCL in deionized water (see Note 2). Make up to 500 mL with deionized water. Check and adjust pH as necessary. Store at room temperature up to 6 month, discard if cloudy.

8. StabilZyme SELECT® Stabilizer (SurModics In Vitro Diagnostic Products, Eden Prairie, MN). Keep refrigerated.

9. Horseradish peroxidase (HRP). We find that a conjugate of HRP with goat anti-mouse immunoglobulin is more stable than purified HRP. (BioRad or similar). Dilute in StabilZyme to make a 0.4% (v/v) stock solution: Add 5 µL HRP conjugate to 1.25 mL StabilZyme Select Stabilizer. Keep refrigerated.

10. Trolox (6-hydroxyl-2, 5, 7, 8-tetramethychroman-2-carboxylic acid), diluted for TAC calibration. Trolox stock solution (100 mM): 0.02503 g dissolved in 1 mL methanol. Store in dark in tightly sealed container no more than 1 week.

3. Methods

Working solutions below must be prepared fresh each day. The Signal Reagent is good for 6–8 h. Samples to be tested may be prepared ahead of time and refrigerated overnight, or frozen at −80°C indefinitely. Semen samples (and other fluids) must be centrifuged at 10,000–15,000×g for 2–5 min to obtain a clarified plasma. Seminal plasma must be diluted before analysis, so if the sample is viscous, it may be diluted two- to fivefold with a physiologic buffer, mixed vigorously to lessen the viscosity, and then centrifuged. Record any dilution made.

This method is designed for use in a 96-well microtiter plate, but the volumes are suitable for small luminometer cuvettes, or may be scaled up for larger tubes. A modification using a luminometer injector is included in Subheading 4.

1. Prepare the Trolox standard curve (see Note 3). Concentrations from 5 to 100 µM may be sufficient, but will depend on the sensitivity of your instrument, range of sample response, and volume of reagents.

2. Prepare Signal Reagent by adding components in this order, with thorough mixing after each addition: to 5 mL 0.1 M Tris–HCl buffer, pH 8.0, add 15 µL of 12 M (30%) fresh hydrogen peroxide, 5 µL of 41.8 mM 4-iodophenol, and 55 µL of 282.2 mM luminol. Cover tube with foil and store immediately at 4°C until use.

3. Prepare HRP Working Solution: Dilute Stock HRP with an equal volume of StabilZyme Stabilizer (see Note 4).

4. Dilute seminal plasma to prepare a 20-fold dilution: Add 25 µL seminal plasma to 475 µL deionized water, mix thoroughly (see Note 5). If the seminal plasma was previously diluted, use a dilution factor that will produce a 20-fold dilution (see Note 6). Other fluids, such as blood plasma, may not need to be diluted

as much. Diluted seminal plasma may be tested and frozen at −80°C for future additional use. Always thoroughly mix thawed samples before use.

5. Set up luminometer for repeated or kinetic measurement, following instrument-specific directions. On a multiwell plate luminometer, we use a repeated measures protocol with 5 s single measurement time, cycle time of 22.5 s (for three wells; this will be longer for more wells), and a total measurement time of 3–10 min, depending on the sample. For a tube luminometer, we use a 2 s count time, interval of 8 s, and total measurement time of 10 min. These values may be changed depending on response conditions (see Note 7). We run our assays at room temperature (25°C).

6. Use low ambient light conditions to prepare the reagents and load wells or tubes (see Note 8). Set up tubes of reagents, pipettors, tips, and samples in a familiar order so that the following steps may be completed in as brief a time as possible.

7. Set up three types of wells or tubes: Total Activity (TA), Nonspecific Background (NSB), and Sample or Standard (S). The TA result may change with time as the Signal Reagent degrades, so it must be included with every few runs, or at regular intervals during the assay. Usually, the NSB can be determined three times per assay. However, we normally run a sample-specific NSB with each sample to account for endogenous activity. Each sample or standard should be repeated to obtain three replicates.

8. For a single-tube luminometer, load reagents into one tube and complete measurements on that tube before loading subsequent tubes. For a multiwell plate luminometer, usually only three wells are loaded per run; these may be each of the three types. Reagent volumes listed below are for microtiter plates, and may be scaled up for tubes.

9. Prepare a TA tube/well to determine the characteristics of the reactive oxygen generating system. As the Signal Reagent degrades with time, this may yield a lower result, and samples run at that time must be analyzed with this new baseline. Add in sequence under low light conditions: 240 µL StabilZyme Stabilizer; 30 µL Signal Reagent, 30 µL HRP Working Solution. Mix well using the pipettor. Quickly place tube or plate into luminometer, and assay for the set time, or longer to determine the extent of a useful (flat, high plateau) response (see Note 9). Collect data from the luminometer, attached computer, or printout.

10. Prepare and analyze an NSB tube/well to determine background signal. Add in sequence under low light conditions: 220 µL StabilZyme Stabilizer, 30 µL Signal Reagent, 50 µL

deionized water, buffer, or diluent used for sample dilution. Or, 50 μL sample may be added instead of latter component (see Note 10). Mix well, assay, and collect data as in step 9.

11. Prepare and analyze each concentration of Trolox standard. Add in sequence under low light conditions: 190 μL StabilZyme Stabilizer, 30 μL Signal Reagent, and 30 μL HRP Working Solution. Mix well. Immediately before placing tube or plate into luminometer, add 50 μL of one concentration of well-mixed Trolox standard and mix with pipettor. Immediately assay and collect data as in step 9 (see Notes 11 and 12). Analyze each dose in triplicate.

12. Prepare and analyze each sample (see Note 12). Add in sequence under low light conditions: 190 μL StabilZyme Stabilizer, 30 μL Signal Reagent, and 30 μL HRP Working Solution. Mix well. Immediately before placing tube or plate into luminometer, add 50 μL of a well-mixed sample and mix with pipettor. Immediately assay and collect data as in step 9. Analyze each sample in triplicate.

13. Download data or enter printed data into a spreadsheet or statistics program (see Note 13). Subtract NSB (preferably, the sample-specific NSB) result from each point, if needed. Create a data set of relative light units (rlu)/s versus time for each inhibition curve ("reverse peak") and the relevant TA "plateau" in the same data file. Plot the TA plateau and the standard or sample data on one graph (Fig. 1).

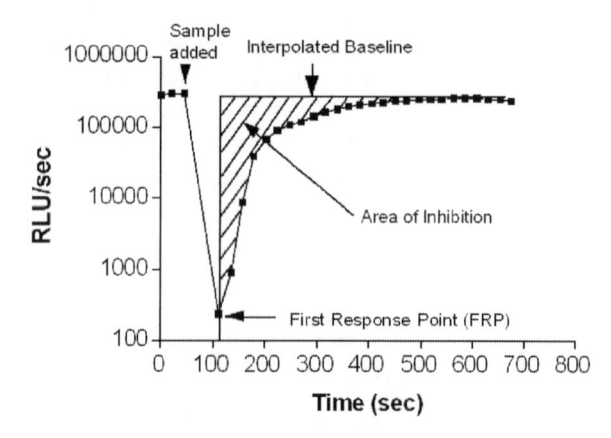

Fig. 1. Example of the combined graph of Total Activity (TA) and the inhibition curve, corrected for NSB. The Area of Inhibition (AOI) is shown (hatched area). The interpolated baseline is the horizontal line drawn from the last TA data point. The *arrowhead* shows the point at which sample is added. No data are collected between the point of sample addition and the first response point. The first response point is indicated, with the first response line (FRL) drawn vertically from it. *RLU/s* relative light units per second, *filled square* data points.

14. There are two data analysis choices. A common method is to calculate the time to 10% recovery of Total Activity. This may be done graphically (manually) or mathematically by interpolating the time point at which the upward recovery line passes through the horizontal at 10% of the TA. The result is then expressed as a time, e.g., 340 s (to 10% recovery). This method is simply calculated, but suffers from errors when the response is low, and from a low slope of the standard curve compared to the alternate method (step 15). The 10% recovery method has an advantage of being usable when the inhibition curve starts to return but does not completely return to the TA plateau.

15. (Alternate method). An "Area of Inhibition" (AOI) may be calculated with a few more analysis steps. A steeper and thus a more accurate slope of the standard curve is obtained. Also, small inhibition curves are easily measured, allowing a greater sensitivity (down to 12.5 μM Trolox) and better precision at the lowest detectable dose (6.25 μM, or 0.3125 nanomoles/well Trolox equivalents) using the multiwell plate luminometer. Although other methods of determining an area under a curve exist, we use the simple approach of image analysis to obtain AOI (therefore, it is critical to keep the graph size and conditions of analysis standard). Copy the graph (Fig. 1) and paste into NIH Image or other image analysis program. Often, the first point for the sample will be below the TA line; draw a vertical line ("first response line," FRL) from this point to the TA line (this may be done in the graphing or image analysis program). If the right side of the inhibition curve does not meet the TA line (see Note 14). In the image analysis program, ensure that all pixels are filled in for the perimeter of the AOI, from the FRL through the inhibition curve, and along the TA (any breaks in the TA may be completed with a straight line, the "interpolated baseline" if, for example, the sample was run longer than the TA well). Select the enclosed space for area calculation; this is the AOI (in pixels).

16. Generate a standard curve for the experiment. Convert or plot Log transformations of the AOI or 10% recovery time measurements for the Trolox standards. Convert Trolox concentrations (μM) to Log Trolox concentrations for the x-axis. Delete high or low points of the standard curve that are not on the linear regression line, and do not use any sample values in those ranges. Calculate the regression equation (Fig. 2).

17. Calculate Trolox equivalents (in μM) from log-transformed data for samples using the regression equation. Compute the antilog of the results to arrive at the Trolox equivalents (in μM) of the samples.

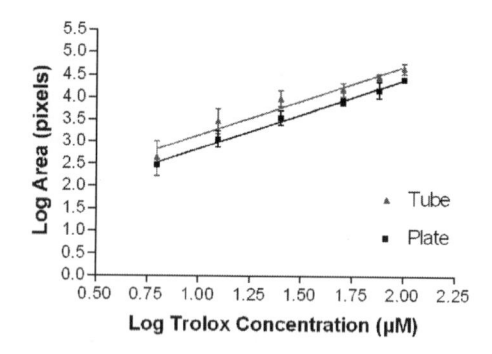

Fig. 2. Trolox standard curve examples. Log AOI for Trolox standards is plotted on the y-axis, and Log Trolox concentration from 6.25 to 100 µM is plotted on the x-axis. The upper line (*triangles*) is from a tube luminometer; the lower line (*squares*) is from a multiwell plate luminometer.

18. Multiply the calculated Trolox equivalents of each sample by the dilution factor used in the analysis. This is the final result.
19. Use of an injector streamlines the assay further, but requires modifications to the method (see Note 15).
20. Validation of the assay for different fluids may be performed by adding known amounts of Trolox to samples, and measuring the difference in TAC from the same sample without Trolox. Calculate the % recovery of added Trolox; we find this to be 90% or higher for seminal plasma.
21. Limitations of this assay are discussed (see Note 16).

4. Notes

1. P-Iodophenol or other phenols are used to enhance luminol chemiluminescence. The effect is dramatic but transient. Thus, careful and consistent timing of all steps is essential.
2. Water should be purified to 18 mega-ohms, and preferably ultrafiltered.
3. The standard curve for Trolox may be made in deionized water (or 0.1 M Tris–HCl buffer, pH 8.0) using the following dilutions starting with the 100 mM stock. Adjust these volumes if your luminometer requires higher volumes. Mix each standard thoroughly before proceeding to the next dilution.

Standard to Make	Vol. of standard		Plus	Vol. diluent	Note
1,000 µM	50 µL	100 mM Stock	+	4,950 µL	Not used in curve
100 µM	200 µL	1,000 µM	+	1,800 µL	
75 µM	750 µL	100 µM	+	250 µL	
50 µM	667 µL	75 µM	+	333 µL	
25 µM	500 µL	50 µM	+	500 µL	
12.5 µM	500 µL	25 µM	+	500 µL	
6.25 µM	500 µL	12.5 µM	+	500 µL	

Example: To make a 100µM Standard, add 200µl of the 1000µM Standard to 1800µl of diluent

4. HRP-conjugated immunoglobulins or other sources of HRP will have different activities of peroxidase. If a new source or lot is to be used, test it first by running a Total Activity tube or well. Make dilutions sufficient to yield 500,000–1,000,000 rlu/s, or other value within the measurement range of your luminometer.

5. Samples may be diluted in deionized water or 0.1 M Tris–HCl buffer; we have not seen any differences. Use of sperm incubation medium (i.e., HTF, Ham's F10, BWW) with or without 0.3–0.5% protein or HEPES also does not significantly affect the assay.

6. Dilution of seminal plasma is required to obtain a response that can be accurately evaluated; e.g., one that does not inhibit the signal to nondetectable levels. However, if the response is too low or too high, the seminal (or other) plasma may be reevaluated at other dilutions. Change the volumes depending on the needs of your luminometer. Addition of protease inhibitors to seminal plasma decreases TAC by about 30%; always treat samples in a consistent way.

7. Measurements must be repeated often enough (at least 10 to over 25 times) to generate an inhibition curve, without missing the peak (low) response or the point at which values return to their high plateau. This is simple in a one-place tube luminometer, since all measurements are collected for only one tube at a time. For multiple tubes or a plate luminometer, it is important not to have too many samples running at once, since the cycle return time may be too long to obtain sufficient points on each curve. Typically, we only run three wells at a time for this reason. We have experimented with using the luminometer's injectors to add HRP or peroxide at the start time of each sample run to avoid these timing issues (see below).

8. We recommend closing window blinds and turning off overhead lights. We set up a low-wattage lamp or dark-room light to provide just enough light to enable work.

9. The Total Activity result is a high level (usually 500,000–1,000,000 rlu/s) plateau that should remain almost flat for enough time (4–10 min) to complete one run of the assay. A slight downward trend is common, and can be compensated during analysis. If the plateau falls dramatically during a 4–10 min span, make fresh Signal Reagent. Be sure the hydrogen peroxide and HRP are fresh.

10. The NSB tube has no HRP added, and the result should be no signal. Spurious light leaks or exposure will cause spikes, which must be corrected before analysis. If the sample is included, endogenous peroxidase (e.g., myeloperoxidase from neutrophils) activity could generate a signal. The NSB result, if not "0," must be subtracted from the TA and all other analyses of the tested components.

11. The generation of a standard curve using Trolox or other antioxidant of choice is essential to quantify the results from the experimental samples. Adjust the range of standards to obtain at least three to four linear points in a dose–response curve. Points on a curve that are out of place, or curves that do not resemble previous ones usually indicate a pipetting or mixing error, or degradation of the standard. If any deviations are noted, prepare a new set of standards and analyze before analyzing samples.

12. Analyzing three standards or samples at a time (for the microplate luminometer without using the injector) may compromise the results due to the delay in time between loading the first sample to loading and measuring the third. We suggest making these three wells for each run: sample, NSB containing the sample, and TA. If loaded in reverse order, there will not be an important delay for the inhibition curve in the first well, and each test will include the nonspecific background for the sample tested. An option is to run the three wells without sample for 3 min (thus obtaining two backgrounds and one TA measurement), opening the lid and quickly adding 50 μL of the same standard or sample to the first two wells, and restarting the run for an additional 5 min. This will yield a sample analysis, a sample NSB, and the continuation of the TA analysis. The two TA analyses (first 3 min, a short break, and then the last 5 min) can be combined to form a complete "baseline."

13. We use Microsoft Excel for the raw data import, adjust for NSB if needed, then import into GraphPad Prism to generate graphs and run regression analyses.

14. If the TA well or tube was not run for as long a time as the sample or standard, a straight line may be drawn from the last point of the TA plateau ("interpolated baseline") to intersect

with the sample curve on the right side. Sometimes, the inhibition is very strong and, even with dilution the inhibition curve does not approach the TA plateau. In these cases, it may be possible to still calculate a 10% recovery time. If AOI is being used, the best solution is to choose a time at which most samples will have returned to the TA plateau (5 min in our experience), and draw a straight vertical line at that time to complete the perimeter of the AOI. We also suggest calculating the percentage of Total Activity reached by the plateau at 5 min. Make a note that this sample completely inhibited oxidant generation.

15. Multiwell luminometers equipped with injector(s) allow the possibility of starting the reaction by injection rather than the laborious and variable technique of stopping mid-way, adding sample, and continuing the analysis. However, alterations are needed to do this. The sample or standards cannot be injected without thoroughly cleaning the system's tubing between each injection. We attempted to inject Signal Reagent plus HRP solution, but found that this mixture degraded too quickly (by 10 min) to use for multiple analyses. We settled on adding the HRP to the sample (seminal plasma contains low amounts of peroxidase already), and injecting a slightly modified Signal Reagent to start the reaction. Prepare Signal Reagent-I (for a 50 μL injection) by adding components in this order, with thorough mixing after each addition: to 8.33 mL 0.1 M Tris–HCl buffer, pH 8.0, add 15 μL of 12 M (30%) fresh hydrogen peroxide, 5 μL of 41.7 mM 4-iodophenol, and 55 μL of 282.1 mM luminol. Cover tube with foil and store immediately at 4°C until use. All other reagents are used as described above, although the HRP can be reduced to 20 μL by making a 0.6% stock solution. Set up a repeated measures protocol with injection at the start (or after a min) of the analysis, total time 10 min, and 5 s single measurement time. Data collection and analysis are the same as above, except there is only one series of data instead of two for each well, and there is no need to draw a vertical FRL line since all data are collected immediately after Signal Reagent addition. Thus, the analysis is simplified. We compared standard curves from the noninjector and injector assays and found them to be identical (Fig. 3). These methods differ primarily in that, at the point of injection, both a stimulation and an inhibition of reactive oxygen species occurs, probably resulting in the lower maximum values seen. In contrast, for the noninjection technique there is an inhibition of an already maximally stimulated response.

16. Every method for measuring antioxidants has limitations, and these are important to point out. The current method, like many others, does not measure enzymatic antioxidants. Specific assays are available for superoxide dismutase, catalase, and glutathione peroxidase, for example. The specific free radicals generated by

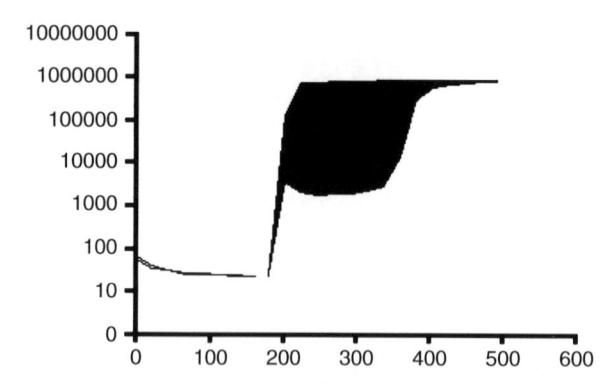

Fig. 3. Example of a graph obtained from data collected using the injector method in the multiplate luminometer. The samples were analyzed 3 min before injection of the Signal Reagent, thus there is a low response line until the point of injection at 180 s. This is a combined graph of TA response (the almost vertical and upper plateau edge of the shaded area) and the sample response (lower and right side of the shaded area). The AOI is filled in for area determination. *Y-axis* rlu/s, *X*-axis seconds. *Shaded area* AOI.

any chosen cell-free generating system will not necessarily match those produced in vivo. We believe that the hydrogen peroxide-HRP system used by us and many others is a reasonable choice, although use of myeloperoxidase or other mammalian radical generating systems would be superior in theory (30–33). Different antioxidants have different effects on scavenging radicals or inhibiting the radical production system, as discussed by Strube et al. (34). Some antioxidants quickly inhibit radical generation and others work slowly (fast vs. slow TRAP). Rhemrev et al. (35) discuss the various components of the seminal plasma nonenzymatic antioxidant system, and suggest that the initial fast response is primarily due to vitamin C, uric acid, tyrosine, proteins, and polyphenolic compounds. The slow TRAP response is partly due to proteins, tyrosine and polyphenolic compounds, but the molecules responsible for about half of the slow TRAP response could not be determined by those authors. In some cases, the inhibition is permanent, and a plateau of lowered radical production develops. This point is of concern, since such a pattern would prevent calculation of area of inhibition, and, more importantly, may represent a different mode of action of the antioxidant. We did find a lowered plateau pattern when we tested certain antioxidants in our system, such as acetylcysteine or butylated hydroxytoluene.

Acknowledgments

The authors thank Ashok Agarwal, Ph.D. of the Cleveland Clinic, Cleveland, Ohio, for sending us his TAC protocol, which we modified for this study. We also thank Professor Seymour Klebanoff,

M.D. Ph.D., of the Division of Allergy and Infectious Diseases, Department of Medicine, University of Washington, for allowing us to use his tube luminometer and for illuminating discussions. The Paul G. Allen Foundation for Medical Research supported this research.

References

1. Ceconi C et al (2003) Oxidative stress in cardiovascular disease: myth or fact? Arch Biochem Biophys 420:217–221
2. Lewis SE et al (1995) Total antioxidant capacity of seminal plasma is different in fertile and infertile men. Fertil Steril 64:868–870
3. Agarwal A et al (2003) Role of reactive oxygen species in the pathophysiology of human reproduction. Fertil Steril 79:829–843
4. Aitken J, Fisher H (1994) Reactive oxygen species generation and human spermatozoa: the balance of benefit and risk. Bioessays 16:259–267
5. Aitken RJ, Sawyer D (2003) The human spermatozoon – not waving but drowning. Adv Exp Med Biol 518:85–98
6. Alkan I et al (1997) Reactive oxygen species production by the spermatozoa of patients with idiopathic infertility: relationship to seminal plasma antioxidants. J Urol 157:140–143
7. Lewis SE et al (1997) Comparison of individual antioxidants of sperm and seminal plasma in fertile and infertile men. Fertil Steril 67:142–147
8. Sharma RK, Agarwal A (1996) Role of reactive oxygen species in male infertility. Urology 48:835–850
9. Sikka SC (2001) Relative impact of oxidative stress on male reproductive function. Curr Med Chem 8:851–862
10. Ramya T et al (2011) Altered levels of seminal nitric oxide, nitric oxide synthase, and enzymatic antioxidants and their association with sperm function in infertile subjects. Fertil Steril 95:135–140
11. Shamsi MB et al (2009) DNA integrity and semen quality in men with low seminal antioxidant levels. Mutat Res 665:29–36
12. Ochsendorf FR (1999) Infections in the male genital tract and reactive oxygen species. Hum Reprod Update 5:399–420
13. Pasqualotto FF et al (2000) Seminal oxidative stress in patients with chronic prostatitis. Urology 55:881–885
14. Potts JM, Pasqualotto FF (2003) Seminal oxidative stress in patients with chronic prostatitis. Andrologia 35:304–308
15. Shahed AR, Shoskes DA (2000) Oxidative stress in prostatic fluid of patients with chronic pelvic pain syndrome: correlation with gram positive bacterial growth and treatment response. J Androl 21:669–675
16. Thomson LK et al (2009) Cryopreservation-induced human sperm DNA damage is predominantly mediated by oxidative stress rather than apoptosis. Hum Reprod 24:2061–2070
17. Kumar D, Jugdutt BI (2003) Apoptosis and oxidants in the heart. J Lab Clin Med 142:288–297
18. Whitehead TP et al (1992) Enhanced chemiluminescent assay for antioxidant capacity in biological fluids. Anal Chim Acta 266:265–277
19. Glazer AN (1990) Phycoerythrin fluorescence-based assay for reactive oxygen species. Methods Enzymol 186:161–168
20. Wayner DD et al (1985) Quantitative measurement of the total, peroxyl radical-trapping antioxidant capability of human blood plasma by controlled peroxidation. The important contribution made by plasma proteins. FEBS Lett 187:33–37
21. Wayner DD et al (1987) The relative contributions of vitamin E, urate, ascorbate and proteins to the total peroxyl radical-trapping antioxidant activity of human blood plasma. Biochim Biophys Acta 924:408–419
22. Harrison D et al (2003) Role of oxidative stress in atherosclerosis. Am J Cardiol 91: 7A–11A
23. Cao G et al (1993) Oxygen-radical absorbance capacity assay for antioxidants. Free Radic Biol Med 14:303–311
24. Cao G et al (1995) Automated assay of oxygen radical absorbance capacity with the COBAS FARA II. Clin Chem 41:1738–1744
25. Cao G, Prior RL (1998) Comparison of different analytical methods for assessing total antioxidant capacity of human serum. Clin Chem 44:1309–1315
26. Benzie IF, Strain JJ (1996) The ferric reducing ability of plasma (FRAP) as a measure of "antioxidant power": the FRAP assay. Anal Biochem 239:70–76
27. Mahfouz R et al (2009) Diagnostic value of the total antioxidant capacity (TAC) in human seminal plasma. Fertil Steril 91: 805–811

28. Said TM et al (2003) Enhanced chemiluminescence assay vs colorimetric assay for measurement of the total antioxidant capacity of human seminal plasma. J Androl 24:676–680
29. Sharma RK et al (1999) The reactive oxygen species-total antioxidant capacity score is a new measure of oxidative stress to predict male infertility. Hum Reprod 14: 2801–2807
30. Heinecke JW (2003) Oxidative stress: new approaches to diagnosis and prognosis in atherosclerosis. Am J Cardiol 91:12A–16A
31. Klebanoff SJ (1968) Myeloperoxidase-halide-hydrogen peroxide antibacterial system. J Bacteriol 95:2131–2138
32. Klebanoff SJ (1999) Myeloperoxidase. Proc Assoc Am Physicians 111:383–389
33. Rosen H, Klebanoff SJ (1976) Chemiluminescence and superoxide production by myeloperoxidase-deficient leukocytes. J Clin Invest 58:50–60
34. Strube M et al (1997) Pitfalls in a method for assessment of total antioxidant capacity. Free Radic Res 26:515–521
35. Rhemrev JP et al (2000) Quantification of the nonenzymatic fast and slow TRAP in a postaddition assay in human seminal plasma and the antioxidant contributions of various seminal compounds. J Androl 21:913–920

Part VIII

Analytical Tools and Methods

Chapter 32

Methods of Sperm DNA Extraction for Genetic and Epigenetic Studies

Jeanine Griffin

Abstract

High quality DNA extractions developed for mammalian somatic cells are ineffective for sperm, due mainly to the high degree of nuclear compaction in sperm. The highly specialized nuclear proteins in sperm create a chromatin structure that is at least six times denser than histone bound DNA. Unlike somatic cells, sperm DNA is highly compacted by the replacement of histones with sperm-specific low molecular weight proteins called protamines. Both the protamines and the disulfide bridges formed within and between protamines inhibit the extraction of sperm DNA by standard techniques used for somatic cells. Here we describe the guanidine thiocyanate method reported by Hossain with additional modifications resulting in high molecular weight DNA of high quality with an A260/280 ratio ranging between 1.8 and 2.0 and an A260/230 ratio of 2.0 and greater. The DNA is efficiently digested with restriction enzymes and amplified by PCR.

Key words: Sperm, Protamines, Guanidinium thiocyanate, DNA, Chaotropic agent, Dithiothreitol, Proteinase K

1. Introduction

A wide variety of methods have been developed to extract DNA from various sources. Despite the many methods reported, there are similarities common among them. DNA isolation techniques used routinely for mammalian somatic cells have proven ineffective for mammalian sperm (1, 2). Bahnak et al. reported a guanidinium protocol for the isolation of mammalian spermatozoa (2). Hossain et al. later modified this protocol making it simple and efficient by replacing lengthy CsCl ultracentrifugation steps with isopropanol and incorporating proteinase K digestion (3). Commercial companies have utilized the guanidine thiocyanate method in kit form for the isolation of RNA, DNA, and protein; however, these kits require the use of toxic solutions such as chloroform and phenol. Other kits, utilize column purification of the DNA following lysis.

Changes to the modified protocol were incorporated to achieve isolation of high molecular weight genomic DNA of high quality, eliminating incomplete protein digestion that appears as a glassy gelatinous glob attached to the DNA strands and the removal of chaotropic salts that can coprecipitate with the DNA.

The lysis of DNA begins with the breakdown of tissues or cells. Lysis is carried out in a salt solution containing detergents to denature proteins or proteases for digestion of proteins, or in some cases both. The components that make up the lysis buffer for the modified guanidine thiocyanate DNA extraction method include guanidine thiocyanate, beta-mercaptoethanol, sarkosyl, sodium citrate, and proteinase K. Here we describe the purpose of each component used in the extraction of human sperm DNA and the changes we incorporated into the modified protocol.

Guanidine thiocyanate is a chaotropic agent and a strong protein denaturant used in most RNA isolation protocols. A 4 M solution of guanidine thiocyanate irreversibly inactivates RNase and DNase, disintegrates cellular membranes, dissolves proteins by disrupting their secondary structures, and dissociates nucleoproteins from nucleic acids allowing digestion of the proteins with proteases. Guanidine thiocyanate enhances the activity of proteinase K making it the protease of choice.

During spermiogenesis, testes-specific transition proteins replace somatic histones in round spermatids. The transitions proteins are later replaced by protamines in the elongated spermatid resulting in highly condensed chromatin (4–9). Soon after synthesis, protamines are phosphorylated. Once the protamines bind to DNA, most of the phosphate groups are removed and cysteine residues are oxidized, forming disulfide bridges that link the protamines together. Additional dissociation of nucleoproteins from nucleic acids is required. Dithiothreitol (DTT) and 2-mercaptoethanol are both strong reducing agents that cleave disulfide bonds, allowing proteins to unfold completely. Although the modified method called for 2-mercaptoethanol, we chose to use DTT, which is more effective, less toxic, quicker and the odor is easily tolerated. Sodium dodecyl sulfate (SDS) is routinely used in DNA extraction methods; however, SDS has very low solubility in high-salt chaotropic solutions. Sodium lauroyl sarcosinate (Sarkosyl), like SDS, is an ionic detergent used to denature proteins and disrupt biological membranes in addition to having excellent solubility in chaotropic solutions, making it the detergent of choice in guanidine buffers.

Prior to DNA extraction, both fresh semen and frozen sperm are washed twice with 150 mM NaCl and 10 mM EDTA solution to remove cellular and noncellular components prior to the addition of lysis buffer. EDTA removes metal ions that may be present in seminal fluid, cryoprotectants or on the cell surface, to reduce contaminants in the final DNA extract.

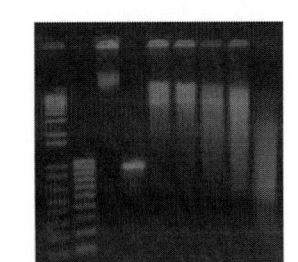

Fig. 1. Electrophoresis of sperm DNA samples on 0.7% agarose gel. Lane 1 = HINDIII lambda DNA marker, Lane 2 = 100 bp DNA size marker, Lane 3 = undigested genomic DNA, Lane 4 = PCR product, Lane 5 = HINDIII digest, Lane 6 = EcoRI digest, Lane 7 = BamHI digest, Lane 8 = DraIII digest, Lane 9 = HinfI digest.

Lysis was found to be complete within 2 h, whereas 3 h was used in the modified protocol. Following lysis, the addition of isopropanol facilitates DNA precipitation. The strands of DNA are transferred to a microcentrifuge tube with 0.1 M sodium citrate in 10% alcohol. By incubating at room temperature for 30 min, any remaining chaotropic salts will go into solution, while the DNA remains precipitated in the presence of high salt and alcohol (sodium ions in the presence of alcohol neutralize the negatively charged phosphate groups on the DNA backbone facilitating DNA precipitation).

This wash is most effective when performed twice. A final wash is performed in 70% alcohol. It is difficult to measure the concentration of high molecular weight genomic DNA because the DNA solution is often heterogeneous and viscous (10).

This extraction method results in high quality, high molecular weight genomic DNA (Fig. 1) with approximately 80% yield, an A260/280 ratio ranging between 1.8 and 2.0, and an A260/230 ratio of 2.0 and greater. The DNA is efficiently digested with restriction enzymes and amplified by PCR.

2. Materials

All procedures are carried out at room temperature unless otherwise specified. All solutions are to be prepared with ultrapure water. All reagents are stored at room temperature with the exception of proteinase K and DTT (stored according to manufacturer's recommendation).

2.1. Reagents

1. 6 M guanidine isothiocyanate from supplier.
2. 30% Sarkosyl from supplier.
3. 5 M NaCl in water.

4. 1 M DTT in water—prepared fresh.
5. Proteinase K (20 mg/mL) from supplier.
6. 0.5 M EDTA in Tris–HCl pH 8.0.
7. Isopropanol.
8. Ethanol.
9. 10 mM Tris–HCl, pH 8.0.

2.2. Solutions

1. 0.1 M Sodium citrate in 10% EtOH.
2. Sperm wash buffer—~20 mL/sample: 150 mM NaCl, 10 mM EDTA, pH 8.0.
3. Extraction buffer—3 mL/sample: 2.12 mL 6 M guanidine thiocyanate (final concentration 4.24 M), 60 µL, 5 M NaCl (final concentration 100 mM), 100 µL, 30% Sarkosyl (final concentration 1%), 450 µL, 1 M DTT (final concentration 150 mM), 30 µL, 20,000 mg/ml (final concentration 200 µg/mL), 240 µL H$_2$O.

3. Methods

Carry out all procedures at room temperature unless otherwise specified.

1. Wash frozen sperm or semen in 10 volumes of sperm wash buffer in a 15 ml conical tube. Centrifuge at 750×*g* for 10 min. Pour off supernatant. Add 10 ml of sperm wash buffer to the pellet and vortex tube resuspending the cells. Centrifuge at 750×*g* for 10 min. Pour off supernatant carefully to avoid disturbing sperm pellet (see Note 1).
2. Vortex pellet briefly to resuspend cells in the small residual volume of buffer remaining in the tube. Add 3 mL of extraction buffer to the tube and gently mix by inverting the tube several times (see Notes 2–4).
3. Place tube in a 56 °C water bath and incubate for 2 h. Invert tube three times halfway through incubation.
4. After the 2 h incubation, remove tube from water bath and let cool to room temperature. Add 2.4 mL isopropanol and mix gently by inverting 25 times or until DNA strands form.
5. Using a Pasteur pipette whose end has been sealed and shaped into a U, spool the DNA out of solution and transfer to a 2 mL microcentrifuge tube containing 2 mL 0.1 M sodium citrate in 10% EtOH. Let sit at room temperature for 30 min with occasional mixing by inverting tube three times.

Carefully remove the buffer using a pipette. Repeat wash once (see Note 5).

6. Wash the DNA pellet twice in 1 mL 70% EtOH. Mix by inverting tube several times. Carefully remove alcohol solution and invert tube to drain. Let dry for 10 min or until no traces of alcohol remain (see Note 6).

7. Rehydrate DNA in 10 mM Tris–HCl, pH 8.0 or 8 mM NaOH (400–600 µL). Place vial in a 65 °C water bath for 30 min to 1 h until DNA is in solution tapping the tube periodically to help disperse the DNA. DNA can be left at room temperature overnight to further rehydrate. If using 8 mM NaOH, once DNA is in solution, neutralize with 0.1 M HEPES to desired pH (see Notes 7–9).

4. Notes

1. If pellet is too loose, centrifuge at $1,000 \times g$. Higher speeds will clump the sperm, making it difficult to resuspend. A swinging bucket centrifuge rotor will collect the sperm at the bottom of the tube, while a fixed-angle centrifuge rotor will collect the sperm up the side of the wall, making it easy to lose sperm when decanting the solution.

2. To avoid shearing the high molecular weight DNA, treat gently and do not vortex from this point on.

3. Prepare lysis buffer fresh each time. DTT becomes unstable in solution. It is important to prepare fresh each time. A stock solution of 1 M DTT prepared in water may be stored in aliquots at −20°C. Do not freeze thaw the stock solution, use only once and then discard.

4. Lyse 30×10^6–120×10^6 cells in 3 mL of lysis buffer. If sperm count is lower, lysis can be carried out in 600 µL in a 2 mL microcentrifuge tube with isopropanol volume reduced to 480 µL. Centrifugation at $5,000 \times g$ for 3 min at room temperature is sufficient to spin down DNA.

5. If the DNA becomes fragmented or there is very little precipitate, collect the DNA precipitate by centrifugation at $5,000 \times g$ for 3 min at room temperature. Caution needs to be taken at this point as the DNA pellet may have a glassy appearance and could easily be lost. Centrifuging DNA at high speeds will compact the DNA and make it difficult to rehydrate, do not exceed $5,000 \times g$.

6. If allowed to dry completely, DNA will be difficult to rehydrate.

7. For 1 mL of 8 mM NaOH add the indicated amounts of 0.1 M HEPES. Prepare fresh 8 mM NaOH each time.

Final pH	0.1 M HEPES (µL)
8.4	86
8.2	93
8.0	101
7.8	117
7.5	159

8. An alternative rehydration buffer of 10 mM TE (pH 8.0) may be used. However, EDTA has been shown to interfere with some restriction enzyme digests. DNA rehydrated in water degrades over time.

9. High molecular weight genomic DNA is frequently nonhomogeneous and very viscous. This makes obtaining an accurate concentration measurement difficult. In such cases, to measure the concentration, pipette a 10–20 µL aliquot of the sample using a yellow tip with the end cut off and dilute in TE, vortex vigorously for 2 min.

References

1. Shiurba R, Nandi S (1979) Isolation and characterization of germ line DNA from mouse sperm. Proc Natl Acad Sci U S A 76:3947–3951
2. Bahnak BR et al (1988) A simple and efficient method for isolating high molecular weight DNA from mammalian sperm. Nucleic Acids Res 16:1208
3. Hossain AM et al (1997) Modified guanidinium thiocyanate method for human sperm DNA isolation. Mol Hum Reprod 3(11): 953–956
4. Hecht NB (1989) Mammalian protamines and their expression. In: Hnilica LS, Stein JL, Stein GS (eds) Histones and other basic nuclear proteins. CRC, Orlando, pp 347–373
5. Hecht NB (1990) Regulation of 'haploid expressed genes' in male germ cells. J Reprod Fertil 88:679–693
6. Oliva R, Dixon GH (1991) Vertebrate protamine genes and the histone-to-protamine replacement reaction. Prog Nucleic Acid Res Mol Biol 40:25–94
7. Dadoune JP (1995) The nuclear status of human sperm cells. Micron 26:323–345
8. Steger K (1999) Transcriptional and translational regulation of gene expression in haploid spermatids. Anat Embryol (Berl) 199: 471–487
9. Brewer L et al (2002) Condensation of DNA by spermatid basic nuclear proteins. J Biol Chem 277:38895–38900
10. Sambrook J, Russell D (2001) Preparation and analysis of eukaryotic genomic DNA. In: Molecular laboratory manual, 3rd edn. Cold Spring Harbor Laboratory Press, New York, pp 6.3–6.64

Chapter 33

Isolating mRNA and Small Noncoding RNAs from Human Sperm

Robert J. Goodrich, Ester Anton, and Stephen A. Krawetz

Abstract

The isolation of spermatozoal RNA is a challenging procedure due to the intrinsic heterogeneous population of cells present in the ejaculate and the small quantity of RNA present in sperm. The transcripts contained within these gametes includes a wide variety of messenger RNAs (mRNAs), small noncoding RNAs (sncRNAs), and highly fragmented ribosomal RNAs (rRNAs).

The protocol described in this chapter to isolate both the mRNA and sncRNA fractions represents years of development towards automation. It combines a guanidinium thiocyanate–phenol–chloroform-based methodology to reduce the content of DNA and a column-based system. Both manual and semi-automated options are described, with preference given to automation for consistent results. A novel quality control procedure has been developed to assess the integrity and purity of the entire population of isolated mRNAs due to the absence of intact rRNAs.

Key words: Human spermatozoa, PureSperm gradient, mRNA and sncRNA extraction, RNeasy Mini kit, RNeasy Minelute kit, QIAcube

1. Introduction

Isolation of the RNA content of human spermatozoa presents a unique set of challenges. First, the intrinsic heterogeneous population of cells present in the ejaculate (1) necessitates the introduction of a purifying step in which only the spermatozoa are isolated. This is necessary to ensure the complete absence of contaminating somatic cells while maximizing the recovery of viable sperm. Second, the small quantity of RNA present in these cells (50 fg of RNA/cell (2) and 0.3 fg of small noncoding RNA (sncRNA)/cell) requires optimization of the RNA extraction protocol to maximize yield. Third, the absence of ribosomal RNA (rRNA) markers hinders quality assessment.

A wide variety of transcripts defining the composition of RNAs in mature spermatozoa that range from the larger messenger RNAs (3) (mRNAs) to the interesting collection of sncRNAs (4). Spermatozoa are transcriptionally and translationally quiescent, thus devoid of intact rRNAs (reviewed by Krawetz (2)) although fragments abound (5). Thus integrity cannot be assessed as a function of the recovery of intact 28S and 18S rRNAs.

Several protocols have been developed and are continuing to be developed for the isolation of spermatozoal RNAs from various sources (6–9). The procedure described in this chapter is outlined in Fig. 1 and combines two of the most common methods used for RNA extraction: a guanidinium thiocyanate–phenol–chloroform-based methodology to remove the bulk of the contaminating DNA and a column-based kit for ease of use. This can be performed manually as described in Subheading 3.2 or in an automated fashion with the use of a QIAcube as described in Subheading 3.3. The latter affords the consistent and simultaneous isolation of both the mRNA and sncRNA component from multiple samples. Quality control, outlined in Subheadings 3.4–3.6, includes DNase treatment, reverse transcription, and PCR amplification with intron-spanning primers to verify the absence of genomic contamination and mRNA integrity. This was now supplemented by genome-wide assessment using microarrays (Subheading 3.7) that has now been superceded by bar-coded multiplexed sequencing.

The protocol described below presents a standardized methodology to obtain good-quality fractions of spermatozoal sncRNAs and mRNAs. These RNAs are well suited to subsequent molecular assays to define the relative distribution of these various transcripts using expression arrays or high-throughput sequencing. The resulting sperm RNA fingerprints have proven a useful means to assess the molecular basis of reproductive competence (10) and the paternal contribution to the zygote and embryo development (11).

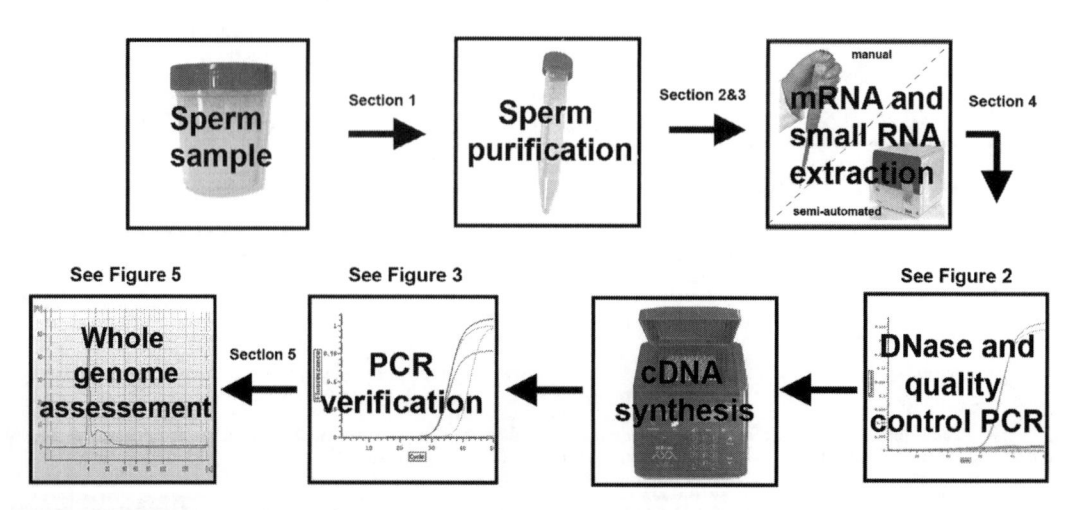

Fig. 1. Flow diagram of sperm RNA extraction and quality control processes.

2. Materials

2.1. Components for Preparing Sample for Extraction

1. PureSperm (Spectrum Technologies).
2. PureSperm buffer (Spectrum Technologies).

2.2. Components for Manual Extraction of RNAs

1. 0.2 mm RNase-free stainless steel beads (NextAdvance).
2. Qiazol (QIAGEN).
3. RNeasy Mini kit (QIAGEN).
4. RNeasy Minelute kit (QIAGEN).
5. Disruptor Genie (Scientific Industries).

2.3. Components for Extraction of RNAs Using QIAcube Custom Protocol

1. QIAcube tips 1,000 µL (QIAGEN).
2. QIAcube tips 200 µL (QIAGEN).
3. QIAcube tubes (QIAGEN).
4. QIAcube Adaptors (QIAGEN).
5. RNeasy Mini kit (QIAGEN).
6. RNeasy Minelute kit (QIAGEN).
7. QIAcube (QIAGEN).
8. Disruptor Genie (Scientific Industries).

2.4. Components for Quality Control

1. RNase Block (Stratagene).
2. Turbo DNA-free (Ambion).
3. Oligo dT primer (Invitrogen).
4. 100 mM dNTP set (Invitrogen).
5. Superscript III RT (Invitrogen).
6. QuantiTect SYBR Green PCR kit (QIAGEN).
7. PRM1 primers (intron spanning).
   Forward: CAGAGCCGGAGCAGATATTAC
   Reverse: ATTTATTGACAGGCGGCATTGTT.

3. Methods

3.1. Preparing Samples for Extraction

1. Dilute PureSperm to 50% with PureSperm buffer. Equilibrate to room temperature (at least 10–15 min). Aliquot 3 mL of 50% PureSperm per sample (1.5 mL PureSperm and 1.5 mL PureSperm buffer) into individual 15 mL conical tubes.

2. Allow fresh ejaculate samples to liquefy at room temperature for 30 min. If liquefied sample was frozen thaw sample rapidly by hand.

3. Transfer the sample to a 15 mL conical tube and add 6–10 mL PureSperm buffer to wash sample by inverting the tube several times.

4. Using a hemocytometer estimate the total number of cells.

5. Centrifuge for 15 min at 4°C at 300×g, and then discard the supernatant into a 10% bleach solution.

6. Resuspend the cell pellet in 1 mL of PureSperm buffer.

7. Overlay the sample onto the 50% PureSperm step-gradient using a transfer pipette being careful not to disturb the interface (see Note 1).

8. Disengage the rapid acceleration and brake options from the centrifuge (see Note 2). Centrifuge the gradient for 20 min at room temperature at 200×g (see Note 3).

9. Remove the pellet from the conical tube by directing the tip of transfer pipette to the bottom of the tube. Extract the sperm pellet with as little liquid as possible.

10. Transfer the sperm pellet to a new conical tube and resuspend in 10 mL PureSperm buffer. Wash the cells by inverting the tube several times.

11. Using a hemocytometer estimate the total number of cells.

12. Centrifuge for 15 min at 4°C at 300×g. Discard the supernatant into a 10% bleach solution.

13. As the purified sperm are pelleting (step 12) prepare the materials required for cell lysis. Prepare one homogenization tube for each sample. To a 2 mL microcentrifuge tube, add 500 µL RLT buffer (QIAGEN RNeasy Kit), 7.5 µL β-mercaptoethanol, and 100 mg of 0.2 mm nuclease-free stainless steel beads.

14. After centrifugation, decant the supernatant and resuspend the sperm pellet in any residual liquid (typically 100–200 µL).

15. Add up to 100 million cells to the homogenization tube prepared in step 13. If the sample is greater than 100 million cells, divide it into equal portions that are less than 100 million cells and treat as separate samples until the extraction is complete.

16. Place the homogenization tube in a Disruptor Genie (Scientific Industries) and shake for 5 min.

17. Add 500 µL of Qiazol (QIAGEN) and shake for 2 min with the Disruptor Genie as above.

18. Add 200 µL of chloroform and rapidly shake by hand for 15 s. Let the mixture sit at room temperature for 3–5 min.

19. Centrifuge for 20 min at 4°C at 12,000×g.
20. Remove the upper aqueous layer using a wide bore pipette tip and place the recovered aqueous layer into a 2 mL tube. Extract RNAs using a modified RNeasy (QIAGEN) protocol outlined in Subheading 3.2. To minimize protocol variation, automate with the use of a QIAcube (QIAGEN).

3.2. Manual Extraction of RNAs

RNeasy MiniKit optimized for the recovery of sperm RNA (see Note 4).

1. Add 360 μL of 100% ethanol for every 500 μL of upper aqueous phase and mix by pipetting up and down.
2. Apply 700 μL of the sample, including any precipitate, to an RNeasy mini column that is placed in a 2 mL collection tube (supplied). Close tube and spin fluid through at ≥8,000×g for 15 s. Remove the flow-through (FT) containing the small RNA fraction and place the small RNA fraction in a 15 mL conical tube. Repeat with the remaining sample.
3. After the entire sample has been centrifuged through the RNEasy column and the FT collected, place the columns containing the large RNAs at 4°C. Continue to step 4 to extract the small RNA first and then complete the extraction of the large RNA (see below Purification of large RNAs (continue at step 12)).
4. From step 3 add 0.65 volumes of 100% ethanol to the 15 mL conical tube containing the small RNA FT, and mix thoroughly by vortexing. Do not centrifuge.
5. Transfer 700 μL of the sncRNA-containing sample to an RNeasy MinElute spin column placed in a 2 mL collection tube. Centrifuge for 15 s at ≥8,000×g. Discard the FT.
6. Repeat **step 5** until the entire sample has passed through the RNeasy MinElute spin column. Discard the FT each time in a guanidinium waste container.
7. Add 500 μL RPE wash buffer to the RNeasy MinElute spin column. Centrifuge for 15 s at ≥8,000×g. Discard the FT.
8. Add 500 μL of 80% ethanol to the RNeasy MinElute spin column. Centrifuge for 15 s at ≥8,000×g. Discard the FT and the collection tube.
9. Place the RNeasy MinElute spin column in a new 2 ml collection tube. Centrifuge for 1 min at ≥8,000×g.
10. Place the RNeasy MinElute spin column in a 1.5 mL collection tube. Add 14 μL of RNase-free water directly to the spin column membrane (see Note 5). Centrifuge for 1 min at ≥8,000×g to elute the small RNA. Repeat with another 14 μL (see Note 6).
11. Add 1 μL of RNase Block (Stratagene) and DTT to a final concentration of 10 mM. Store at −80°C.

Purification of large RNAs (continued):

12. Remove the columns containing the large RNAs from the temporary 4°C storage. Add 700 μL RW1 buffer to the RNeasy column. Centrifuge at ≥8,000×g. Discard FT and collection tube.
13. Transfer the RNeasy column to a new collection tube (supplied). Pipette 500 μL RPE buffer onto the RNeasy column. Centrifuge at ≥8,000×g, for 15 s. Discard FT.
14. Load another 500 μL RPE buffer onto the RNeasy column. Centrifuge for 2 min at ≥8,000×g to dry the RNeasy silica-gel membrane.
15. Transfer the RNeasy column to another tube (not supplied) and spin again as above for 1 min to remove any traces of ethanol.
16. Transfer the RNeasy column to the final 1.5 mL collection tube (supplied).
17. Elute the large RNA fraction with 50 μL of RNase-free H_2O (see Note 5). Centrifuge at ≥8,000×g for 1 min and then repeat elution with an additional 50 μL of RNase-free H_2O (see Note 6).
18. Add 1 μL of RNase Block (Stratagene) and add DTT (final concentration of 10 mM) to the 100 μL final RNA sample. Store at −80°C.

3.3. Extraction of RNAs Using QIAcube Custom Protocol (Preferred)

Binding large RNA to the RNeasy Mini Column (see Note 7).

1. For each 500 μL of upper aqueous phase (step 20, Subheading 3.1) add 360 μL 100% ethanol and mix.
2. Transfer 430 μL from the middle position of the rotor adapter onto the RNeasy Mini column (remove snap caps before loading).
3. Centrifuge for 1 min at maximum speed.
4. Mix and load 430 μL from the middle position of the rotor adapter onto the RNeasy Mini column.
5. Centrifuge for 1 min at maximum speed.
6. Recover the flow-through from the middle position and save for small RNA isolation.
7. Repeat steps 1–6 until all of the upper aqueous phase has been loaded onto the column.
8. Store RNeasy Mini columns containing the bound large RNA at 4°C until step 22.

Isolation of Small RNAs:

9. Place fresh RNeasy MinElute columns in the wash positions.

10. Add 540 μL of 100% ethanol to the 860 μL FT that resides in the middle position of the rotor adapter.
11. Mix and transfer 700 μL of the ethanol FT to the RNeasy MinElute column that is placed in the wash position.
12. Centrifuge for 1 min at maximum speed.
13. Transfer the remaining 700 μL of the ethanol FT to the RNeasy MinElute column that is placed in the wash position.
14. Centrifuge for 1 min at maximum speed.
15. Repeat steps 9–13 until all of the small RNA FT has been passed through the column.
16. Wash the column with 500 μL RPE buffer by centrifugation for 1 min at $8,000 \times g$.
17. Wash with 500 μL of 80% ethanol by centrifugation at maximum speed for 2 min.
18. Transfer the RNeasy MinElute column to the elution position.
19. Add 14 μL of RNase-free water and then incubate for 5 min (see Note 8).
20. Recover the eluted RNA by centrifugation for 2 min at maximum speed.
21. Repeat steps 18 and 19, and then to the combined eluent add 1 μL of RNase Block (Stratagene) and DTT (final concentration of 10 mM) to your sample. Store at −80°C.

Isolation of large RNAs:

22. Place a new rotor adapter with the RNeasy Mini columns containing the bound large RNA that was stored in step 8 at 4°C in the rotor wash position.
23. Wash with 700 μL of RW1 buffer by centrifugation for 1 min at maximum speed.
24. Wash with 500 μL of RPE buffer by centrifugation for 1 min at $8,000 \times g$.
25. Wash with 500 μL of RPE buffer by centrifugation for 2 min at maximum speed.
26. Transfer the washed RNeasy Mini column to the elution position.
27. Add 50 μL of RNase-free water and then incubate for 5 min (see Note 9).
28. Recover the eluted RNA by centrifugation for 1 min at maximum speed.
29. Repeat steps 27 and 28, and then to the combined eluent add 1 μL of RNase Block (Stratagene) and DTT (final concentration of 10 mM) to your sample. Store at −80°C.

3.4. Quality Control: Removing Residual DNA

1. Add 0.1 vol. 10× Turbo DNase buffer and 2 μL Turbo DNase (Ambion) (see Note 10). Incubate for 20–30 min at 37°C. Following incubation add 0.1 volumes of the inactivation reagent and then incubate for 5 min at room temperature mixing occasionally. Centrifuge at 10,000×g for 1.5 min and transfer the RNA to a new tube.

2. Residual DNA contamination is assessed by QuantiTect SYBR green real-time PCR with PRM1 (annealing temperature [Ta]=65.5°C) primers (see Note 12). The DNase-treated samples should be void of amplification (Fig. 2). If products are apparent, perform a second DNase treatment using DNase buffer 2 comprising 200 mM Tris–HCl, buffer, pH 7.5, 100 mM $MgCl_2$, plus 5 mM $CaCl_2$, prepared with nuclease-free H_2O.

3. Quantify RNA in samples devoid of DNA (see Note 11).

3.5. Quality Control: Reverse Transcriptase PCR

1. Perform reverse transcription PCR using equal amount RNA from all samples. Add the following components to a nuclease-free microcentrifuge tube:

   1 μL of oligo(dT).

   10 pg to 5 μg of total RNA.

   1 μL 10 mM dNTP mix.

   Up to 13 μL nuclease-free water.

2. Heat mixture to 65°C for 5 min and then incubate on ice for at least 1 min.

3. Collect the contents of the tube by brief centrifugation and then add:

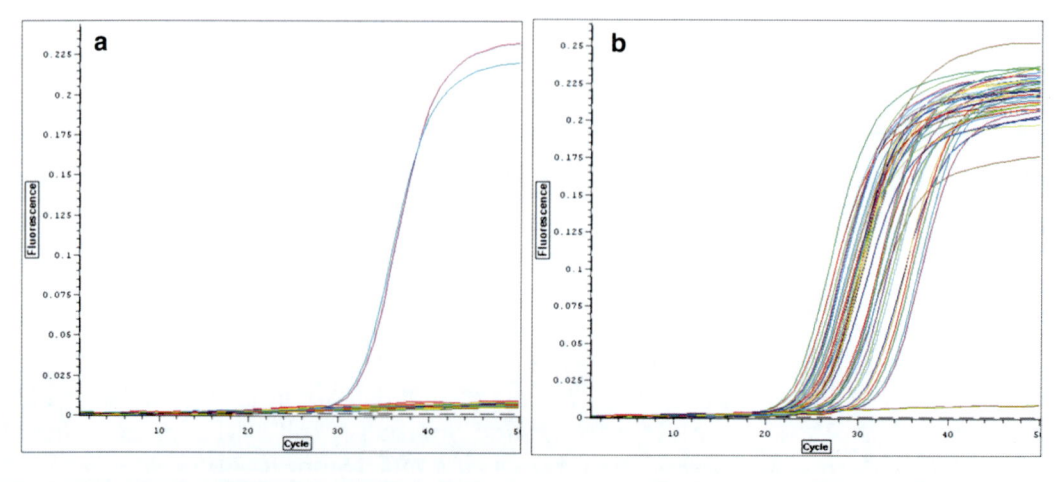

Fig. 2. PCR assessment of DNA contamination and RNA integrity. (**a**) Only human genomic controls amplified. All samples were free of contaminating DNA. (**b**) Subsequent to DNase treatment, all samples successfully amplified PRM1 mRNA.

4 µL 5× First-Strand Buffer.

1 µL 0.1 M DTT.

1 µL RNase Block.

1 µL of SuperScript™ III RT (200 units/µL).

4. Incubate at 50°C for 30–60 min.
5. Inactivate the reaction by heating at 70°C for 15 min.

3.6. Quality Control: Assessing RNA Integrity by PCR

1. Following reverse transcription, perform Real-Time PCR with PRM1 (Ta = 65.5°C) primers to authenticate cDNA from spermatozoal RNAs. A typical 20 µL reaction contains 10 µL Quantitect SYBR green mix, 7 µL nuclease-free water, 1 µL of primer [10 µM], and 2 µL of template. Products from cDNA amplification will be smaller than products from DNA amplification (Fig. 3) because the PRM1 primers chosen are intron spanning.

3.7. Optional Quality Control Assessment

1. The Bioanalyzer (Agilent) with the small RNA chip kit can be used as a quality control measure and to quantify sncRNA as a function of the profile. As shown in Fig. 4b, the presence of sncRNAs is detected as a rise in fluorescent units above background between 6 and 30 nucleotides. The concentrations of the small RNA and the miRNA fraction for each sample can then be determined (see Note 13).

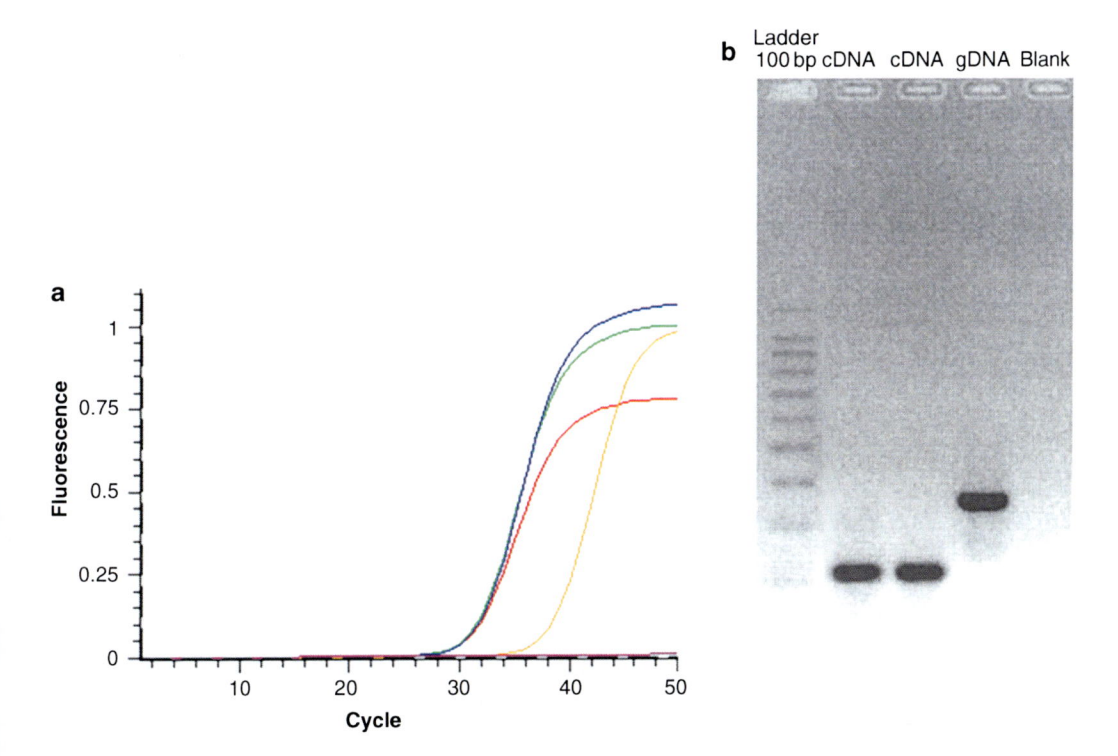

Fig. 3. PCR assessment of cDNA synthesis. (**a**) cDNA samples amplified. (**b**) Gel image showing PCR products from cDNA templates are smaller than products from genomic DNA when amplified with intron spanning primers.

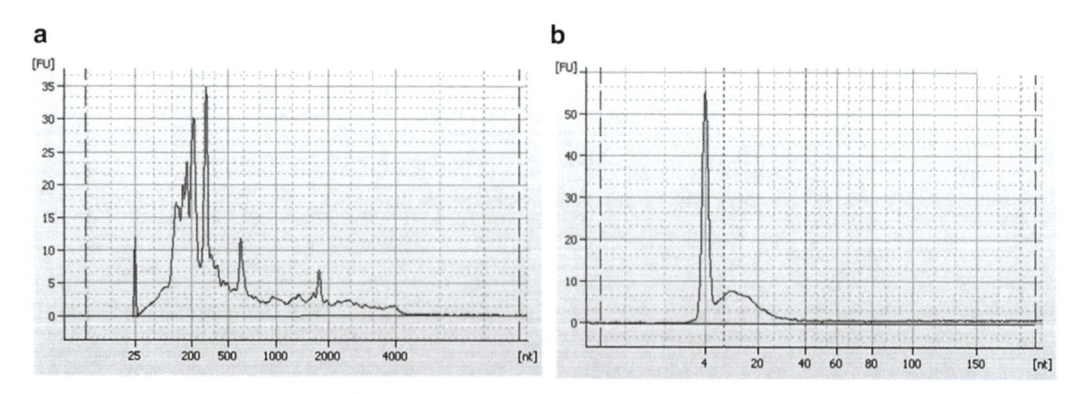

Fig. 4. Agilent Bioanalyzer profiles. (**a**) Total RNA extracted from sperm. Degraded 18S and 28S rRNAs are indistinguishable. (**b**) sncRNA extracted from sperm. Note that the sncRNAs are 6–30 nucleotides in length.

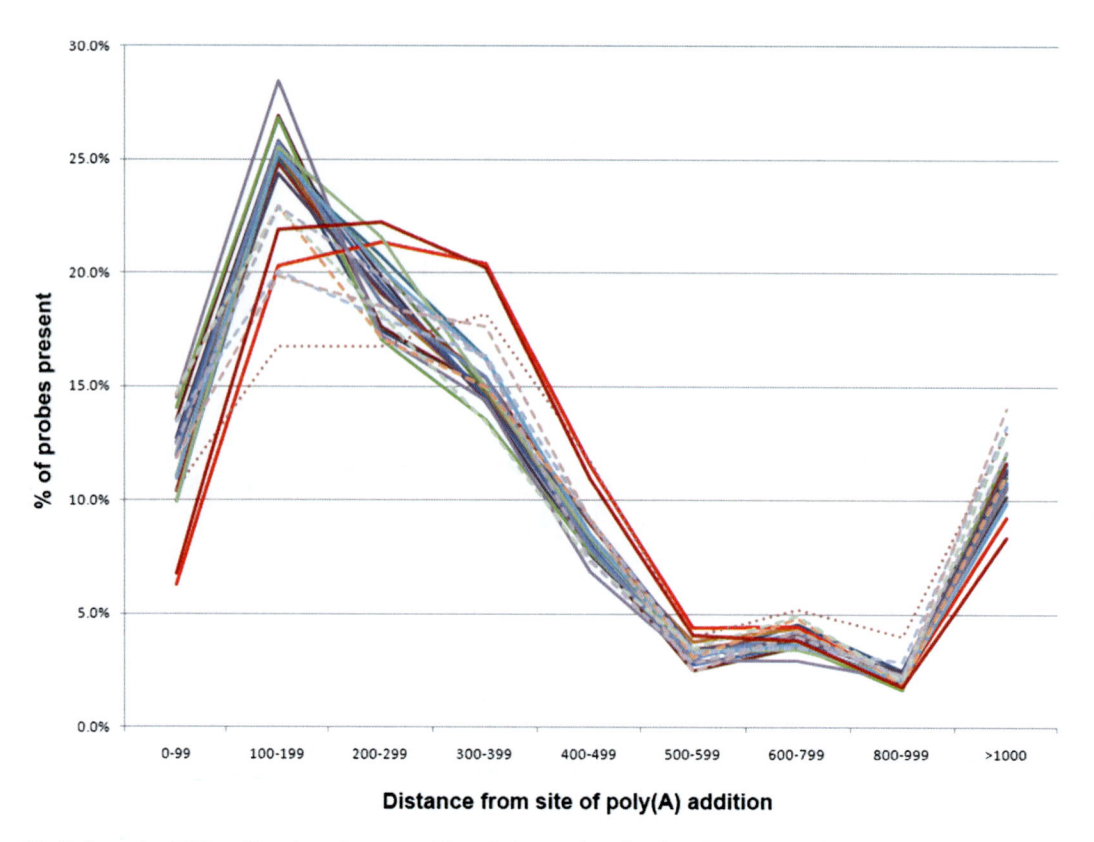

Fig. 5. Assessing RNA quality using microarrays. The relative number of probes detected as a function of distance from the 3′ end of each transcript was calculated. Testes RNAs (*solid red lines*) have a slightly broader peak than the sperm RNA samples indicating greater quantity of intact mRNAs.

2. Transcript length from the 3′ end determined via microarray analysis can be used as an indicator of mRNA integrity. The results are shown in Fig. 5. The complete sequences of each human mRNA queried on the HT12 Illumina array were obtained from GenBank. HT12 probe array sequences were

obtained from the Illumina annotation file. The position of each probe relative to the site of poly(A) addition (3′ end) was calculated by mapping the Illumina probe sequence to the transcript sequence. The probes called as present with a *P*-value of detection of *P* < 0.01 for each sample were retrieved. The present probes were grouped into bins based on probe distance from the 3′ end of the corresponding transcript. Each bin contained the probes within a 100 base multiple of the distance from the site of poly(A) addition. The percentage of present probes in each bin was calculated as a function of total probes detected and their distance from the 3′ end of the transcript.

4. Notes

1. It may be helpful to practice the PureSperm overlay. To do this use colored water in place of the sample. Practice with 50% PureSperm gradient until you are comfortable with the procedure.
2. When centrifuging the PureSperm gradient it is very important to use a centrifuge with minimal acceleration and without a brake. Sudden changes in rotation speed may allow somatic cells to penetrate through the gradient.
3. The PureSperm gradient should be maintained at room temperature. Centrifuging at cold temperatures may change the density and compromise the gradient.
4. Some steps are modified from QIAGEN manual to accommodate the different homogenization protocol.
5. Heating RNase-free H_2O to 65°C may maximize RNA yield during column elution.
6. If more than one column is used to extract the small or large RNA from one sample, the eluates can be pooled prior to stabilization with RNase Block.
7. Extraction of RNAs using a QIAcube requires the purchase of a custom protocol as described in Subheading 3.
8. The elution volume of small RNAs using the QIAcube can be adjusted in 2 μL increments to a total of 30 μL.
9. The elution volume of large RNAs using the QIAcube can be adjusted in 5 μL increments to a total of 100 μL.
10. Treatment of small RNAs with DNase I is optional and is usually not necessary.
11. RiboGreen has proven an effective assay to quantify RNA (12).

12. Dilute PRM1 forward and PRM1 reverse primers to 10 μM in one tube. Divide into several aliquots. Use a single aliquot as your working primer solution. The others can be used as required.
13. Sperm RNA, like somatic cells, is 80% rRNA though it is degraded (5) and thus undetectable by Bioanalyzer analysis (Fig. 4).

Acknowledgements

The authors would like to acknowledge KLAB alumni for their contributions to the development of this protocol. E.A. is recipient of a mobility grant (Programa José Castillejo, ref. JC 2010-0247) from the Ministerio de Educación (Gobierno de España).

This work was supported by the Charlotte B. Failing Professorship Fetal Therapy/Diagnosis to S.A.K.

References

1. Lewis SE (2007) Is sperm evaluation useful in predicting human fertility? Reproduction 134:31–40
2. Krawetz SA (2005) Paternal contribution: new insights and future challenges. Nat Rev Genet 6:633–642
3. Ostermeier GC et al (2002) Spermatozoal RNA profiles of normal fertile men. Lancet 360:772–777
4. Ostermeier GC et al (2005) A suite of novel human spermatozoal RNAs. J Androl 26: 70–74
5. Johnson GD et al (2011) Cleavage of rRNA ensures translational cessation in sperm at fertilization. Mol Hum Reprod 17(12):721–726
6. Cappallo-Obermann H et al (2011) Highly purified spermatozoal RNA obtained by a novel method indicates an unusual 28S/18S rRNA ratio and suggests impaired ribosome assembly. Hum Mol Reprod 17:669–678
7. Das PJ et al (2010) Total RNA isolation from stallion sperm and testis biopsies. Theriogenology 74:1099–1106
8. Feugang JM et al (2010) Transcriptome analysis of bull spermatozoa: implications for male fertility. Reprod Biomed Online 21:312–324
9. Goodrich R et al (2007) The preparation of human spermatozoal RNA for clinical analysis. Arch Androl 53:161–167
10. Platts AE et al (2007) Success and failure in human spermatogenesis as revealed by teratozoospermic RNAs. Hum Mol Genet 16: 763–773
11. Ostermeier GC et al (2004) Reproductive biology: delivering spermatozoan RNA to the oocyte. Nature 429:154
12. Goodrich RJ et al (2003) Multitasking with molecular dynamics Typhoon: quantifying nucleic acids and autoradiographs. Biotechnol Lett 25:1061–1065

Chapter 34

A Review of Genome-Wide Approaches to Study the Genetic Basis for Spermatogenic Defects

Kenneth I. Aston and Donald F. Conrad

Abstract

Rapidly advancing tools for genetic analysis on a genome-wide scale have been instrumental in identifying the genetic bases for many complex diseases. About half of male infertility cases are of unknown etiology in spite of tremendous efforts to characterize the genetic basis for the disorder. Advancing our understanding of the genetic basis for male infertility will require the application of established and emerging genomic tools. This chapter introduces many of the tools available for genetic studies on a genome-wide scale along with principles of study design and data analysis.

Key words: GWA, Whole genome, Next-generation sequencing, Sperm, Male infertility, Microarray

1. Introduction

Normal spermatogenesis requires the precise regulation of hundreds or even thousands of genes (1, 2). Disruption of gene pathways involved in endocrine regulation, meiosis, development, differentiation, spermatogenesis, or sperm function can result in male infertility. Because of the complexity of events required for the production of viable and functional sperm, it is not surprising that the underlying cause for male factor infertility is not identified in about 50% of cases (3). It is likewise not known what proportion of idiopathic male infertility cases has a genetic basis. Given the evolutionary consequences of male infertility, there is certainly strong selective pressure against the transmission of common single gene events that give rise to spermatogenic defects. Based on this premise (and mounting genetic evidence supports this notion) genetic-based male infertility is likely a multigenic disorder, or a result of rare *de novo* events that lead to the phenotype.

The two best-documented and most common genetic causes for male infertility are azoospermia factor (AZF) deletions of the

Y chromosome, which account for 10–15% of azoospermia and severe oligozoospermia cases (4), and Klinefelter syndrome, in which an extra copy of the X chromosome in men results in severe hypospermatogenesis or azoospermia. Klinefelter syndrome is the underlying cause for spermatogenic defects in about 3–4% of infertile men (5). In addition, chromosomal translocations and inversions are more frequently observed in infertile men than in the general population (6).

In spite of enormous efforts by researchers worldwide, little progress has been made in the identification of novel genetic variants associated with male infertility. In most cases, targeted approaches for the identification of genetic features associated with male infertility have proven an ineffective, and cumulatively a very costly, strategy.

Despite the disappointingly slow progress in understanding the genetic basis for idiopathic male infertility, great advancements on several fronts have positioned the field for what will hopefully be rapid progress in this area in the years to come. Completion of the Human Genome Project and the associated annotation of genes across the genome were important early steps in understanding the organization of the human genome (7). The application of genomic tools to characterize the transcriptome of the developing testis gave us greater appreciation for the complexity of spermatogenesis. In addition, the systematic manipulation or knockout of several thousand genes in the mouse by a large number of researchers has yielded a wealth of information about the genes required for normal spermatogenesis and male fertility.

While comprehensive genomic studies targeting male infertility have lagged behind other more prominent and better funded diseases, the lessons learned from the study of other complex diseases will also serve to accelerate male infertility genetic research. Additionally, great advances in the development of tools to interrogate the genome through high-resolution array technologies and high-throughput genomic sequencing, along with the concomitant rapid decline in the cost of these technologies, will enable the study of the genetic basis for male infertility at a level not previously imagined.

Clearly genomic tools will be necessary to unravel the complex genetic background for male infertility. This chapter will serve as a brief primer for the effective design and implementation of genome-wide studies. Cohort selection, assay options, and analysis considerations are addressed.

2. Genomic Features

The types of genomic features that can be interrogated using genome-wide approaches depend largely on the platform utilized (Table 1). The primary features of interest that are evaluated in

Table 1

Comparison of whole genome technologies for characterizing genetic variation. The numbers listed reflect approximate values for a typical Caucasian genome; numbers will vary depending on the population origin of the genome, and the specific details of the platform and analysis pipeline being used. CNVs can in principle be genotyped from CGH arrays, but the exact number depends on the design of the array (e.g., untargeted whole genome versus targeted to regions of known CNV). In principle complex structural variants such as inversions and translocations can be discovered and genotyped by exome sequencing, but success will be highly contingent on the breakpoint locations of these rare forms of variation

Technology	Cost per sample	What can be discovered	What can be genotyped	# of variants discovered	# of variants genotyped
SNP arrays	$200–$500	CNVs	SNPs, CNVs	10–500 CNVs	300,000–5 M SNPs, 10–500 CNVs
CGH arrays	$200–$500	CNVs	CNVs	10–500	Varies
Whole exome sequencing	$500	CNVs, SNPs, indels, inversions, translocations	CNVs, SNPs, indels, inversions, translocations	25,000–50,000 SNPs, 800–1,600 indels, 10–50 CNVs	25,000–50,000 SNPs, 800–1,600 indels, 10–50 CNVs
Whole genome sequencing	$3,000–$10,000	CNVs, SNPs, indels, inversions, translocations	CNVs, SNPs, indels, inversions, translocations	>3 M SNPs, >100,000 indels, 1,000 CNVs, 10–50 inversions and translocations	>3 M SNPs, >100,000 indels, 1,000 CNVs, 10–50 inversions and translocations

the context of genome-wide studies include point mutations, single-nucleotide polymorphisms (SNPs), copy number variations (CNVs), and regions of homozygosity (ROHs) (8–11). Point mutations and SNPs both involve variation of a single nucleotide, and represent different spaces along a continuum of allele frequencies. SNPs are arbitrarily defined as single base changes with a frequency greater than 1% in the population (although sometimes a threshold 5% is used), while point mutations refer to rarer, often de novo changes. Both can result in changes in gene dosage or function, or they can be silent with no functional effect. ROHs are long stretches of DNA sequence homozygous for all SNPs genotyped and typically indicate recent, often cryptic, consanguinity in the genealogical history of an individual (9). CNVs involve duplications, deletions, or complex rearrangements of genomic regions generally larger than 1 kb in length (12). They can likewise be common or rare and can have no apparent effect, or can result in complete gene disruption, or increased or decreased gene dosage or function. As discussed below, different genome-wide analysis platforms are appropriate for the assessment of some features and not suitable for others.

3. Genome-Wide Approaches for the Study of Complex Diseases

The predominant strategy for performing genome-wide genetic studies for the past decade has employed microarrays (13). Microarray technologies represented a huge advance over previous technologies for genetic analysis in terms of the amount of genetic information that could be ascertained in a single experiment, which conferred a significant cost and time advantage. While microarrays are still widely used, they do have limitations in that they only provide information for a portion of the genome (albeit a significant portion).

More recently, technologies have been developed that enable the sequencing of essentially the entire genome in a high-throughput and cost-effective manner (14). Variations of this technology enable the selective capture and subsequent sequencing of a region of interest or of the various RNA species within a specific tissue or a cell type.

3.1. Genomic Microarrays

DNA microarrays exist in two primary forms: (1) comparative genomic hybridization (CGH) arrays (15) and (2) SNP arrays (13). While the two types of arrays yield some of the same information, there are profound differences in the way the arrays are constructed and the data generated by each.

Genomic microarrays are composed of several thousand to more than one million sets of probes designed to be complementary to the genome of interest. The probe sets are arranged in a matrix configuration so that each point on a microarray slide corresponds with one small region of the genome. The probes can be composed of synthetic oligonucleotides as short as 20 nucleotides in length or long nucleotides derived from bacterial artificial chromosome (BAC) libraries.

As the name implies, CGH arrays are used to compare two genomes (15). This is accomplished by competitively hybridizing to an array, a reference genome labeled with one fluorophore along with a test genome labeled with another fluorophore. CGH arrays are useful for determining differences in copy number based on the intensity of hybridization of the test sample compared with the control at a given locus. While CGH arrays are well suited for the assessment of CNVs, they do not provide SNP genotype information. Agilent Technologies (Santa Clara, CA, USA) and Roche NimbleGen (Madison, WI, USA) are the largest producers of CGH microarrays.

For analysis with SNP arrays, a single labeled test sample is applied to the microarray. Probes on the array are designed to be complementary to a region containing an SNP. Two sets of probes for each locus hybridize preferentially with one SNP allele or the other, so based on hybridization patterns genotype can be assessed for each sample. In addition to genotype, with proper data normalization the intensity of hybridization can be used to determine relative copy number of each locus across the genome. Likewise, the identification of long homozygous stretches coupled with intensity data can be useful in inferring single allele deletions or regions of copy-neutral homozygosity. Many modern SNP arrays actually contain a mixture of two probe types: traditional SNP probe sets that hybridize to known SNPs in an allelic-specific manner, and "copy number" probes, nonredundant probes tiled in regions of known CNV and in areas of low SNP coverage to give better sensitivity to copy number changes. The largest manufacturers of SNP genotyping arrays are Illumina (San Diego, CA, USA) and Affymetrix (Santa Clara, CA, USA), although both Agilent and NimbleGen have recently begun offering combined CGH/SNP arrays.

3.2. Next-Generation Sequencing

A more recently developed strategy for genome-wide genetic assessment is whole genome sequencing (16, 17). While Sanger-based DNA sequencing strategies have been successfully employed for several decades, only in the past few years, with the development of second- and third-generation next-generation sequencing (NGS) using non-Sanger-based methods, have sequencing capacity and cost been such that genome-wide sequencing has been a realistic option for basic research. Sanger sequencing was greatly limited by the number of reactions that could be run in parallel, with a capacity of 96 wells per sequencer.

Second-generation sequencing greatly increases capacity by arraying several hundred thousand sequencing templates on a solid surface enabling massively parallel sequencing (16). The three commercially available second-generation sequencing systems that employ variations of the NGS technology are the Roche 454 genome analyzer, the Illumina HiSeq sequencers, and the Life Technologies (Foster City, CA, USA) SOLiD sequencer. Complete Genomics (Mountain View, CA, USA) has developed its own NGS platform, but operates it as a sequencing service rather than selling the machines.

Most recently, third-generation single-molecule sequencing platforms are beginning to emerge that promise longer read lengths, reduced cost, and faster run times (17). Developers of third-generation sequencing include Ion Torrent (Life Technologies, Carlsbad, CA, USA), Pacific Biosciences (Menlo, CA, USA), and Oxford Nanopore (Oxford, UK).

In addition to whole genome sequencing emerging strategies for capturing specific regions of the genome, for example a single chromosome or all exonic sequence within the genome (exome), further improve the affordability of high-throughput sequencing studies.

The first human genome sequence was generated over a 13-year period at a cost of $2.7 billion (http://www.genome.gov). With current NGS technologies an entire human genome can be sequenced in a week or less at a cost of less than $5,000, and the efficiency of sequencing in terms of time requirements and cost continues to drop sharply. With the increasing availability and affordability of whole genome sequencing, generating the sequence data is no longer a significant issue. The management and analysis of whole genome sequence data is rapidly becoming the bottleneck and the area requiring the greatest amount of technical expertise.

4. Study Design Considerations

Thoughtful and thorough study design is critical to the success of any study, but given the enormous costs, magnitude of data generated, and special considerations required for a genome-wide study, careful study design is of utmost importance. Study design considerations that are discussed here in the context of genome-wide studies include sample size, sample selection, data quality, and study replication. Several excellent reviews have been written to address design considerations (18–21). This chapter provides an introductory primer to the topic.

4.1. Sample Size

In general, suitably powered GWAS include several hundred to several thousand cases and controls. Determination of optimal sample size is not a simple matter, as the majority of initial studies lack the data required for the necessary power calculations. Power calculations are crucially dependent on the genetic architecture of the trait under analysis, that is, the number of loci modulating the trait, the number of risk alleles at each locus, their frequencies and effect sizes, and the effect of interactions between risk alleles. These parameters are essentially unknowable from first principles. However, data generated from numerous GWAS studies targeting other complex diseases serves as a good reference for predicting sample size requirements. The number of samples required is negatively correlated with the frequency of the causal or risk variants and the effect size of each variant. Common polymorphisms present in >20% of a population will generally confer minimal risk (odds ratio [OR] of 1.1–1.5). Less common variants with frequencies in the range of 5–20% may have ORs up to 3.0 (21). As a general rule, at least 1,000 cases and 1,000 controls are required for an 80% power to detect significantly associated common variants with ORs in the range of 1.5 (21). Common variants that confer greater risk may be detected with fewer samples, but in general, studies on the order of 1,000 or more samples have been required to detect the majority of common variants so far implicated in complex disease. Obviously the identification of rare variants, particularly those that confer minimal risk, will require manyfold larger studies.

4.2. Sample Selection

Sample selection is a critical component of a GWAS in order to maximize the power of the study to detect true variants and to avoid false positives arising from selection bias or confounding factors. As SNP and CNV frequencies can vary widely across different populations, ethnic homogeneity between groups is absolutely necessary (18, 21). As with all case–control-type experiments, groups should be matched as closely as possible, and potential confounders should be considered and avoided (18, 21). It is usually not possible to completely avoid issues such as population stratification as self-reported ethnic ancestry is sometimes vague, unavailable, or subjective. With the large amount of genetic information generated during a genome-wide study, genetic population structure can be addressed empirically, and if the differences between groups are not too great, the data can be corrected (18). The ability to correct association test statistics for artifacts introduced by the experiment, be it in sample selection or genotyping, represents the most important statistical advantage of GWAS over the candidate gene approach. Many QC protocols suggest that approximately 50,000 unlinked markers are required for robust assessment of population structure (22); it is currently an open question as to how well population structure can be detected using exome sequencing data, which generates 700–1,000 variants per sample (23). The most widely used approach to detect population structure today is called *principal components analysis*, which measures covariation in genotypes among individuals, and is implemented in the package EIGENSOFT (http://genepath.med.harvard.edu/~reich/Software.htm).

Another important assumption of standard GWAS design is that all individuals in the cohort are unrelated. Inclusion of large numbers of individuals from a single family can create a case- or control-specific enrichment of an allele that is unrelated to disease phenotype; this effect is especially pernicious in the analysis of rare variants. A common QC step is thus to identify relative pairs or duplicate samples prior to analysis by examining the number of identical SNP genotypes between all pairs of samples in the cohort.

4.3. Data Quality

Quality of the data generated in a GWAS is likewise critical in detecting true associations. Considerations include DNA purity and integrity, plate layout, and sample processing by microarray or sequencing. Obviously DNA purity and integrity are important in any type of genetic study. In array studies care should be given to ensure that closely matched DNA quantities are applied to each chip. DNA quantification using a DNA-specific assay such as PicoGreen (Invitrogen, Carlsbad, California, USA) will generally give more precise results than simple 260 and 280 nm absorbance measurements. Another consideration when plating samples for analysis is sample distribution. Cases and controls should be randomly distributed across plates to avoid spurious associations arising from systematic batch effects (18). After the data are generated, it is essential that data quality be assessed on both a per-locus and per-sample basis. For each locus (e.g., an SNP) the number of individuals confidently assigned a genotype, referred to as the "call rate," is a standard QC metric; sites with low call rates should be removed. Other commonly used marker QC steps are (A) testing for departure from Hardy–Weinberg Equilibrium, (B) testing for differential missingness between cases and controls, and (C) removing extremely low-frequency variants that are difficult to genotype (the latter is applied to SNPs rather than CNVs).

Similarly, sample quality should be assessed after observing the genotype data from the experiment. Samples that have a large proportion of sites without a genotype call (high "no call" rate) or samples that have an unusually large proportion of sites with a particular genotype call (e.g., an excessive proportion of heterozygotes) should be removed. In the analysis of CNV, samples that have excessive number of CNVs compared to the rest of the population are usually excluded.

5. Analysis

5.1. Common Variants

Over the past decade, GWA studies have relied upon genotyping panels of known common polymorphisms that are incomplete representations of the genomes under consideration. While each of these common polymorphisms is directly tested for association, GWA studies are capturing indirect information about the effects of many additional ungenotyped variants in the cohort that are linked to the marker panel.

In the simplest form, testing for association at common variants can be reduced to a test of independence between the columns of a contingency table, that is, a Chi-square test comparing the allele counts (a 2 × 2 table) or genotype counts (a 2 × 3 table) between the case and control cohorts. A more sophisticated approach is to use a logistic regression modeling approach, which can accommodate covariates known to alter disease risk, such as gender, age, and environmental exposures. A number of options for association testing are provided in the widely used software package PLINK (http://pngu.mgh.harvard.edu/~purcell/plink/).

Visualizing the results of association testing is an essential step in the interpretation of a common variant GWA study. Two key types of summary plots used in a GWA study are the quantile–quantile plot ("qq plot") and the "Manhattan" plot. The QQ plot is generated by plotting the observed distribution of association test p-values against the distribution of p-values expected under the null hypothesis that there is no association between genotype and phenotype (Fig. 1). A healthy QQ plot would show a strong fit to the line $y = x$. Minor deviations in the upper tail of the p-value distribution are expected when a rare, significant association signal is present in the data. More general deviation along the length of the

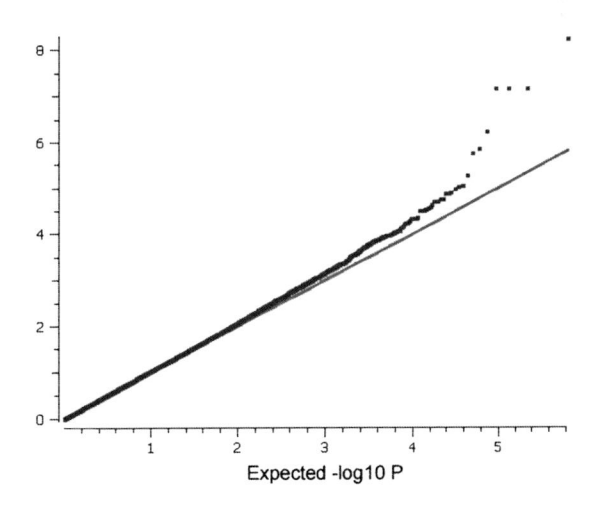

Fig. 1. The QQ plot is generated by plotting the observed distribution of association test p-values against the distribution of p-values expected under the null hypothesis that there is no association between genotype and phenotype. A healthy QQ plot would show a strong fit to the line $y = x$. Minor deviations in the upper tail of the p-value distribution indicate that a significant association signal is present in the data. More general deviation along the length of the entire distribution is a sign of test statistic inflation, usually the result of experimental batch effects or population structure.

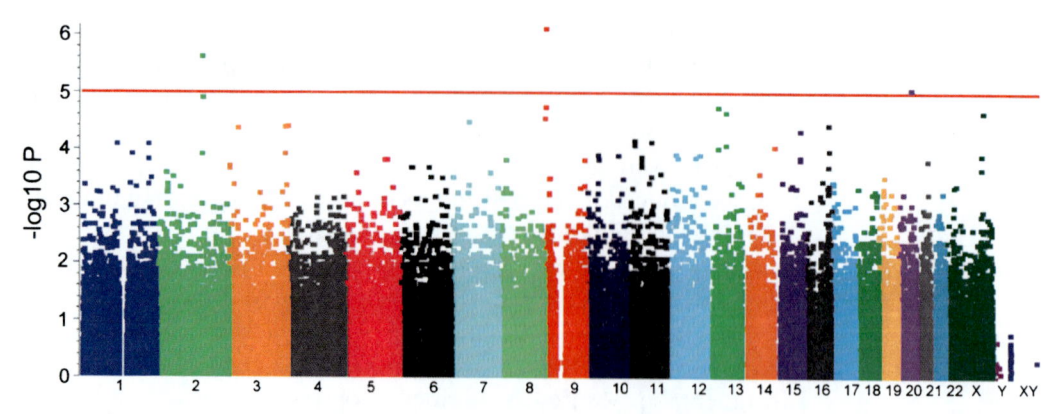

Fig. 2. The Manhattan plot is a plot of −log10 *p*-value against genomic position. Each bar represents one chromosome from 1 to 22 followed by X, Y, and the XY pseudoautosomal region. Each point represents one SNP on the microarray chip. The most robust associations will often appear as clustered peaks of low *p*-values; this clustering is a result of shared genealogical history among closely linked SNPs, a phenomenon referred to as linkage disequilibrium (LD).

entire distribution is a sign of test statistic inflation, usually the result of experimental batch effects or population structure. The Manhattan plot is a plot of −log10 *p*-value against genomic position (Fig. 2). The most robust associations will often appear as clustered peaks of low *p*-values; this clustering is a result of shared genealogical history among closely linked SNPs, a phenomenon referred to as linkage disequilibrium (LD). Numerous small *p*-values that appear as isolated events (e.g., with no linked SNPs of similar association strength) are a hallmark of batch effects (24). In summary, both QQ plots and Manhattan plots are important for assessing the overall quality of the data and the evidence for association in a GWA study.

Multiple testing is clearly the chief concern when assessing the significance of an association test from a GWA study. There exists a multitude of approaches to addressing multiple testing. These range from conservative penalty-based approaches that rescale reported p-values by the number of tests conducted (e.g., Bonferroni and Sidak methods) to methods that estimate the proportion of true positive tests from the observed distribution of *p*-values (25) to the computationally intensive but accurate family of permutation tests. The value of significance required for publication will depend on the journal, and some workers claim that a threshold of 5×10^{-7} is sufficient for publication (26) while others have recommended 5×10^{-8} (27).

Special considerations are needed for testing common CNVs in an association study. The technology and statistical methods for identifying CNVs differ from SNP genotyping. CNVs are often *discovered* from the same experimental data that are used for doing the association study. SNP genotypes, on the other hand, are only generated for sites of known common polymorphism targeted by an SNP array. CNV discovery algorithms usually are applied on one sample at a time, and they often return the location of a CNV but not a reliable estimate of the CNV genotype for the sample. Further, as these CNV discovery algorithms have incomplete power to discover CNVs (discovery power is a function of probe density, among other things), the classification error rate among samples can be quite high, especially for a common CNV. It is thus inappropriate to directly test the CNV calls from a CNV locus with greater than 5% allele frequency, using the standard association testing frameworks mentioned above. A second, related consideration is *batch effects* that produce systematic differences in probe intensity among sets of samples, which can also produce differential CNV call rates among cases and controls. Robust association methods that simultaneously assign CNV genotypes and test for association are strongly recommended for testing common CNVs (28).

5.2. Rare Variants

The standard protocols for testing common variants derive power from large sample sizes, as large samples are needed to precisely estimate the frequency of a single allele in both case and control populations. If cohorts of 1,000 cases are necessary to generate enough precision for a 3% frequency polymorphism, then what are our prospects for assessing mutations that are much rarer than 1% or even unique? The most popular classes of rare-variant analyses can be described as the "burden" approach and the "pathway" approach. These strategies should only be applied to rare CNVs (<1% frequency) for the technical considerations discussed in the previous section. Rare SNPs should be interpreted with caution if they are generated by array genotyping or by sequencing.

The "burden" approach was initially pioneered for the analysis of rare CNVs in neuropsychiatric disease (29). "Genomic burden" analysis is simply comparing the total number of CNV calls per sample in cases and controls, using for instance a Mann–Whitney U-test or permutation strategy. While each CNV in the analysis may be individually too rare to definitively attribute a risk estimate, one can assign risk to rare CNVs as a class using this approach. Many related derivatives of this approach exist, which involve aggregating rare CNVs or SNPs into groups, based on their location within a genomic region or a gene set that operates as a biological pathway (e.g., testing for an enrichment of CNVs in neurogenesis genes in cases of schizophrenia). Pathway definitions can be obtained from numerous sources; one of the more popular ones is the Gene Ontology project (http://www.geneontology.org).

Another rare genetic event detectable by SNP arrays are large chromosome segments that are inherited identical-by-descent (HBD regions or HBDRs). These large ROHs are likely to have shared a common ancestor sometime in the recent past (i.e., they are nearly identical or identical), increasing the chances that a rare recessive mutation may be homozygous within the HBDR.

The use of HBD, when applied to consanguineous families, has been an unequivocal success in pinpointing the location of causal variants for simple recessive monogenic diseases. HBD analysis can also be used to screen for the location of rare variants in common disease case–control studies of unrelated individuals, either using a single-locus association testing framework or by testing for a homozygosity "burden", an enrichment of size or functional impact of HBD regions aggregated across the genome. This novel approach has produced results for a growing list of common diseases, including schizophrenia (30) and Alzheimer's disease (31). There are a handful of methods available for identifying HBD regions from SNP data; the most accurate ones will use an explicit population genetic model to determine the significance of runs of homozygosity, which occur frequently in normal outbred populations. An example of such a method is the one employed by the software package BEAGLE (32). HBD analyses are sensitive to population structure, so matching cases and controls on ethnicity is essential.

5.3. Study Validation and Replication

Most genome-wide studies aim to detect extremely rare or weak biological signals. The extremely large number of tests conducted in such studies increases the probability that such rare events are observed spuriously by chance. Validation and replication are two follow-up study techniques that are key to obtaining reliable and publishable results from a genome-wide scan. Almost always, validation and replication entail a focused follow-up using locus-specific assays. In the case of SNPs, there are numerous single-locus assays available such as mass spectrometry (as offered by Sequenom Corporation) or TaqMan qPCR (Life Technologies). These two platforms also work for CNVs, which can also be interrogated by simple PCR assays and multiplex ligation-dependent probe amplification (MLPA) assays, among others.

Validation refers to confirmation of the initial finding, reproducing the same result in the same samples using a second experimental technique. For instance, a standard validation step in an array-based CNV analysis is to validate deletion calls of interest using PCR, or a second, custom high-density targeted array, prior to attempting further experimental follow-up in additional samples. Biologically interesting CNV calls are often enriched for false positives.

A replication study attempts to reproduce the biological signal of interest using a complete new set of samples, and perhaps a new technology. As replication may require large sample sizes, it is not unusual for an investigator to opportunistically evaluate an existing large GWA dataset in an independent cohort to replicate their finding. Alternatively, locus-specific assays can be applied to a second cohort for replicating the initial finding. Especially for GWA studies, replication is a critical step for publication in a high-impact journal.

6. Conclusion

Genome-wide approaches can be powerful tools for identifying the genetic basis for complex diseases or traits. In order to achieve the most statistical power, careful detail must be paid to proper study design, sample phenotyping, assay performance, and data analysis. With careful study design and execution, GWA studies have proven very effective in identifying novel variants associated with, or causal of, numerous disease states. Growing interest in applying these technologies to the study of male infertility will likely result in numerous advances in our understanding of the underlying etiologies of male infertility, which will ultimately result in more complete diagnoses and improved treatments.

References

1. Iguchi N et al (2006) Expression profiling reveals meiotic male germ cell mRNAs that are translationally up- and down-regulated. Proc Natl Acad Sci USA 103:7712–7717
2. Schultz N et al (2003) A multitude of genes expressed solely in meiotic or postmeiotic spermatogenic cells offers a myriad of contraceptive targets. Proc Natl Acad Sci USA 100: 12201–12206
3. Dohle GR et al (2005) EAU guidelines on male infertility. Eur Urol 48:703–711
4. Poongothai J et al (2009) Genetics of human male infertility. Singapore Med J 50:336–347
5. Paduch DA et al (2009) Reproduction in men with Klinefelter syndrome: the past, the present, and the future. Semin Reprod Med 27:137–148
6. Martin RH (2008) Cytogenetic determinants of male fertility. Hum Reprod Update 14: 379–390
7. Lander ES et al (2001) Initial sequencing and analysis of the human genome. Nature 409: 860–921
8. Almal SH, Padh H (2011) Implications of gene copy-number variation in health and diseases. J Hum Genet 57(1):6–13
9. Ku CS et al (2011) Regions of homozygosity and their impact on complex diseases and traits. Hum Genet 129:1–15
10. Ku CS et al (2010) The pursuit of genome-wide association studies: where are we now? J Hum Genet 55:195–206
11. Conrad DF et al (2010) Origins and functional impact of copy number variation in the human genome. Nature 464:704–712
12. Stankiewicz P, Lupski JR (2010) Structural variation in the human genome and its role in disease. Annu Rev Med 61:437–455
13. Grant SF, Hakonarson H (2008) Microarray technology and applications in the arena of genome-wide association. Clin Chem 54: 1116–1124
14. Davey JW et al (2011) Genome-wide genetic marker discovery and genotyping using next-generation sequencing. Nat Rev Genet 12: 499–510
15. Oostlander AE et al (2004) Microarray-based comparative genomic hybridization and its applications in human genetics. Clin Genet 66:488–495
16. Schuster SC (2008) Next-generation sequencing transforms today's biology. Nat Methods 5:16–18
17. Schadt EE et al (2010) A window into third-generation sequencing. Hum Mol Genet 19: R227–R240
18. Glessner JT, Hakonarson H (2011) Genome-wide association: from confounded to confident. Neuroscientist 17:174–184
19. Kim SY et al (2010) Design of association studies with pooled or un-pooled next-generation sequencing data. Genet Epidemiol 34: 479–491
20. Pahl R et al (2009) Optimal multistage designs – a general framework for efficient genome-wide association studies. Biostatistics 10:297–309
21. Zondervan KT, Cardon LR (2007) Designing candidate gene and genome-wide case-control association studies. Nat Protoc 2:2492–2501
22. Anderson CA et al (2010) Data quality control in genetic case-control association studies. Nat Protoc 5:1564–1573
23. (2010) A map of human genome variation from population-scale sequencing. Nature 467: 1061–1073

24. Sebastiani P et al (2011) Retraction. Science 333:404
25. Storey JD (2003) The positive false discovery rate: a Bayesian interpretation and the q-value. Ann Statist 31:2013–2035
26. Clarke GM et al (2011) Basic statistical analysis in genetic case-control studies. Nat Protoc 6:121–133
27. Pe'er I et al (2008) Estimation of the multiple testing burden for genomewide association studies of nearly all common variants. Genet Epidemiol 32:381–385
28. Barnes C et al (2008) A robust statistical method for case-control association testing with copy number variation. Nat Genet 40:1245–1252
29. International Schizophrenia Consortium (2008) Rare chromosomal deletions and duplications increase risk of schizophrenia. Nature 455:237–241
30. Lencz T et al (2007) Runs of homozygosity reveal highly penetrant recessive loci in schizophrenia. Proc Natl Acad Sci USA 104:19942–19947
31. Nalls MA et al (2009) Extended tracts of homozygosity identify novel candidate genes associated with late-onset Alzheimer's disease. Neurogenetics 10:183–190
32. Browning SR, Browning BL (2010) High-resolution detection of identity by descent in unrelated individuals. Am J Hum Genet 86:526–539

Chapter 35

Methods for the Analysis of the Sperm Proteome

Sara de Mateo, Josep Maria Estanyol, and Rafael Oliva

Abstract

Proteomics is the study of the proteins of cells or tissues. Sperm proteomics aims at the identification of the proteins that compose the sperm cell and the study of their function. The recent developments in mass spectrometry (MS) have markedly increased the throughput for the identification and study of the sperm proteins. Catalogues of spermatozoal proteins in human and in model species are becoming available laying the groundwork for subsequent research, diagnostic applications, and the development of patient-specific treatments. A wide range of MS techniques is also rapidly becoming available for researchers. This chapter describes a methodological option to study the sperm cell using MS and provides a detailed protocol to identify the proteins extracted from a Percoll-purified human sperm population and separated by two-dimensional polyacrylamide gel electrophoresis (2D-PAGE) using LC-MS/MS.

Key words: Proteome, Proteomics, Spermatozoa, Protamine, Epigenetics, Mass spectrometry

1. Introduction

Proteomics is defined as the systematic analysis of all the proteins in a tissue or cell and aims to identify the expression levels of all proteins of one functional state of a biological system (1, 2). Mass spectrometry (MS) has become a method of choice for the analysis of complex protein samples, and it is enabled by the availability of genome sequence databases and protein ionisation methods (1, 3). Proteomic technologies have improved significantly and have proven to be very valuable in the study of reproductive biology and medicine (4).

Essentially, two different approaches have been applied to the proteomic study of the sperm cell using mass spectrometry (MS): (1) two-dimensional (2D) separation of proteins followed by peptide generation and identification through matrix-assisted laser desorption ionisation MS (MALDI-MS) or liquid chromatography followed by tandem MS (LC-MS/MS) and (2) initial digestion of proteins to generate peptides, followed by LC-MS/MS analysis (Fig. 1; refs. 5, 6). The first approach usually involves the separation of proteins using isoelectric focussing followed by a polyacrylamide gel electrophoresis separation in the presence of SDS (SDS-PAGE) to separate the proteins in a second dimension according to molecular weight (Fig. 1). This approach has been widely used in the past leading to the identification of many proteins present in the sperm cell (7–13).

The other mass spectrometry approach (LC-MS/MS) combines the solute separation power of HPLC with the detection power of a mass spectrometer (Fig. 1). HPLC can separate proteins or peptides on the basis of a number of unique properties such as charge, size, hydrophobicity, and the presence of specific tags or amino acids. Because the HPLC is coupled to a tandem mass spectrometer with an interface it is possible to rapidly separate a complex protein mixture and identify its components. LC-MS/MS experiments usually generate peptide primary structure (sequence) information from each peptide in the initial mix. In addition, different approaches to identify protein abundance differences are also available (14).

A final comment must be devoted to the protamines, the most abundant nuclear proteins in sperm (15–19). Because of their peculiar chemical nature (over 50% of positively charged amino acids) it is extremely difficult to identify them using mass spectrometry. When using a 2D approach, because of their highly positive charge, they are focused at one end of the IEF strip, thus usually out of the usual isoelectric point (pI) range of the first dimension. In addition, protamines are insoluble in the presence of SDS, so little protein can be expected to enter the second SDS-PAGE dimension. An approach to specifically separate the human protamines using 2D-PAGE followed by their identification has been described (20). However, a convenient option to analyse protamines and the proteome from a single sample is to process independent aliquots and to analyse them separately using a 1D-acidic-PAGE for protamines and a conventional 2D-PAGE or LC-MS/MS for the remainder of the proteins (Fig. 1) (13, 19, 21–25).

The present chapter describes the procedures for purification of the spermatozoa using density gradient centrifugation, extraction of sperm proteins, their separation using two-dimensional gel electrophoresis, the identification of the proteins from excised spots after trypsin digestion, and peptide separation and identification using liquid chromatography followed by tandem mass spectrometry (LC-MS/MS). The LC-MS/MS procedure described here can also be applied to identify mixtures of proteins present in excised gel bands from one-dimensional SDS polyacrylamide gel electrophoresis separations (24).

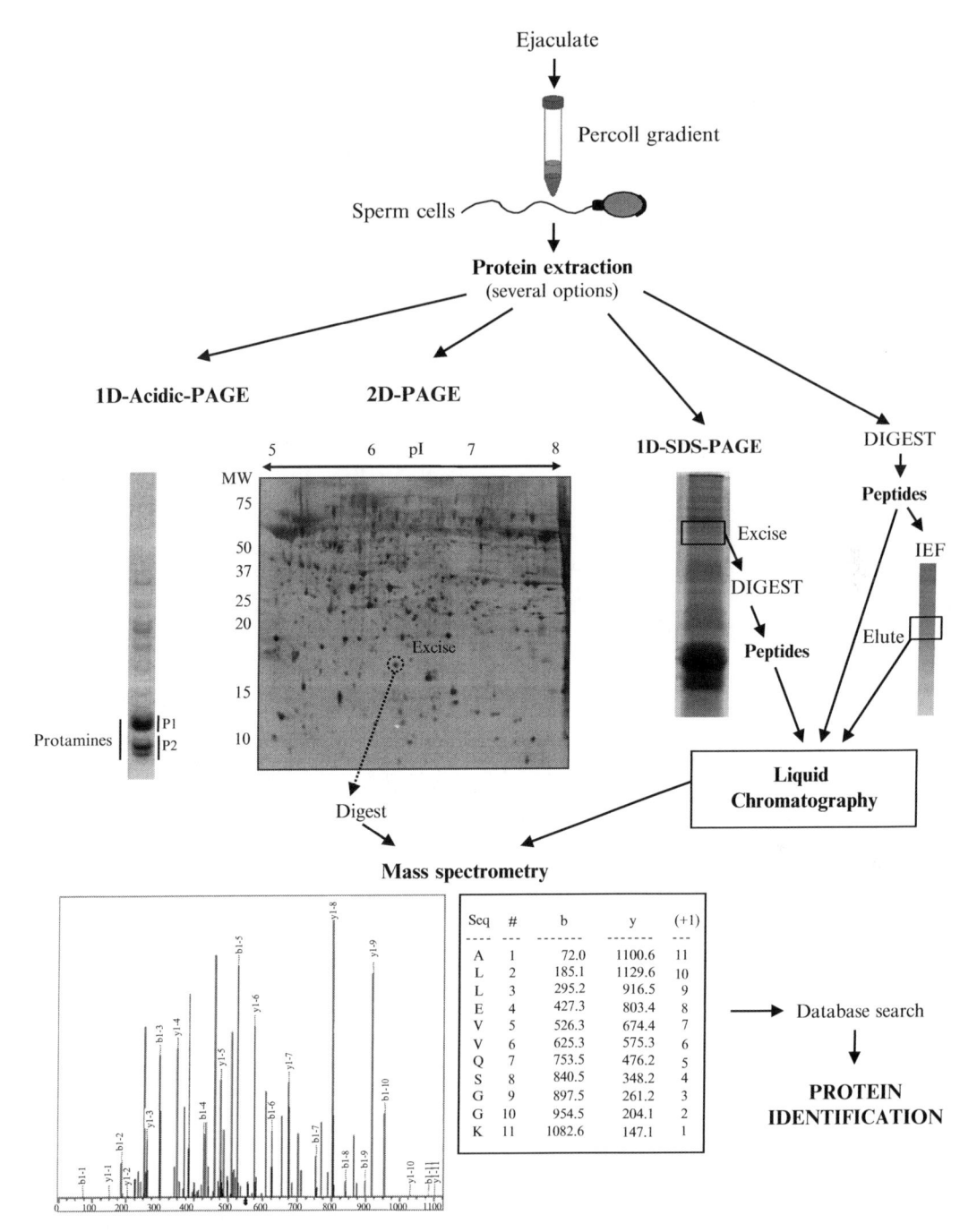

Fig. 1. Extraction and analysis of proteins from spermatozoa. The nuclear protamines can be extracted and analysed using conventional electrophoresis on acidic gels (*left*) using conventional procedures (13, 19, 21–25). In order to analyse the rest of the sperm proteome using mass spectrometry, an aliquot of the sperm cells can be analysed using 2D-PAGE (*centre*) or 1D-SDS-PAGE followed by excision of protein spots/bands from the 2D and 1D gels and identification through mass spectrometry (either MALDI-MS or LC-MS/MS for 2D-PAGE and LC-MS/MS for 1D-SDS-PAGE). In addition, a powerful procedure involves digestion of the initial protein extract into peptides followed by separation of the peptides thought LC and finally protein identification using MS/MS (*right*) (5, 6, 24). A LC-MS/MS spectrum is shown at the *bottom left*, and the peptide sequence is also indicated. A search of protein databases for the peptides identified from each sample gives rise to the positive identification of the protein analysed.

2. Materials

Prepare all solutions using ultrapure water (milli-Q water; 18 MΩ cm at 25°C). Prepare and store all reagents at the temperature stated. Follow manufacturer's instructions for solution handling and disposal.

2.1. Percoll Gradient

1. 100% Isotonic Percoll: 87% Percoll®, 10% Ham's F10 10×, 0.25% NaHCO$_3$. Mix 8.7 mL Percoll®, 1 mL Ham's F10 10×, and 300 µL 7.5% NaHCO$_3$. Prepare freshly or store shortly at 4°C.

2. 50% Isotonic Percoll (50% Percoll, 47% Ham's F10 1×, and 0.25% NaHCO$_3$): Mix 5 mL 100% isotonic Percoll (prepared in step 1), 4.7 mL Ham's F10 1×, and 300 µL 7.5% NaHCO$_3$. Prepare freshly or store shortly at 4°C.

2.2. 2D-PAGE Components

1. Lysis buffer (7 M Urea, 2 M Thiourea, 1% CHAPS, 1% *n*-octyl-glucopyranoside, 0.5% IPG buffer, 18 mM DTT, 2.4 mM PMSF): Prepare lysis buffer stock (store at −20°C) without DTT and PMSF. Add 4.2 g urea, 1.52 g thiourea, 0.1 g CHAPS, 0.1 g *n*-octyl-glucopyranoside, and 50 µL IPG buffer (Bio-Rad®) to a 15 mL tube. Add milli-Q water until 9.5 mL, mix gently for 5–10 min at room temperature and add additional milli-Q water until 10 mL. Just before protein extraction: add 18 µL 1 M DTT and 24 µL 100 mM PMSF for every 1 mL of lysis buffer (see Note 1).

2. Equilibration stock buffer (6 M urea, 0.375 M Tris–HCl pH 8.8, 20% glycerol, and 2% SDS): Prepare equilibration stock buffer freshly by mixing 9.01 g urea, 12.5 mL 1.5 M Tris–HCl pH 8.8, 5 mL glycerol, and 0.5 g SDS. Add milli-Q water until 25 mL. Just before the equilibration step prepare the equilibration buffers I and II.

3. Equilibration buffer I (2% DTT): Add 0.24 g DTT to 12.5 mL equilibration stock buffer.

4. Equilibration buffer II (2.5% iodoacetamide): Add 0.3 g iodoacetamide to 12.5 mL equilibration stock buffer.

5. SDS gel buffer (1.5 M Tris–HCl, pH 8.8). Weigh 181.7 g of Tris base, place it on a 1 L graduated cylinder, and add milli-Q water until 900 mL. Mix and adjust the pH with HCl then make up to 1 L with water. Store at room temperature.

6. 10% Ammonium persulfate (APS): Mix 0.1 g APS with 1 mL milli-Q water. Prepare fresh.

7. Running buffer TGS (25 mM Tris, 192 mM glycine, and 0.1% SDS): Make a TGS 10× and store it at room temperature. For TGS 10× preparation, mix 60 g Tris with 288.4 g glycine and 20 g of SDS. Make up to 2 L with milli-Q water.

8. Agarose solution (1%): Mix 0.25 g agarose with 25 mL milli-Q water in a 50 mL tube and add 0.001 g of bromophenol blue. Warm the solution in the microwave oven: place the tube in a glass beaker and open the cap slightly. Let the agarose warm in the microwave oven until it starts boiling. Close the tube and place the agarose solution at 4°C until use.

9. Fixing solution for EZBlue staining (50% methanol/10% ethanol): Mix 500 mL methanol with 100 mL ethanol and add milli-Q water to 1 L final volume. Store at room temperature.

2.3. LC-MS/MS Analysis Components

1. Trypsin solution: 12.5 ng/μL in 25 mM amonium bicarbonate. Store at −20°C until use.
2. Liquid chromatography gradient Buffer A (3% acetonitrile and 0.1% formic acid): Mix 15 mL acetonitrile with 500 μL of formic acid and add milli-Q water to 500 mL.
3. Liquid chromatography gradient Buffer B (97% acetonitrile and 0.1% formic acid). Mix 485 mL acetonitrile with 500 μL of formic acid and add milli-Q water to 500 mL.

3. Methods

3.1. Preparation of the Sperm Cells Prior to Proteomic Analysis (See Note 2)

1. Remove seminal plasma from sperm cells by adding two volumes of Ham's F10 1×, 0.25% NaHCO$_3$ the ejaculate and spinning at 3,000×g for 10 min at 4°C.
2. Remove the supernatant, add 1 mL of Ham's F10 1× and count the sperm cells with a Makler chamber to determine the concentration of spermatozoa (millions of spermatozoa/mL).
3. Prepare 1 mL of 50% Percoll for approximately every 50 million spermatozoa (see Note 3).
4. Add Ham's F10 1×, 0.25% NaHCO$_3$ medium to the washed sperm sample to achieve a final concentration of 50 million spermatozoa/mL.
5. Prepare the gradients: Place 1 mL of 50% Percoll in a 15 mL tube for each gradient. Prepare one tube for every 50 million sperm cells.
6. Add 1 mL of washed sperm sample. To accomplish this hold the tube at an angle of approximately 45°, introduce the micropipette tip into the tube at a distance of approximately 2 cm away from the Percoll gradient (do not touch the Percoll gradient). Carefully place (very slowly) 1 mL of the sperm sample on top of the 50% Percoll gradient (see Note 4). At this stage two clear phases should be visible, the gradient on the bottom and the sample on the top, divided by a sharp interface.
7. Place each tube on ice carefully and let the gradient settle for 10 min.

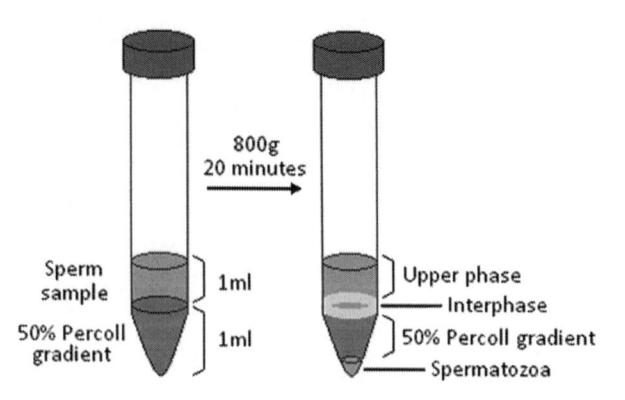

Fig. 2. Purification of the spermatozoa prior to proteomic analysis. The sperm sample is layered on the top of a Percoll gradient and spun at 800 × g for 20 min at 4°C to recover the purified spermatozoa from the pellet. The different layers obtained after centrifugation are also shown. Varying Percoll concentrations can be used to obtain more or less stringent selection of cells (see Notes 2 and 3).

8. Centrifuge the gradients at $800 \times g$ for 20 min at 4°C.
9. Place the tube on ice carefully. A representation of the gradient before and after the centrifugation step is shown in Fig. 2. Discard the upper phase and the 50% gradient phase and keep the interphase and spermatozoa. To remove the upper phase do not tip the tube, hold it upright and remove the top part of the gradient (debris and other cells) down to the interphase with a micropipette. Start from the top and lower the pipette as the phase is aspirated.
10. Use a micropipette to aspirate the interphase layer. Keep this phase in a new microcentrifuge tube for use as a control. The components of this layer should be other cells, debris and some aberrant sperm cells.
11. Carefully aspirate the 50% Percoll layer using a micropipette. Take care to avoid aspirating the spermatozoa present in the pellet.
12. To recover the spermatozoa from the pellet, pipette 1 mL of Ham's F10 1×, 0.25% $NaHCO_3$ medium and dispense approximately a fourth part of it onto the pellet. Pipette up and down only this fourth part to resuspend the pellet and place it in a new microcentrifuge tube. Repeat this step with the rest of the Ham's F10 1×, 0.25% $NaHCO_3$ until the entire pellet is resuspended (see Note 5).
13. Check the retrieved interphase layer and the resuspended pellet using phase contrast microscopy. Count the sperm cells from each phase with a Makler counting chamber (million spermatozoa/mL) (see Note 6).
14. Remove the residual Percoll from the purified sperm cells by adding Ham's F10 1×, 0.25% $NaHCO_3$ and spinning at $3,000 \times g$ for 5 min at 4°C. Remove the supernatant and

combine all the spermatozoa from each gradient into a single microcentrifuge tube with 1 mL Ham's F10 1×, 0.25% NaHCO$_3$. Wash the sperm cells a second time. At this point the spermatozoa should be depleted of potentially contaminating cells present in the original semen and therefore ready to proceed to the proteomic study (see Note 7).

3.2. Protein Extraction and Separation Through Two-Dimensional Gel Electrophoresis (2D-PAGE)

This protocol is adapted for use with a Bio-Rad vertical electrophoresis system. Other systems may require slightly different setups.

1. Centrifuge the sperm cells at 3,000×g for 5 min at 4°C. Remove the supernatant and add the appropriate volume of lysis buffer to the purified sperm pellet to achieve a concentration of approximately 330 million spermatozoa/mL (see Note 7).

2. Extract the sperm proteins for 1 h at room temperature with gentle shaking.

3. Centrifuge the protein extract at 3,000×g for 5 min at 4°C. Place the supernatant containing the solubilized proteins in a new microcentrifuge tube. Keep the pellet and supernatant at −20°C for potential future protein studies (see Note 8).

4. To run the first dimension (protein separation according to its isoelectric point; pI) carefully load the recommended 300 µL of sperm proteins into a focusing tray. Distribute the sample all along one well of the tray from one electrode to the other.

5. Remove the 17 cm ReadyStrip IPG strips (Bio-Rad) pH 3–10 from the freezer, and let them warm at RT for a few minutes (see Note 9). Always handle the strips with forceps previously soaked with ethanol, never with the hands even if wearing gloves. Hold the strip with one pair of forceps, and peel off the cover sheet with another pair of forceps. Place the strip in the well where the sample is spread with the acrylamide side of the strip facing the sample and the anion end oriented toward the anode end of the tray.

6. Let the strip soak with the sample for 10 min and then add mineral oil on the top to cover the strip (approximately 6 mL). The purpose of the mineral oil is to prevent the strip from drying and burning since the strip will warm up during the focusing procedure.

7. Set the focusing program for human sperm cell proteins as follows (see Note 10):
   – Passive rehydratation step for 12 h.
   – Focusing program.
   Step 1: Slow ramp, 250 V for 20 min.
   Step 2: Slow ramp, 10,000 V for 2 h and 30 min.
   Step 3: Rapid ramp, 10,000 V/h until 40,000 V.
   Step 4: Rapid ramp, 500 V for 10 h.

8. The focused strips can be stored at −20°C in a capped tube until the second dimension (protein separation according to its molecular mass) is performed.

9. To run the second dimension, prepare the SDS-PAGE gels: clean the outer and inner glass plates with ethanol (22.3 × 20 cm and 20 × 20 cm, respectively). Assemble the gel cassettes with the outer and inner plates, spacers, and sandwich clamps in the slab gel casting stand.

10. For one gel, mix 17.2 mL milli-Q water, 11.04 mL gel buffer, 480 µL 10% SDS, 9.6 mL 30% acrylamide/bisacrylamide (29:1), 9.6 mL 30% Duracryl, 62.4 µL TEMED, and 224 µL 10% APS. Insert a gel comb with one reference well immediately after loading the gel taking care to avoid introducing air bubbles. Let the gel solidify for 2 h at room temperature or overnight at 4°C.

11. Equilibrate the strip prior to loading it on the top of the SDS-PAGE in two equilibration steps in order to minimize electroendosmotic effects, which are responsible for reduced protein transfer from the first to the second dimension. First equilibration: place the focused strip in a rehydratation tray, add 6 mL of equilibration buffer I and equilibrate the strip for 15 min. Remove the equilibration buffer I, add 6 mL of equilibration buffer II and equilibrate the strip for 15 min more (see Note 11).

12. Dilute the TGS 10× to yield 4 L of TGS 1× (400 mL TGS 10× and 3,600 mL milli-Q water). Warm the agarose solution in the microwave until it starts to boil, and let it cold at room temperature for 5–10 min. Do not allow the agarose solidify again.

13. Soak the equilibrated strip in TGS 1× to remove the free DTT or iodoacetamide. Place the strip between the two glass plates (the acrylamide surface of the strip should be placed towards one glass, and the anode end (+) of the strip should be oriented to the left, close to the reference well). Press carefully all along the strip with a comb to ensure the strip is in contact with the SDS-PAGE gel.

14. Seal the IPG strip with the agarose solution and let the agarose solidify. Be careful not to pour very hot agarose on top of the strip. Also be careful not to cover the reference well with the melted agarose.

15. Place the gel into the central cooling core and transfer the apparatus to the buffer tank. Add 4 L of TGS 1×. Load 3 µL of molecular weight standard in the reference lane at the left.

16. Run the gel at 300 V for 3 h or until the dye front (bromophenol blue present in the agarose solution) has reached the bottom end of the gel.

17. Disassemble the gel from the central cooling core, and proceed with the EZBlue (Invitrogen®) staining protocol: wash the gel

three times with milli-Q water with gentle shaking. Remove water and fix the gel with fixing solution for 15 min with gentle shaking. Remove fixing solution and add EZBlue for protein staining for 45 min. Destain the gel with milli-Q water for 10 min or until the background is removed.

3.3. Excision of Gel Spots and LC-MS/MS Analysis

1. Scan the gels in transmission scan mode using a high-resolution calibrated scanner (e.g., GS-800, Bio-Rad).

2. Print the gel image and number every spot that it is going to be excised on the basis of high intensity and resolution.

3. Place the gel on a wet glass precleaned with ethanol. Use new gloves soaked in ethanol and prepare enough microcentrifuge tubes for all the spots that are going to be excised. Prepare a box of autoclaved tips and a clean and ethanol soaked scalpel.

4. Cut the tip with the scalpel to create a hole similar in size to that of the spot that is going to be excised. Excise all the spots by punching with the micropipette tip and place them in the corresponding numbered microcentrifuge tubes.

5. Proceed to LC-MS/MS analysis. Clean the excised spots by adding 2 volumes of 25 mM ammonium bicarbonate and incubating at room temperature for 15 min. After that discard the liquid and add the same volume of acetonitrile (ACN) for 15 min. Discard the liquid and repeat 2 times the above 2 steps. After that add the same volume of acetonitrile and incubate for an additional 15 min. Following this incubation, remove the ACN and trypsinize the gel fragments with 125 ng of trypsin at 37°C overnight.

6. Liquid chromatography: Inject 10 µL of tryptic peptides into a nanoLC (Proxeon EASY-nLC) at a flow rate of 500 nL/min. Clean and trap the sample through a trap column of C18-PepMap100, 5 µm, 100 A, 300 µm ID 5 mm. Peptides can then be separated through an analytical column of C18 PepMap 3 µm, 100 A, 75 µm ID, 150 mm from Proxeon with the following gradient: 0–40 min (0–60% Buffer B; see Note 12), 40–45 min (60–70% Buffer B), 45–55 min (70–100% Buffer B), and 55–60 min (100% Buffer B).

7. Analyze the peptides on a Velos LTQ-Orbitrap with a Proxeon/stainless steal emitter ion source (Thermo Fisher Scientific) and a precursor ion selection in the Orbitrap at a resolution of 30,000 following 15 IT MS/MS with a collision energy of 35 in positive mode.

8. Acquire the MS/MS data using Xcalibur 2.1 (Thermo Fisher Scientific) and submit the results to the SEQUEST software using the HUMAN UniProt SwissProt database search with the following thresholds: 25 ppm of peptide mass tolerance and 0.8 Da of fragment ion tolerance. Use the following criteria to accept identifications: sequest Xcore (1.5 for +1, 2 for

+2, 2.5 for +3, 2.75 for +4, 3 for more than +4), probabilistic Protein Prophet index (>0.99) using the 4.3 version, the extent of sequence coverage and the number of peptides matched.

4. Notes

1. The ampholytes from the IPG Buffer should correspond to the IPG strips (Immobilized pH gradient) pH range.
2. The vast majority of the cells present in a normal sperm sample (ejaculate) are spermatozoa. However, a variable proportion of additional cells such as immature spermatids, aberrant spermatozoa, somatic cells (leukocytes and male tract epithelial cells) and bacteria are also present. In abnormal sperm samples the proportion of these types of additional cells present in the semen can be substantially increased. Thus, it is very important to take this fact into account and to perform appropriate methods to eliminate these contaminating cells. A common procedure used is to perform density gradient purification of the sperm cells (11, 26). In addition leukocytes may also be eliminated using CD45 affinity (27). The procedure for sperm cell purification using density gradient centrifugation is described here.
3. 50% Percoll can be used instead of a higher percentage of Percoll so that the resulting sperm cells are representative of the average spermatozoon present in the normozoospermic sperm samples, rather than representing a highly selected subpopulation of the fittest spermatozoa (26). Other types of density media such as PureSperm may also be used instead of Percoll.
4. To ease the placement of the sperm sample on the top of the 50% Percoll gradient a track can be created during the addition of the 50% gradient in the 15 mL tube by adding the gradient into the tube following the procedure described in step 3 of Subheading 3.1 (tilt tube, recline the pipette at approximately 5 cm away from the bottom of the tube and add the gradient carefully down the side of the tube).
5. The use of a fourth part of 1 mL of Ham's F10 1× for the resuspension of the purified sperm cell pellet is intended to avoid putting in contact the fresh Ham's F10 1× with the walls of the tube after gradient centrifugation (which may be contaminated from the gradient itself and other undesirable semen components).
6. If the interphase has a considerable amount of spermatozoa it is possible that the gradient has not been properly prepared and that a valuable amount of sperm cells are being lost in this fraction. In this case, use the interphase material and repeat steps 3–13 of Subheading 3.1 to try to recover as many sperm cells as possible.

7. The centrifugation speed indicated is OK for the purpose used here but it affects mobility and cell viability. Therefore lower centrifugation speeds may be required for other types of experiments. The indicated sperm cell concentration is enough to recover sufficient protein to identify the resulting 2D-PAGE spots using MS. As stated in step 4 of Subheading 3.2, it is recommended that 300 μL of sample be loaded onto each IPG strip.

8. Proteins that are not solubilized with the urea-thiourea lysis buffer can be further extracted with HCl for acid–urea gel protein separation and MS protein identification.

9. Most of the components specified here are components from Bio-Rad®, but other suppliers can be used. Strip length and pH range can be selected according to the goals of the proteome study.

10. To set up the focusing program follow manufacturer's instructions (Protean IEF Cell, Bio-Rad). For samples with different characteristics from the proteins extracted from human spermatozoa modify the settings to allow the proteins to focus properly.

11. Shorter equilibration times can be used, but some proteins may not migrate out of the IPG gel strip and enter the SDS-PAGE gel. It is recommended that the IPG strip be stained after SDS-PAGE to verify that all proteins have migrated out of the IPG strip.

12. Different conditions and longer times must be used if bands excised from mono-dimensional gels are used instead of spots excised from two-dimensional gels.

Acknowledgment

Supported by grants from the Ministerio de Educación y Ciencia (BFU2009-07118), fondos FEDER. The authors acknowledge the critical revision by Dr. Alexandra Amaral.

References

1. Domon B, Aebersold R (2006) Mass spectrometry and protein analysis. Science 312:212–217
2. Cox J, Mann M (2007) Is proteomics the new genomics? Cell 130:395–398
3. Aebersold R, Mann M (2003) Mass spectrometry-based proteomics. Nature 422:198–207
4. Brewis IA, Gadella BM (2010) Sperm surface proteomics: from protein lists to biological function. Mol Hum Reprod 16:68–79
5. Oliva R et al (2009) Sperm cell proteomics. Proteomics 9:1004–1017
6. Baker MA, Aitken RJ (2009) Proteomic insights into spermatozoa: critiques, comments and concerns. Exp Rev Proteomics 6:691–705
7. Com E et al (2003) New insights into the rat spermatogonial proteome: identification of 156 additional proteins. Mol Cell Proteomics 2:248–261

8. Pixton KL et al (2004) Sperm proteome mapping of a patient who experienced failed fertilization at IVF reveals altered expression of at least 20 proteins compared with fertile donors: case report. Hum Reprod 19:1438–1447
9. Chu DS et al (2006) Sperm chromatin proteomics identifies evolutionarily conserved fertility factors. Nature 443:101–105
10. Dorus S et al (2006) Genomic and functional evolution of the Drosophila melanogaster sperm proteome. Nat Genet 38:1440–1445
11. Martinez-Heredia J et al (2006) Proteomic identification of human sperm proteins. Proteomics 6:4356–4369
12. Martinez-Heredia J et al (2008) Identification of proteomic differences in asthenozoospermic sperm samples. Hum Reprod 23:783–791
13. de Mateo S et al (2007) Marked correlations in protein expression identified by proteomic analysis of human spermatozoa. Proteomics 7:4264–4277
14. Baker MA et al (2010) Analysis of proteomic changes associated with sperm capacitation through the combined use of IPG-strip pre-fractionation followed by RP chromatography LC-MS/MS analysis. Proteomics 10:482–495
15. Oliva R, Dixon GH (1991) Vertebrate protamine genes and the histone-to-protamine replacement reaction. Prog Nucleic Acid Res Mol Biol 40:25–94
16. de Yebra L et al (1993) Complete selective absence of protamine P2 in humans. J Biol Chem 268:10553–10557
17. de Yebra L, Oliva R (1993) Rapid analysis of mammalian sperm nuclear proteins. Anal Biochem 209:201–203
18. Oliva R (2006) Protamines and male infertility. Hum Reprod Update 12:417–435
19. de Mateo S et al (2009) Protamine 2 precursors (Pre-P2), protamine 1 to protamine 2 ratio (P1/P2), and assisted reproduction outcome. Fertil Steril 91:715–722
20. Yoshii T et al (2005) Fine resolution of human sperm nucleoproteins by two-dimensional electrophoresis. Mol Hum Reprod 11:677–681
21. Oliva R et al (2008) Proteomics in the study of the sperm cell composition, differentiation and function. Syst Biol Reprod Med 54:23–36
22. de Mateo S et al (2011) Improvement in chromatin maturity of human spermatozoa selected through density gradient centrifugation. Int J Androl 34:256–267
23. de Mateo S et al (2011) Protamine 2 precursors and processing. Protein Pept Lett 18:778–785
24. de Mateo S et al (2011) Proteomic characterization of the human sperm nucleus. Proteomics 11:2714–2726
25. Castillo J et al (2011) Protamine/DNA ratios and DNA damage in native and density gradient centrifuged sperm from infertile patients. J Androl 32:324–332
26. Mengual L et al (2003) Marked differences in protamine content and P1/P2 ratios in sperm cells from percoll fractions between patients and controls. J Androl 24:438–447
27. Koppers AJ et al (2011) Phosphoinositide 3-kinase signalling pathway involvement in a truncated apoptotic cascade associated with motility loss and oxidative DNA damage in human spermatozoa. Biochem J 436:687–698

Part IX

Methodologies to Evaluate Sperm Epigenetics

Chapter 36

Methodology of Aniline Blue Staining of Chromatin and the Assessment of the Associated Nuclear and Cytoplasmic Attributes in Human Sperm

Leyla Sati and Gabor Huszar

Abstract

In this chapter, the laboratory methods for detection of sperm biomarkers that are aimed at identifying arrested sperm development are summarized. These probes include sperm staining with aniline blue for persistent histones, representing a break in the histone-transition protein-protamine sequence, immunocytochemistry with cytoplasmic sperm proteins, highlighting cytoplasmic retention during spermiogenesis, DNA nick translation testing for DNA chain fragmentation due to various reasons, for instance low HspA2 chaperone protein levels, and consequential diminished DNA repair. Finally, we briefly provide references on our work on sperm hyaluronan binding, abnormal Tybergerg sperm morphology, and the increased levels of chromosomal aneuploidies in sperm with developmental arrest. A very interesting aspect of the biomarker field is the discovery (Sati et al, Reprod Biomed Online 16:570–579, 2008) that the various nuclear and cytoplasmic defects detected by the biomarkers are related, and may simultaneously occur within the same spermatozoa as evidenced by a combination of biomarkers, such as aniline blue staining (persistent histones) coupled with cytoplasmic retention, DNA fragmentation, Caspase-3, Tygerberg abnormal morphology, and increased levels of chromosomal aneuploidies. We show examples of this >80% overlap in staining patterns within the same spermatozoa.

Key words: Human sperm biomarkers, Chromatin development, Cytoplasmic retention, Sperm DNA integrity/fragmentation, Paternal contribution

1. Introduction

In the past 20 years, major progress has occurred in the Sperm Physiology Laboratory at the Department of Obstetrics and Gynecology at Yale School of Medicine. The ultimate goal was the development of objective biomarkers that would predict sperm fertilizing potential, independently from sperm concentration and motility in the semen of a man.

Douglas T. Carrell and Kenneth I. Aston (eds.), *Spermatogenesis: Methods and Protocols*, Methods in Molecular Biology, vol. 927,
DOI 10.1007/978-1-62703-038-0_36, © Springer Science+Business Media, LLC 2013

The major advances were the description and role of the sperm creatine kinase (CK) activity that reflects arrested sperm development at the level of cytoplasmic extrusion.

In a clinical study, it was demonstrated that in couples with oligozoospermic husbands treated with intrauterine insemination, the occurrence of pregnancies correlated with sperm creatine kinase levels, whereas sperm concentration and motility were noncontributory (2). In another sperm creatine kinase study of 84 couples treated with IVF, in husbands with high sperm creatine kinase levels, there were no pregnancies in spite of normal sperm concentration and motility (3). These two findings led to the formation of the concept of unexplained male infertility (men with diminished fertilizing potential or diminished sperm paternal contribution in spite of normal sperm concentration and motility).

The next step in the sperm biomarker studies was the assessment of sperm chromatin maturity via aniline blue staining. Aniline blue stains persistent histones in the sperm nucleus. Increased levels of persistent histones in turn indicate a break in the developmental sequence of histones-transition proteins-protamines, which significantly affects DNA chain folding and vulnerability for increased DNA fragmentation. The DNA chain fragmentation causes a decline in sperm fertilizing potential and in the paternal contribution of sperm to the developing embryo (4, 5).

Other biomarkers we have identified that are contribute sperm fertility and function is the HspA2 chaperone protein, which exhibits various functions. For instance HspA2 is a component of the synaptonemal complex and thus supports meiosis. Due to the transport role of HspA2, the chaperone also supports cellular processes including DNA repair (6). Sperm specimens from men which contain low HspA2 levels show increased levels of chromosomal aneuploidies, DNA fragmentation and a decline in fertility as demonstrated in a study of IVF husbands ($N=135$ men) (7) who did not establish IVF pregnancy and displayed low sperm HspA2 levels, although their semen parameters were well into the normozoospermic range. Low HspA2 levels are also associated with increased levels of DNA fragmentation, most likely due to a diminished level of DNA chain repair activity.

Also of interest is the finding that the attributes of incomplete or arrested development such as CK, aniline blue, DNA fragmentation, low HspA2, abnormal morphology, and apoptosis appear to be common and related in individual spermatozoa. This is illustrated in Figs. 1, 2, 3, and 4, in which sperm are double stained with aniline blue and a second biomarker. It is noteworthy that aniline blue staining and the other biomarkers, including cytoplasmic markers (Fig. 2), DNA fragmentation markers (Fig. 3), and apoptotic markers (Fig. 4), show similar patterns when evaluated within the same cell (8). There is >80% similarity in the sperm biomarker staining patterns, indicating that once spermatozoa

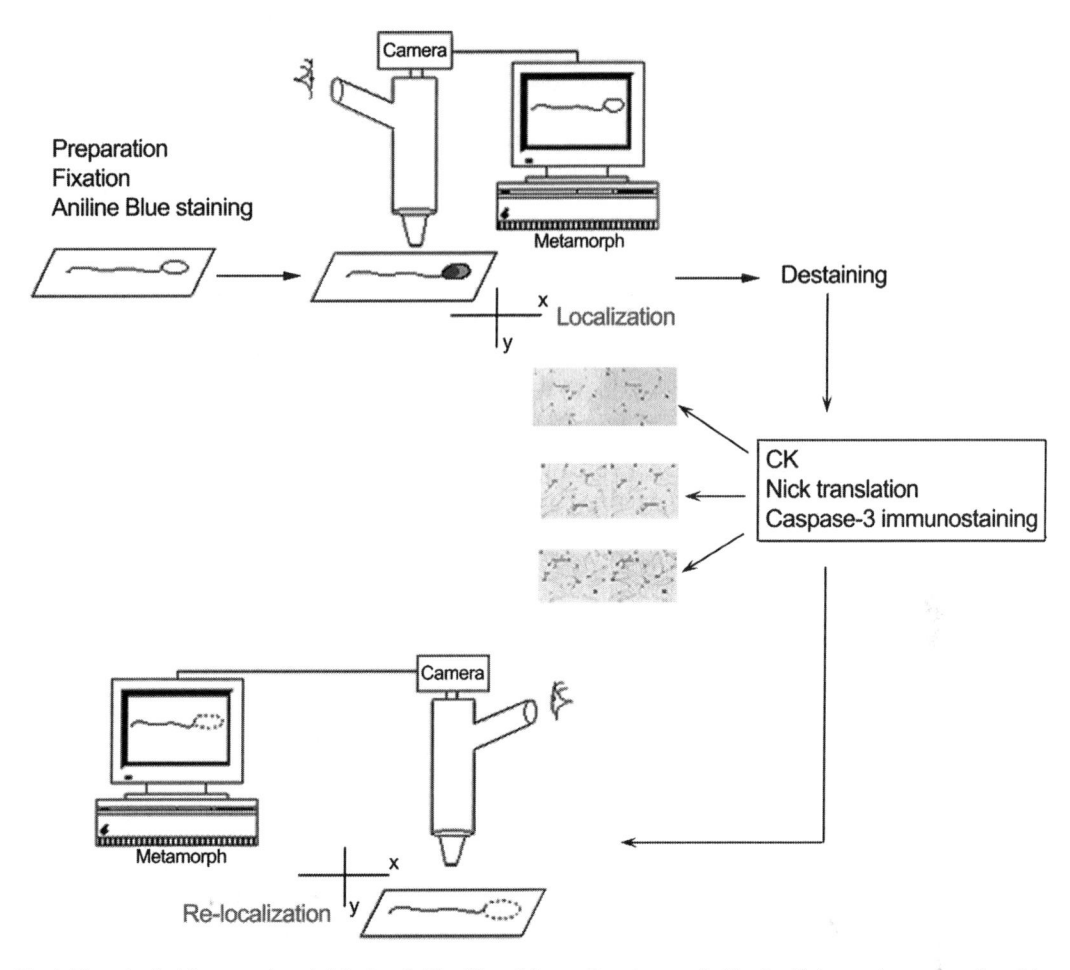

Fig. 1. Flow chart of the experimental design. Aniline blue staing pattern is recorded by the Metamorph program, the slides are destained from aniline blue, the slide is restained with a second probe, and the same sperm field treated with the two probes is compared and the staining is quantified as light, intermediate, and dark (1).

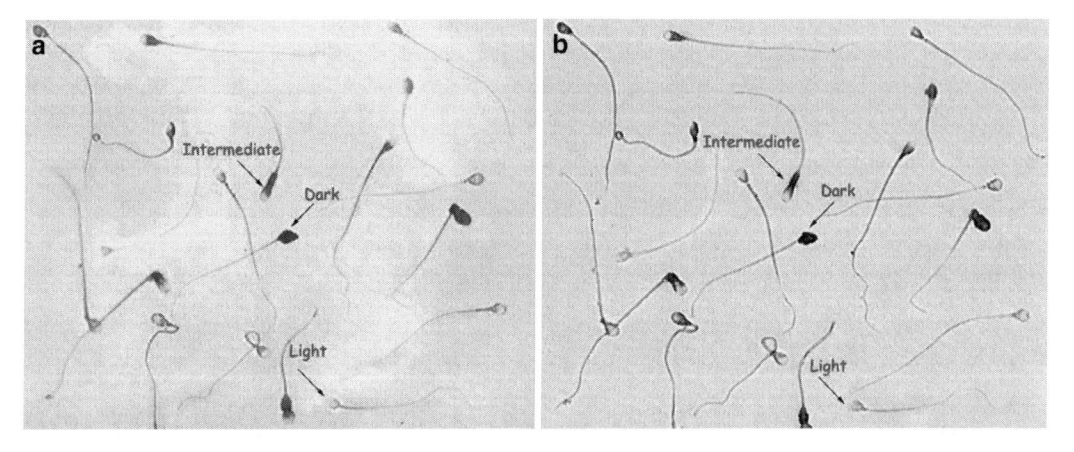

Fig. 2. (a) Aniline blue staining and (b) creatine kinase immunostaining of the same spermatozoa. Note the substantial degree of similarity in the light-, intermediate-, and dark-staining patterns with aniline blue and CK.

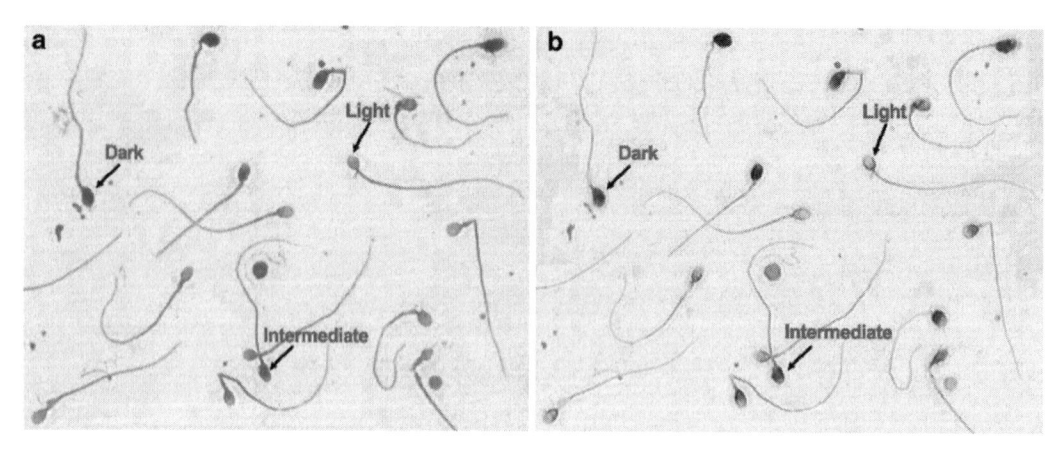

Fig. 3. (a) Aniline blue staining and (b) DNA Nick translation of the same spermatozoa. Note the substantial degree of similarity in the light-, intermediate-, and dark-staining patterns with the two methods.

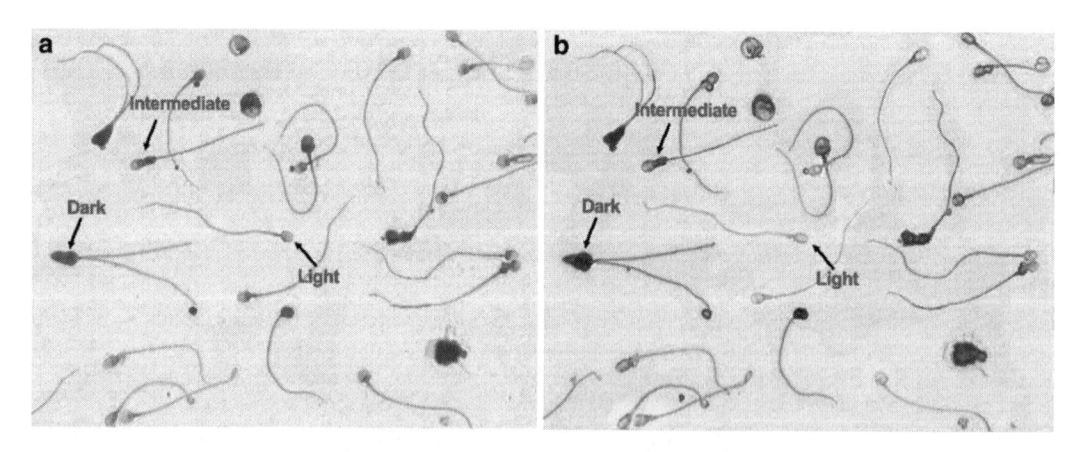

Fig. 4. (a) Aniline blue staining and (b) Caspase-3 immunostaining of the same spermatozoa. Note the similarity in the light-, intermediate-, and dark-staining patterns of the aniline blue and Caspase-3 panels. Also, Caspase-3 immunostaining is present in the mid-piece of intermediate-type spermatozoa, whereas in dark spermatozoa with more extensive maturity arrest both the head and the mid-piece are stained (a and b).

undergoes a developmental arrest, this event structurally and functionally affects multiple cytoplasmic and nuclear attributes within the same sperm (1).

2. Materials

Prepare all solutions using double-distilled water.

2.1. Solutions for Sperm Preparation

1. 1 M Imidazole stock solution: Weigh 68.1 g imidazole and dissolve in 1 l of double-distilled water. Adjust the pH to 7.0. Store at 4°C (see Note 1).

2. Saline/imidazole sperm washing medium (SAIM) containing 0.3% bovine serum (30 mM imidazole/0.15 M NaCl): Combine 15 ml 1 M imidazole stock solution (at 4°C) and 4.5 g NaCl. Add double-distilled water to 500 ml.

3. We use saline/imidazole solution for washing and resuspending sperm before preparing slides. PBS solution may also be used for this step. 0.5 M sodium phosphate stock solution, pH 7 (see Note 2): this solution is prepared by mixing together 0.5 M Na$_2$HPO$_4$ (dibasic) and 0.5 M NaH$_2$PO$_4$ (monobasic) solutions. For 0.5 M dibasic, dissolve 71 g of anhydrous Na$_2$HPO$_4$ in 1 l of double-distilled water. For 0.5 M monobasic, dissolve 79 g NaH$_2$PO$_4$ monohydrate in double-distilled water. Once these solutions are prepared, 0.5 M sodium phosphate buffer is prepared by mixing approximately 296 ml monobasic solution and 104 ml dibasic solution for a pH of 7 (see Note 3).

2.2. Solutions for Aniline Blue Staining

1. Aniline blue staining solution: Weigh 0.5 g aniline blue and add 48 ml double-distilled water and 2 ml of glacial acetic acid.

2.3. Solutions for In Situ DNA Nick Translation

1. PBS: see Subheading 2.1, step 3.

2. Fixative for slides: Methanol and glacial acetic acid in a ratio of one part acetic acid to three parts methanol (for example, 50 ml of acetic acid and 150 ml of methanol to make 200 ml fixative).

3. 1 M Tris–HCl, pH 7.2: Add 60 g Tris to 500 ml double-distilled water. Adjust pH with concentrated HCl to 7.2.

4. 0.1 M Tris–HCl, pH 7.2: Add 10 ml 1 M Tris–HCl, pH 7.2 to 90 ml double-distilled water.

5. 20 mM lithium 3,5-diiodosalicylate (LIS) stock solution: Add 3.95 g LIS to 500 ml double-distilled water (see Note 4).

6. 1 M dithiothreitol (DTT): Add double-distilled water to 1.5 g DTT to a final volume of 10 ml (see Note 5).

7. 0.1 M DTT: Dilute one part of 1 M DTT with nine parts of double-distilled water (see Note 5).

8. Nucleotide mix: 1 mM mix of dGTP, dCTP, dATP, and dTTP. We make aliquots and keep them in the freezer.

9. Biotin: Prepare a 1% solution from 50 mg biotin stock by adding all of the contents of this tube to 5 ml PBS. Freeze this concentrated biotin solution in 0.5 ml aliquots. To make the working solution (0.01%), dilute at a ratio of 1 part 1% biotin to 99 parts double-distilled water (see Note 6).

10. 1 M magnesium sulfate (see Note 7).

11. Coomassie Blue Counterstain: Add 0.04 g Coomassie Blue to 25 ml isopropanol, and 10 ml glacial acetic acid. Add double-distilled water to reach a final volume of 100 ml.

2.3.1. DNA Solution

1. 90.5 μl distilled water.
2. 1.0 μl 0.1 M DTT.
3. 1.0 μl 1 M magnesium sulfate.
4. 5.0 μl 1 M Tris–HCl, pH 7.2.
5. 1.0 μl 1 mM biotin-16-dUTP (as supplied by Roche).
6. 1.0 μl 1 mM mix of dGTP, dCTP, dATP, and dTTP.
7. 0.5 μl DNA polymerase I at concentration supplied by Roche.

2.4. Solutions for Immunostaining of Sperm Slides for Cytoplasmic Retention and Apoptosis (CK and Caspase-3)

PB-sucrose: Add 22.5 g sucrose to 20 ml 0.5 M sodium phosphate buffer, pH 7 (as prepared above), and add double-distilled water to 500 ml.

3. Methods

Carry out all procedures at room temperature unless otherwise specified.

3.1. Sperm Preparation for Aniline Blue Staining and/or In Situ DNA Nick Translation Studies

1. Centrifuge the semen samples through 1.5 ml of a 40% single-phase isolate gradient prepared in HTF media at $1,600 \times g$ for 10 min at room temperature. Resuspend the pellet in 2–4 ml of HTF.
2. Centrifugation sample a second time at $800 \times g$ for 10 min at room temperature. Resuspended the sperm pellet fraction in the SAIM or HTF (no albumin) prior to use.

3.2. Sperm Preparation for Immunostaining Studies

1. Assess aliquots of liquefied semen for sperm concentration and motility, then dilute the semen samples with physiological saline containing 0.3% bovine serum albumin and 30 mmol/l imidazole pH 7.2 (SAIM) up to a final volume of 10 ml.
2. Centrifuge the semen samples at $500 \times g$ for 18 min at room temperature. Discard the supernatant, and resuspend each sperm pellet in the SAIM solution to a concentration of $10-25 \times 10^6$ sperm/ml (see Note 8).

3.3. Aniline Blue Staining of Sperm Chromatin

1. Place one drop of the concentrated sperm suspension in SAIM or HTF (no albumin) on a clean microscope slide and spread over the surface of the slide (see Note 9). Allow the slide to dry. Slides may be stored at this point under refrigeration or used immediately for staining.

2. Place in a Coplin jar containing acidic aniline blue stain for 5 min.

3. Rinse in distilled water, dry, and mount using Cytoseal.

3.4. CK-Immunocytochemistry for Cytoplasmic Retention

1. Add several drops of PB-sucrose to a polylysine slide.

2. Add 3–5 µl of the resuspended sperm suspension to the PB-sucrose, and mix carefully using the pipette tip. The final sperm concentration should ideally be between 10 and 15 million/ml. Allow sperm to settle onto polylysine-treated slides overnight in a humidified chamber at 4°C (see Note 10).

3. The next morning, carefully aspirate the residual fluid so as not to disturb the attached sperm and add 3.7% formalin in PB-sucrose. After 20 min at room temperature, again carefully aspirate the fluid from the slide and allow to dry. Slide may be stored at this point under refrigeration or used immediately for immunostaining.

4. Prior to immunostaining, block the fixed sperm in PB-sucrose containing 3% BSA for 1 h at room temperature, followed by a quick subsequent wash with PB-sucrose.

5. For CK immunostaining, apply the first antibody (our CK, raised in rabbit) to the fixed sperm at a 1:1,000 dilution in PB-sucrose containing 0.1% BSA. Maintain the slide in a humidified chamber for 2 h at room temperature.

6. Wash the slide three times with PB-sucrose and apply the second (biotinylated) rabbit antibody at a 1:1,000 dilution. Hold the slide at room temperature for 1.5 h. 30 min before the end of incubation, prepare AB (avidin-biotin) complex (ABC; Vector) in PB-sucrose according to manufacturer's instructions.

7. After three washes in PB-sucrose, apply the diluted AB complex to the slide and hold at room temperature for ½ h.

8. Again wash the slide three times in PB-sucrose, then apply DAB solution. Allow color to develop for 5–10 min at room temperature.

9. Wash the slide thoroughly three times with PB-sucrose, dehydrate through an ethanol series: 70, 85, and 100% ethanol quickly, immerse in xylene twice for 5 min each, and mount using Cytoseal.

3.5. Strict Morphology

Our methods are summarized in ref. 8.

3.6. Nick Translation for Sperm DNA Integrity (9)

1. After processing sperm, smear resuspended sperm on a clean glass slide and allow to air-dry. As soon as the smear is dry, fix in 3:1 methanol/acetic acid for 15 min (see Note 11).

2. Dehydrate through an ethanol series: 70, 85, and 100% ethanol (5 min each). Air dry. Slides may be stored at 4°C after dehydration.

3. Place slides in a Coplin jar or small tray containing 10 mM DTT in 0.1 M Tris, pH 7.2. Hold at room temperature for ½ h (see Note 12).

4. Wash twice with 0.1 M Tris, pH 7.2, and add 10 mM LIS in 0.1 M Tris, pH 7.2. Leave at room temperature 2–3 h (store LIS as a 20 mM solution in distilled water) (see Note 13).

5. Wash three times with PBS. Add biotin (a 1:10 dilution of a 0.01% stock, kept at −20°C) to completely cover sperm smear. Leave at room temperature for 20 min.

6. Wash three times with PBS. Add avidin (a 1:10 dilution of stock as supplied by Vector, kept at 4°C) (see Note 14). Leave at room temperature for 20 min.

7. Wash three times with PBS. Add DNA polymerase solution.

8. Float a plastic coverslip on the drop of polymerase mixture, and leave at room temperature for 30 min.

9. While the slide is incubating with the DNA polymerase solution, make up ABC solution: Add 9 µl each of components A and B from Vector ABC peroxidase kit to 1 ml PBS.

10. Wash slide three times in PBS. Add ABC solution, and leave at room temperature for ½ h.

11. Wash three times with PBS, and add DAB solution (prepared as for anti-CK staining). Leave 10–15 min, and then wash thoroughly with distilled water. Check slide under the microscope to ensure that sperm are stained before counterstaining (see Note 15).

12. Counterstain with Coomassie Blue (0.04% Coomassie Blue powder in 25% isopropanol, 10% glacial acetic acid) for 10–20 s. Wash thoroughly with distilled water and observe to ensure that tails are sufficiently stained. If not, repeat Coomassie Blue staining.

13. Allow slide to air dry and coverslip with Cytoseal.

3.7. Fluorescent In Situ Hybridization (FISH)

Our methods are summarized in refs. 10, 11.

3.8. Active Caspase-3 Immunostaining

1. Caspase-3 immunocytochemistry is carried out similar to the methods described for CK immunostaining (1, 12). However, after the spermatozoa is exposed to the 3% BSA blocking solution, treat the slides with a 1:300 dilution of active Caspase-3 antibody overnight at 4°C.

2. Further, process the slides with 1:1,000 dilution of a biotinylated anti-rabbit second antibody.

3.9. Double Probing of Spermatozoa with Aniline Blue and Several Nuclear and Cytoplasmic Markers of Sperm Immaturity

The double probing studies in the same spermatozoa are carried out as outlined in Fig. 1. The steps are detailed as follows.

1. Smear spermatozoa onto glass slides, and fix with methanol–acetic acid, then stain with aniline blue to detect persistent histones as described above (1, 13).

2. Capture the images of sperm fields and individual spermatozoa carefully using the Metamorph™ imaging program (Universal Imaging Co., Downingtown, PA, USA).

3. Subsequently, with the slides still on the microscope platform, record the X–Y coordinates of the fields in order to facilitate the re-localization of the same sperm fields.

4. Destain the sperm slides from aniline blue by overnight incubation in the fixative solutions that are appropriate for the second biochemical markers (methanol for DNA nick translation, and paraformaldehyde for the CK or Caspase-3 immunocytochemistry) (1). Carry out all the destaining procedures on a shaking platform at room temperature.

5. Treat destained spermatozoa with one of the second biochemical probes: (1) CK immunocytochemistry to demonstrate cytoplasmic retention in diminished-maturity spermatozoa (14); (2) DNA nick translation for the detection of DNA chain breaks (15); and (3) Caspase-3 immunostaining to detect the apoptotic process in spermatozoa (12).

6. As illustrated in Figs. 2, 3, and 4, there is a substantial overlap between the staining patterns of the same spermatozoa (assessed as light, intermediate, and dark) with the various cytoplasmic and nuclear biomarkers; the similarity in staining intensity is over 70%. This suggests that when there is an arrest in sperm development, the various sperm compartments and developmental processes are interrelated and commonly affected (1, 4).

3.10. Destaining for CK or Caspase-3 Immunocytochemistry

1. After recording the sperm fields, use 0.5% paraformaldehyde solution in PB-sucrose overnight at room temperature to destain the aniline blue-stained sperm cells.

2. Remove formaldehyde by three washes with PB-sucrose and air dry the slides.

3. Start the immunocytochemistry protocol by exposing to a 3% bovine serum albumin (BSA) blocking solution in PB-sucrose at room temperature. All the subsequent steps are the same as those described for sperm immunocytochemistry procedure.

3.11. Destaining for In Situ DNA Nick Translation

1. After recording of the fields, use 30% methanol solution overnight to destain the aniline blue-stained sperm cells.

2. Cover the slides with 20 mmol/l of imidazole buffer pH: 7.0 for 1 h, and apply 30% methanol for 15 min, followed by air drying.

3. Treat the slides with methanol–glacial acetic acid (3:1) for 15 min.
4. After exposure to a dehydrating ethanol series (70, 85, and 100%), air-dry the slides and treat with DTT and LIS to initiate DNA decondensation. All the subsequent steps are the same as those described for the in situ DNA Nick translation procedure.

4. Notes

1. If you accidentally add too much HCl and the pH goes below 7, add dissolved NaOH (a few pellets of NaOH dissolved in a small amount of double-distilled water) until the pH rises to 7.
2. Keep this solution at room temperature, since it will precipitate at 4°C.
3. The pH should be checked and adjusted if necessary. If the pH is higher than 7, add more monobasic solution to bring it to pH 7. If the pH is lower than 7, add more dibasic solution to bring it to pH 7.
4. Store the solution in a dark or foil covered bottle at 4°C.
5. Keep frozen in aliquots and do not refreeze. It is best to thaw a new aliquot each time.
6. Aliquot this 0.01% solution. It is this 0.01% solution that is then further diluted 1:9 in PBS before using for slides.
7. Since only 1 µl of this is needed for 0.1 ml polymerase mixture, there should be enough once prepared to last as long as you will need it.
8. When resuspending the pellet, take care to avoid making any bubbles since they can cause a stress factor on sperm and induce DNA strand breaks. Therefore gentle treatment is required for this step.
9. Acidic aniline blue staining is used to detect chromatin defects of sperm nuclei related to their nucleoprotein content. However, even if you can stain the semen smears directly with aniline blue, it is difficult to evaluate the slides due to dark background staining. Therefore, it is best to rinse the semen sample with wash solution or dilute the semen to be able to analyze the slides.
10. It is possible to perform this step either within the chambers of a multichambered slide or within the borders of a pap-pen circle. All subsequent steps are carried out in a humidified chamber also.

11. This fixation step is very crucial. Carefully watch the smear while it is drying out, and as soon as the sperm is dried start the fixation with methanol–acetic acid in 3:1 ratio.

12. Once you thaw stock DTT solution for this step, save some of it as you also need to use 1 μl of 0.1 M DTT in the polymerase mixture for step 7. Slides should not be incubated for more than 30 min in DTT solution. DTT is a strong agent that reduces disulfide bridges and prevents the formation of intra- and intermolecular bridges between residues in proteins, thus destroying their tertiary structure. The reducing capacity of DTT is limited to pH values above 7. Therefore, a slightly alkaline TRIS buffer solution is used.

13. LIS incubation time for swelling of sperm heads may differ between samples. Therefore, check the decondensation status under the microscope every ½ h in the dark. Over-incubation results in blowing out of the sperm heads. In addition, LIS incubation should be carried out in the dark. Therefore, cover the Coplin jar with aluminum foil.

14. It is sufficient to add only 25–50 μl and float a plastic coverslip over the drop.

15. If sperm are sufficiently immunostained, allow to air dry. If not, restain with DAB for 10–15 additional minutes, then wash and dry as before.

References

1. Sati L et al (2008) Double probing of human spermatozoa for persistent histones, surplus cytoplasm, apoptosis and DNA fragmentation. Reprod Biomed Online 16:570–579
2. Huszar G, Vigue L, Corrales M (1990) Sperm creatine kinase activity in fertile and infertile oligospermic men. J Androl 11:40–46
3. Huszar G, Vigue L, Morshedi M (1992) Sperm creatine phosphokinase M-isoform ratios and fertilizing potential of men: a blinded study of 84 couples treated with in vitro fertilization. Fertil Steril 57:882–888
4. Huszar G et al (2007) Fertility testing and ICSI sperm selection by hyaluronic acid binding: clinical and genetic aspects. Reprod Biomed Online 14:650–663
5. Agarwal A, Said TM (2003) Role of sperm chromatin abnormalities and DNA damage in male infertility. Hum Reprod Update 9:331–345
6. Huszar G et al (2000) Putative creatine kinase M-isoform in human sperm is identified as the 70-kilodalton heat shock protein HspA2. Biol Reprod 63:925–932
7. Ergur AR et al (2002) Sperm maturity and treatment choice of in vitro fertilization (IVF) or intracytoplasmic sperm injection: diminished sperm HspA2 chaperone levels predict IVF failure. Fertil Steril 77:910–918
8. Prinosilova P et al (2009) Selectivity of hyaluronic acid binding for spermatozoa with normal Tygerberg strict morphology. Reprod Biomed Online 18:177–183
9. Rigby PW et al (1977) Labeling deoxyribonucleic acid to high specific activity in vitro by nick translation with DNA polymerase I. J Mol Biol 113:237–251
10. Jakab A et al (2005) Intracytoplasmic sperm injection: a novel selection method for sperm with normal frequency of chromosomal aneuploidies. Fertil Steril 84:1665–1673
11. Kovanci E et al (2001) FISH assessment of aneuploidy frequencies in mature and immature human spermatozoa classified by the absence or presence of cytoplasmic retention. Hum Reprod 16:1209–1217
12. Cayli S et al (2004) Cellular maturity and apoptosis in human sperm: creatine kinase, caspase-3

and Bcl-XL levels in mature and diminished maturity sperm. Mol Hum Reprod 10:365–372

13. Sati L et al (2004) Persistent histones in immature sperm are associated with DNA fragmentation and affect paternal contribution of sperm: a study of aniline blue staining, fluorescence in situ hybridization and DNA nick translation. In: Annual Meeting of the American Society for Reproduction Medicine, Philadelphia, PA

14. Huszar G, Vigue L (1993) Incomplete development of human spermatozoa is associated with increased creatine phosphokinase concentration and abnormal head morphology. Mol Reprod Dev 34:292–298

15. Irvine DS et al (2000) DNA integrity in human spermatozoa: relationships with semen quality. J Androl 21:33–44

Chapter 37

Isolation of Sperm Nuclei and Nuclear Matrices from the Mouse, and Other Rodents

W. Steven Ward

Abstract

The isolation of mammalian sperm heads from their tails is complicated by the relatively high density of the tails, but facilitated by the fact that protamine condensation of the sperm chromatin and the insolubility of the perinuclear theca make the sperm nucleus stable in sodium dodecyl sulfate. Two methods are described for the isolation of rodent sperm nuclei using sucrose step gradients in which the sperm nuclei are only centrifuged one time, minimizing potential damage by mechanical stress.

Key words: Sperm nuclei, Sperm nuclear matrix, Sperm nuclear halo, Sperm DNA

1. Introduction

The mammalian sperm nucleus is a unique model for the study of chromatin for several reasons. Firstly, the DNA is packaged by proteins found only in that cell type, the protamines (1). Secondly, this nucleus has nuclear matrix that is unusually stable to mechanical stress so that it can be experimentally manipulated with much greater ease than most somatic cell nuclear matrices (2, 3). Thirdly, sperm DNA contains sex-specific methylation patterns (4) as well as specific histone modifications (5) that contribute to proper embryogenesis.

The structure of the sperm cell presents peculiar challenges as well as interesting advantages that make the isolation of the nuclei different from most other cells. The isolation of mammalian sperm heads from their tails is complicated by the relatively high density of the tails which contains the overwhelming majority of the mass of the entire cell (6). This separation is made more difficult by the high density and low viscotic resistance of the tails, which makes traditional sucrose density gradients less useful. On the other hand, the high degree of condensation of the sperm DNA by the protamines and the intermolecular disulfide bonds of the protamines stabilize the sperm nuclei to such a degree that they can survive sonication (7) and even washing with the ionic detergent sodium dodecyl sulfate (SDS) (8).

The most popular method for isolation of nuclei from cells was developed by Blobel and Potter in 1966 (9). The basis for the isolation was the high sedimentation velocity of nuclei relative to all other components in the somatic cell. Cells were lysed, and then centrifuged through a high concentration of cold sucrose resulting in a high yield of nuclei. This method was modified for mouse sperm by adding a discontinuous sucrose gradient with three different concentrations (7). Our longstanding interest in the sperm nucleus has been the organization of DNA by the sperm nuclear matrix. We found that washing spermatozoa with 0.5% SDS was crucial for the isolation of sperm nuclear matrices from hamster (2), mouse (10), and human (11). If sperm nuclei are washed with the nonionic detergent Triton X-100 (TX100), the sperm nuclei decondense completely when the protamines are extracted. But when the sperm are washed with 0.5% SDS, salt extraction of the protamines results in nuclear matrices that retain the original shape of the sperm nucleus with DNA emanating from the matrices in loops to form a halo around the nucleus.

In this chapter, two methods of isolation for the sperm nuclear matrix are presented. The methods were originally developed for isolation of sperm nuclei and nuclear matrices from the hamster, but have since been modified for the mouse. In the first method, spermatozoa are washed with SDS immediately upon extraction (SDS Nuclei) (12). The advantage of this method is that nucleases and some proteases are denatured in the first step, possibly protecting the nuclear matrices and DNA. The disadvantage is that the SDS removes the lipid bilayer from the tails, which are not fully dissolved by SDS alone, making them even less buoyant in sucrose density gradients, and they have a tendency to contaminate the sperm head fraction. In the second method (adapted from (2)), the sperm nuclei are washed with SDS after they are separated from the sperm tails (T-SDS Nuclei).

2. Materials

2.1. SDS-Nuclei and Nuclear Matrix

1. 50 mM Tris–HCl, pH 7.4, 0.5% SDS, 5 mM $MgCl_2$ (see Note 1).
2. 2 M Sucrose, 50 mM Tris–HCl, pH 7.4, 5 mM $MgCl_2$ (see Note 2).
3. 0.075 g/ml CsCl, 2 M sucrose 25 mM Tris–HCl, pH 7.4 (see Note 3).

37 Isolation of Sperm Nuclei and Nuclear Matrices from the Mouse, and Other Rodents

2.2. T-SDS-Nuclei

4. 50 mM Tris–HCl, pH 7.4.
5. 0.45 g/ml CsCl, 25 mM Tris–HCl, pH 7.4, and 0.25% TX100 (see Note 4).
6. Sonicator fitted with microtip.

2.3. Common to Both Methods

7. 2 M NaCl, 25 mM Tris–HCl, pH 7.4.
8. 1 M dithiothreitol (DTT) stock (see Note 5).
9. DNAse I (see Note 6).
10. 1 M stock solution of $MgCl_2$.
11. Ultracentrifuge with a swing bucket rotor that holds 30–40 ml (such as a Beckman SW28).
12. 100 µg/ml ethidium bromide (EtBr) or DAPI.

3. Methods

3.1. SDS-Nuclei: Direct SDS Method for Mouse Sperm Nuclear Isolation

1. Place the ultracentrifuge rotor in the ultracentrifuge and start the vacuum and refrigeration to 4°C, as described by the manufacturer (see Note 7).
2. Place the swing buckets on ice and the centrifuge tubes that fit inside them on ice.
3. Dissect epididymides from up to ten mice. Cut them open with scissors and tease out the sperm with small forceps, putting the sperm into 20 ml of 50 mM Tris–HCl, pH 7.4, 0.5% SDS, and 5 mM $MgCl_2$ (see Note 8). Mix gently, and place the suspension on ice.
4. Set up a triple step gradient as follows in the next three steps. All solutions should be at 4°C. Put 5 ml of 0.075 g/ml CsCl, 2 M sucrose 25 mM Tris–HCl, pH 7.4, into two ultracentrifuge tubes for a swing bucket rotor with a 35 ml capacity. The centrifuge tubes should be on ice. This is most easily done with a 10 ml syringe (see Note 9).
5. Carefully overlay 10 ml of 2 M sucrose, 50 mM Tris–HCl, and 5 mM $MgCl_2$ using a 60 ml syringe to dispense the solution.
6. Finally, carefully layer the 20 ml of the sperm suspension onto one of the 10 ml sucrose cushions. This is done by tilting the tube carefully so as not to disturb the sucrose gradient and slowly pouring the sperm suspension over the sucrose. As the sperm suspension fills, the tube is slowly up-righted. The final triple gradient should have 5 ml of CsCl/sucrose solution at the bottom, with 10 ml of 2 M sucrose solution over that, and 20 ml of the sperm suspension at the top.
7. Centrifuge the triple gradient at 25,000 rpm ($113,000 \times g$) for 1.5 h at 4°C. Carefully remove the tubes from the swing bucket

holders, always keeping the tubes vertical. Occasionally some of the solution spills out over the tube creating suction. These tubes can be removed using a small hemostat.

8. Aspirate the three layers using a short glass pipette attached to a tube connected to a vacuum source with a large Erlenmeyer flask to collect the aspirate. The sperm heads should be at the bottom (see Note 10).

9. If the goal is to isolate sperm heads, stop here and resuspend the pellet in the desired solution. If the goal is to make nuclear matrices, or nuclear halos, proceed to step 10.

10. Take up the sperm heads with 1 ml of 2 M NaCl, 25 mM HCl, pH 7.4. Mix gently and avoid excessive mechanical stress, as sperm nuclei are susceptible to lysis.

11. Add 2 mM DTT and incubate at room temperature for 10–30 min. Check for nuclear halos by placing 3 µl of the suspension on a slide, and then mixing with 3 µl of 100 µg/ml EtBr in 2 M NaCl by gently pipetting up and down. Place a coverslip over the slide and examine under a fluorescent microscope for halos. Fully extracted halos have a bright halo of fluorescence surrounding a nuclear matrix that retains the shape of the sperm head (see Note 11). When halos are formed, place the suspension back on ice.

12. If the goal is to isolate nuclear halos, stop here. Sperm nuclear halos are very stable and can be centrifuged to remove the salt, but they will clump. An alternative for other procedures, such as restriction endonuclease digestion, is to dilute the sperm halos or even to dialyze the salt.

13. If the goal is to isolate sperm nuclear matrices, add DNAse I to 50 µg/ml, and $MgCl_2$ to 30 mM, and incubate at 37°C for 1 h. DNAse I will digest DNA in 2 M NaCl if the magnesium concentration is high enough. View nuclear matrices under a fluorescent microscope as in step 11 to ensure that the DNA has been digested. Nuclear matrices will appear hook shaped, as the original sperm heads, but will be very condensed because the DNA has been removed.

14. Centrifuge the nuclear matrices at 2,000 rpm ($1,327 \times g$) for 10 min. Wash twice with 2 M NaCl to remove DNAse I, and then resuspend in the desired solution.

3.2. T-SDS Nuclei: Triton X-100 then SDS Method for Mouse Sperm Nuclear Isolation

1. Set up the rotor and centrifuge tubes as in steps 1 and 2, above.

2. Dissect epididymides from up to ten mice. Cut them open with scissors and tease out the sperm with small forceps, putting the sperm into 20 ml of 50 mM Tris–HCl, pH 7.4.

3. Sonicate the sperm suspension with the mini-probe sonicator at a setting 70% of maximum on ice for 2 min. This separates the sperm heads and tails and breaks up the tails into smaller

fragments. Check the suspension with a phase microscope to ensure that the tails are separated and fragmented. Dilute the suspension with 20 ml 2 M sucrose, 50 mM Tris–HCl, and 5 mM $MgCl_2$.

4. Set up a triple step gradient as follows: Put 5 ml of 0.45 g/ml CsCl, 25 mM Tris–HCl, pH 7.4, and 0.25% NP-40 into each ultracentrifuge tube. Carefully overlay 10 ml of 2 M sucrose, 50 mM Tris–HCl, and 5 mM $MgCl_2$ using a 60 ml syringe to dispense the solution. Finally, carefully layer about 28 ml of the sonicated sperm suspension onto each of the 10 ml sucrose cushions. The final triple gradient should have 5 ml of CsCl solution at the bottom, with 10 ml of 2 M sucrose solution over that, and 25 ml of the sonicated sperm suspension at the top (see Note 12).

5. Centrifuge the triple gradient at 20,500 rpm ($75,600 \times g$) for 45 min at 4°C.

6. Aspirate the solutions with care to remove the broken tails at both interfaces (see Note 13).

7. If the goal is to isolate sperm nuclei that have not been washed in SDS, the pellet should be resuspended at the final desired solution. If the goal is to make nuclear matrices, continue to step 22.

8. Resuspend the pelleted nuclei in step 20 in 50 ml of 50 mM Tris–HCl, pH 7.4, and spin down at 2,000 rpm ($1,327 \times g$) for 10 min (this is necessary since even a small amount of CsCl will precipitate the SDS). Resuspend in 50 ml of 50 mM Tris–HCl, pH 7.4, and add 0.5% SDS. Centrifuge at 2,000 rpm ($1,327 \times g$) for 10 min.

9. If the goal is to isolate SDS washed sperm nuclei, stop here. If the goal is to isolate sperm halos or sperm nuclear matrices, continue exactly as described in steps 11–14, above.

4. Notes

1. The $MgCl_2$ is a holdover from the original Blobel and Potter nuclear isolation method (9). Magnesium condenses chromatin, and consequently the nuclei themselves, making them easier to centrifuge through sucrose. This is not required for sperm nuclei. However, another effect of magnesium is to make nuclei less adherent to each other, and this may also be important for sperm nuclei. $MgCl_2$ is optional in all solutions.

2. The 2 M sucrose solutions are difficult to make. The sucrose should be added to a beaker that is larger than the final solution (a 2 L beaker for a 1 L solution) with graduations. Then, water

should be added to about 90% of the final volume. The sucrose will take up most of the volume. This should be placed on a heated magnetic stirrer, and the suspension left to stir with the heater set on warm until sucrose is dissolved. The $MgCl_2$ and the Tris–HCl can be added from 1 M stock solutions. Then the solution can be brought up to the final volume.

3. The rationale for the CsCl was to separate the sperm heads from the tails not only by sedimentation velocity through the sucrose but by chemical density as well. The DNA in the sperm heads has a density of 1.7 g/ml compared to that of an average protein which is about 1.3 g/ml. The actual concentrations of CsCl in both solutions in these methods were determined by trial and error.

4. As in Note 3, the rationale for the CsCl was to separate sperm heads based on the density of the DNA they contain. In this solution, the TX100 removes the lipid membrane from the sperm heads as they enter this lowest portion of the step gradient (see Subheading 3), increasing the total density of the sperm heads. Adding the TX100 to this step, rather than an earlier step, allows the tails to retain their lipid membranes, reducing their sedimentation velocity through sucrose.

5. The 1 M DTT stock solution should be kept in small aliquots in a freezer. Aliquots can be thawed several times, but excessive freeze/thaw cycles result in oxidation.

6. There are two ways to make DNAse I stocks. The first is to buy a bovine pancreatic DNAse I in solution, and store it as recommended. A less expensive method is to purchase crystallized DNAse I and make a stock solution of 1 mg/ml in 10 mM Tris–HCl, pH 7.4, freeze it in small aliquots, and then thaw the aliquots as needed. The final concentration of DNase I required for the last digestion step is 10 mg/ml.

7. Most ultracentrifuges require that the vacuum pump be started and the temperature set before placing the rotor with the material to be centrifuged inside. The centrifugation will be run at 4°C.

8. For the mouse, the total amount of sperm nuclei can be increased if one also includes the sperm in the vas deferens. We have found that nucleases exist in the vas deferens fluid that may digest sperm DNA (13), but these are washed away by the procedures that are described (unpublished observations). Make sure to minimize mechanical stress by mixing the epididymal extracts gently with the 50 mM Tris–HCl solution at the start. The SDS will separate the heads from the tails, and it is a good idea to make sure that this has occurred by examining the suspension by phase microscopy. For the hamster, it is possible to use too many sperm. If so, all the heads will be trapped by the tails at the interface. If this happens, repeat with fewer sperm.

9. All sucrose solutions in this method should be put into the centrifuge tubes with a syringe rather than a pipette. The solutions are extremely viscous, especially when kept cold, and dispensing them with a syringe is the only method for doing this.

10. This aspiration is delicate. Most of the sperm tails will be caught at the interface between the first (50 mM Tris–HCl) and second (2.0 M Sucrose) solutions. Try to aspirate this interface first to avoid contamination of the sperm heads with tails. Next, try to aspirate the top solution. As you approach the bottom, you may want to leave a small amount of solution so that you do not aspirate the heads. One can often see the nuclear pellet, and if it is loose stop the aspiration before you get to the bottom.

11. If the halos are not very large, continue the incubation. If the halos never become large, you can try incubation at 37°C for 5 or 10 min, although this is not usually necessary for mice. Hamster sperm nuclei do require a more elevated temperature. Note that this solution should NOT be viscous even when the halos are fully formed. If the suspension is viscous, it means that the sperm heads have completely decondensed and the nuclear matrices were not stable. Occasionally the sperm heads decondense and lose their original shape usually by the apical ends expanding. This means that the spermatozoa were treated too roughly at the beginning and the preparation is not very good. Less mechanical disruption at the beginning will help with this problem. Also, note that if EtBr is used to visualize the halos, they should appear very small when first seen but grow larger in the 10–30 s while examining the sperm halos. This is because EtBr supercoils the DNA loops making them small, but in the presence of UV light, EtBr catalyzes single-strand breaks in the DNA allowing them to uncoil.

12. The CsCl solution is not viscous at all, and it is sometimes not easy to pour the sucrose onto the top of this solution. An alternative method I have used is to put the 2.0 M sucrose solution in first, and then inject the CsCl solution with a syringe that has a small gauge tube attached to the opening underneath the sucrose solution. With either method, the final sperm suspension can be poured on top by gently tilting the tube because the viscous sucrose helps keep the gradient stable.

13. This aspiration is very different from the one in step 8 and Note 9. The top and bottom layers are not viscous, and aspirate quickly. The interfaces should be aspirated first, as in Note 9. The pelleted nuclei in this method are much looser, and care must be taken not to aspirate them.

Acknowledgment

This work was supported by NIH Grant HD060722.

References

1. Balhorn R et al (1977) Mouse sperm chromatin proteins: quantitative isolation and partial characterization. Biochemistry 16:4074–4080
2. Ward WS et al (1989) DNA loop domains in mammalian spermatozoa. Chromosoma 98:153–159
3. Choudhary SK et al (1995) A haploid expressed gene cluster exists as a single chromatin domain in human sperm. J Biol Chem 270:8755–8762
4. Surani MA et al (1984) Development of reconstituted mouse eggs suggests imprinting of the genome during gametogenesis. Nature 308:548–550
5. van der Heijden GW et al (2008) Sperm-derived histones contribute to zygotic chromatin in humans. BMC Dev Biol 8:34
6. Yanagimachi R (1994) Mammalian fertilization. In: Knobil E, Neill JD (eds) The physiology of reproduction, 2nd edn. Raven, New York, pp 189–317
7. Mayer JF Jr et al (1981) Spermatogenesis in the mouse 2. Amino acid incorporation into basic nucleoproteins of mouse spermatids and spermatozoa. Biol Reprod 25:1041–1051
8. Kvist U (1980) Importance of spermatozoal zinc as temporary inhibitor of sperm nuclear chromatin decondensation ability in man. Acta Physiol Scand 109:79–84
9. Blobel G, Potter VR (1966) Nuclei from rat liver: isolation method that combines purity with high yield. Science 154:1662–1665
10. Mohar I et al (2002) Sperm nuclear halos can transform into normal chromosomes after injection into oocytes. Mol Reprod Dev 62:416–420
11. Barone JG et al (1994) DNA organization in human spermatozoa. J Androl 15:139–144
12. Ward WS, Coffey DS (1989) Identification of a sperm nuclear annulus: a sperm DNA anchor. Biol Reprod 41:361–370
13. Boaz SM et al (2008) Mouse spermatozoa contain a nuclease that is activated by pretreatment with EGTA and subsequent calcium incubation. J Cell Biochem 103:1636–1645

Chapter 38

Protamine Extraction and Analysis of Human Sperm Protamine 1/Protamine 2 Ratio Using Acid Gel Electrophoresis

Lihua Liu, Kenneth I. Aston, and Douglas T. Carrell

Abstract

Protamines, sperm-specific nuclear proteins, are essential for sperm chromatin condensation and DNA stabilization. They are small, highly basic, and rich in disulfide bonds. Under reducing conditions, protamines, along with other basic proteins, are soluble in acid solutions. Because of their small and similar molecular weights, SDS-PAGE cannot resolve protamine 1 and protamine 2 well. Urea-acid gel electrophoresis separates proteins based on the level of the positive charge and is thus a suitable method for resolving protamines 1 and 2. Here, we describe the commonly used protamine extraction method and the Urea-acid gel electrophoresis for assessment of protamine 1/protamine 2 ratio.

Key words: P1/P2 ratio, Protamine, Chromatin, Nuclear proteins

1. Introduction

Protamines are small, basic sperm-specific nuclear proteins that are rich in arginine and cysteine. Human protamines include protamine 1 and protamine 2. The latter can be subdivided into protamine 2, 3, and 4, which vary by minor changes at the N-terminus. Protamines 2, 3, and 4 are 57, 54, and 58 amino acids in length, respectively, and protamine 1 has 51 amino acids. For all the human protamines, their arginine ratios are similar at about 48%, but protamine 1 has a higher ratio of cysteine (12%) than protamine 2 (9%) (1, 2).

Under reducing conditions, protamines, along with other basic proteins, are soluble in acid solutions. Acid extraction under reducing condition is the most commonly used method for protamine extraction.

Because of their similarities in size and amino acid content, it is difficult to separate protamines 1 and 2 with SDS-PAGE based

Douglas T. Carrell and Kenneth I. Aston (eds.), *Spermatogenesis: Methods and Protocols*, Methods in Molecular Biology, vol. 927, DOI 10.1007/978-1-62703-038-0_38, © Springer Science+Business Media, LLC 2013

solely on molecular weight. Acid-Urea gel separates proteins based both on molecular weight and cumulative charge. Since both protamines are highly positively charged, they are easily separated from other proteins. Protamine 2 is slightly more positive than protamine 1, as it contains more lysine and histidine; however protamine 1 contains one more cysteine than protamine 2. As a result of this cysteine content difference, a step to induce an alkylation modification of the cysteine is requisite to yield an effective separation of protamine 1 and protamine 2 on acid-urea gel electrophoresis.

2. Materials

Prepare all solutions using ultrapure water (Milli-Q UF Plus). Store reagents following manufacturers' instructions. Unless otherwise noted, all the procedures should be processed at room temperature. Carefully follow all waste disposal regulations.

2.1. Protamine Extraction

1. 1 mM phenylmethanesulfonylfluoride (PMSF) in H$_2$O (Sigma) (see Note 1).
2. 20 mM ethylenediaminetetraacetic acid (EDTA), 1 mM PMSF, 100 mM Tris–HCl (pH 8.0) (see Note 1).
3. 6 M guanidine and 575 mM dithiothreitol (DTT): 5.73 g guanidine, 0.886 g DTT in 10 ml H$_2$O (see Notes 2 and 3).
4. 522 mM sodium iodoacetate: 1.147 g in 10 ml H$_2$O (see Note 4).
5. 100% Ethanol.
6. 0.5 M HCl.
7. 100% TCA: 100 g TCA in 100 ml water (w/v).
8. 1% 2-mercaptoethanol in acetone (see Note 5).
9. Microcentrifuge.
10. 1.5 ml or 2.0 ml Eppendorf tubes.

2.2. Acid-Urea Polyacrylamide Gel Electrophoresis

1. 40% acrylamide/bis solution, 19:1 from supplier (BIO-RAD) (see Note 6).
2. 30% acrylamide/bis solution, 37.5:1 from supplier (ProtoGel; National Diagnostics).
3. 5 M Urea and 6.67M Urea.
4. Ammonium persulfate (AP) (see Note 7) (Pharmacia Biotech).
5. Tetramethylethylenediamine (TEMED) (Pharmacia Biotech).
6. 43.1 ml glacial acetic acid in 56.9 ml water.
7. 3 M potassium acetate (pH 4.0) titered with Acetic acid.

8. Sample loading buffer: 0.375 M potassium acetate, pH 4.0, 15% sucrose, 0.05% Pyronin Y.
9. 40 ml water-saturated N-Butanol (see Note 8).
10. Electrophoresis running buffer: 0.9 N acetic acid.
11. Electrophoresis apparatus: Multiphor II Electrophoresis System (see Note 9).
12. Coomassie blue solution: Dissolve 2.5 g brilliant blue R250 in 450 ml of methanol, then add 450 ml H_2O and 100 ml acetic acid.
13. Destain solution: 10% acetic acid, 30% methanol, and 60% H_2O.

3. Methods

Carry out all procedures at room temperature unless otherwise noted. As a quality control measure, an aliquot of the same-pooled semen should be used for each protamine extraction and urea-acid gel electrophoresis.

3.1. Protamine Extraction (3)

1. Use the proper amount of semen, which depends on the sperm concentration. $10-20 \times 10^6$ sperm are required for the assay, with 20×10^6 being preferable. An extraction control should also be performed at the same time as extraction of experimental samples (see Note 10).
2. Centrifuge semen at $4,000 \times g$ for 5 min, and remove the supernatant.
3. Resuspend sperm in 1 mM PMSF in H_2O, centrifuge at $4,000 \times g$ for 5 min, and remove the supernatant.
4. Resuspend sperm in 100 μl of 20 mM EDTA, 1 mM PMSF, and 100 mM Tris–HCL (pH 8.0). Vortex to mix.
5. Add 100 μl of 6 M guanidine and 575 mM DTT to the mix above, and vortex briefly.
6. Add 200 μl of 522 mM sodium iodoacetate to the mixture, and vortex briefly.
7. Incubate the mixture in the dark at room temperature for 30 min.
8. Add 1 ml of ice-cold Ethanol, mix, and incubate at −20°C for 1 min. Centrifuge at $14,000 \times g$ for 8 min. Save the precipitate. Repeat Ethanol wash once.
9. Resuspend the precipitate in 0.8 ml of 0.5 M HCl with vortex. Incubate at 37°C for 10 min.

10. Centrifuge at 10,000×g for 10 min, and transfer the supernatant to another Eppendorf tube.
11. To the supernatant, add 200 µl of TCA (100%), and mix it by vortexing. Hold at –20°C for 3 min. Centrifuge at 14,000×g for 10 min. Carefully remove the supernatant taking care not to disturb the precipitate. Sometimes, the precipitate is not visible.
12. Add 1 ml of 1% 2-mercaptoethanol in acetone to the precipitate, and vigorously vortex. Following vortexing precipitate can generally be seen. Centrifuge at 14,000×g for 10 min, and remove as much supernatant as possible, being careful not to disturb the pellet. Perform this procedure twice.
13. Allow samples to dry overnight under a fume hood with caps open. Samples are ready to be applied to an acid-urea polyacrylamide gel.

3.2. Acid-Urea Polyacrylamide Gel Electrophoresis

3.2.1. Gel Preparation (4)

1. Separation gel: Mix 8 ml 40% acrylamide with 6 ml 6.67 M urea, and degas with vacuum. Add 2 ml of 1.6% AP in 43.1% acetic acid (freshly prepared) and 70 µl TEMED to above, and gently mix being careful to avoid producing air bubble in the mixture. Load the gel to the gel cassette with a syringe and needle (see Note 11). Cover the top of the gel with the water-saturated N-butenol. Prepare the stacking Gel after the separation gel has polymerized. Remove the N-butanol, and wash with water once.
2. Stacking gel: Mix 2 ml 30% acrylamide, 4 ml 5 M urea, and 1 ml 3 M potassium acetate (pH 4.0), and then degas with vacuum. Then add 1 ml of 1.6% of AP in H_2O and 70 µl TEMED, and mix without creating bubbles. Load the gel on the top of the separation gel.

3.2.2. Sample Preparation

Resuspend the dried protamine-extraction pellet in 40 µl of sample loading buffer. Briefly spin to pellet down the insoluble fraction.

3.2.3. Electrophoresis (See Note 9)

1. Fill both of the buffer chambers of the electrophoresis apparatus with 1 L of electrophoresis running buffer each.
2. Place the polymerized gel into the electrophoresis apparatus.
3. Program the electrophoresis with the following parameters:

Pre-run (prior to loading the samples):	200 V	30 mA	15 W	100 VH
Slow-run (after loading samples):	50 V	20 mA	15 W	100 VH
Separation run:	200 V	300 mA	15 W	800 VH

4. Initiate the pre-run.

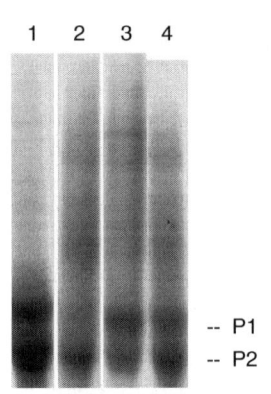

Fig. 1. Image of a urea-acid gel containing protamine extraction samples. Lane 1 shows the extraction control sample. Lane 2 shows a sample with a decreased P1/P2 ratio. Lane 3 shows a sample with an increased P1/P2 ratio. Lane 4 shows a sample with a normal P1/P2 ratio. *P1* protamine 1. *P2* protamine 2.

5. Pause the program, and load the protamine-samples to the gel after completing the pre-run.
6. Continue the program until the run is complete.

3.2.4. Coomassie Blue Stain Gel

Stain the gel in the Coomassie blue solution for at least 1 h, and then destain in the destain solution until the background of the gel is clear (see Fig. 1).

3.2.5. Human Protamine P1/P2 Ratio Analysis

Scan the gel (wet) with any reasonable scanner. Protamine 2 migrates faster than protamine 1. Measure the densities of the protamine bands with the program Image J, which can be downloaded from NIH Web site. Calculate the protamine 1/protamine 2 ratio based on the densities of P1 and P2 bands.

4. Notes

1. The reagents containing PMSF need to be freshly prepared. PMSF serves to inhibit proteinase activity.
2. 6 M guanidine and 575 mM DTT should be aliquoted and stored at −20°C.
3. Any reagents that contain DTT or 2-mercaptoethanol should be processed in a fume hood.
4. 522 mM sodium iodoacetate should be stored at 4°C and protected from light.
5. 2-mercaptoethanol should be added to acetone freshly.
6. 40% acrylamide/Bis solution should be stored at 4°C. Each time before using, the solution should be shaken to mix. Otherwise the concentration of acrylamide may be inaccurate.

7. Ammonium persulfate solution needs to be prepared right before use.
8. To 40 ml N-butanol, add 5 ml ddH$_2$O, and shake vigorously to mix. Let bottle sit for several minutes, and unabsorbed water will settle to the bottom. The top layer will be water-saturated N-butanol. This is performed to prevent N-butanol from drawing water from the gel when used as an overlay. Only two to three drops of the N-butanol are required for each gel.
9. Other electrophoresis systems are also suitable for acid-urea PAGE, but the electrophoresis conditions must be optimized for the system used.
10. Samples utilized for the extraction control are composed of a pool of >10 semen samples and serve to evaluate the quality of sample extractions. The pooled semen sample can be aliquoted into tubes of 20×10^6 sperm each. One tube should be extracted and run on the gel each time the assay is performed.
11. Aspirate the gel mix with the syringe without a needle to avoid drawing air bubbles, and then affix the needle to load the gel mix into the cassette.

References

1. Gusse M et al (1986) Purification and characterization of nuclear basic proteins of human sperm. Biochim Biophys Acta 884:124–134
2. McKay DJ et al (1986) Human sperm protamines. Amino-acid sequences of two forms of protamine P2. Eur J Biochem 156:5–8
3. de Yebra L, Oliva R (1993) Rapid analysis of mammalian sperm nuclear proteins. Anal Biochem 209:201–203
4. Spiker S (1980) A modification of the acetic acid-urea system for use in microslab polyacrylamide gel electrophoresis. Anal Biochem 108:263–265

Chapter 39

Analysis of Gene-Specific and Genome-Wide Sperm DNA Methylation

Saher Sue Hammoud, Bradley R. Cairns, and Douglas T. Carrell

Abstract

Epigenetic modifications on the DNA sequence (DNA methylation) or on chromatin-associated proteins (i.e., histones) comprise the "cellular epigenome"; together these modifications play an important role in the regulation of gene expression. Unlike the genome, the epigenome is highly variable between cells and is dynamic and plastic in response to cellular stress and environmental cues. The role of the epigenome, specifically, the methylome has been increasingly highlighted and has been implicated in many cellular and developmental processes such as embryonic reprogramming, cellular differentiation, imprinting, X chromosome inactivation, genomic stability, and complex diseases such as cancer. Over the past decade several methods have been developed and applied to characterize DNA methylation at gene-specific loci (using either traditional bisulfite sequencing or pyrosequencing) or its genome-wide distribution (microarray analysis following methylated DNA immunoprecipitation (MeDIP-chip), analysis by sequencing (MeDIP-seq), reduced representation bisulfite sequencing (RRBS), or shotgun bisulfite sequencing). This chapter reviews traditional bisulfite sequencing and shotgun bisulfite sequencing approaches, with a greater emphasis on shotgun bisulfite sequencing methods and data analysis.

Key words: DNA methylation, Bisulfite sequencing, Sperm, Shotgun bisulfite sequencing, MeDIP

1. Introduction

DNA methylation in mammals, established by DNA methyltransferases (also known as DNMTs), is a post-replication modification that is predominantly found on cytosines of the CpG dinucleotide sequence (1, 2). In mammals, CpG methylation has been shown to be a primary chromatin attribute associated with the main mode of repression at transposons and retroviral elements (3–7), essential for development (1), genomic imprinting, and X-chromosome inactivation (8–10).

Although it was once thought that DNA methylation in eukaryotes could act as a stably inherited modification affecting gene regulation and cellular differentiation (11), a number of reports suggested otherwise. DNA methylation profiles are more dynamic than previously anticipated. (1) DNA methylation profiles are reset twice during development (Fertilization and PGCs); however, the extent of reprogramming is unclear (12). (2) More recently genetic and biochemical data in plants and mammals indicate that genomic DNA methylation patterns in differentiated cells can be reshaped in part by active demethylation mediated by a family of 5-methylcytosine (m^{5}C) glycosylases (13–15). These data indicate the existence of a more dynamic cellular methylome.

The evolution of next-generation sequencing technologies has provided unprecedented opportunities for high-throughput functional genomic research and new avenues to explore the methylation dynamics during development, germ cell differentiation, cellular differentiation, somatic cell reprogramming, and cancer. The four most frequently used sequencing-based technologies are the bisulfite-based methods MethylC-seq and reduced representation bisulfite sequencing (RRBS); and the enrichment-based techniques methylated DNA immunoprecipitation sequencing (MeDIP-seq) and methylated DNA binding domain sequencing (MBD-seq). Here, we review both tested and optimized bisulfite-based methods for human and mouse sperm. Bisulfite treatment of genomic DNA deaminates unmethylated cytosine, causing its chemical conversion to uracil upon alkaline desulfonation, but does not modify methylated cytosines. After PCR and sequencing of DNA amplicons or library preparations, the methylation status of every m^{5}C is detected in a targeted region of interest (16, 17) or genome-wide (18–21).

2. Materials

2.1. Candidate Bisulfite Sequencing

2.1.1. Kits

1. Several bisulfite kits are commercially available, however, we use the EpiTect Bisulfite kit (Qiagen).
2. Zero Blunt® TOPO® PCR Cloning Kit.
3. QIAGEN Plasmid Mini Kit.

2.1.2. PCR Reagents

1. PFU 10× Buffer.
2. 10 mM DNTPs.
3. 10 µM stock Forward Primer.
4. 10 µM stock Reverse Primer.
5. PFU Taq DNA polymerase.
6. Bisulfite-treated genomic DNA.
7. DNAse-free RNAse-free H$_2$O.

2.1.3. Cloning Solutions	All solutions are provided in the Invitrogen kit except for the Laurel Broth Growth-promoting media and plates, antibiotic selection marker and X-GAL for blue and white colony screening (Kanomyocin final concentration 50 μg/ml, X-GAL 25 μg/ml).	
2.2. Shotgun Bisulfite Sequencing	DNA methylation analysis at base pair resolution via shotgun sequencing in human sperm has not been published, but has been successfully performed by our lab (unpublished data). Here we provide our latest protocol.	

1. Epitect Bisulfite Kit.
2. Sonicator (preferred is Covaris due to high reproducibility).
3. Illumina TruSeq DNA Sample Preparation kit.
4. MiniElute PCR purification kit (Qiagen) or Agencourt AMPure XP beads.
5. Illumina TruSeq DNA Sample Preparation kit.
6. Illumina HiSeq2000 machine.

3. Methods

3.1. Candidate Bisulfite Sequencing

1. Extract DNA from sperm.
2. Up to 2 μg genomic may be used for bisulfite treatment in the Epitect bisulfite kit. Carry out bisulfite conversion as described with minor changes in the PCR conditions (Table 1; see Note 1).

Table 1
Modified PCR conditions

Denaturation	98°C	10 min
Incubation	60°C	25 min
Denaturation	98°C	10 min
Incubation	60°C	85 min
Denaturation	98°C	10 min
Incubation	60°C	175 min
Denaturation	98°C	10 min
Incubation-added step necessary to achieve >99% conversion	60°C	120 min
Hold	20°C	Indefinite

3. After bisulfite treatment clean samples following kit protocols, with a minor exception: perform two 30 μl elutions rather than one 50 μl elution.

4. After DNA clean-up, samples are ready for PCR. PCR conditions will vary depending on the primers and template. Typically our primers are 27–30 bp long with melting temperatures of 57–60°C. CpGs are usually avoided in the primer region but if necessary substitute CG for YG in order to avoid methylation amplification bias. With this primer design strategy, we have had >80% success using the following PCR reaction mix and conditions.

5. Prepare PCR mix with the following components:
   5 μl of PFU 10× Buffer.
   1 μl of 10 mM DNTPs.
   2.5 μl of 10 μM stock Forward Primer.
   2.5 μl of 10 μM stock Reverse Primer.
   0.5 μl of PFU Taq DNA polymerase.
   1–2 μl of bisulfite treated genomic DNA.
   36.5–37.5 μl of DNAse-free RNAse-free H_2O.

6. Perform PCR under the following conditions (see Note 2):

Cycles	Temperature (°C)	Time
1	94	4 min
35	94	45 s
	55	45 s (~3°C below primer temp)
	72	45 s
1	72	10 min
1	4	Final hold

7. After amplification run 5 μl of the PCR product on a 1% agarose gel. If a single product is obtained, then the sample can be cloned directly. If multiple products are present or a significant quantity of primer dimer is detected, then we recommend gel-purifying the band of interest.

8. Cloning can be performed using traditional restriction enzyme digestion techniques and T4 ligations, however we use topoisomerase blunt TA cloning kits. To minimize cost 3 μl reactions can be performed.

9. Transfect all 3 μl of the cloning reaction into Top 10 chemically competent cells (made in house) or commercially available with the cloning kit.

10. Heat shock cells for 45 s and set on ice for 1 min.

11. Allow cells to recover in 400 μl SOC (provided in the TA cloning kit) for 1 h at 37°C prior to plating.

12. Plate cells on LB agar plates with antibiotic selection marker and X-Gal (25 μg/ml) and incubate overnight (12–16 h). If using the Zero Blunt TOPO kit blue and white screening is not necessary.

13. The following morning, 10–20 colonies are screened using Colony PCR: Suspend each colony in 100 μl LB + antibiotic. Then use 1.5 μl of the suspension in a 50 μl PCR reaction. The PCR reaction is the same as above, except the initial denaturation is longer (94°C for 10 min). The resuspended bacterial cells remain viable for approximately 1 week at 4°C.

14. Reculture cells containing the correct insert in 5 ml LB + Antibiotic (for high copy plasmids) at 37°C rotating overnight. If using low copy plasmids larger cultures maybe needed to obtain sufficient DNA.

15. Perform plasmid purification using the QIAGEN Plasmid Mini Kit.

16. Submit 10–20 colonies (or more if deemed appropriate for a particular study) for sequencing with a commercial or core facility.

3.2. Shotgun Bisulfite Sequencing

1. Spike 3 μg of genomic DNA with 1% unmethylated lambda genomic DNA (~30 ng) (promega). Lambda is used as internal control to determine conversion efficiency and later also used by bioinformatics packages to estimate % mC false discovery rate (see Note 3).

2. Shear the genomic DNA with a sonicator. We have used Covaris (Covaris, Inc., Woburn, MA) and perform shearing according to their published protocols, which yields DNA fragments ranging from 100 to 400 bp with a peak at 200 bp.

3. After the DNA is sheared, clean using the MiniElute PCR purification kit (Qiagen) and measure the DNA size range using a bioanalyzer.

4. Repair DNA ends and add an adenosine base to the 3′ end of the DNA fragments according to manufacturer instructions (Illumina, Inc.).

5. After end repair, ligate methylated adaptors to DNA fragments using the TruSeq DNA Sample Preparation Kit according to manufacturer instructions and purify DNA using the QIAquick PCR purification Kit (see Note 4).

6. At this point DNA is ready for bisulfite conversion. For bisulfite conversion, we use and follow the EpiTect Bisulphite Kit (Qiagen) with a total of 1 μg of sheared genomic DNA with slight modification. The thermal cycler conditions for all DNA denaturation steps should be adjusted to 98°C for 10 min, and

Table 2
Incubation conditions for bisulfite conversion

Denaturation	98°C	10 min
Incubation	60°C	25 min
Denaturation	98°C	10 min
Incubation	60°C	85 min
Denaturation	98°C	10 min
Incubation	60°C	175 min
Denaturation	98°C	10 min
Incubation-added step necessary to achieve >99% conversion	60°C	120 min
Hold	20°C	Indefinite

an additional 2 h of incubation at 60°C should be added at the end of the procedure to ensure complete conversion (Table 2).

7. Subsequently, apply high-fidelity Phusion DNA polymerase for strand replication of all bisulfite-converted DNA according to instructions in the Illumina TruSeq kit. (~10 PCR cycles).

8. Purify samples with a PCR purification kit (Qiagen), then separate DNA fragments on a 1% agarose gel or with the Pippin Prep DNA Size Selection System. Excise products in the 300–400 bp size range.

9. Sequence the resulting library of bisulfite-converted DNA fragments with an Illumina HiSeq2000 machine using the 101-base paired end format for ease of mapping and optimal read numbers.

3.3. Shotgun Bisulfite Sequencing Analysis

Currently, there is not a single established pipeline for the analysis of shotgun bisulfite sequence data; however, we have developed our own pipeline. All of the software we are using in processing and analyzing the bisulfite alignment data is open source and freely available from the USeq project site (http://useq.sourceforge.net).

1. First, aligh the Fastq Illumina sequencing data using Novocraft's novoalign aligner (http://www.novocraft.com) in bisulfite mode with the following prameters, "-rRandom -t120 -h120 -b2." Add a chrLambda sequence to the index. This enables one to measure the bisulfite conversion efficiency. Add a chrPhiX sequence to estimate the read quality. Lastly, add a chrAdapter sequence containing all permutations of the adapter sequences to remove these artifacts from the data.

2. Process the novoalignments using NovoalignbisulfiteParser (http://useq.sourceforge.net/cmdLnMenus.html#NovoalignBisulfiteParser). This application parses the text-based novoalignments into "PointData": converted Cs (Ts—non-methylCs) and non-converted Cs (methylCs) at each reference C sequenced in the genome for both the plus and the minus strands.

3. The Point Data can be used to assess methylation at genes, Cpg islands, etc. Details for programs available and usage guides for USEQ and methylation can be found at (http://useq.sourceforge.net/cmdLnMenus.html).

4. Notes

1. Although the Epitect Bisulfite Kit suggests up to 2 μg of DNA, we typically use 1 μg. Using too much DNA will result in incomplete bisulfite conversion. One way to QC bisulfite conversion efficiency is to spike the DNA samples with unmethylated lambda DNA (Promega), then treat DNA with sodium bisulfite, use lambda primers to assess bisulfite conversion. Alternatively, if working with mammalian cells, then imprinted loci can be evaluated to estimate conversion efficiency.

2. Extension times for amplifying deaminated DNA are longer because of the presence of uracil, which decreases the rate of DNA polymerization.

3. Lambda reads are parsed from the dataset and examined for C to T conversion rates. Following the PCR modifications discussed here with the Epitect Bisulfite Kit consistently and reproducibly results in >99% conversion with minimal degradation/fragmentation of DNA.

4. The methylated cytosines in the adaptors resist conversion in the subsequent bisulfite step.

References

1. Okano M et al (1999) DNA methyltransferases Dnmt3a and Dnmt3b are essential for de novo methylation and mammalian development. Cell 99:247–257
2. Li E et al (1992) Targeted mutation of the DNA methyltransferase gene results in embryonic lethality. Cell 69:915–926
3. Bird AP, Wolffe AP (1999) Methylation-induced repression–belts, braces, and chromatin. Cell 99:451–454
4. Walsh CP et al (1998) Transcription of IAP endogenous retroviruses is constrained by cytosine methylation. Nat Genet 20:116–117
5. Dodge JE et al (2005) Inactivation of Dnmt3b in mouse embryonic fibroblasts results in DNA hypomethylation, chromosomal instability, and spontaneous immortalization. J Biol Chem 280:17986–17991
6. Karpf AR, Matsui S (2005) Genetic disruption of cytosine DNA methyltransferase enzymes

induces chromosomal instability in human cancer cells. Cancer Res 65:8635–8639
7. Chen T, Li E (2004) Structure and function of eukaryotic DNA methyltransferases. Curr Top Dev Biol 60:55–89
8. Jaenisch R, Jahner D (1984) Methylation, expression and chromosomal position of genes in mammals. Biochim Biophys Acta 782:1–9
9. Surani MA (1998) Imprinting and the initiation of gene silencing in the germ line. Cell 93:309–312
10. Ng HH, Bird A (1999) DNA methylation and chromatin modification. Curr Opin Genet Dev 9:158–163
11. Holliday R, Pugh JE (1975) DNA modification mechanisms and gene activity during development. Science 187:226–232
12. Reik W et al (2003) Mammalian epigenomics: reprogramming the genome for development and therapy. Theriogenology 59:21–32
13. Metivier R et al (2008) Cyclical DNA methylation of a transcriptionally active promoter. Nature 452:45–50
14. Gehring M et al (2006) DEMETER DNA glycosylase establishes MEDEA polycomb gene self-imprinting by allele-specific demethylation. Cell 124:495–506
15. Kangaspeska S et al (2008) Transient cyclical methylation of promoter DNA. Nature 452:112–115
16. Frommer M et al (1992) A genomic sequencing protocol that yields a positive display of 5-methylcytosine residues in individual DNA strands. Proc Natl Acad Sci U S A 89:1827–1831
17. Clark SJ et al (1994) High sensitivity mapping of methylated cytosines. Nucleic Acids Res 22:2990–2997
18. Lister R et al (2008) Highly integrated single-base resolution maps of the epigenome in Arabidopsis. Cell 133:523–536
19. Lister R et al (2009) Human DNA methylomes at base resolution show widespread epigenomic differences. Nature 462:315–322
20. Cokus SJ et al (2008) Shotgun bisulphite sequencing of the Arabidopsis genome reveals DNA methylation patterning. Nature 452:215–219
21. Laurent L et al (2010) Dynamic changes in the human methylome during differentiation. Genome Res 20:320–331

Chapter 40

Evaluating the Localization and DNA Binding Complexity of Histones in Mature Sperm

David Miller and Agnieszka Paradowska

Abstract

The paternal genome in many animal taxa is efficiently packaged into the sperm nucleus by switching from a histone (nucleosome)-based chromatin configuration to one using predominantly protamines. Nonetheless, various studies have shown that some nucleosomes, often containing modified histones are retained in mature sperm and bind DNA with distinct sequence compositions. Considering the significance of histone modifications in epigenetic phenomena and the fact that sperm histones and their bound DNA must be carried into the oocyte, this chapter describes methods aimed at examining and analysing the histone composition of sperm chromatin. The focus is on both microscopic visualisation and evaluation of sequence composition of histones and histone-bound DNA in human and mouse spermatozoa. However, similar methods may be applicable to the sperm of other mammalian and even non-mammalian classes.

Key words: Sperm chromatin, Histones, Immunocytochemistry, Endonuclease digestion, ChIP-chip

1. Introduction

As a terminally differentiated cell, the spermatozoon is exquisitely specialised to deliver the paternal genome to the egg. Its chromatin is at least tenfold more condensed than that found in most somatic cells and in many species this exceptional level of compaction is achieved by substituting histones with protamines during the later, haploid stages of spermatogenesis (1, 2). In mammals studied to date, substitution begins with the incorporation of sperm-specific histones including H3.3 into nucleosomes during spermatogenesis (3). Acetylation of canonical H4 is one histone modification that follows until the general displacement of histones by transition proteins occurs and finally their displacement by protamines (4, 5). Nonetheless, it has long been known that histones and nucleosomes are not completely absent from the sperm nucleus (6–8). More recently, we among others have begun to reveal what histones are present in the sperm nucleus, where they might be located, what modifications they carry (methylation, acetylation, etc.), and what DNA sequences they associate with (7, 9–11). The data from such studies suggest that sperm histones are there by design and not by accident. They paint a rather different and far more elaborate picture of the paternal genome than is currently in vogue. The most recent thinking is that paternally derived histones and the way the DNA is differentially packaged in the sperm nucleus conveys important epigenetic information to the zygote that can affect subsequent development. By studying these phenomena more closely, we shall begin to understand what those effects may be.

This chapter describes approaches for the study of sperm chromatin on two levels. The first deals with the microscopic localisation of histones in sperm by immunocytochemistry and the second with the isolation and analysis of histone (and protamine)-bound DNA.

1.1. Visualisation of Sperm Histones

Sperm chromatin is so highly condensed that it is often impossible for antibodies to gain access to proteins that may lie deeply buried within the compacted chromatin framework. It may therefore be necessary to carry out a chromatin decondensation step before applying any antibodies; otherwise false negatives may result. A number of protocols are available for achieving this, although we use the procedure described by Ramos (12) for human sperm (see decondensation Subheadings 2.2 and 3.2). Before beginning work, it is important to ensure that all solutions and working reagents are fresh and that reagents are of the highest grade possible. Ultrapure, deionised water (18 MΩ or greater) should be used throughout, and if the integrity of RNA is important, the water should be treated with DEPC. The protocol described uses Tris-buffered solutions but others have used phosphate-buffered solutions with equal success.

1.2. Selective and Differential Recovery of Sperm Chromatin for Histone Localization Experiments

For simplicity's sake, we refer to nucleosomes rather histones hereafter because nucleosomes are what package DNA in the absence or alongside protamines. Reference will be made to their constituent histones where applicable.

There are two principle choices available for examining the DNA sequences binding to sperm nucleosomes. The older, classical method is adapted from nuclease or salt-based methods (or combinations thereof) developed originally to probe the nucleus for clues about how chromatin configuration responds to differentiation, gene expression, and silencing (13, 14). It is predicated on the assumption that these biological changes are dynamic and require active chromatin reconfiguration into more relaxed domains that facilitates access to salts and digestive enzymes. The sperm nucleus is essentially in a "frozen" state where transcription of the haploid genome is highly unlikely. Instead, sperm chromatin seems to be in a poised state that may be a preparation for activation events that can only ensue following its introduction to the öoplasm (15). Nevertheless, the structure and organisation of sperm chromatin is as good a template for these classical techniques as any somatic cell nucleus. The advantages and disadvantages of these techniques over the newer ChIP-based procedures are discussed below.

Chromatin ImmunoPrecipitation (ChIP) is a relatively recent technique for accessing specific DNA binding sites in nuclei. Its specificity can be highly discriminating, depending almost entirely on the exquisitely selective recognition of particular target. ChIP is becoming an indispensible tool in epigenetics where characterising the posttranslational modifications to nucleosomal histones is so important (16). For example, acetylation/deactylation and methylation/demethylation of lysine residues on H3 and H4 are of particular importance in the condensation and relaxation of chromatin in response to the demands of gene expression, and the body of information relating to ChIP-derived DNA and RNA sequences in numerous cells and tissues is set to explode in the coming years. So it is perhaps not surprising that such investigations should be undertaken on sperm chromatin. Yet, histones modified with methyl and acetyl groups are also found in mammalian spermatozoa and must be transmitted to the öocyte on fertilisation (6, 17). The absence of gene expression in mature spermatozoa indicates an alternative function for and retention of their modified histones.

In comparison with classical techniques, ChIP is far more target and gives a much narrower "view" of DNA binding sites. High specificity is its main advantage, allowing one to dissect out modified nucleosomes and examine their attendant DNAs with a high degree of accuracy. However, if one requires a more "global" approach to exploring the architectural differences in chromatin, the high specificity of ChIP may be a disadvantage, particularly in sperm where nucleosomes coexist with protamines but as a minor proportion of the chromatin. Both approaches have been used to investigate sperm chromatin, and they can be applied in tandem relying on classical techniques to probe architectural differences followed by ChIP with specific antibodies to dissect out subdomains bound by nucleosomes containing modified histones.

2. Materials

2.1. Preparation of Sperm Nuclei

1. *Prepare a sperm washing solution*, such as Physiological Tris-buffered saline (TBS: 0.15 M NaCl, 25 mM Tris-HCl, pH 7.2) containing 1%w/v human or bovine serum albumin (TBS-A; see Note 1).
2. *Prepare 80% and 40%v/v stocks of PureSperm (Nidacon) or Percoll (Sigma)* following the manufacturer's instructions (Nidacon) or an established procedure (see Note 2) (18).

2.2. Decondensation of Sperm Chromatin (See Note 3)

1. Physiological Tris (or phosphate)-buffered saline: 10× stock containing 1.5 M NaCl.
2. Dithiothreatol: 100 mM DTT in water. Store in 0.1 ml aliquots at −80°C until use.
3. Heparin: 5,000 U/ml in water. Store in 200 μl aliquots at −80°C until use.
4. Paraformaldehyde: 4%w/v in 1× TBS (2 g PFA + 50 ml TBS) or pre-chilled (−20°C) methanol: acetone 1:1 (see Note 4).
5. Prepare 100 ml 0.2%v/v Triton X-100 (10.0 ml × 10 TBS + 200 μl TX-100 to 50 ml with water); use a cut down pipette tip to dispense TX-100 and dissolve thoroughly before use as TX-100 is highly viscous.
6. Decondensing solution (Make fresh every time): Add 2.0 ml of 0.2% TX-100, 250 μl DTT (100 mM), and 200 μl heparin (5,000 U/ml) to 7.55 ml water and mix thoroughly. The final preparation should be 2.5 mM DTT, 0.2% Triton X-100, 100 U/ml Heparin Sulphate in 10 ml of TBS solution (see Note 3).

2.3. General Micrococcal Nuclease Digestion of Sperm Chromatin (See Note 5)

1. Sperm washing buffer 1 (SWB1) = HTF (HEPES-buffered Synthetic Oviductal Fluid): 10 mM HEPES, 80 mM NaCl, 3.1 mM KCl, 0.3 mM NaH$_2$PO$_4$, pH 7.3 (see Note 6).
2. Sperm washing buffer 2 (SWB2): 50 mM Tris–HCl, pH 8.0, 2 mM phenyl methyl sulphonyl fluoride (PMSF; from 20 mM stock in ethanol), 1% (w/v) mixed alkyltrimethylammonium bromide in water (ATAB; Sigma).
3. Pre-digestion buffer (PDB): 0.5% (v/v) Triton X-100, 2 mM PMSF in Na$^+$ and Mg^{2+} free PBS.
4. Digestion buffer (DB): 2 mM PMSF, 10 mM DTT in (Na$^+$ and Mg^{2+} free) PBS.
5. TE buffer: 10 mM Tris–HCl, 1 mM EDTA, pH 7.5.

2.4. Selective ChIP-chip of Sperm Chromatin

1. Dilution Buffer: 0.01% SDS, 1.1% Triton X-100, 1.2 mM EDTA, 16.7 mM Tris–HCl pH 8.1, 167 mM NaCl.
2. Low Salt Buffer: 0.1% SDS, 1% Triton X-100, 2 mM EDTA, 20 mM Tris–HCl, pH 8.1, 150 mM NaCl.
3. High Salt Buffer: 0.1% SDS, 1% Triton X-100, 2 mM EDTA, 20 mM Tris–HCl, pH 8.1, 500 mM NaCl.
4. LiCl Buffer: 0.25 M LiCl, 1% IGEPAL-CA630, 1% deoxycholic acid (sodium salt), 1 M EDTA, 10 mM Tris–HCl, pH 8.1.
5. TE Buffer: 1 mM EDTA, 10 mM Tris–HCl, pH 8.0.
6. Lysis Buffer: 1% SDS, 10 mM EDTA, 50 mM Tris–HCl, pH 8.0.
7. Elution Buffer: 1% SDS, 0.1 M NaHCO$_3$.
8. Protein A Agarose/Salmon Sperm DNA suspended in TE buffer (Millipore, USA).

3. Methods

3.1. Preparation of Sperm Nuclei

1. To obtain working sperm concentrations for microscopy, resuspend pelleted and (TBS-A or HTF) washed sperm in a small (50–100 µl) volume of TBS-A or HTF and dilute 1:50 in the same buffer before counting the sperm in a hemocytometer or other suitable counting graticule (see Note 7).
2. Add a few microlitres of formalin to stop motility and aid in counting.
3. Once the concentration of sperm in the sample is known, dilute sperm to 10^6/ml in the same buffer (*without formalin*) to provide sufficient sperm for microscopy.
4. Spot 10 µl samples over a 1 cm area, in pairs spaced 2 cm apart onto cleaned, uncoated acid washed (5 N HCl overnight followed by rinsing in ultrapure water) glass slides (~10,000 sperm).
5. Allow slides to dry in a warm oven for at least 2 h prior to further processing.
6. Circle the spots with a wax or diamond marker. This ensures that you can locate their positions at all times and minimise antibody use in all subsequent steps.

3.2. Decondensation of Sperm Chromatin

1. Wash (water) air-dried slides carefully twice in a coplin jar and carefully blot dry.
2. Place in humidified chamber in warm cabinet (30°C). Leave for 10 min to give slides time to warm up.

3. Carefully pipette 100 µl decondensing solution onto spots and incubate in the humid chamber.

4. Remove slides from the chamber every 2 min and observe under phase contrast optics. Within 5 min, you should see a loss of birefringence as the chromatin begins to decondense. This should be allowed to continue until birefringence fades altogether but without losing head shape. If shape is lost, decondensation has gone on for too long. The process should take no longer than 15 min but will vary depending on the sperm source and species (see Note 3). At this stage, it is wise to check again by staining a few slides with hematoxylin to verify that the extent of decondensation is as homogeneous as possible from sperm to sperm. These slides can be discarded.

5. Following decondensation, immediately fix the cells for 15 min by placing the slides in a coplin jar filled with 4% PFA or pre-chilled (−20°C) methanol:acetone (1:1) (see Note 4).

6. Wash the slides twice in TBS for 5 min.

7. Allow the slides to air dry and retain in a dry box (with desiccant) at 4°C until further use.

3.3. Antigen Localisation

There are many procedures available to the user for the visualisation and localisation of antigens by immuno-microscopy and undoubtedly, many readers of this chapter will have their own preferences in this regard. The range of good, commercially available antibodies to histones is now quite wide and expanding with several manufacturers offering reagents recognising core histones comprising all major subtypes (H2A, H2B, H3, and H4) as well as histones modified on various residues with methyl, acetyl, and sumoyl groups. We recommend that you begin with either a core-histone antibody or one recognising a histone-subtype regardless of modification. To give you an idea of what to expect, rather than describe our own IF procedure, which is essentially adapted from standard protocols, we have compiled a portfolio of pictures (Fig. 1) showing data from several laboratories (including our own). The procedure described is applicable to any primary antibody that is deemed suitable for this purpose by the manufacturer.

We have found that of all species tested so far, decondensation of human sperm chromatin is the most difficult to achieve with a reasonable degree of homogeneity. Figure 1a, b shows ejaculate and epididymal human and mouse sperm labelled with antibodies recognising H4K5Ac, K8Ac, K12Ac, and K16Ac (Calbiochem; green) and protamine 2 (kind gift of Rod Balhorn; red). It is clear that human and mouse sperm give good signals, but the variation in human sperm is far higher. This may be a consequence of the ejaculatory source of the former where seminal proteins can coat and mask internal antigens. However, labelling of the posterior nucleus is clear from the majority of these cells. The same antibody

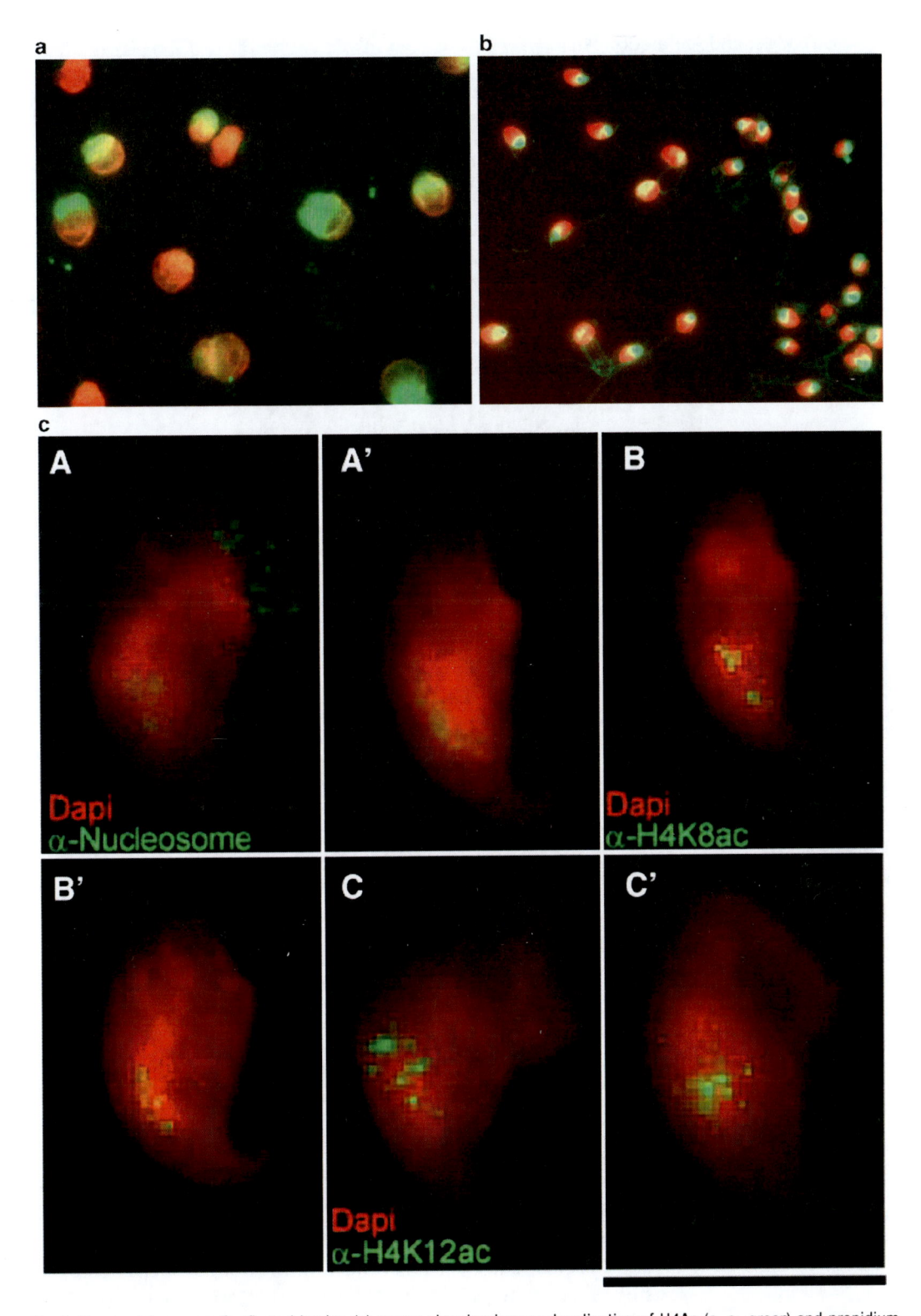

Fig. 1. Human (a), mouse (b–d), and bovine (e) sperm showing immunolocalisation of H4Ac (a–c; *green*) and propidium iodide staining of DNA (*red* in a, b, and f and DAPI *pseudocoloured red* in c). (d) A group of mouse sperm labelled with DNA probes to endonuclease sensitive (*green*) and resistant (*red*) chromatin. DAPI staining is in *blue*. (e) Immunolocalisation of CENPA in bovine sperm. (f) H4 (*green*) in marsupial sperm; DNA in *red*. Decondensation of sperm chromatin was achieved before labelling in all cases except the marsupial. (a), (b), and (d). Miller et al., unpublished work. (c) from Van der Heiden et al. (6); (e) from Palmer et al. (19); (f) from Soon et al. (20). Upper panels (e and f) show CENPA only (e) and histone + DNA (f). Lower panels (e and f) show CENPA + DNA (e) and DNA only (f).

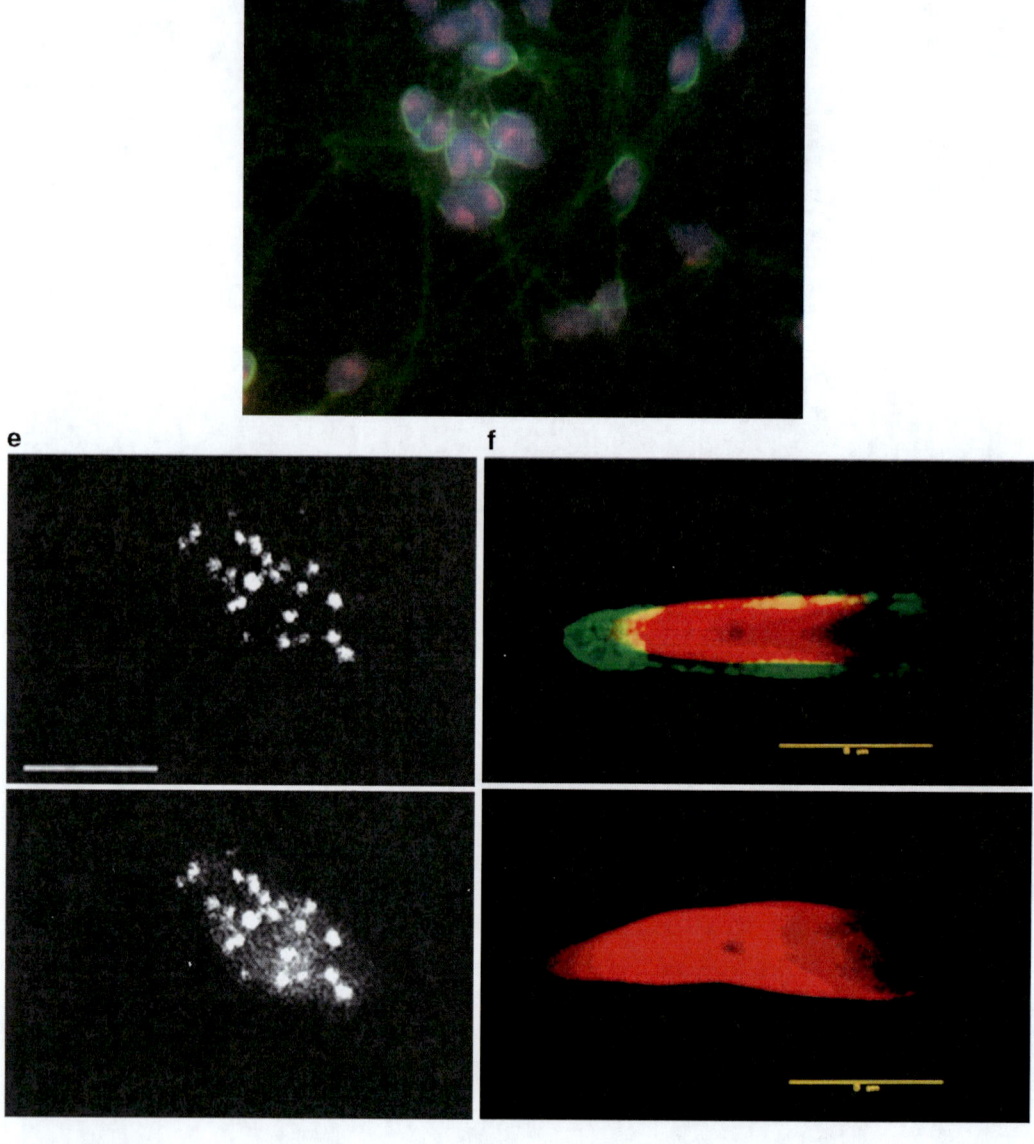

Fig. 1. (continued)

gives a striking sub-acrosomal location in mouse sperm and protamine 2 is located throughout the nucleus in both species (*unpublished data). A similar location for human sperm histones has been reported elsewhere (21), while our mouse data resembles that reported for H4K8Ac and K12Ac (Fig. 1c; (6)). The H3 variant CENPA is restricted to centromeric DNA, and an antibody to it stains discrete foci in bovine sperm (Fig. 1e; (19)). Perhaps the most bizarre distribution of sperm histones reported so far is that of the dasyurid marsupial, which retains a large proportion of its chromatin in a peripherally located nucleosomal configuration (Fig. 1f; (20)). The location of nuclease sensitive (green) and resistant (red) chromatin in mouse sperm chromatin is shown in Fig. 1d; (10), DAPI in blue. This labelling resembles that reported for the location of H1 in mouse sperm (22)).

3.4. General Micrococcal Nuclease Digestion of Sperm Chromatin

The following protocol is suitable for preparing 10^7 to 10^8 human sperm for digestion. These numbers of cells typically yield between 30 and 300 µg of DNA (based on ~3 pg per sperm) and will require anywhere from one to ten or more complete ejaculate samples. Apart from the density gradient step, which is carried out at room temperature, keep everything cold until the digestion procedure itself.

1. Prepare fresh or frozen (liquefied) semen using a discontinuous density gradient as above. HTF can be used instead of physiological tris-buffered saline (PTBS) but is not strictly necessary.

2. Recover pellets, resuspend in ice-cold SWB2 (0.1 ml), and immediately count a diluted (×10 in PTBS or HTF) aliquot to determine sperm concentration.

3. Incubate the remainder of the resuspended pellet on ice for 10 min, manually resuspending every minute or so to prevent cell clumping. At the end of the incubation period, make up to 5 ml with ice-cold HTF and centrifuge at $500 \times g$ for 10 min at 4°C.

4. Discard the supernatant and resuspend the pellet in 1 ml of ice-cold PDB; incubate for 10 min at 4°C.

5. Centrifuge as before and resuspend the pellet in 1 ml of DB. Incubate for 30 min at 37°C, removing from the incubator every 5 min or so to manually resuspend the pellet.

6. Add $CaCl_2$ to a final concentration of 0.6 mM, and five to ten units of micrococcal nuclease per 10^8 sperm nuclei (sufficient for ~0.3 mg DNA). Digestion time (37°C) should be determined empirically by running post-digested, deproteinated supernatants on agarose gels and must be repeated for every batch of micrococcal nuclease purchased. Decondensation dynamics of murine sperm will also require monitoring for optimisation and repeating for each batch purchased (see Note 8).

7. Stop digestion by the addition of EDTA to 10 mM and continue incubation at 37°C for an additional 20 min to allow digested chromatin to leach from the nuclei. Resuspend pellets manually every few minutes to prevent clumping.

8. Centrifuge for 10 min at $5,000 \times g$ (4°C) and recover the supernatant and pellets.

9. Add proteinase K (final concentration of 200 μg/ml from 2 mg/ml stock) and SDS (0.5% w/v from 10% w/v stock) to the recovered supernatants.
10. Resuspend the pellets in DB and add proteinase K as in step 9.
11. Incubate both supernatants and pellets for 10 h at 55°C with gentle agitation.
12. Add RNAse A (10 mg/ml) to both supernatants and pellets and incubate for 1 h at 37°C.
13. Isolate DNA by extraction in phenol chloroform (1:1) followed by ethanol precipitation. Alternatively, use DNA purification spin columns.
14. Dissolve the washed (70% ethanol), precipitated DNA in 100 μl of TE buffer and measure the DNA concentration with a Nanodrop device or equivalent. Store long term at −80°C and short term at 4°C.

3.5. Selective ChIP-chip of Sperm Chromatin

The protocol for the chromatin immunoprecipitation in combination with microarray (ChIP-chip) of (human) sperm chromatin can be divided into the following essential steps and is based on the protocol described by Steilmann et al. (23): Preparation of spermatozoa from ejaculate, crosslinking of remaining histones to sperm DNA, chromatin shearing by sonication, immunoprecipitation with specific antibodies against sperm nuclear proteins, reverse crosslinking of histone/DNA complex and isolation of histone-bound DNA, and amplification and hybridisation of DNA on your microarray of choice (see Fig. 2).

1. Prepare at least 10^7 sperm cells (see Subheading 3.1).
2. Wash the pellet twice in 1× Dulbecco's PBS (DPBS) supplemented with protease inhibitors (Complete, Roche), dilute pellet with 10 ml of DPBS and place on ice.
3. Perform crosslinking of histones to sperm DNA by the addition of 37% formaldehyde to a final concentration of 1% (270 μl of 37% formaldehyde in 10 ml of cell suspension). Incubate for 10 min on a rotating platform at room temperature (see Note 9).
4. Add glycine to 125 mM in order to quench the crosslinking reaction. Centrifuge 10 min at 500×g, discard supernatant, and wash pellet twice in DPBS.
5. Resuspend the cell pellet in 200 μl of ice-cold DPBS and homogenise with 20 strokes in an Ultra Turrax homogeniser for separation of sperm heads from sperm tails. Add 0.8 ml of Lysis Buffer to the cell suspension and incubate for 20 min on ice.

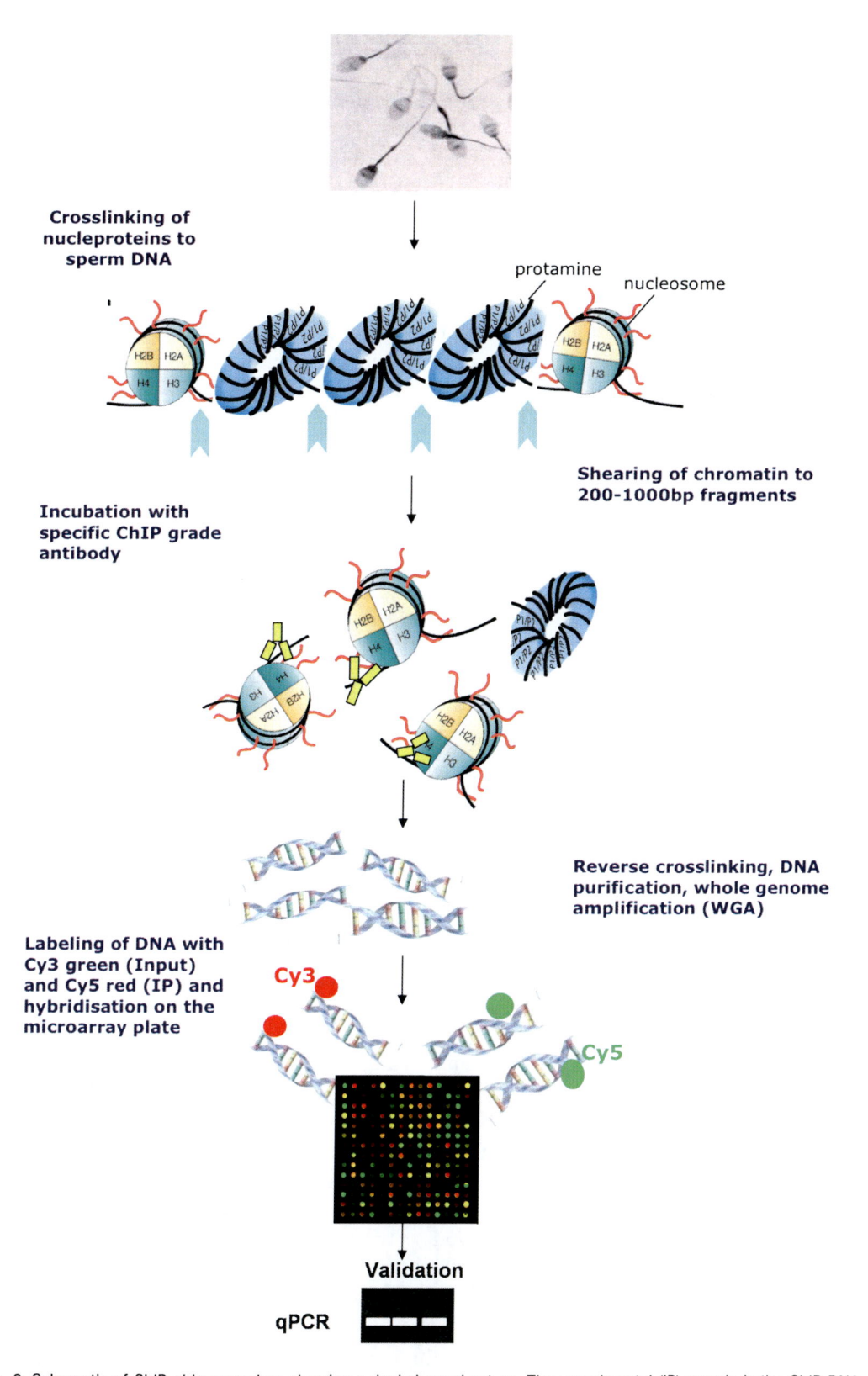

Fig. 2. Schematic of ChIP-chip procedure showing a dual channel set-up. The experimental (IP) sample is the ChIP DNA target (channel 1). The reference target is normally total DNA (channel 2). The targets following endonuclease digestion of sperm chromatin are labelled in the same way.

6. Sonicate the suspension (ten times) on ice with a Branson 250 sonificator (or equivalent) on setting 3, duty cycle 50% for 30 s to shear the DNA and obtain an average fragment length of approximately 200–1,000 bp. Monitoring of the shearing efficiency is essential for the ChIP results.

7. Remove 200 μl of sonicated material, add 8 μl of 5 M NaCl, and incubate 4 h to reverse crosslinks. Recover DNA by standard phenol/chloroform extraction or DNA mini isolation kit (Qiagen) and subsequently analyse DNA fragmentation on an agarose gel. The remaining suspension (0.8 ml) can be frozen at −20°C for further applications provided that sonication of DNA was carried out successfully.

8. Dilute the sonicated sperm cell supernatant tenfold with ChIP Dilution Buffer supplemented with protease inhibitors (Complete tablets, Roche). The final volume of 8 ml can be divided into four aliquots of 2 ml each for incubation with specific antibodies. Usually 1% of the ChIP sample (20 μl) should be saved as a no antibody control to quantify the amount of input DNA in every ChIP reaction.

9. Reverse crosslinks in input material by adding 1 μl of 5 M NaCl and incubating for 4 h at 65°C for subsequent DNA extraction.

10. Before incubation with antibodies, pre-clear the sonicated sperm sample with 75 μl of salmon sperm DNA/protein A agarose solution for 30 min at 4°C. Pellet agarose and collect supernatant for incubation with antibodies.

11. Incubate with specific antibodies of interest at 4°C overnight with constant rotation. The dilutions may vary per antibody and should be optimised according to the manufacturer instructions. The use of nonspecific rabbit polyclonal IgG ChIP grade antibody (Abcam ab46540) as a negative control and antibodies against abundant unmodified histone H3 (Abcam ab179) or unmodified histone H4 as a positive control for histone fraction is highly recommended. ChIP assay with anti-protamine antibody would also be an appropriate control if a ChIP grade protamine antibody becomes available (see Note 10).

12. To collect the antibody/histone complex, add 60 μl of salmon sperm DNA/protein A agarose solution and incubate for 1 h at 4°C. Briefly centrifuge and wash the immunoprecipitated agarose/antibody/histone complex on a rotating platform with the following buffers for 5 min each: Low Salt Buffer twice, High Salt Buffer twice, LiCl Buffer twice, and TE Buffer twice.

13. Add 250 μl of Elution Buffer to release the immunoprecipitated material from the agarose beads. This can be done with 15 min incubation on a rotating platform at RT. Repeat this step and combine eluates.

14. Add 20 μl of NaCl to the ChIP samples and 1 μl NaCl to the previously saved input material and incubate at 65°C for 4 h to dissolve histone/DNA crosslinks (see Note 11).

15. Add 10 μl of 0.5 M EDTA, 20 μl 1 M Tris–HCl, pH 6.5 and 2 μl of 10 mg/ml Proteinase K to ChIP samples and incubate for 1 h at 45°C. DNA can be recovered by standard phenol/chloroform extraction or by using DNA Mini Kit (Qiagen).

16. Following DNA recovery ChIP samples, input and controls can be analysed by quantitative real-time PCR (qPCR). Primer pairs should be designed for the sequence interacting with modified histones and control primers to the DNA sequence located 2 kb upstream or downstream from the estimated histone binding sites. DNA sequences of interest, interacting with modified histones in human and mouse sperm can be downloaded from the GEO Database (7, 9, 11). Enrichment of the immunoprecipitated sample compared to input material is calculated as follows: $\Delta Ct = Ct$ (input) $- Ct$ (immunoprecipitated sample) and % total $= 2^{\Delta Ct} \times 10$ (according to 10% input chromatin of total immunoprecipitated chromatin).

3.6. Genome Wide Analysis Using Microarrays

At least 4 μg of DNA is required for hybridisation to most array platforms. Hence, the amount of ChIP-IP DNA from sperm is frequently limiting. To overcome this shortcoming, whole genome amplification (WGA) can be applied using less than 10 ng of input material. There are a number of kits and protocols for achieving WGA and we have used the Sigma GenomePlex WGA kit (http://www.sigmaldrich.com) as described by O'Geen et al. (24) for this purpose (see Notes 12 and 13). A second round of WGA amplification (re-amplification) may be necessary.

The microarray platform of choice depends on the aims and objectives of the individual study. Several different arrays can be considered for analysing the DNA binding sites of sperm nuclear proteins. Agilent, and Affymetrix, for example, offer whole genome arrays (cover all genomic region with high resolution), promoter arrays (cover all regulatory elements and promoters of available RefSeq genes), ENCODE array and targeted arrays (interrogating focused regions of the genome, e.g., 375 coding regions described by the Encyclopaedia of Coding Elements), and custom design arrays cover sequences defined by the researcher.

3.7. Data Analysis and Validation

A comprehensive discussion of data analysis of ChIP-chip results is beyond the scope of this chapter, but analysis can be carried out using following resources.

1. The Bioconductor package array QualityMetrics (25).
2. ChIPpeakAnno Bioconductor package (26).
3. Gene ontology classification web tool (http://david.abcc.ncifcrf.gov/).

To validate microarray results by qPCR using ChIP samples from different individuals, design primer pairs overlapping the sequence corresponding to peak start and peak end of enriched region. For the controls, design primers 2 kb downstream or upstream away from the identified binding site. Calculate the enrichment of a specific binding site against the input material control.

4. Notes

1. More complex balanced solutions such as Human Tubal Fluid F (HTF; (18)) can also be used although we find no advantage over simple albumin-supplemented TBS for these purposes. PBS can cause sperm tails to curl, which affects sedimentation rates in density gradient media. The choice of buffer for density gradient centrifugation and washing of semen may be more critical for downstream nuclease digestion than for immunocytochemistry, with HTF being preferred by many andrologists regardless. Use PBS throughout if decondensation solution and antibodies are being dissolved in PBS.

2. Sperm preparation is a matter of individual choice and depends on your preference for swim-up, discontinuous density gradient or cell lysis methods to obtain relatively pure populations of (human) spermatozoal nuclei. All methods can give satisfactory yields while minimising contamination with round cells although for those with strict requirements for pure sperm populations, a combination of these procedures may be required (density gradient followed by round cell lysis, for example). For human sperm, density gradients prepared from polyvinylpyrrolidone (PVP)-free colloidal silica are preferred over alternatives like Percoll. We use Puresperm™ from Nidacon but it is not strictly necessary for experimental work with human sperm or for sperm from other species. There are many well-established protocols to follow for Percoll-based sperm preparation (18), but one advantage of PureSperm is that it is supplied ready for use. A general buffer containing a suitable detergent capable of preferentially lysing round cells in a human ejaculate can be used instead of density gradient centrifugation or swim-up. The method described originally by Jeffreys et al. (27) can be used for semen samples with few round cells. A more rigorous process can be followed if heavy contamination with round cells is noted (28).

3. Mammalian sperm chromatin is far more compact than that found in a typical somatic cell nucleus. Consequently, antibodies and other probes have a more difficult and frequently impossible task of finding their targets in sperm nuclei. The

best option that with care will preserve spatial and architectural features is some form of chromatin decondensation. Gentle decondensation is also the only option for successful immunocytology because it helps to retain the target molecules. Salt washing will remove histones and many other positively charged proteins from the nucleus and so should be avoided unless this is your intention. The procedure described in Subheading 3.2 works well for human and mouse sperm and is based on that reported originally by Ramos et al. (12). It is highly recommended that you watch the decondensation process in real time using a good phase contrast microscope to help you achieve the optimal time of exposure. Too short, and decondensation will be insufficient; too long and you risk dispersing the chromatin altogether. This will require calibrating for the sperm of each species you are interested in examining. If possible, use a heated stage to give a constant temperature throughout (suggest 30°C). The decondensing step is the same regardless of whether you choose immunofluorescence (IF) or brightfield microscopy for downstream analysis. We prefer IF for visualising sperm histones because it provides greater sensitivity than brightfield with chromagenic agents such as BCIP or DAB and gives more satisfactory signal to noise ratios, essential in the detection of "rare" components. For all incubations, including decondensing, prepare a humidified box for the slides. Soaking paper towels with water and placing them on the bottom of a large plastic container with the slides mounted above is satisfactory.

4. The choice of paraformaldehyde is again dependent on personal fixative preference and the primary antibodies you wish to use. We strongly advise that you follow any recommendations provided by suppliers on fixative choice while experimenting with your own. We have obtained good results with both PFA and methanol–acetone fixation protocols.

5. The micrococcal nuclease digestion procedure is based on a modified version of that described originally by Zalenskaya et al. (29) for human sperm. The same procedure can be used for mouse sperm although the dynamics of decondensation may differ.

6. For SWB1, HTF (not HEPES buffered) can be used instead.

7. The 90:45 interface, where immature sperm and somatic cells are frequently found can also be harvested for downstream analysis.

8. Digestion time with MNase will depend on cell density and what downstream processing is required. Because of labelling and hybridization dynamics, fragment sizes of approximately 200–500 bp are considered optimal for microarray analysis. Hence, one should aim to obtain a typical nucleosome ladder

from MNase digestion products (see Fig. 2). This will only be seen after isolating the DNA using, for example, a phenol–chloroform, ethanol precipitation extraction step. The post-digestion processing for microarrays often includes re-digestion with the frequent cutting enzymes Rsa1 and Alu1 to help achieve this goal. Digestion to mononucleosomes is suited to microarray analysis provided the tiling on the array is no greater than the fragment sizes to be probed. Mononucleosomes package ~150 bp DNA. It is well suited to downstream ChIP-seq strategies using monospecific antibodies.

9. Because histones bind DNA so strongly, it may not be necessary to carry out cross linking under the conditions described in step 3 of Subheading 3.5.

10. Not every antibody is suitable for ChIP procedure. Commonly offered, so-called "ChIP grade" antibodies are those that have been successfully tested for ChIP applications. A critical step for discriminating a good from a poor ChIP grade antibody is the washing step with lithium chloride.

11. While the high salt concentration should guarantee the effective removal of nonspecific chromatin interactions with the agarose beads, the protein–antibody interactions should not be affected.

12. The latest whole genome amplification (WGA) techniques are now good enough to maintain the complexity of sequence composition with a reasonably high fidelity while providing up to 1,000 times the quantity of original input DNA. WGA is now used routinely in ChIP-chip procedures to help overcome the sample size limitations of microarrays. WGA of small DNA samples has proved to be a robust and reliable alternative to sample pooling in order to generate sufficient DNA for downstream analysis and should be considered if semen samples are a limiting factor. WGA may be compatible with DNA sequencing strategies but outputs should be interpreted with some caution.

13. The initial random amplification step of the WGA protocol should be omitted due to prior fragmentation of sperm chromatin by sonication.

Acknowledgements

The authors gratefully acknowledge the assistance of the British Biotechnology and Science Research Council (BBSRC) and the German Research Foundation (DFG), Project 1 of the Clinical Research Unit KFO 181/2 for supporting the work on which this chapter in based.

References

1. Rathke C et al (2007) Transition from a nucleosome-based to a protamine-based chromatin configuration during spermiogenesis in Drosophila. J Cell Sci 120:1689–1700
2. Dadoune JP (2003) Expression of mammalian spermatozoal nucleoproteins. Microsc Res Tech 61:56–75
3. Orsi GA et al (2009) Epigenetic and replacement roles of histone variant H3.3 in reproduction and development. Int J Dev Biol 53:231–243
4. Grimes SR Jr, Henderson N (1984) Hyperacetylation of histone H4 in rat testis spermatids. Exp Cell Res 152:91–97
5. Christensen ME et al (1984) Hyperacetylation of histone H4 promotes chromatin decondensation prior to histone replacement by protamines during spermatogenesis in rainbow trout. Nucleic Acids Res 12:4575–4592
6. van der Heijden GW et al (2006) Transmission of modified nucleosomes from the mouse male germline to the zygote and subsequent remodeling of paternal chromatin. Dev Biol 298:458–469
7. Hammoud SS et al (2009) Distinctive chromatin in human sperm packages genes for embryo development. Nature 460:473–478
8. Gatewood JM et al (1990) Isolation of four core histones from human sperm chromatin representing a minor subset of somatic histones. J Biol Chem 265:20662–20666
9. Brykczynska U et al (2010) Repressive and active histone methylation mark distinct promoters in human and mouse spermatozoa. Nat Struct Mol Biol 17:679–687
10. Saida M et al (2011) Key gene regulatory sequences with distinctive ontological signatures associate with differentially endonuclease accessible mouse sperm chromatin. Reproduction 142(1):73–86
11. Arpanahi A et al (2009) Endonuclease-sensitive regions of human spermatozoal chromatin are highly enriched in promoter and CTCF binding sequences. Genome Res 19:1338–1349
12. Ramos L et al (2008) Incomplete nuclear transformation of human spermatozoa in oligo-astheno-teratospermia: characterization by indirect immunofluorescence of chromatin and thiol status. Hum Reprod 23:259–270
13. Sanders MM (1978) Fractionation of nucleosomes by salt elution from micrococcal nuclease-digested nuclei. J Cell Biol 79:97–109
14. Henikoff S et al (2009) Genome-wide profiling of salt fractions maps physical properties of chromatin. Genome Res 19:460–469
15. Kramer JA et al (2000) Human spermatogenesis as a model to examine gene potentiation. Mol Reprod Dev 56:254–258
16. Fischer JJ et al (2008) Combinatorial effects of four histone modifications in transcription and differentiation. Genomics 91:41–51
17. van der Heijden GW et al (2008) Sperm-derived histones contribute to zygotic chromatin in humans. BMC Dev Biol 8:34
18. Mortimer D (1994) Sperm recovery techniques to maximize fertilizing capacity. Reprod Fertil Dev 6:25–31
19. Palmer DK et al (1990) The centromere specific histone CENP-A is selectively retained in discrete foci in mammalian sperm nuclei. Chromosoma 100:32–36
20. Soon LL et al (1997) Isolation of histones and related chromatin structures from spermatozoa nuclei of a dasyurid marsupial Sminthopsis crassicaudata. J Exp Zool 278:322–332
21. Li Y et al (2008) Characterization of nucleohistone and nucleoprotamine components in the mature human sperm nucleus. Asian J Androl 10:535–541
22. Pittoggi C et al (1999) A fraction of mouse sperm chromatin is organized in nucleosomal hypersensitive domains enriched in retroposon DNA. J Cell Sci 112(Pt 20):3537–3548
23. Steilmann C et al (2010) The interaction of modified histones with the bromodomain testis-specific (BRDT) gene and its mRNA level in sperm of fertile donors and subfertile men. Reproduction 140:435–443
24. O'Geen H et al (2006) Comparison of sample preparation methods for ChIP-chip assays. Biotechniques 41:577–580
25. Toedling J, Huber W (2008) Analyzing ChIP-chip data using bioconductor. PLoS Comput Biol 4:e1000227
26. Zhu LJ et al (2010) ChIPpeakAnno: a Bioconductor package to annotate ChIP-seq and ChIP-chip data. BMC Bioinformatics 11:237
27. Jeffreys AJ et al (1994) Complex gene conversion events in germline mutation at human minisatellites. Nat Genet 6:136–145
28. Goodrich R et al (2007) The preparation of human spermatozoal RNA for clinical analysis. Arch Androl 53:161–167
29. Zalenskaya IA et al (2000) Chromatin structure of telomere domain in human sperm. Biochem Biophys Res Commun 279:213–218

Part X

In Vitro and Ex Vivo Spermatogenesis Models

Chapter 41

In Vitro Spermatogenesis Using an Organ Culture Technique

Tetsuhiro Yokonishi, Takuya Sato, Kumiko Katagiri, and Takehiko Ogawa

Abstract

Research on in vitro spermatogenesis has a long history and remained to be an unaccomplished task until very recently. In 2010, we succeeded in producing murine sperm from primitive spermatogonia using an organ culture method. The fertility of the sperm or haploid spermatids was demonstrated by microinsemination. This organ culture technique uses the classical air–liquid interphase method and is based on conditions extensively examined by Steinbergers in 1960s. Among adaptations in the new culture system, application of serum-free media was the most important. The system is very simple and easy to follow.

Key words: In vitro spermatogenesis, Spermatogonial stem cells, Sperm, Spermatids, Organ culture, Cryopreservation

1. Introduction

The research on in vitro spermatogenesis has a long history beginning in the early twentieth century (1, 2). In its early phase, the organ culture method was extensively examined and proved to be an effective method to analyze conditions of spermatogenesis (3). However, spermatogenesis did not proceed beyond the meiotic pachytene stage. The organ culture method was therefore abandoned, and other cell culture methods were then adopted by many researchers. In fact, new culture techniques were developed mostly based on the emerging molecular biology methodologies. Such techniques included the establishment of immortalized cell lines derived from germ cells or Sertoli cells (4, 5). Nonetheless, it remained impossible to reliably induce complete spermatogenesis from spermatogonia or spermatognial stem cells in vitro. Little progress had been made in the last couples of decades.

In 2007, we began studying in vitro spermatogenesis mainly by reexamining the method and results obtained by Steinbergers in 1960s (6). We found that serum-free medium, supplemented with knockout serum replacement (KSR), is effective in inducing spermatogenesis from neonatal mouse testes (7). Our method has several advantages as follows (1) it is very simple and easy to repeat, (2) it can reliably produce haploid cells, (3) in vitro spermatogenesis can be maintained for more than 2 months, (4) the resultant haploids are fertile, and (5) cryopreservation of testis tissue fragments is possible and thawed fragmented remain viable and suitable for in vitro sperm production. In this chapter, we introduce our organ culture method as a powerful tool for the evaluation of spermatogenesis in vitro.

2. Materials

2.1. Culture Medium

1. α-Minimum Essential Medium (α-MEM) (Invitrogen, Carlsbad, CA, USA).
2. Knockout Serum Replacement (KSR) (Invitrogen, Carlsbad, CA, USA).
3. AlbuMAX I (40 mg/ml) (Invitrogen, Carlsbad, CA, USA).
4. Penicillin–Streptomycin (×100; 10,000 μ/ml penicillin, 10 mg/ml streptomycin).
5. Sodium bicarbonate (7 %) in water.
6. Organ culture medium (α-MEM + 10%KSR): Dissolve α-MEM (10.1 g) in 500 ml of double-deionized water (DDW) to make a double-concentrated α-MEM. To 100 ml of double-concentrated α-MEM, add 20 ml of KSR, 2 ml of Penicillin–Streptomycin, and 5.2 ml of sodium bicarbonate (7 %). Then add DDW to a total volume of 200 ml. Alternatively, organ culture medium (α-MEM + AlbuMAX) can be made by adding 8 g of AlbuMAX I instead of 20 ml of KSR. Sterilize by filtration through a 0.22 micro-meter Millipore membrane. Store the culture media in refrigerator.

2.2. Agarose Gel Preparation

1. Erlenmeyer flask (200 ml size).
2. 10 cm dishes.
3. Agarose: Place 1.5 g of agarose and 100 ml of DDW in a 200 ml flask, then shake the flask briefly to suspend the agarose. Sterilize the agarose solution by autoclaving (2 atm, 121 °C, 20 min).
4. Agarose dish preparation: Pour 33 ml of agarose solution into three 10-cm dishes. The depth of the solution will be about

Fig. 1. The process of cutting the agarose gel (**a**) into hexahedrins and transferring (**b**) them to 6-well culture plates (**c**).

5 mm. The agarose solution will solidify around 50°C. Leave the dishes for about 2 h at room temperature until agarose gels become solid. These prepared dishes can be preserved by sealing and storing in a refrigerator for several weeks.

2.3. Culture Dish Preparation

1. GF Millipore Express plus membrane, 0.22 micro-meter (MILLIPORE, USA).
2. Erlenmeyer flask (200ml size).
3. 10-cm Cell culture dish, 100 mmre Expstyle.
4. Scalpel blade or metal spatula.
5. 6-Well culture plate.
6. Cut gel into hexahedrons of about 10 × 10 × 5 mm in size with a scalpel blade or spatula (Fig. 1a).
7. Place three to four hexahedrons into each well of 6-well plate (Fig. 1b). Add culture medium (α-MEM + 10%KSR or α-MEM + AlbuMAX) to soak gels completely (Fig. 1c). Keep them in an incubator overnight or longer in order to replace water in the agarose with the medium. At the initiation of a culture experiment, remove old medium by aspiration, and add new culture medium the well to the height of half or up to four fifths the height of the agarose gel pieces.

2.4. Testis Tissue Preparation

1. Mice aged between 0.5 and 7.5 days postpartum (dpp) depending on research objectives (see Note 1).
2. Forceps for microsurgical use.
3. Stereomicroscope.
4. CO_2 incubator.

2.5. Analysis of Spermatogenesis

1. Bouin's fixative.
2. Phosphate-buffered saline (PBS).
3. 4% Paraformaldehyde in PBS.
4. Sucrose solutions: 10%, 15%, and 20% (w/v).
5. Tissue-Tek OCT compound (Sakura Finetechnical, Tokyo, Japan).
6. Cryostat.
7. Adhesive glass slides.
8. Dryer.
9. 0.2 % Triton-X100 in PBS (0.2% PBT).
10. 5 % Bovine serum albumin (BSA) in 0.2% PBT.
11. Rabbit anti-SYCP1 primary antibody (Novus Biologicals, Colorado, USA: NB300-229): Dilute 1:600 in 0.2% PBT supplemented with 5% BSA.
12. Rabbit anti-SYCP3 primary antibody (Abcam, Cambridge, UK): Dilute 1:400 in 0.2% PBT supplemented with 5% BSA.
13. Mouse anti-mouse sperm protein SP56 monoclonal primary antibody (QED Bioscience, Inc., California, USA): Dilute 1:100 in 0.2% PBT supplemented with 5% BSA.
14. Rat anti-GFP primary antibody (Nakalai tesque, Inc., Kyoto, Japan): Dilute 1:1,000 in 0.2% PBT supplemented with 5% BSA.
15. Rabbit anti-GFP Alexa Fluor 488-conjugate primary antibody (Invitrogen, Carlsbad, CA): Dilute 1:50 in 0.2% PBT supplemented with 5% BSA.
16. Alexa Fluor 555-conjugated goat anti-rabbit IgG secondary antibody (Invitrogen, Carlsbad, CA): Dilute 1:200 in 0.2% PBT supplemented with 5% BSA.
17. Alexa Fluor 555-conjugated goat anti-mouse IgG secondary antibody (Invitrogen, Carlsbad, CA): Dilute 1:200 in 0.2% PBT supplemented with 5% BSA.
18. ProLong Gold Antifade Reagent (Invitrogen).
19. Hoechst 33342 dye.

2.6. Cryopreservation of Testis Tissue Fragments

1. Liquid nitrogen storage tank.
2. Serum tube (Sumitomo Bakelite Co., Ltd., MS-4501 W, Japan).
3. Cell freezing medium (see Note 2): Cell Banker1 (Juji Field, Inc., Tokyo, Japan) or TC-Protector cell freezing medium (DS pharma biomedical).
4. −80°C Freezer.

3. Methods

3.1. Culture of Mouse Testis Tissues

1. After euthanasia of mouse pup, remove testes and immediately place them in 3.5-cm dishes containing culture medium (Fig. 2). With two fine forceps, hold the tunica albuginea at two close sites, and pull them apart to tear the tunica and expose seminiferous tubules inside. After fully exposing the seminiferous tubules, remove the tunica. Collect the mass of seminiferous tubules into a dish containing fresh culture medium cooled on ice (Fig. 3).

2. Separate the testis tissue into smaller pieces of seminiferous tubules using forceps (Fig. 4). The size of the pieces is arbitrary, but they should be approximately 1 mg in weight or

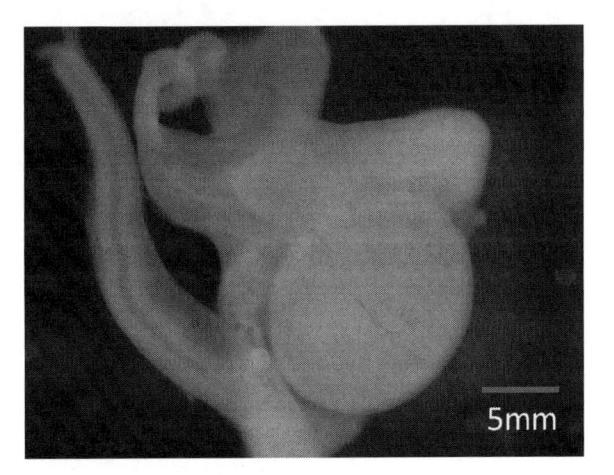

Fig. 2. A testis removed from a pup mouse.

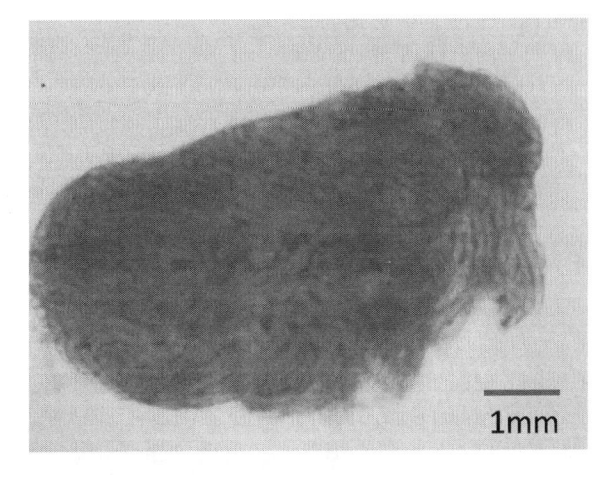

Fig. 3. A testis following removal of the tunica albuginea.

Fig. 4. Dissociation of the seminiferous tubules into fragments suitable for tissue culture or cryopreservation.

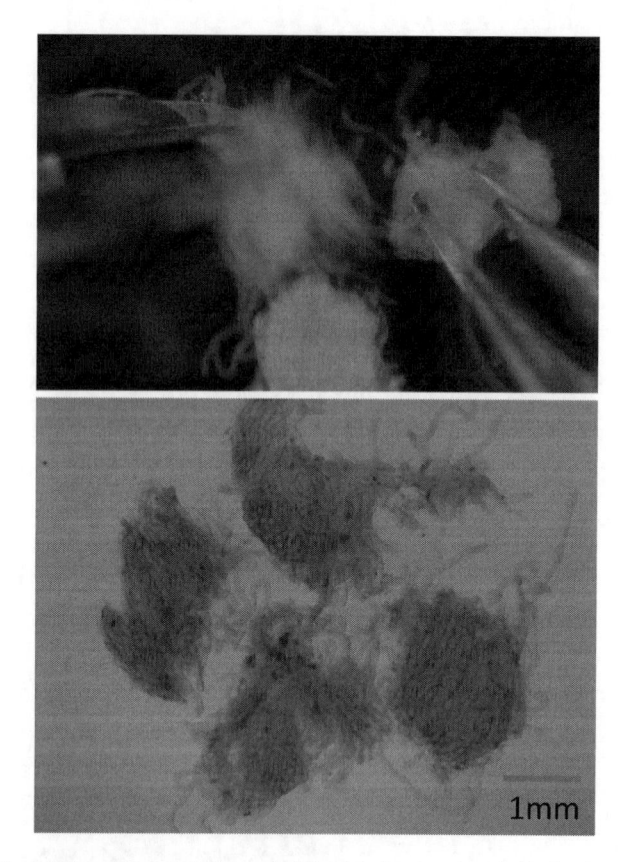

  1 mm³ in size when compacted. Maintain the pieces in culture medium.
 3. Hold each tissue fragment gently using forceps, or hold them with a micropipette by applying mild negative pressure. Transfer one to three of the testis tissue fragments to the hexahedrons (Fig. 5).
 4. Incubate the tissues by placing the 6-well plates in a culture incubator. The culture incubator should be supplied with 5 % CO_2 in air and maintained at 34 °C. Change medium once a week. For medium change, remove the old medium in a well by aspiration, then add the same amount of fresh culture medium (see Note 4).

3.2. Analysis of the Progression of Spermatogenesis

Spermatogenesis in organ culture progresses at almost the same pace as occurs in vivo.

 1. Histological evaluation (H&E, PAS stain): Fix the specimens with Bouinns wiis in and embed them in paraffin. Thin sections should be made in the horizontal direction to obtain the largest

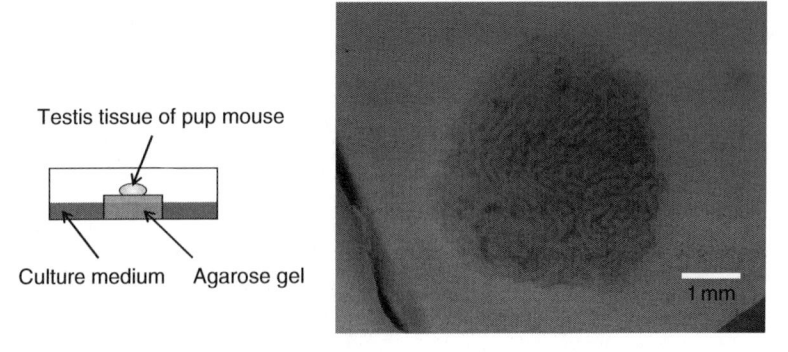

Fig. 5. Schematic and photomicrograph of the organ culture method for testis tissue culture.

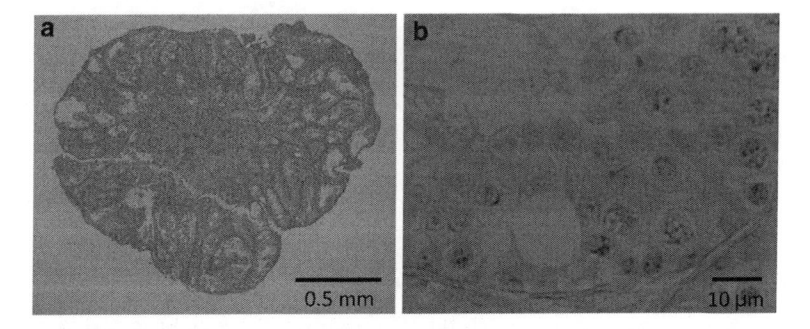

Fig. 6. Photomicrographs of tissue sections following culture. (a) H&E-stained section demonstrating spermatogenesis at the periphery and degeneration toward the center. (b) PAS-stained sections demonstrating intensely stained acrosomes.

cut surface (Fig. 6a). The PAS stain is useful to identify round spermatids because the acrosomal dots or caps stain red (Fig. 6b).

2. Slide preparation for immunohistochemical evaluation: To prepare frozen sections, fix testes tissue fragments with 4% PBS at 4°C for 6–16 h, and wash them with PBS three times, then dehydrate in 10%, 15%, and 20% (w/v) sucrose for 4 h, 4 h, and overnight, respectively. Embed the testis tissues in OCT compound and freeze them in liquid nitrogen. Make sections of about 7 mM in thickness using a cryostat. Place the tissue sections onto an adhesive slide. Dry the tissue sections for 15 min with dryer. Sections can be stored −80°C for later use.

3. Immunohistochemical evaluation: Wash the sections with PBS for 5 min then wash them in 0.2% PBT four times, 10 min each. Incubate slides for blocking in 5% BSA in 0.2% PBT for 60 min, then incubate them with primary antibodies overnight at 4°C. Wash the sections with 0.2% PBT four times, 10 min each. Apply the appropriate secondary antibodies for 60 min

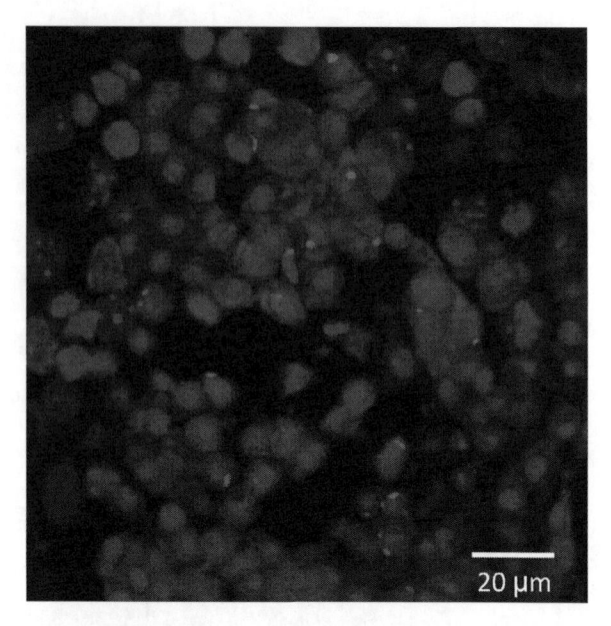

Fig. 7. Cultured testis tissue of a Gsg2-GFP Tg mouse labeled with SP56 antibodies (*red*), GFP (*green*), and Hoechst (*blue*).

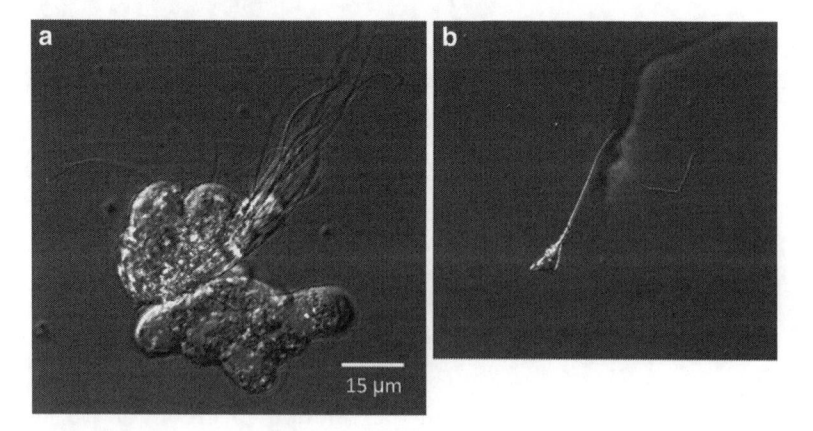

Fig. 8. (**a**) Flagellated spermatids whose heads are bound by Sertoli cells. (**b**) A single sperm obtained from cultured testis tissue.

and wash them with 0.2 % PBT four times, 10 min each. Add one or two drops of ProLong Gold Antifade Reagent and mount a cover glass (Fig. 7).

4. Sperm identification: Place a tissue fragment on a glass slide and add several drops of PBS. Dissociate cells from the tissue using two 26 G needles by tearing and chopping up, then pipette the suspension with micropipetter. Observe the sample with a microscope. Without any staining, flagella are most easily identified. For nuclear staining, add Hoechst dye to the cell suspension (1–2 %v/v). This is helpful to identify the head of sperm or elongated spermatids (Fig. 8).

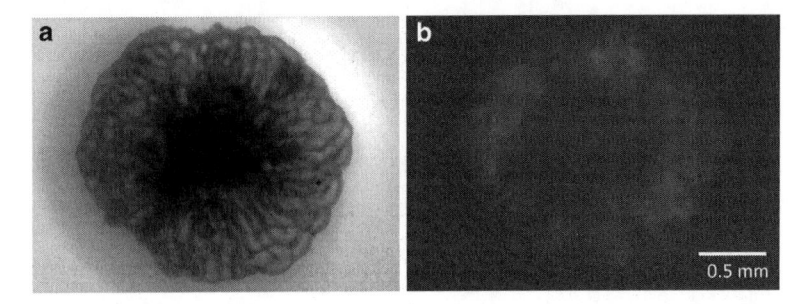

Fig. 9. (a) Testis tissue obtained from a Gsp2/Haspin-GFP mouse. (b) Active spermatogenesis is readily apparent with fluorescent illumination for GFP.

5. Evaluation using transgenic mice: Transgenic mice expressing meiosis- or haploid-specific GFP are very useful in this type of culture study. Such transgenic mice include Acr-GFP (8) and Gsg2/Haspin-GFP (9). The GFP expression can be recognized under a stereomicroscope equipped with fluorescent illumination for GFP (Fig. 9). For the observation of Acr-GFP acrosomes, dissociate the cultured tissues mechanically using needles to release cells into PBS. Observe the cell suspension with a microscope under GFP excitation light.

3.3. Cryopreservation and Thawing of Testis Tissue Fragments

1. To cryopreserve testis tissue, prepare testis tissue fragments in the same manner for culturing as described in Subheading 3.1. Hold each of the tissue fragments gently with micro-forceps and transfer them into sterile serum tubes containing 0.5–1.0 ml of cryoprotectant solution. Three to ten fragments can be cryopreserved in a single tube. Place the tubes in a –80 °C freezer overnight, then, plunge them in liquid nitrogen for storage.

2. Thawing and initiation of culture: Remove tubes from the liquid nitrogen. Place the tubes on a clean bench and let them thaw at room temperature. After thawing, recover the testis tissues using forceps or a micropipette and place them in a dish containing the culture medium to rinse the tissues. Place the tissues on agarose gel for culturing.

4. Notes

1. The choice of mouse age may depend on the purpose of the experiments. As spermatogenic meiosis in mice starts around day 7–8 after birth, we usually use mice younger than that age to enable evaluation meiotic progression in vitro.

2. We have used two commercially available cryoprotectants, Cell Banker1 and TC-Protector cell freezing medium, and have not found a significant difference between them. A more simple DMSO solution is also suitable as cryoprotectant for the testis tissues.

3. Incubation temperature is critical for in vitro spermatogenesis using this system. In our experience, 34 °C was better than 32 °C for promoting spermatogenesis, and 37 °C was detrimental. We usually change medium once a week, which is generally sufficient to maintain the cultured tissue architecture and spermatogenesis for 5–10 weeks or more. However, more frequent (twice a week) medium changes may be advantageous.

Acknowledgments

This work was supported by a Grant-in-Aid for Scientific Research on Innovative Areas, "Regulatory Mechanism of Gamete Stem Cells" (#20116005); a Grant-in-Aid for Scientific Research (C) (#21592080) from the Ministry of Education, Culture, Sports, Science, and Technology, Japan; a grant from the Yokohama Foundation for Advancement of Medical Science; and a grant for Research and Development Project II (No.S2116) of Yokohama City University, Japan, to TO.

References

1. Champy C (1920) Quelques resultats de la methode de culture des tissues. Arch Zool Exp Gen 60:461–500
2. Martinovitch PN (1937) Development in vitro of the mammalian gonad. Nature 139:413
3. Steinberger A et al (1964) Mammalian testes in organ culture. Exp Cell Res 36:19–27
4. Feng LX et al (2002) Generation and in vitro differentiation of a spermatogonial cell line. Science 297:392–395
5. Rassoulzadegan M et al (1993) Transmeiotic differentiation of male germ cells in culture. Cell 75:997–1006
6. Gohbara A et al (2010) In vitro murine spermatogenesis in an organ culture system. Biol Reprod 83:261–267
7. Sato T et al (2011) In vitro production of functional sperm in cultured neonatal mouse testes. Nature 471:504–507
8. Nakanishi T et al (1999) Real-time observation of acrosomal dispersal from mouse sperm using GFP as a marker protein. FEBS Lett 449:277–283
9. Tanaka H et al (1999) Identification and characterization of a haploid germ cell-specific nuclear protein kinase (Haspin) in spermatid nuclei and its effects on somatic cells. J Biol Chem 274:17049–17057

Chapter 42

Testicular Tissue Grafting and Male Germ Cell Transplantation

Jose R. Rodriguez-Sosa, Lin Tang, and Ina Dobrinski

Abstract

Testicular tissue grafting and male germ cell transplantation are techniques that offer unprecedented opportunities to study testicular function and development. While testicular tissue grafting allows recapitulation of testis development and spermatogenesis from immature males of different mammalian species in recipient mice, germ cell transplantation results in donor-derived spermatogenesis in recipient testes.

Testicular tissue grafting results in spermatogenesis from a wide variety of large animal donor species and is therefore an attractive way to study testis development and spermatogenesis and preserve fertility of immature males. Germ cell transplantation represents a functional reconstitution assay for identification of spermatogonial stem cells (SSCs) in a given donor cell population and serves as a valuable tool to study stem cell biology and spermatogenesis. In this chapter we provide detailed methodology to successfully perform both techniques.

Key words: Testicular tissue grafting, Testis development, Male germ cell transplantation, Spermatogenesis, Spermatogonial stem cells

1. Introduction

In the last two decades, important discoveries have provided unprecedented opportunities to explore testicular function and development. In 2002, we first reported that small fragments of testicular tissue from immature males, transplanted under the dorsal skin of immunodeficient mice, were able to survive and undergo full development with the production of sperm (1). Since then, testis tissue xenografting has been shown to be successful in a wide variety of species and emerged as a valuable alternative to study testis development and spermatogenesis of large animals in mice (2).

Testis tissue xenografting is particularly attractive for research as several mice can be transplanted with tissue from a single donor and then assigned to different treatment groups and collection time points. This allows the study of testis tissue from the target species with the advantages of working in a rodent model. For example, this approach reduces the number of large donor animals that are needed for study and the cost associated with it. Moreover, since the resulting sperm are capable of fertilizing and triggering normal embryo development, this approach can also be used to preserve fertility in immature males when sperm collection is not an option (3).

In 1994, Dr. Ralph Brinster and colleagues at the University of Pennsylvania showed that microinjection of germ cells from fertile donor mice into the seminiferous tubules of infertile recipient mice results in donor-derived spermatogenesis and sperm production by the recipient animal (4, 5). The use of transgenic donors carrying the bacterial β-galactosidase gene easily allowed identification of donor-derived spermatogenesis and transmission of the donor haplotype to the offspring by recipient animals. Through the tracing of donor cells, it was revealed that donor germ cells, after being transplanted into the lumen of the seminiferous tubules, were able to move from the luminal compartment to the basement membrane where spermatogonia are located (6). It is generally accepted that only SSCs are able to colonize the stem cell niche in the seminiferous epithelium of recipient mice and initiate spermatogenesis.

Germ cell transplantation provides a functional approach to study the stem cell niche in the testis and to evaluate the stem cell potential of specific populations of germ cells. Moreover, germ cell transplantation can be used to elucidate basic stem cell biology (6–9), to produce transgenic animals through genetic manipulation of germ cells prior to transplantation (10, 11), and to study Sertoli cell–germ cell interaction (12, 13).

2. Materials

2.1. Surgical Instruments

1. Suture, Ethicon 6-0 silk.
2. Wound clips.
3. #5 Dumont Forceps.
4. Hemostat.
5. Extra Fine Graefe Forceps—0.5 mm Tip, Curved.
6. Extra Thin Iris Scissors.

2.2. Testicular Tissue Grafting

1. Phosphate-buffered saline (PBS).
2. Dulbecco's modified Eagle's medium (DMEM).
3. Disposable scalpels, No. 10.
4. Tissue culture dishes (60 × 15 mm).

2.3. Germ Cell Transplantation

1. Collagenase, type IV: dissolve 7 mg of collagenase powder in 7 ml of DMEM to make a working solution of 1 mg/ml. Filter-sterilize the solution (see Note 1).
2. Trypsin–EDTA: prepare 4 ml (0.25 % trypsin plus 1 mM EDTA) (see Note 1).
3. DNase I: dissolve 14 mg of DNase I in 2 ml of DMEM to make a working solution of 7 mg/ml. Filter-sterilize the solution (see Note 1).
4. DMEM.
5. Trypan blue.
6. Nylon mesh cell strainer, 40 and 70 μm (BD biosciences).
7. Busulfan: dissolve Busulfan powder in DMSO and then add an equal volume of sterile distilled water to make a final concentration of 4 mg/ml. Keep solution warm at 35–40 °C prior to use to avoid precipitation of Busulfan (see Note 2).
8. Thin-wall glass capillary tubes: used as the injection needle.
9. Polyethylene tubing: attach the tubing to a 1 ml syringe to make a simple injection apparatus.
10. Sigmacote (Sigma).
11. 1 ml Syringe.

2.4. X-Gal Staining

1. X-gal: Dissolve X-gal powder in *N,N*-dimethyl formamide to give a stock solution of 10 mg/ml. Wrap in aluminum foil (light sensitive) and stored at −20 °C.
2. Potassium Ferrocyanide: make 500 mM stock in water.
3. Potassium Ferricyanide: make 500 mM stock in water.
4. Magnesium chloride: make 1 M stock in water.
5. Sodium deoxycholate: make 1 % stock solution in water.
6. *N,N*-dimethylformamide: used to dissolve X-gal powder.
7. Igepal CA-630 (Sigma): make 10 % stock solution in water.
8. LacZ rinse buffer: 0.2 M sodium phosphate, pH 7.3; 2 mM magnesium chloride; 0.02 % Igepal CA-630; 0.01 % sodium deoxycholate; make up to 200 ml with PBS.
9. LacZ staining buffer: 5 mM potassium ferricyanide, 5 mM potassium ferrocyanide, 1 mg/ml X-gal, make up to 10 ml with LacZ rinse buffer.

3. Methods

3.1. Testicular Tissue Grafting in Mice

From Dobrinski and Rathi (14) and Rodriguez-Sosa et al. (15).

3.1.1. Collection of Donor Tissue

1. Obtain testis tissue by castration or biopsy from a donor male.
2. Place testis in PBS or biopsies into culture medium, maintaining sterile conditions.
3. Keep the collected tissue on ice and transport to the laboratory (see Note 3).
4. For preparation of donor tissue, perform in the tissue culture hood to maintain sterility. Wash each testis in ice-cold PBS containing antibiotics two to three times before transferring into a culture dish with PBS. In the case of biopsies, wash testis fragments two to three times with ice-cold culture medium containing antibiotics by centrifugation at $150 \times g$ for 2 min and resuspend in fresh ice-cold culture medium.
5. If intact testes are collected, remove tunica vaginalis by making an incision along the surface and extrude the testis. Remove from testis all annex structures (spermatic cord, epididymis, connective tissue). Wash testes once in cold PBS and transfer into a culture dish with PBS. Carefully remove the tunica albuginea of the testis by using a scalpel blade and a pair of scissors. If the testis is very small, the tunica can be removed by squeezing the testicular tissue out of the tunica through a small incision made on one end while holding the tunica with a pair of small forceps on the other end.
6. Depending on the size of the testis either the whole testis tissue can be cut into small pieces of around 1–2 mm³ in size using a disposable No. 10 scalpel or large pieces of testis tissue can first be removed from the testis and then cut into smaller pieces. All this should be done in ice-cold culture medium and under sterile conditions in a small culture dish.
7. Transfer the prepared tissue fragments to ice-cold culture medium in small culture dishes on ice until grafting.

3.1.2. Castration of Recipient Mouse

1. Anesthetize recipient mouse following approved protocols and prepare sterile surgical field by clipping the hair (not necessary in nude mice), and wiping with 70 % ethanol and betadine solution.
2. Make a 0.5–1 cm ventral midline skin incision to expose the abdominal wall (see Note 4).
3. Carefully expose the testis, the testicular artery and epididymis, and detach the tail of the epididymis from the gubernaculum by blunt dissection (Fig. 1a).

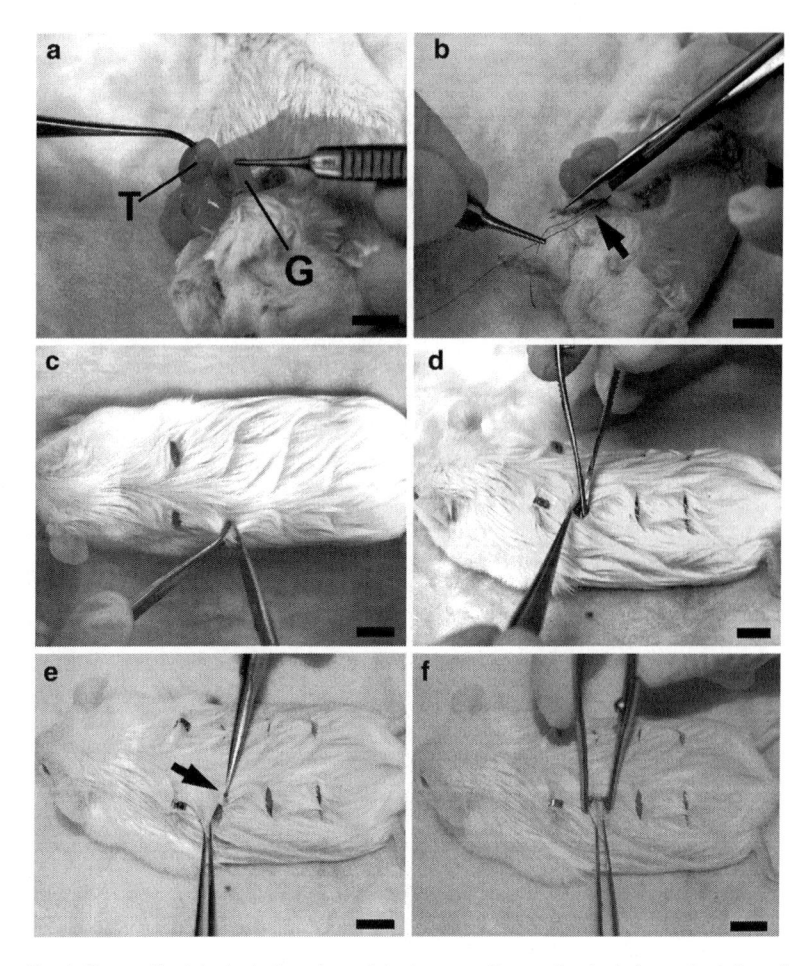

Fig. 1. Key methodological steps in recipient preparation and ectopic transplantation of testicular tissue. (**a**) Photograph illustrating the detachment of the testis (T) from the gubernaculum (G) prior to ligation of the testicular artery and annexed structures. (**b**) Once the testis has been detached from the gubernaculum, the testicular artery and vas deferens are ligated. In this photograph the ligature has already been placed (*arrow*) and the sectioning of the ligated structures is about to be performed. (**c**) Once the mouse has been castrated and its back has been aseptically prepared, 0.5–1 cm incisions are made in the skin to introduce each testis fragment. (**d**) In order to place each testis fragment under the dorsal skin, the subcutaneous tissue is teased apart with small scissors to produce a small cavity. (**e**) The testis fragment is placed with fine forceps deep into the subcutaneous cavity by holding the border of the skin incision with small forceps to expose the cavity. (**f**) Once the testis fragment has been placed, the incision is closed with a Michel clip. Bars = 1 cm.

4. Ligate the testicular artery and the vas deferens together with the blood vessel with silk, and section the ligated structures by cutting between the testis and the ligature (Fig. 1b).
5. Repeat the procedure for the second testis.
6. Suture the abdominal wall with one or two surgical stitches.
7. Close the skin incision with one or two Michel clips.

494   J.R. Rodriguez-Sosa et al.

3.1.3. Ectopic Grafting
(Summarized in Fig. 1c–f)

1. Position the mouse in ventral recumbency and prepare a sterile surgical field on its back as above.
2. Depending on how many grafts are to be inserted (generally 4–8/mouse), make ~0.5 cm long skin incisions on each side of the midline of the back of the mouse (see Note 5).
3. Use forceps to hold a border of the skin incision, and make a subcutaneous cavity by teasing apart the connective tissue using scissors.

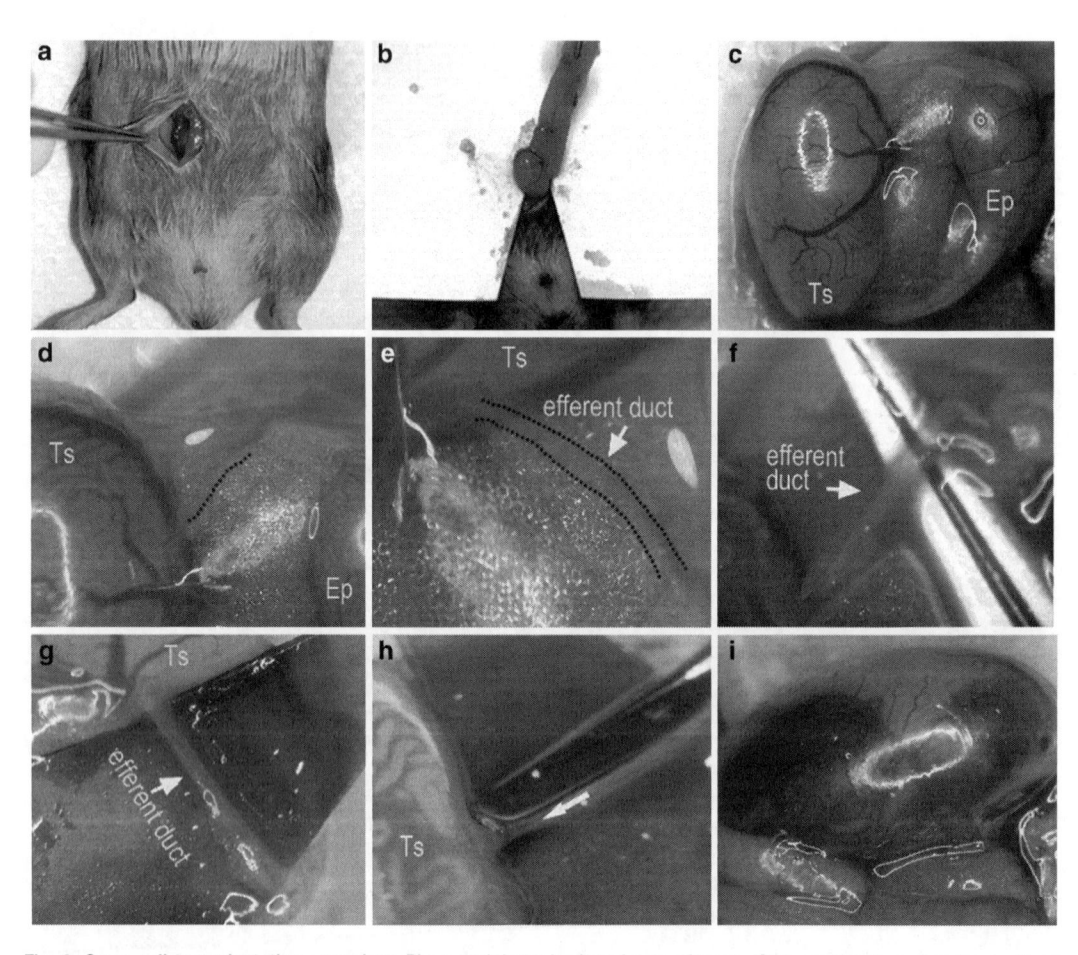

Fig. 2. Germ cell transplantation procedure. Place recipients in dorsal recumbence after anesthesia and make a ~1 cm midline abdominal incision (**a**). Expose testis by withdrawing the fat pad attached to the epididymis and testis and place a thin sterile drape underneath the fat pad/testis for better visual identification (**b**). Position the testis and epididymis so that the efferent ducts buried in the fat pad are discernible (**c–e**). Identify the efferent ducts and gently remove fat tissue around the ducts. A piece of colored paper or plastic can be placed underneath the ducts for better visualization (**g**). Break or grind the pipette tip according to the size of the ducts and load cell suspension into the pipette (**h**). Carefully insert the pipette into a duct in the bundle of efferent ducts, gently thread a few millimeters toward the testis (*the arrow* in H shows the direction of pipette injection and threading). A testis with successful injection into the seminiferous tubules is shown (**i**). *Ts* Testis, *Ep* epididymis.

42 Testicular Tissue Grafting and Male Germ Cell Transplantation

4. Using an iris forceps, place a piece of testis tissue deep into the subcutaneous cavity, holding the border of the skin incision with another iris forceps.
5. Close the skin incision with one Michel clip and keep mouse on heating pad until it starts to recover from anesthesia.
6. Transfer the mouse to a cage with additional insulation and cover and monitor until mice are fully recovered (see Note 6).

3.1.4. Collection of Testicular Grafts for Analysis and Sperm Harvesting

1. Sacrifice the host mouse according to animal care and use guidelines and make a midline skin incision on the back skin running from the tail to the neck, and open the skin. This exposes the grafts which can be located either on the subcutaneous tissue or attached to the skin (see Note 7).
2. Carefully remove the grafts using a pair of forceps and a pair of scissors.
3. Record the number of grafts recovered, along with the size and weight of individual grafts.
4. Retrieve the seminal vesicles from the abdomen of the mouse and record their weight as an indication of testosterone production by the grafted tissue (see Note 8).
5. For histological analysis, suspend xenografts into a sample vial containing Bouin's solution (or other fixative) in a volume ~10 times that of the xenograft, and label the vial appropriately. Incubate overnight in the refrigerator followed by washing at least three times in 70 % ethanol preferably at intervals of 24 h, and proceed for processing and embedding in paraffin (see Note 9).
6. For sperm harvesting, wash xenografts by spinning them down at $300 \times g$ for 1 min and resuspend them in culture medium containing antibiotics. Cut grafts into small pieces and mince carefully with the forceps in a tissue culture dish containing 3–5 ml of culture medium. Finally, filter minced tissue through the 40-μm cell strainer.

3.2. Male Germ Cell Transplantation

3.2.1. Preparation of Recipient Mice with Busulfan

1. Recipients should be immunologically tolerant (either genetically matched to donors or immunodeficient) to the donor testis cells.
2. Recipients should be either naturally devoid of spermatogenesis (e.g., W/W^v mice) or depleted of endogenous germ cells. Germ cell depletion can be achieved by irradiation or chemotherapeutic drugs such as Busulfan. In this protocol, we describe the method of preparing recipient mice with Busulfan.
3. Treat recipients at 4–6 week of age through intraperitoneal injection of Busulfan (see Note 10). The optimal dose of Busulfan is strain dependent. In commonly used recipient strains, a dose of 40–50 mg/kg is sufficient to deplete endogenous

germ cells (e.g., 44 mg/kg for nude mice, 50 mg/kg for B6/129 F1 recipients).

4. Dissolve Busulfan powder in DMSO and then add an equal volume of sterile distilled water to make a final concentration of 4 mg/ml. Keep solution warm at 35–40 °C before use to avoid precipitation of Busulfan. Discard solution if precipitation is observed.

5. After Busulfan treatment, allow at least 1 month before using recipients. Recipients can be used between 1 and 3 months post-Busulfan treatment.

3.2.2. Preparation of Microinjection Pipettes

1. Choose borosilicate glass pipettes (capillary tubes) with a 1.0 mm outer diameter, a 0.75 mm inner diameter, and a length of 3 in.

2. Siliconize glass pipettes with Sigmacote, rinse pipettes with methanol and blow dry.

3. Pull the pipettes using a pipette puller (see Note 11).

4. Break the pipette tips under a dissecting microscope prior to use to achieve a diameter of approximately 50 μm at the tip.

3.2.3. Preparation of Donor Cells for Transplantation

1. Choice of donor strain is dependent on the experimental question studied. If quantification of spermatogenesis is used as the endpoint, use donors with an easily identifiable genetic marker such as Lac-Z (e.g., B6.129 S7-Gt(ROSA)26Sor/J from Jackson Laboratory) or GFP (e.g., C57BL/6-Tg(CAG-EGFP)1Osb/J from Jackson Laboratory).

2. Prepare enzyme solutions (see Note 1).

3. Collect testes aseptically and remove the tunica albuginea. Spread seminiferous tubules gently with fine forceps to facilitate enzymatic digestion.

4. Transfer tubules into collagenase solution, incubate at 37 °C for 5–10 min, and agitate frequently.

5. Wash tubules twice by spinning (200–300 × g for 3 min) and resuspend in PBS w/o Ca^{2+}.

6. Resuspend tubules in trypsin–EDTA and shake until they become sticky and cloudy.

7. Monitor the digestion of tubules at 37 °C as it should occur within 1–2 min.

8. Add DNase solution and shake well. Incubate for 1–2 min.

9. Add 3 ml of FBS to stop the action of enzymes.

10. Filter the cell suspension using a nylon mesh with 40–70 μm pore size to remove cell/tissue chunks.

11. Collect cells at 600 × g for 5 min and resuspend cells in a small amount of DMEM (<100 μl).

12. Count cells and adjust the volume to a desired final concentration (usually 100×10^6). Keep cell suspension on ice prior to use.

13. Each Busulfan-treated mouse testis will be filled with only 10–15 µl of cell suspension. Due to potential waste and leakage, aim to prepare 30–50 µl cell suspension per testis.

3.2.4. Transplantation Procedure (Summarized in Fig. 2)

1. Anesthetize recipients. Place in dorsal recumbency and surgically prepare the abdominal area. Make a ~1 cm midline abdominal incision to expose the abdominal wall.

2. Lift abdominal wall by using a small forceps at the point of the white line to avoid accidentally injuring abdominal organs, and proceed to make a ~0.5 cm incision at the midline of the abdominal wall to expose the peritoneal cavity.

3. Use one iris forceps to hold the abdominal wall and use another pair of iris forceps to search for the fat pads attached to the epididymis and testis in the peritoneal cavity.

4. Gently pull the fat pad out until the testis is exteriorized and the testicular artery and epididymis are clearly visible. Work on one testis at a time.

5. Place a thin sterile drape made from autoclaved index cards underneath the fat pad/testis for better visual identification (optional). The drape also works to absorb fluid.

6. Add a drop of Trypan blue dye into the cell suspension and carefully load the cell suspension into the polyethylene tubing connected to a 1 ml syringe. Attach the pulled pipette into the tubing and gently force the cell suspension into the pipette by applying pressure to the syringe (see Note 12).

7. Identify the efferent ducts (that connect the testis to the epididymis) and gently remove fat tissue around the ducts. Work carefully as the ducts and the membrane around them are translucent.

8. Carefully insert the pipette into a duct in the bundle of efferent ducts; gently thread a few millimeters toward the testis.

9. Hold the injection pipette in place with one hand, and use the other hand to reach for the syringe plunger. Gently depress the plunger to ensure that the suspension flows into the rete testis and the seminiferous tubules begin to fill.

10. The injection rate and flow of cell suspension is regulated by thumb pressure. Avoid sudden increase in pressure; monitor the movement of suspension in tubules. Stop the injection when almost all surface tubules have been filled before the testis starts to become ischemic (see Note 13).

11. Return the testis to the abdominal cavity. Repeat procedure on the contralateral testis. Suture the abdominal wall with 6-0 silk suture and close the skin with metal wound clips. Monitor mice on a warming pad until full recovery.

3.2.5. Analysis of Recipient Testes

1. Allow 2–4 months after transplantation before analysis.
2. Sacrifice the recipient mouse according to animal care and use guidelines and collect the testis and epididymis into PBS. When a Lac-Z transgenic donor strain has been used, the epididymis can be used as a positive staining control as it has endogenous beta-galactosidase activity.
3. For detection of donor cells by Lac-Z staining, fix the testis for 1–2 h at 4 °C in 4 % paraformaldehyde (PFA) in PBS. Rinse 3×30 min at room temperature in lacZ rinse buffer.
4. Incubate overnight at 37 °C in lacZ staining solution. Testis can be stained as a whole or after dispersing the tubules.
5. Fix and store in 10 % neutral-buffered formalin. If sections are analyzed use neutral fast red as counterstain.

4. Notes

1. The amount described is for digesting two donor mouse testes. Scale the solution accordingly if digestion of more or fewer testes is intended.
2. Handle Busulfan with caution. Measure and dissolve Busulfan power in a chemical hood. Allow Busulfan to complete dissolve in DMSO before adding water. Add water very slowing to the Busulfan/DMSO solution. Keep Busulfan solution warm in a preheated block during intraperitoneal injection and discard solution if precipitation is observed.
3. Testes can be stored overnight at 4 °C prior to preparation of pieces for transplantation without affecting the ability of testis to survive and develop after grafting.
4. Castration can also be performed through scrotal incisions.
5. Spermatogenesis and development is variable between xenografts. Generally, xenografts located closest to the limbs show the best grafting outcome.
6. In some cases gonadotropin supplementation to recipient mice may be required to improve development and germ cell differentiation of xenografts. Horse and monkey xenografts show higher development and germ cell differentiation when hCG is applied to recipient mice (16, 17).
7. Individual grafts can be removed surgically by anesthetizing the mouse and removing the graft through a skin incision. However, this may affect the endocrine balance similar to the situation in hemi-castration.

8. The seminal vesicles are androgen-dependent accessory sex glands. Their weight is an indicator of the levels of bioactive testosterone in the host mouse. In intact male mice, seminal vesicles weigh 100–300 mg, while in castrated male mice the weight is <10 mg.

9. If the grafts are large (weighing over 100 mg), they should be cut into small pieces to assure sufficient penetration of the fixative.

10. For transplantation to work efficiently, choice and treatment of recipient animals is important. Recipients should be immunologically tolerant (either genetically matched to donors or immunodeficient) to the donor testis cells. Recipients should be depleted of endogenous germ cells either by irradiation or by chemotherapeutic drugs such as Busulfan. Alternatively, recipients that are naturally devoid of spermatogenesis (e.g., W/W^v mice) can be used without germ cell depletion.

11. Different pipette puller machines have different settings, therefore, test a few settings. Generally, a setting similar to that used for making enucleation pipettes will work.

12. Break the pulled pipette tip under a dissecting microscope before attaching it to the tubing. Make sure the diameter of the broken tip does not exceed the diameter of the ducts. After loading the pipette, test and adjust the flow of cell suspension on a piece of sterile absorbent paper to ensure that there is no solution leakage without pressure applied. Also, make sure that the pipette is not blocked by cell aggregates.

13. Watch the speed of tubule filling closely under a dissecting microscope. Depending on the total cell number injected and the number of stem cells present in the injected cell suspension, injection volume can be customized. With good practice, a greater than 70 % filling of the tubules can be easily achieved.

References

1. Honaramooz A et al (2002) Sperm from neonatal testes grafted in mice. Nature 418:778–781
2. Rodriguez-Sosa JR, Dobrinski I (2009) Recent developments in testis tissue xenografting Reproduction 138:187–194
3. Schlatt S, Rodriguez-Sosa JR, Dobrinski I (2011) Testicular xenografting. In: Orwig K, Hermann B (eds) Male germline stem cells: developmental and regenerative potential. Springer/Humana Press, New York, pp 205–225, Chapter 10
4. Brinster RL, Avarbock MR (1994) Germline transmission of donor haplotype following spermatogonial transplantation. Proc Natl Acad Sci USA 91:11303–11307
5. Brinster RL, Zimmermann JW (1994) Spermatogenesis following male germ-cell transplantation. Proc Natl Acad Sci USA 91:11298–11302
6. Nagano M, Avarbock MR, Brinster RL (1999) Pattern and kinetics of mouse donor spermatogonial stem cell colonization in recipient testes. Biol Reprod 60:1429–1436

7. Nagano MC (2003) Homing efficiency and proliferation kinetics of male germ line stem cells following transplantation in mice. Biol Reprod 69:701–707
8. Kubota H, Avarbock MR, Brinster RL (2004) Growth factors essential for self-renewal and expansion of mouse spermatogonial stem cells. Proc Natl Acad Sci USA 101:16489–16494
9. Oatley JM et al (2006) Identifying genes important for spermatogonial stem cell self-renewal and survival. Proc Natl Acad Sci USA 103:9524–9529
10. Nagano M et al (2001) Transgenic mice produced by retroviral transduction of male germ-line stem cells. Proc Natl Acad Sci USA 98:13090–13095
11. Ryu BY et al (2007) Efficient generation of transgenic rats through the male germline using lentiviral transduction and transplantation of spermatogonial stem cells. J Androl 28:353–360
12. Hess RA et al (2006) Mechanistic insights into the regulation of the spermatogonial stem cell niche. Cell Cycle 5:1164–1170
13. Oatley MJ, Racicot KE, Oatley JM (2011) Sertoli cells dictate spermatogonial stem cell niches in the mouse testis. Biol Reprod 84:639–645
14. Dobrinski I, Rathi R (2008) Ectopic grafting of mammalian testis tissue into mouse hosts. In: Xou SX, Singh SR (eds) Methods in molecular biology, germ line stem stem cells. Humana Press, New Jersey, pp 139–148, Chapter 10
15. Rodriguez-Sosa JR, Schlatt S, Dobrinski I (2011) Testicular tissue transplantation for fertility preservation. In: Seli E, Agarwal A (eds) Fertility preservation: emerging technologies and clinical applications. Springer, New York, Chapter 25
16. Rathi R et al (2006) Germ cell development in equine testis tissue xenografted into mice. Reproduction 131:1091–1098
17. Rathi R et al (2008) Maturation of testicular tissue from infant monkeys after xenografting into mice. Endocrinology 149:5288–5296

Part XI

Transgenic Techniques

Chapter 43

Transgenic Modification of Spermatogonial Stem Cells Using Lentiviral Vectors

Christina Tenenhaus Dann

Abstract

The continuous production of spermatazoa throughout the reproductive lifetime of a male depends on the maintenance of a pool of progenitor cells called spermatogonial stem cells (SSCs). SSCs represent a very small fraction of the cellular population in the testes and lack definitive molecular markers for their identification. The discovery of conditions that allow one to propagate mouse SSCs in vitro essentially indefinitely has truly facilitated studies of the molecular mechanisms regulating SSC function. While multiple conditions for culturing SSCs have now been described, here we detail a method for culturing SSCs that uses a simpler medium than the original formulation. As with numerous other primary and stem cell cultures, it is difficult to introduce DNA into cultured SSCs using standard transfection approaches. However, VSV-G pseudotyped lentivirus efficiently infects cultured SSCs with minimal toxicity. Here we present protocols for producing lentivirus and stably modifying the genome of cultured SSCs using lentiviral vectors.

Key words: Spermatogonial stem cell, GS cell, Germline stem cell, Transgene, Lentivirus, Tissue culture, Germ cell

1. Introduction

Spermatogonial stem cells (SSCs) have the developmental capacity to give rise to an unlimited number of spermatazoa, ensuring continued fertility throughout the reproductive lifetime of males. SSCs are also the key to reestablishment of spermatogenesis following tissue damage and by transplanting SSCs into a sterile animal one can restore fertility. Aside from clinical applications in reproductive medicine, it has been proposed that SSCs could also serve as a source of cells for a variety of other types of cell therapy because SSCs exhibit remarkable plasticity (1, 2). For instance, SSCs have the capacity for "dedifferentiation" into pluripotent ES-like cells in vitro without introducing exogenous factors into the cells (3–10).

Douglas T. Carrell and Kenneth I. Aston (eds.), *Spermatogenesis: Methods and Protocols*, Methods in Molecular Biology, vol. 927,
DOI 10.1007/978-1-62703-038-0_43, © Springer Science+Business Media, LLC 2013

Intriguingly, SSCs/spermatogonia can even transdifferentiate directly into certain somatic cell types, such as prostatic epithelial cells, when introduced into an appropriate microenvironment (11). Determining how regulatory factors within the SSCs function to ensure the continuous, yet controlled production of spermatogenic progenitors, without the generation of pluripotent or tumorogenic cells is an important question, particularly if the therapeutic potential of SSCs is to be realized.

SSCs are a very rare cellular population dispersed amongst millions of somatic and germ cells in the adult male testis (e.g., 0.01% of 30 million testicular cells in the mouse) (12, 13). Further, SSCs cannot be distinguished from other undifferentiated spermatogonia in vivo because definitive markers for the SSCs are lacking (14). For these reasons, elucidation of the molecular details governing SSC development has been challenging. Fortunately, a major breakthrough came in 2003 when conditions were discovered for propagating mouse SSCs in long-term in vitro cultures, termed "germline stem" (GS) cells (15). The method begins with the derivation of a primary cell culture from the testes. During the initial days of culture the germ cells are supported by somatic cells from the testes. Subsequently, the germ cells are cultured on mitotically inactivated mouse embryonic feeder cells in a complex medium containing basic fibroblast growth factor (bFGF) and glial cell line-derived neurotrophic factor (GDNF), critical growth factors that stimulate proliferation and maintenance of the SSCs (16, 17). GS cells proliferate to form clusters of interconnected cells, including SSCs and differentiating transit-amplifying cells, on the surface of the feeder layer (15).

The original formulation of GS cell growth medium consists of Stem Pro, which is a proprietary base medium and supplement from Life Technologies, plus an additional 22 ingredients. This complete medium currently costs more than $450 per 500 mL. Also, maintaining numerous stock solutions and preparing fresh GS growth medium frequently requires a considerable investment of time. To save time and money, simpler formulations of media for maintenance of GS cells have been described. However, when GS cells are grown in these media the germ cell clusters exhibit reduced adherence to the feeder cells (C.T.D. unpublished data and (18)). All things considered, GS cells require somewhat unusual handling compared to standard fibroblast cell cultures. Preparation of a "simple" medium formulation (F12GFB) for GS cell growth and the method for routine GS cell maintenance using F12GFB are described in detail here (Table 1 and Subheading 3.3). Despite poor adherence, GS cell cultures propagated in F12GFB have been shown to maintain a population of functional SSCs based on colony formation in a transplantation assay, the gold standard for measuring SSC self-renewal (C.T.D. unpublished data). Kanatsu-Shinohara et al. (2006) have also reported that adherence to MEFs is not directly required for SSC self-renewal (19).

Table 1
F12GFB medium components

Step	Component	Final concentration
1. Base media		
	DMEM/nutrient mixture F12-HAM	1×
2. Add nongrowth factor components		
	L-Glutamine	2 mM
	100× Antibiotic–Antimycotic	1×
	Fetal bovine serum (see Note 14)	1%
	2-Mercaptoethanol (see Note 15)	55 µM
	50× B27 Supplement Minus Vitamin A	1×
3. Add growth factor components		
	Recombinant human basic FGF (see Note 16)	10 ng/mL
	Recombinant rat GDNF (see Note 17)	10 ng/mL

In order to use the GS cell culture model system to study the mechanisms regulating SSC self-renewal and differentiation, it is desirable to have a method for genetically manipulating GS cells. For instance, RNA interference is a useful tool for assessing gene function and can be accomplished through introduction of a transgene that codes for small hairpin RNA (shRNA) expression in cells. However, similar to other primary and stem cells, GS cells are difficult to transfect by standard means such as lipofection or electroporation (20). On the other hand, an effective tool for introducing foreign DNA stably into the genome of GS cells with minimal toxicity is lentivirus (21, 22). Lentiviral transgenes are advantageous in that they are stably expressed, unlike other retroviral vectors that seem to be prone to silencing (23, 24). A method for preparation and application of lentivirus to genetically modify GS cells is presented here (see Subheadings 3.4–3.7). The shRNA expression vector, pLLNeo-Oct3, which can be used to knockdown rat or mouse *Pou5f1* (Oct4) expression, is used as an example throughout (Fig. 1).

2. Materials

2.1. Plating of Mitotically Inactivated MEFs

1. MEF medium: Dulbecco's Modified Eagle's Medium with 4,500 mg/L glucose containing 1.5 g/L sodium bicarbonate, 15% fetal bovine serum (e.g., Hyclone Characterized), and 1% penicillin/streptomycin. Store at 4°C.

2. 0.2% Gelatin from porcine skin Type A; Add 1.0 g gelatin powder to 500 mL ultra-purified water and autoclave to dissolve and sterilize. Store at room temperature.

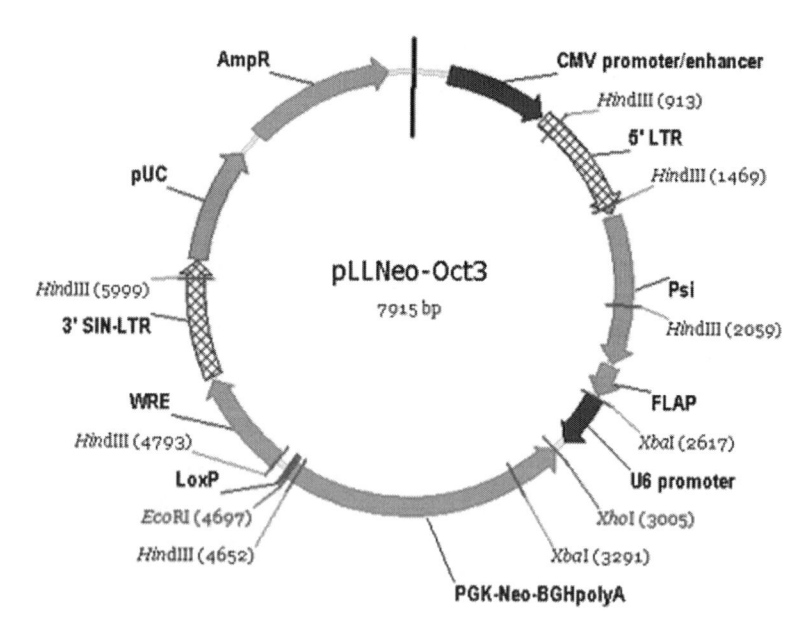

Fig. 1. pLLNeo-Oct3 vector. Elements important for lentiviral production include self-inactivating long-terminal repeats (5′ LTR and 3′ SIN-LTR), Psi HIV packaging signal, central polypurine track (FLAP) and Woodchuck hepatitis viral response element (WRE). The Polymerase III U6 promoter drives expression of an shRNA that targets *Pou5f1* (not labeled). The Phosphoglycerate Kinase promoter drives expression of a Neomycin Resistance gene, which is followed by a bovine growth hormone poly A addition site (PGK-Neo-BGHpolyA); the entire PGK-Neo-BGHpolyA cassette is flanked on either side by unique *Eco*RI and *Xho*I restriction enzyme sites.

3. Cryovial of MEFs; Store in vapor phase of liquid nitrogen upon delivery from supplier. Mitotically inactivated MEFs can be purchased from a limited number of suppliers. MEFs derived from the DR4 strain are well established to support GS cell growth and are resistant to neomycin, hygromycin, puromycin, and 6-thioguanine, important traits relevant when using them as a feeder layer during antibiotic selection for establishing a stably transgenic cell line. However, DR4 MEFs are also relatively difficult to produce and are therefore costly and have limited availability. As an alternative, we have found that the Neomycin resistant MEFs sold by Applied Stem Cell also support GS cell growth.

2.2. Recovery of Cryopreserved GS Cells

1. F12GFB medium (Table 1).
2. Mitotically inactivated MEFs plated in tissue culture dishes as needed.
3. Cryovial of GS cells.

2.3. GS Cell Propagation Using F12GFB Medium

1. F12GFB medium (Table 1).
2. Mitotically inactivated MEFs plated in tissue culture dishes as needed.

3. 0.05% Trypsin.
4. Dulbecco's Phosphate-Buffered Saline (DPBS) without calcium or magnesium.
5. 0.2% Gelatin from porcine skin Type A.

2.4. Preparation of Lentivirus

Portions of this section were reproduced from "Germline Stem Cells" (2008) with permission from Humana Press, a part of Springer Science + Business Media (25).

1. Plasmid DNAs (pMD2G, pMDLg/pRRE, pRSV-REV, pLL-Neo-Oct3 or derivative) (26, 27) prepared with a transfection-grade plasmid DNA isolation kit (e.g., Qiagen EndoFree Plasmid Maxi Kit) and resuspended in sterile TE (pH 7.5) at 0.5–2 mg/mL (see Note 1).
2. 0.2% Gelatin from porcine skin Type A.
3. Low passage number (<20) 293 FT cells (Invitrogen) maintained in T150 flasks or 10 cm plates.
4. 293 medium: Dulbecco's Modified Eagles' Medium with 4,500 mg/L glucose and 3.7 g/L sodium bicarbonate, and supplemented with 10% FBS, 2 mM glutamine and 1% penicillin/streptomycin. Add 500 µg/mL Geneticin (G418) to medium prior to use for cell maintenance.
5. 0.5 M $CaCl_2$ in ultrapure (Millipore) water; filter sterilize and store at 4°C for 2 months.
6. 2× HeBS pH 7.0: 0.28 M NaCl, 0.05 M HEPES, 1.5 mM Na_2HPO_4 anhydrous in ultrapure (Millipore) water; filter sterilize and store at 4°C for 2 months.
7. BW: 2.5 mM HEPES pH 7.3 in ultrapure (Millipore) water; filter sterilize and store aliquots at −20°C.
8. 40 mM Caffeine (see Note 2).
9. 50 mL Steriflip-HV filter unit with 0.45 µm PVDF (Millipore).
10. 30 mL Polyallomer Konical tube (Beckman) and adapters (Beckman, 358156).
11. SW-28 Rotor and Ultracentrifuge (Beckman).

2.5. Determination of Relative Infectious Titer Based on Antibiotic Resistance

1. COS-7 cells (ATCC) or other adherent cell line (see Note 3).
2. 293 Medium (see Subheading 2.4) containing 500 µg/mL G418.
3. 6 mg/mL polybrene (Hexadimethrine Bromide) in water; filter sterilize and store aliquots at −20°C. Add polybrene to medium (6 µg/mL) just prior to use.
4. 1% Crystal violet in 100% ethanol.
5. Phosphate-buffered saline (PBS).

2.6. Transduction of GS Cells with Lentivirus

1. F12GFB medium (Table 1).
2. Titered lentivirus.

2.7. Establishment of Stably Transgenic GS Cell Lines

1. 96-Well plate, 60 mm dish, multi-well plates for expansion.
2. Stereo dissecting microscope.
3. 9 in. glass pasteur pipet pulled over a bunsen burner to a fine capillary.
4. Aspirator tube mouthpiece (e.g., Drummond).
5. Latex tubing to connect mouthpiece to pasteur pipet.
6. Syringe filter (25 mm, 0.2 mM polyethersulfone) connected by latex tubing in line between mouthpiece and pasteur pipet.

3. Methods

3.1. Plating of Mitotically Inactivated MEFs

1. Pipet a sufficient volume of 0.2% gelatin to cover the bottom of wells of culture dishes in which the MEFs will be plated.
2. Incubate the plate(s) at 37°C for at least 10 min and until step 10 (see Note 4).
3. Prepare a 50 mL tube with 9 mL pre-warmed MEF medium.
4. Remove the vial of frozen cells from liquid nitrogen and thaw quickly in a 37°C water bath by holding the vial partly submersed and swirling it gently.
5. When only a small volume of ice remains (about 45 s), decontaminate the outside of the vial by spraying it with 70% ethanol and wipe dry.
6. Use a 2 mL pipette to transfer the cells to the 50 mL tube with 9 mL MEF medium.
7. Centrifuge at $270 \times g$ for 5 min and discard supernatant.
8. Resuspend the pellet in 6 mL MEF medium.
9. Determine the cell concentration using a hemocytometer.
10. Add an appropriate volume of MEF medium to obtain a final cell concentration of 2×10^5 cells/mL.
11. Aspirate the gelatin from plates. The plates do not need to be dry prior to plating the MEFs.
12. Plate out diluted MEFs to multi-well plates at a concentration of 4.2×10^4 cells per cm^2 (e.g., 2 mL per well of 6-well plate).
13. Incubate the plate at 37°C. MEFs should be given time (at least 4 h) to adhere prior to using as a feeder layer for GS cells. The medium on the plated MEFs should be replaced every 3–4 days until they are used for GS cell growth (see Note 5).

3.2. Recovery of Cryopreserved GS Cells

1. Ensure that a well (12-well size) of high-quality mitotically inactivated MEFs is prepared (see Note 5).
2. Remove the vial of frozen cells from liquid nitrogen and thaw quickly in a 37°C water bath by holding the vial partly submersed and swirling it gently.
3. When only a small volume of ice remains (about 45 s), decontaminate the outside of the vial by spraying it with 70% ethanol and wiping dry.
4. Using a 2 mL pipette, transfer the cells to a 15 mL tube containing 5 mL F12GFB.
5. Centrifuge at $270 \times g$ for 5 min and discard supernatant.
6. Resuspend the pellet in 5 mL F12GFB and repeat Step 5.
7. Discard medium from the well of MEFs, rinse briefly with F12GFB and discard.
8. Resuspend the pellet in 1 mL F12GFB and transfer to the well containing rinsed MEFs.
9. Incubate the plate overnight at 37°C.
10. After an overnight incubation most of the living germ cells should be loosely attached to the MEFs. Discard the medium containing cell debris and replace with 1 mL of fresh F12GFB.

3.3. GS Cell Propagation Using F12GFB Medium

During the first 1–2 days following thawing or passaging, GS cells will be single, isolated cells on the MEFs. Subsequently the cells will begin to proliferate to form clusters of interconnected cells (see Note 6). Once clusters comprise approximately 50 or more cells, or no more than 10–14 days after the cells were plated, the culture should be passaged. The way in which one passages GS cells depends on the appearance of the culture and the desired downstream application. GS cell cultures grown in F12GFB include MEFs, floating germ cell clusters and loosely adherent germ cell clusters, and the ratio of floating and adherent clusters varies widely for unknown reasons. Further, the ratio of GS cells to MEFs depends on several factors including the number of GS cells plated and the amount of GS cell proliferation and cell death. In this section, two methods for harvesting GS cells are described that differ in the way that the floating clusters and MEFs are handled. In method 1, the floating clusters are discarded; this method is similar to the routine passaging method used for adherent somatic cell lines. It is appropriate to use method 1 for routine passaging of GS cells if a significant fraction of spermatogonia is attached to the feeder layer. An optional gelatin subtraction procedure at the end of method 1 can be useful for certain applications in which the separation of spermatogonia from MEFs is required; however, gelatin subtraction is not essential during routine passaging of robustly growing cultures. Method 2 presents steps for harvesting a maximal yield of germ cells without

MEFs by trypsinization of both the floating and loosely adherent germ cell clusters. Both methods are presented assuming a single well of a 6-well plate is being processed, but the method can be scaled according to the investigator's needs.

3.3.1. Passaging of GS Cells: Method 1

1. Ensure that a well of healthy MEFs is prepared and available before beginning.
2. Ensure that a significant fraction of germ cells is attached to the MEFs.
3. Remove medium with floating cells and discard.
4. Gently pipet 1 mL PBS (for 6-well) into the well and then remove and discard.
5. Pipet 1 mL trypsin into the well.
6. Incubate the plate at 37°C for 5 min.
7. Pipet trypsin up and down in the well to mechanically disrupt cells and transfer to a conical tube.
8. Pipet 1 mL F12GFB medium into the well and wash off remaining cells. Transfer the supernatant to the tube with the rest of the cells/trypsin. Mix vigorously by pipeting.
9. Centrifuge at $270 \times g$ for 5 min and discard supernatant.
10. Resuspend cells in 1 mL F12GFB medium.
11. Determine the concentration of spermatogonia using a hemocytometer (see Note 7).
12. Plate out 4×10^5 spermatogonia per well of a 6-well plate. For routine passaging this typically corresponds to replating 20% (1:5) of the cells harvested from one well to a new well of the same size. It is expected that the culture will need to be passaged about once per week.

Steps 13–17: Optional gelatin subtraction for separation of spermatogonial fraction from MEFs.

13. Replate the cell suspension from step 10 back on to the original well.
14. Incubate the plate at 37°C for 1.5–2 h. MEFs will be partially adhering to the plastic and germ cells will be floating or loosely attached to the MEFs.
15. Transfer the medium with floating cells into a conical tube.
16. Using a 2 mL serological pipet, very gently pipet an additional 2 mL F12GFB up and down on the well to remove loosely attached germ cells and pool with the cells from step 15. Pay particular attention to the edges of the well, where a significant fraction of the clusters may be found.
17. Visually inspect the well with an inverted microscope to check whether the GS cells are being effectively harvested. If needed, continue to rinse the well very gently with F12GFB to collect all of the germ cells.

3.3.2. Passaging of GS Cells: Method 2

1. Collect the medium with floating clusters and dispense into a conical tube.
2. Pipet 2 mL PBS on to the well (for a 6-well).
3. Wash the loosely adhering clusters off as follows. Pipet the PBS up and down while moving the tip in a swirling motion across the well. The tip should hover closely over the surface of the plate but should not scrape the surface of the plate. Continue to pipet until the germ cell clusters are removed, as determined by visual inspection with an inverted microscope. Pay particular attention to the edges of the well, where a significant fraction of the clusters may remain attached.
4. Transfer the 2 mL of PBS and cells to the conical tube.
5. Wash the remaining cells off with another 1 mL of PBS and add to the tube.
6. Centrifuge at $270 \times g$ for 5 min.
7. Aspirate the supernatant by pipeting and discard.
8. Resuspend cells in 1 mL of 0.05% trypsin by pipeting vigorously up and down.
9. Incubate cells in trypsin for 5 min at room temperature.
10. Add 1 mL F12GFB to cells in trypsin and pipet vigorously up and down.
11. Centrifuge at $270 \times g$ for 5 min and discard supernatant.
12. Resuspend cells in 1 mL of F12GFB by pipeting vigorously up and down.
13. Plate cells out as described in Step 12 of Method 1 (see Subheading 3.3.1).

3.4. Preparation of Lentivirus

The following protocol is written for a single 10 cm plate but is typically scaled up to a minimum of four plates. Portions of this section reproduced from "Germline Stem Cells" with permission from Humana Press, a part of Springer Science + Business Media (25).

1. Apply 5 mL 0.2% gelatin to each 10 cm plate and incubate at 37°C for 10 min. Using gelatin-coated plates improves adhesion of the cells.
2. Remove gelatin and apply 3×10^6 293 FT cells in 10 mL medium (without G418) to each plate.
3. Gently rock the plate to ensure an even distribution of cells and allow cells to adhere ~16 h (overnight) at 37°C (see Note 8).
4. Mix DNAs (3 μg pMD2G, 5 μg pMDLG/pRRE, 2.5 μg pRS-REV, and 10 μg pLLNeo-Oct3 or derivative) with BW to give a final volume of 250 μL (26, 27).
5. Add 250 μL 0.5 M $CaCl_2$ to DNA/BW and mix well.

6. Add the DNA/BW/CaCl$_2$ mixture slowly drop-wise (1 drop every 2 s) to 500 μL 2× HeBS in a 15 mL tube while vortexing at a medium speed.

7. Allow transfection complexes to form undisturbed at room temperature for exactly 30 min.

8. Using a 2 mL serological pipet, slowly apply transfection complexes one drop at a time over the entire plate by placing the tip of the pipette just above the surface of the medium in the plate to prevent disruption of cell adhesion.

9. Incubate ~16 h (overnight) at 37°C.

10. Refresh the transfection complexes with 9.5 mL fresh medium (without G418) (see Note 9). Add 0.5 mL 40 mM caffeine and swirl gently to mix. Use biosafety level 2 guidelines (see Note 10) for this and all subsequent steps involving lentivirus.

11. Incubate ~24 h (overnight) at 37°C.

12. Collect the supernatant medium from the plate ("A" supernatant on cells from 24 to 48 h post-transfection) (see Note 11) into a 50 mL conical tube and replace with 10 mL fresh medium (without G418). Store supernatant medium "A" at 4°C overnight.

13. Incubate ~24 h (overnight) at 37°C.

14. Collect the supernatant medium from the plate ("B" supernatant on cells from 48 to 72 h post-transfection) into a 50 mL conical tube.

15. Cover the surface of the plate with bleach to kill the cells, transfer the bleach to a waste beaker, and discard the plate (see Note 12).

16. Centrifuge tubes of collected supernatant medium at 2,500 × g for 15 min at 4°C to remove cellular debris.

17. Filter supernatant medium using a SteriFlip (0.45 μM) or similar filtration unit.

18. Place conical bottom tubes into the rotor tubes fitted with adapters (SW28 rotor with Beckman Ultracentrifuge) and pipette supernatant into each tube.

19. Centrifuge at 82,705 × g for 90 min at 4°C to pellet lentivirus particles.

20. Pour off supernatant medium into the bleach waste (see Note 13) and use a pipette and/or Kimwipes to remove as much residual liquid as possible from the side of the tube.

21. Apply 40 μL (or 10 μL per 10 cm plate of lentivirus supernatant) of F12GFB to the pellet and incubate for 2–16 h at 4°C.

22. Resuspend the pellet by pipetting up and down 20 times and aliquot (5–20 μL each) prior to storing at −80°C.

3.5. Determination of Relative Infectious Titer Based on Antibiotic Resistance

1. Apply 1×10^5 COS7 cells to each well of a 6-well plate and allow cells to adhere ~16 h (overnight) at 37°C (see Note 3).
2. Aliquot 1 mL of 293 medium containing 6 µg/mL polybrene into 1.5 mL tubes.
3. Add virus (typically 1–5 µL of lentivirus diluted 1:50) to each tube. Test multiple quantities of virus, perform duplicates for each amount of virus being tested and also include one well with no virus.
4. Replace medium in each well of the 6-well plate with medium containing virus.
5. Incubate ~24 h (overnight) at 37°C.
6. Replace medium containing virus with fresh medium.
7. Incubate ~24 h (overnight) at 37°C.
8. Replace medium with fresh medium containing 500 µg/mL of G418.
9. Continue to feed cells every 3–4 days with medium containing G418 until the negative control well contains no living cells (~10–12 days of G418 selection).
10. To stain cells with crystal violet and count colonies, first rinse the wells two times with PBS.
11. Pipet 1 mL crystal violet solution per well and incubate at room temperature for 10 min.
12. Remove crystal violet solution and wash away excess stain by rinsing the wells multiple times with PBS until all excess dye has been rinsed out.
13. The number of colonies obtained will depend not only on the number of viral particles present but also on the activity of the promoter used to drive expression of the Neomycin resistance gene in COS7 cells. As a reference point, transduction with 1 µL of 1:50 diluted lentiviral particles from the pLLNeo-Oct3 vector (see "A" supernatant from Subheading 3.4) would be expected to yield 50–100 colonies; transduction with 5 µL of 1:50 would be expected to lead to confluent cell growth (see Note 12).

3.6. Transduction of GS Cells with Lentivirus

1. Prepare a suspension of GS cells by tryspinization, as in Subheading 3.3.2.
2. Pipet 1×10^5 cells into a 1.5 mL tube, per transduction.
3. Centrifuge at $1,000 \times g$ for 5 min in a microcentrifuge and discard supernatant.
4. Resuspend cells in lentivirus (typically 2–30 µL) and F12GFB totaling 100 µL, mix by pipeting gently up and down, and transfer mixture of cells and lentivirus to a well in a 96-well plate.
5. Incubate ~16 h (overnight) at 37°C.

6. Discard the medium with virus and replace with 100 µL fresh medium.

7. Expression from a lentiviral vector is delayed by about 1 day compared to expression from transfected plasmid DNA and should be evident on the second day after transduction.

3.7. Establishment of Stably Transgenic GS Cell Lines

Note that pLLNeo-Oct3 leads to knockdown of *Pou5f1*, a gene that is required for SSC self-renewal. Therefore, robust growth of G418-resistant clones does not actually occur using this particular vector.

1. Transduce cells as described in Subheading 3.6.

2. Three days after transduction trypsinize the transduced cells (see Subheading 3.3) and plate onto a 60 mm dish (see Note 14). Plate the cells in F12GFB containing 200 µg/mL G418.

3. Feed cells every 3–4 days with F12GFB containing 200 µg/mL G418. G418-resistant clusters should become evident and continue to expand in size. Each cluster should be well isolated from other clusters.

4. Once clusters contain 100+ cells, generally after 10–20 days of growth in G418-containing medium, individual clusters can be isolated manually as follows.

5. Create a fine capillary pipet by heating the middle of the tip of a glass pasteur pipet over a Bunsen burner flame until it is soft. Remove from flame and quickly yet gently pull straight outwards on either end until it is stretched to at least twice the original length. After the glass has cooled (about 3 s) then pull sharply outwards until the narrow region in the middle snaps. Retain the main part of the pipet while discarding the other end. Expect that the diameter of the capillary opening will be much larger than a germ cell cluster; however, if the capillary is too thin it will not be stiff enough to resist surface tension in the culture dish.

6. Using the aspirator mouthpiece and tube assembly (see Subheading 2.7), remove individual clusters from the feeder layer while viewing the clusters under low magnification (about 30× is sufficient) using a dissecting microscope. Transfer each cluster into a well of a 96-well plate containing F12GFB medium with 200 µg/mL G418. Collect at least 10 clusters/clones. The process of transferring a cluster to an individual well will cause it to disintegrate into smaller clusters.

7. Maintain individual clusters in a 96-well plate for 7–14 days by feeding with F12GFB medium with 200 µg/mL G418 every 3–4 days, and then trypsinize and transfer to a larger well (48-well). Continue to expand each clone until sufficient cells are obtained for cryopreservation. Analyze the resulting cell lines as desired.

4. Notes

1. The viral packaging plasmids and their sequences can be obtained at Addgene.org.

2. Addition of caffeine during lentiviral production has been shown to improve titers by three- to eightfold (28, 29). Caffeine has limited solubility in aqueous medium. Add caffeine powder to DMEM base medium containing 1× penicillin/streptomycin to obtain 40 mM final concentration. Vortex and/or mix at room temperature on a rocker until the caffeine has dissolved. Filter sterilize and store at 4°C.

3. Any adherent cell line that is not already resistant to G418 (or the antibiotic to be used for selection) and is amenable to lentivirus transduction may be used for titering (see Note 11).

4. Throughout this chapter incubations at 37°C are to be done in a humidified incubator with 5% CO_2 unless noted otherwise.

5. High-quality MEFs are characterized by homogeneous and complete coverage of the plastic. It is important that MEFs appear as a confluent lawn throughout the well to ensure that GS cell growth is optimal. GS cells can be plated on DR4 MEFs up to 14 days after plating the MEFs; however, Neomycin resistant MEFs (from Applied Stem Cell) lose integrity after 7–10 days.

6. In the first 1 or 2 days following thawing and plating of GS cells, one may have difficulty in distinguishing the germ cells from other cells, such as dead MEFs. However, after about 3 days doublets and larger clusters of spermatogonia should become visible. Generally speaking, cluster size should approximately double each day during the first several days following plating of individualized cells.

7. When using a hemocytometer to count GS cell cultures, spermatogonia are readily distinguishable from MEFs because the spermatogonia have a small, round morphology while MEFs are larger and more irregularly shaped. Also, as the GS cells proliferate the ratio of spermatogonia to MEFs will increase.

8. 293FT cells should be ~70% confluent on the day of transfection, nearly 100% confluent the morning after transfection and may exhibit a decline in confluency thereafter during lentivirus production (depending on the plasmid being packaged).

9. The volume per plate may be varied from 7.5 to 10 mL such that the final supernatant volume to be concentrated is 30 mL, which is the volume constraint of the centrifuge tubes used in ultracentrifugation.

10. The use of lentiviral vectors described here is considered biosafety level 2. The biosafety office at your institution should be

notified prior to using these reagents. The production of virus and handling of concentrated viral stocks should be performed in a biosafety cabinet (tissue culture hood) while wearing two pairs of gloves and a lab coat. All pipettes and plates exposed to infectious virus should be decontaminated with bleach prior to disposal as standard biohazardous material. This maybe accomplished by placing a 500 mL beaker with about 300 mL 50% bleach (in tap water) in the tissue culture hood and using this as a waste beaker as well as a source of bleach for decontaminating plates or pipettes.

11. Generally, the "A" collection of lentivirus tends to have a higher titer than the "B" collection, although this may vary between lentiviral vectors or preparations. Although "A" and "B" may be pooled at any point after their collection, to obtain a maximal concentration of virus it is recommended that the "A" and "B" collections are treated as distinct preparations through the titering procedure. Following titering one may choose to utilize only the higher titer "A" lentivirus and discard the "B" lentivirus.

12. The goal of titering in COS7 cells by antibiotic selection is to determine relative titer, for instance of multiple virus preparations. The method presented will not be useful for obtaining an absolute titer (such as viral particles per volume of virus) because infection rates and activity of the promoter driving Neomycin expression may vary between GS cells and other cell types. Nonetheless, titering in COS7 cells can be useful for confirming production of infectious lentivirus and determining relative titers. For example, one can compare lentivirus preparations of vectors for target shRNA expression and scrambled control shRNA expression to ensure similar infection rates are achieved once vectors are applied in GS cells.

13. As an alternative to antibiotic selection, if the lentiviral vector includes a fluorescence reporter then a flow cytometer can be used to sort transgenic cells and establish a cell line. To generate clonal cell lines by flow cytometry, cells can be sorted directly into wells of a 96-well dish (10 cells per well yields an average of one surviving cluster-forming cell per well). To generate a polyclonal cell line, 10,000 cells sorted into a single well of a 96-well plate will effectively and quickly yield a GS cell line composed of cells containing various transgene integration sites in the genome.

14. Fetal bovine serum (FBS): Heat inactivate FBS prior to use by incubating thawed and mixed FBS at 56°C for 30 min. Aliquot and store at −20°C.

15. 2-Mercaptoethanol stock: Dilute to 10 mM in DMEM/F12 base media immediately prior to use.

16. bFGF: Reconstitute bFGF at a concentration of 25 μg/mL in sterile DPBS containing 0.1% BSA. Aliquot and store at −20°C.

17. GDNF: Reconstitute GDNF at a concentration of 100 μg/mL in sterile DPBS containing 0.1% BSA. Aliquot and store at −20°C.

Portions of this chapter are adapted from a chapter previously published by Christina Dann and David Garbers in Methods in Molecular Biology series: Germline stem cell (Vol 450, pgs 193–209)

Acknowledgments

Thank you to Danielle Fanslow for assistance in preparing the manuscript and Hye-Won Song for comments on the manuscript.

References

1. Kubota H, Brinster RL (2006) Technology insight: in vitro culture of spermatogonial stem cells and their potential therapeutic uses. Nat Clin Pract 2:99–108
2. de Rooij DG, Mizrak SC (2008) Deriving multipotent stem cells from mouse spermatogonial stem cells: a new tool for developmental and clinical research. Development 135:2207–2213
3. Seandel M et al (2007) Generation of functional multipotent adult stem cells from GPR125+ germline progenitors. Nature 449:346–350
4. Mizrak SC et al (2010) Embryonic stem cell-like cells derived from adult human testis. Hum Reprod 25:158–167
5. Kossack N et al (2009) Isolation and characterization of pluripotent human spermatogonial stem cell-derived cells. Stem Cells 27:138–149
6. Ko K et al (2009) Induction of pluripotency in adult unipotent germline stem cells. Cell Stem Cell 5:87–96
7. Kanatsu-Shinohara M et al (2008) Pluripotency of a single spermatogonial stem cell in mice. Biol Reprod 78:681–687
8. Guan K et al (2006) Pluripotency of spermatogonial stem cells from adult mouse testis. Nature 440:1199–1203
9. Golestaneh N et al (2009) Pluripotent stem cells derived from adult human testes. Stem Cells Dev 18:1115–1126
10. Conrad S et al (2008) Generation of pluripotent stem cells from adult human testis. Nature 456:344–349
11. Simon L et al (2009) Direct transdifferentiation of stem/progenitor spermatogonia into reproductive and nonreproductive tissues of all germ layers. Stem Cells 27:1666–1675
12. Tegelenbosch RA, de Rooij DG (1993) A quantitative study of spermatogonial multiplication and stem cell renewal in the C3H/101 F1 hybrid mouse. Mutat Res 290:193–200
13. Nagano MC (2003) Homing efficiency and proliferation kinetics of male germ line stem cells following transplantation in mice. Biol Reprod 69:701–707
14. Oatley JM, Brinster RL (2008) Regulation of spermatogonial stem cell self-renewal in mammals. Annu Rev Cell Dev Biol 24:263–286
15. Kanatsu-Shinohara M et al (2003) Long-term proliferation in culture and germline transmission of mouse male germline stem cells. Biol Reprod 69:612–616
16. Lee J et al (2007) Akt mediates self-renewal division of mouse spermatogonial stem cells. Development 134:1853–1859
17. Kubota H et al (2004) Growth factors essential for self-renewal and expansion of mouse spermatogonial stem cells. Proc Natl Acad Sci USA 101:16489–16494
18. Wu Z et al (2009) Spermatogonial culture medium: an effective and efficient nutrient mixture for culturing rat spermatogonial stem cells. Biol Reprod 81:77–86
19. Kanatsu-Shinohara M et al (2006) Anchorage-independent growth of mouse male germline stem cells in vitro. Biol Reprod 74:522–529

20. Kanatsu-Shinohara M et al (2005) Genetic selection of mouse male germline stem cells in vitro: offspring from single stem cells. Biol Reprod 72:236–240
21. Hamra FK et al (2002) Production of transgenic rats by lentiviral transduction of male germ-line stem cells. Proc Natl Acad Sci USA 99:14931–14936
22. Dann CT et al (2008) Spermatogonial stem cell self-renewal requires OCT4, a factor downregulated during retinoic acid-induced differentiation. Stem Cells 26:2928–2937
23. Pfeifer A et al (2002) Transgenesis by lentiviral vectors: lack of gene silencing in mammalian embryonic stem cells and preimplantation embryos. Proc Natl Acad Sci USA 99: 2140–2145
24. Ikawa M et al (2003) Generation of transgenic mice using lentiviral vectors: a novel preclinical assessment of lentiviral vectors for gene therapy. Mol Ther 8:666–673
25. Dann CT, Garbers DL (2008) Production of knockdown rats by lentiviral transduction of embryos with short hairpin RNA transgenes. Methods Mol Biol 450: 193–209
26. Dann CT et al (2006) Heritable and stable gene knockdown in rats. Proc Natl Acad Sci USA 103:11246–11251
27. Dull T et al (1998) A third-generation lentivirus vector with a conditional packaging system. J Virol 72:8463–8471
28. Ellis BL et al (2011) Creating higher titer lentivirus with caffeine. Hum Gene Ther 22:93–100
29. Ellis BL et al (2011) Creating higher titer lentivirus with caffeine. Hum Gene Ther 22: 93–100

Chapter 44

Methods for Sperm-Mediated Gene Transfer

Marialuisa Lavitrano, Roberto Giovannoni, and Maria Grazia Cerrito

Abstract

The transgenic technologies represent potent biotechnological tools that allow the generation of genetically modified animals useful for basic research and for biomedical, veterinary, and agricultural applications. Among transgenic techniques, we describe here the sperm-mediated gene transfer methods that is gene transfer based on the spontaneous ability of sperm cells to bind and internalize exogenous DNA and to carry it to oocyte during fertilization, producing genetically modified animals with high efficiency.

Key words: Spermatozoa, Transgenic animals, SMGT, Transgenesis, Mouse, Pig

1. Introduction

Transgenic techniques. The transgenic technologies represent potent biotechnological tools that allow the generation of genetically modified animals useful for basic research and for biomedical, veterinary, and agricultural applications. Since the production of the first genetically modified mouse in 1980 (1) and of the first genetically modified livestock in 1985 (2) by DNA microinjection into the zygote's male pronucleus, several methods for producing transgenic animals have been developed (3). To date, the most widely used methods for the production of transgenic animals are (1) direct microinjection of exogenous DNA into the pronucleus of fertilized oocytes (DNA pronuclear microinjection, DNA-PMI), (2) Embryonic Stem (ES) cell transfer, (3) nuclear transfer using genetically modified embryonic or somatic donor cells (Somatic Cell Nuclear Transfer, SCNT), (4) the use of viral-derived DNA sequences as vectors for the introduction of exogenous DNA into embryos (4–6), (5) sperm-mediated gene transfer (SMGT) (7–9).

1.1. Direct Microinjection of Exogenous DNA into the Pronucleus of Fertilized Oocytes

The DNA-PMI was the first transgenesis technique successful in producing transgenic rodents (1) and farm animals (2). This method is based on the microinjection of exogenous DNA into the male pronucleus of a single-cell embryo. The technique is well established, however high embryo-lethality associated with the extensive embryonic manipulation is still a limitation. In pigs, transgenic efficiency of DNA-PMI is reduced due to a very high lipid content of the porcine oocyte requiring the centrifugation of the zygote to visualize the male pronucleus (2).

1.2. Embryonic Stem Cell Transfer

ES cells have been extensively used to transfer genetic modifications to embryos for the production of genetically modified animals with site-specific modifications into the genome. The technique is based on the integration of DNA sequences into the ES cell genome by homologous recombination in vitro and subsequent injection of selected clonal populations of genetically modified ES cells into the blastocoel cavity of a preimplantation embryo, originating in a chimeric animal and, after appropriate breeding, transgenic offspring (10, 11).

1.3. Somatic Cell Nuclear Transfer

The SCNT, i.e., the transfer of a somatic cell nucleus carrying the desired genetic modification into enucleated oocytes, also allows site-specific modification of an animal genome, inserting targeted mutations or deleting endogenous genes (i.e., knock-out of porcine GGTA1 gene) (12, 13) as well as randomly introducing exogenous genes into the genome with the possibility of selecting positive cells generating transgenic animals. SCNT has been demonstrated to be applicable to large mammals, with better efficiency in pigs (14).

1.4. Viral-Derived DNA Sequences as Vectors for the Introduction of Exogenous DNA into Embryos

Viral-derived DNA sequences are used as vectors for the introduction of exogenous DNA into embryos via infection (15). Pioneering experiments using retroviruses to infect murine embryos (16, 17) to produce genetically modified mice date back to the mid-1970s. Today, lentiviral sequences are mostly used to allow genetic modification of preimplantation embryos, with the expression of the transgene and its stable transmission to the progeny (15) both in rodents and in farm animals (18).

1.5. Sperm-Mediated Gene Transfer

The SMGT method is based on the spontaneous ability of sperm cells to bind and internalize exogenous DNA and to carry it to oocyte during fertilization, producing genetically modified animals with high efficiency (7). SMGT may, in principle, be successfully applied to all animal species whose reproduction is mediated by gametes. In the last 20 years, it has been demonstrated that the ability of sperm cells to bind foreign DNA is not restricted to one or few species but represents instead a widespread feature of spermatozoa from all species, ranging from echinoids to mammals.

The first report of mammalian sperm cells being able to act as vectors for foreign DNA was published in 1971 by Brackett et al. (19) but was ignored, only to be rediscovered after the publication of the first SMGT paper in 1989 (7). Since 1989 successful in vitro uptake of exogenous DNA by sperm cells of different animal species followed in most cases by post-fertilization transfer and maintenance of the transgene have been reported (9). The SMGT procedure has been optimized in large mammals (8, 20, 21) and applied to several animal species from silkworm (22, 23) to fish (24, 25), from chicken (26) to rabbit (27), and from goat (28, 29) to bovine (30).

Initial observations indicates that (1) sperm cells from virtually all animal species are able to take up exogenous DNA, (2) DNA uptake can be improved by electroporation or by lipofection, (3) exogenous DNA binding occurs in the subacrosomal segment of the sperm head of different species, and (4) the interaction is ionic, reversible, and sequence-independent.

It has been demonstrated that sperm/DNA interaction is not a random event, but rather the uptake of exogenous DNA by sperm cells is a complex mechanism involving specific factors, which occurs in three steps: (1) binding of exogenous DNA to sperm cells, (2) nuclear internalization of exogenous DNA, and (3) integration of exogenous DNA into sperm chromatin.

More specifically, the exogenous DNA interacts with DNA-binding protein(s) (DBP) of 30–35 kDa on the sperm cell surface, leading to the formation of a DNA/DBP complex which is dependent on the expression of MHC class II and triggers CD4-mediated internalization (31, 32). DNA/DBP/CD4 penetrates inside the nucleus and reaches the nuclear matrix. Subsequently, a small fraction of exogenous DNA undergoes recombination with the sperm chromosomal genome at few selected sites of "accessible" chromatin. Alternatively the exogenous DNA is degraded by a sperm endogenous nuclease, the activity of which is triggered in a dose-dependent manner upon interaction of spermatozoa with foreign molecules (33).

The integration of the exogenous DNA has been reported to occur in a unique site or in more sites of integration within the genome (8, 32), particularly when a combination of different transgenes has been used for SMGT experiments (34).

There are two major powerful natural barriers to antagonize the spontaneous undesired intrusion of exogenous molecules: an inhibitory factor IF-1 abundant in mammalian seminal fluid or bound to the spermatozoal membrane in marine animals, that prevents the binding of foreign molecules, and a nuclease activity present in spermatozoa or in the fish seminal fluid (32, 35). Thus, the removal of seminal fluid from the sperm samples is a crucial step for a successful SMGT protocol. In fact, it has been demonstrated that only epididymal and ejaculated spermatozoa, depleted of trace amount of seminal fluid by sequential washing steps, are able to take up exogenous DNA (20, 33).

Selection of sperm donors and optimization of DNA uptake have been demonstrated to be the key steps for the successful outcome of SMGT.

The nominal parameters that a sperm donor should possess to serve as a good vector for exogenous DNA are the quality of semen based on standard parameters used in conventional animal breeding programs (volume, concentration, presence of abnormal sperm cells, motility at time of collection, and high progressive motility after 2 h) and the ability of the sperm cells to take up and internalize exogenous DNA.

To optimize the protocol for generating transgenic animals, it is necessary to establish when, for how long and in what quantity DNA must be added to sperm.

However, it is not possible to establish a unique SMGT protocol suitable for all species, because of the wide differences in the environmental requirements for fertilization in different species, i.e., water or seawater for amphibians and fish and the female genital tract for mammals. Therefore, differential protocols adapted to different experimental conditions must be developed for each species.

2. Materials

All the chemicals and media are from Sigma Aldrich, unless otherwise indicated. Prepare all the solutions using ultrapure water and analytical grade reagents. Prepare and store all the reagents at room temperature, unless otherwise indicated. All media are filtered through 0.22 μm filters and stored at 4°C under sterile conditions for no more than 4 weeks. Diligently follow all waste disposal regulations when disposing waste materials. All of the procedures involving animals must be performed in accordance to local and national regulations and approved by an ethical committee.

2.1. Mouse: Preparation of Gametes; Sperm/DNA Uptake; In Vitro Fertilization; Embryo Implantation into Foster Mothers

1. Fertilization Medium (FM) is based on Whittingham's Tyrode solution (36), from which sodium lactate, penicillin, and streptomycin are omitted; NaH_2PO_4 is replaced by 0.15 mM Na_2HPO_4 and NaCl is increased to 120 mM: Add about 800 mL of water to a 1-L graduated cylinder. Weight and transfer to the cylinder 6.97 g of NaCl, 2.106 g of $NaHCO_3$, 1 g of Glucose, 0.201 g of KCl, 0.027 g of $Na_2HPO_4 \cdot 2\ H_2O$, 0.264 g of $CaCl_2 \cdot 2H_2O$, 0.102 g of $MgCl_2 \cdot 6H_2O$, and 0.055 g of Na Pyruvate. Mix and add water to 1 L.

2. FM/BSA: Add 4 mg/mL Bovine Serum Albumin Fraction V (BSA) to the FM solution. Store at 4°C until use.
3. Ham's F10 medium: Hypoxanthine-free, supplemented with $NaHCO_3$ (2.106 g/L) and 4 mg/mL BSA. Osmolarity should be between 275 and 290.
4. Falcon 3653 dishes for squeezing epididymis, collecting eggs, fertilizations, and embryo cultures.
5. Silicon oil to cover embryo cultures.
6. Exogenous DNA.
7. Animals: Mouse CD1 and BDF1 strains (Charles River). CD1 males are used as sperm donors. Vasectomized CD1, tested for sterility, are mated with CD1 recipients 8–12 h before implanting the embryos. CD1 recipients are retired females. BDF1 females 3–10 weeks old are used as egg donors.
8. Pregnant Mare Serum Gonadotropin (PMSG).
9. Human chorionic gonadotropin (hCG; Intervet International B.V.).
10. Anesthetic: Concentrated 40× avertine anesthetic stock is prepared as follows: 10 g tribromoethanol are dissolved in 10 mL of *tert*-Amyl alcohol and stored at 4°C. The working solution is obtained by diluting the stock 1:40 with physiological solution.
11. Surgical set.

2.2. Swine: Selection of Donor Boars; Sperm Samples Preparation; Sperm/DNA Uptake; Fertilization

1. Boars of proven fertility.
2. Swine Fertilization Medium (SFM): Add about 800 mL of water to a 1-L graduated cylinder. Weight and transfer to the cylinder 11.25 g of glucose, 10 g of Na citrate $2H_2O$, 4.7 g of EDTA·$2H_2O$, 3.25 g of citric acid H_2O, and 6.5 g of Trizma. Mix and adjust pH to 6.8. Add water to 1 L and store at 4°C until use. Before use pre-warm the solution to 37°C.
3. SFM/BSA: Add 6 mg/mL Bovine Serum Albumin Fraction V (BSA) to the SFM solution. Store at 4°C until use. Before use pre-warm the solution to 25°C.
4. Exogenous DNA.
5. eCG (1,250 IU; Folligon1 Intervet International B.V., Boxmeer, The Netherlands).
6. hCG (750 IU; Corulon1 Intervet International B.V.).
7. Set for artificial insemination.
8. Surgical set for laparoscopic insemination.

3. Methods

3.1. Mouse: Preparation of Gametes

1. Preparation of mouse epididymal spermatozoa. Collect the spermatozoa from the cauda epididymis of fertile males (that have abstained for at least 3 days but no longer than 7 days). Prepare the spermatozoa by squeezing the terminal part of the vas deferens and puncturing the middle part of the epididymis into 1 mL of FM supplemented with 4 mg/mL of BSA overlaid with silicon oil. Allow the sperm suspension to disperse by incubating the drop for 30 min at 37°C in 7% CO_2 in air (see Note 1) (35).

2. Preparation of mouse oocytes. Induce superovulation in mouse females by intraperitoneal injection of 5 IU of PMSG followed by 5 IU of hCG 48 h later. 13 h after hCG injection sacrifice the females by cervical dislocation, remove the oviducts, and place them in a PBS-containing dish. Using a 1-mL syringe squeeze out the oocytes and transfer them to a dish containing FM/BSA, overlaid with silicon oil. Transfer 30–70 eggs to a new dish containing FM/BSA for in vitro fertilization (IVF).

3.2. Mouse: Sperm/DNA Uptake (See Note 2)

1. Count the sperm cells by diluting 10 μl of sperm suspension with 990 μl of HCl 0.1 N, place 10 μl in a hemocytometric chamber and count the cells. Dilute the sperm suspension to a final concentration of 1–2 million cells/mL (7, 35).

2. Add the exogenous DNA (see Note 3), at a concentration ranging from 0.01 to 1 μg/10^6 sperm cells, to the sperm suspension and incubate for 30 min at 37°C in 7% CO_2 in air.

3.3. Mouse: In Vitro Fertilization

1. Transfer $1-2 \times 10^6$ spermatozoa from the sperm/DNA incubation experiment to the egg-containing dishes and place the dish in the incubator for 5 h.

2. Transfer non-degenerated eggs into dishes containing Ham's F10 and incubate overnight (see Note 4). After 26–28 h evaluate the fertilization rate by determining the percentage of two-cell stage embryos (see Note 5).

3.4. Mouse: Embryo Implantation into Foster Mothers

1. Transfer two-cell stage embryos to a pre-warmed dish containing PBS supplemented with 1 mg/mL of BSA and overlaid with silicon oil.

2. Collect 15–20 embryos in PBS with a micropipette and surgically implant them into the oviduct of plugged CD1 mouse females mated with vasectomized males 8–12 h before. The procedure is performed under anesthesia (avertin).

3.5. Mouse: Birth and Genotyping of Offsprings

1. Let the deliveries go to term.
2. Collect biopsies from tail and ear of newborn mice.
3. Detect the presence of the transgenic constructs in the biopsies following standard procedures (37).

3.6. Swine: Selection of Donor Boars and Optimization of Sperm/DNA Uptake (See Note 6)

1. Collect the semen from donor boars of proven fertility in a sterile plastic bag placed in a thermostatic container pre-warmed to 37°C to avoid temperature shock for spermatozoa. Discard the first fraction of the ejaculate (see Note 7) and collect the initial 40% of the second fraction, filtered through sterile gauze.

2. Evaluate the quality of semen on a slide pre-warmed to 37°C by measuring sperm concentration, sperm vitality, sperm motility. Reevaluate sperm motility after the washing procedures described in Subheading 3.6 (see Note 8). Among the boars evaluated, select the ones with the best scores in semen parameters.

3. The semen of the selected boars should then be evaluated on the basis of the capacity to uptake exogenous DNA after the removal of seminal fluid. Prepare time-course experiments of sperm/DNA uptake to test the amount of DNA (ranging from 20 ng to 1 µg/10^6 spermatozoa) to be added to sperm, time and temperature of incubation, as described in Subheadings 3.6–3.8. Assess the sperm/DNA uptake by incubating $5-10 \times 10^6$ washed sperm cells/mL (see Subheading 3.6) with linearized, radiolabeled DNA at different temperatures (time-course experiments at 17, 20, 25, or 37°C). At specific time-points withdraw an aliquot (10^6 sperm cells) from the incubation solution, dilute it in 1 mL of SFM and wash twice by centrifuging at $1,500 \times g$ for 5 min in a microfuge. Prepare nuclei as previously described (38) and analyse them by scintillation counting.

3.7. Swine: Sperm Samples Preparation

1. Semen collection, refer to Subheading 3.6.

2. Remove the seminal fluid by carefully washing the sperm. Transfer aliquots (5 mL) of semen in 50-mL Falcon tubes (Becton Dickinson, Milan, Italy) and dilute the sperm suspension 1:10 with SFM pre-warmed to 37°C. Centrifuge the sperm/SFM ($800 \times g$) for 10 min at 25°C. Remove and discard the supernatant by aspiration without disturbing the pellet. Resuspend the pellet in approximately 45 mL of SFM/BSA, pre-warmed to 25°C. Centrifuge the sperm/SFM/BSA ($800 \times g$) for 10 min at 17°C. Remove and discard the supernatant by aspiration without disturbing the pellet. Count the spermatozoa in a hemocytometric chamber and resuspend the cells at desired working dilutions (see Note 9).

3.8. Swine: Sperm/DNA Uptake

1. Add the proper amount of DNA, resuspended in SFM/BSA, to washed sperm cells and incubate the sperm/DNA solution for up to 2 h at the proper temperature.

2. Gently invert the tube every 20 min to avoid sedimentation of sperm cells.

3.9. Swine: Fertilization

Several fertilization procedures can be used.

1. Artificial insemination can be performed in normally cycling or in artificially induced sows.
2. Induce synchronization in prepubertal gilts by intramuscular injection of eCG, followed 60 h later by hCG. Check the signs of oestrous every 6 h beginning 32 h after hCG injection.
3. 36 h After hCG injection perform insemination (10^9 sperm incubated with DNA) by using an inseminating pipette (Melrose Catheter) according to standard procedures.
4. Alternatively, 36 h after hCG injection perform laparoscopic insemination at the utero-tubal junction with 5×10^8 DNA-incubated spermatozoa per uterine horn (see Fantinati et al. for details) (39).
5. Let the deliveries go to term.
6. Collect biopsies from tail and ear of newborn piglets.
7. Detect the presence of the transgenic constructs in the biopsies following standard procedures (37).

4. Notes

1. It is recommended to pool sperm cells squeezed from the epididymis of two males.
2. A modification of SMGT method is based on the removal of the sperm membrane before DNA/sperm cell incubation, followed by the injection of the DNA-treated sperm head within the oocyte's cytoplasm (Intra-Cytoplasmic Sperm Injection, ICSI). This so-called ICSI-SMGT method has been successfully applied to mouse (40) and large mammals (41).
3. Plasmid DNA can be used either in circular form or linearized. In case of linear DNA, the plasmid is usually linearized using restriction enzymes that separate the eukaryotic expression cassette from the vector backbone sequences.
4. The transfer of non-degenerated eggs from FM/BSA medium to Ham's F10 medium, outside the incubator, should be performed as quickly as possible to reduce embryo damage.
5. The epididymal spermatozoa routinely give a fertilization rate of 60–100% under optimal conditions and the incubation with DNA does not usually affect this efficiency.
6. Selection of sperm donors is based on (1) the evaluation of the quality of semen based on standard parameters used in

conventional animal breeding programs, (2) the ability of the sperm cells to take up and internalize exogenous DNA. To optimize the sperm/DNA interaction, timing, duration, and quantity of DNA to be added to sperm should be established. There is a window of opportunity that coincides with the early stage of capacitation, DNA is ideally added within 30 min after washing the sperm and capacitation time should be modulated to allow for complete interaction between the sperm and the DNA (2–4 h). Since capacitation is calcium dependent after removal of seminal fluid, a calcium-free medium (SFM) in which the capacitation process is slowed down is recommended. Calcium-free medium is also important to avoid the activation of calcium-dependent endogenous endonucleases. To allow capacitation in the absence of seminal fluid and calcium, it is possible to vary the temperature (from 17 to 37°C) and the BSA concentration (from 6 to 30 g/L SFM). DNA should be added to sperm in sufficient amount so as to be taken up by the majority of sperm cells and nuclei without overloading, to prevent endogenous nucleases activation.

7. The initial fraction of ejaculate contains a small amount of spermatozoa and the cells are less viable and have reduced motility.

8. The selection of the semen donor and the quality of semen are crucial factors for the success of the SMGT procedure (20). The quality of the semen collected for the SMGT experiment could influence the entire experiment, so it is very important to perform the SMGT procedure on semen that has been properly selected. Good semen parameters are vitality of at least 90%, motility not below 80% initially and 65% after the washing procedure, concentration >10^6 spermatozoa/mL, and cells with abnormal morphology less than 15% (39).

9. To set up the experimental conditions for sperm/DNA uptake, it would be useful to test the best combination of DNA concentration, time and temperature of sperm/DNA incubation, as described in Subheading 3.6. Incubation solution of 10^8 cells per mL is recommended and also allows time-course experiments for testing the above-mentioned experimental conditions. In the SMGT procedure, 10^9 spermatozoa are incubated with the proper amount of DNA in 10 mL SFM/BSA. At the end of the incubation time, for laparoscopic insemination, the sperm/DNA solution is split into two aliquot of 5 mL per uterine horn while for artificial insemination the incubation solution is diluted to 120 mL (39).

Acknowledgments

This work was supported by the European Commission's Sixth Framework Programme, under the priority thematic area "Life Sciences, Genomics and Biotechnology for Health," contract no. LSHB-CT-2006-037377, Xenome, and by the Italian Minister of Research and University (FIRB, RBAP06LAHL).

References

1. Gordon JW et al (1980) Genetic transformation of mouse embryos by microinjection of purified DNA. Proc Natl Acad Sci USA 77:7380–7384
2. Hammer RE et al (1985) Production of transgenic rabbits, sheep and pigs by microinjection. Nature 315:680–683
3. Wall RJ (2002) New gene transfer methods. Theriogenology 57:189–201
4. Kues WA, Niemann H (2004) The contribution of farm animals to human health. Trends Biotechnol 22:286–294
5. Aigner B et al (2010) Transgenic pigs as models for translational biomedical research. J Mol Med 88:653–664
6. Sachs DH, Galli C (2009) Genetic manipulation in pigs. Curr Opin Organ Transplant 14: 148–153
7. Lavitrano M et al (1989) Sperm cells as vectors for introducing foreign DNA into eggs: genetic transformation of mice. Cell 57:717–723
8. Lavitrano M et al (2002) Efficient production by sperm-mediated gene transfer of human decay accelerating factor (hDAF) transgenic pigs for xenotransplantation. Proc Natl Acad Sci USA 99:14230–14235
9. Lavitrano M et al (2006) Sperm-mediated gene transfer. Reprod Fertil Dev 18:19–23
10. Capecchi MR (1989) Altering the genome by homologous recombination. Science 244: 1288–1292
11. Koller BH, Smithies O (1992) Altering Genes in Animals by Gene Targeting. Annu Rev Immunol 10:705–730
12. Tseng Y-L et al (2005) alpha1,3-Galactosyltransferase gene-knockout pig heart transplantation in baboons with survival approaching 6 months. Transplantation 80:1493–1500
13. Yamada K et al (2005) Marked prolongation of porcine renal xenograft survival in baboons through the use of alpha1,3-galactosyltransferase gene-knockout donors and the cotransplantation of vascularized thymic tissue. Nat Med 11:32–34
14. Lagutina I et al (2007) Comparative aspects of somatic cell nuclear transfer with conventional and zona-free method in cattle, horse, pig and sheep. Theriogenology 67:90–98
15. Pfeifer A (2004) Lentiviral transgenesis. Transgenic Res 13:513–522
16. Jaenisch R, Fan H, Croker B (1975) Infection of preimplantation mouse embryos and of newborn mice with leukemia virus: tissue distribution of viral DNA and RNA and leukemogenesis in the adult animal. Proc Natl Acad Sci USA 72:4008–4012
17. Jaenisch R (1976) Germ line integration and Mendelian transmission of the exogenous Moloney leukemia virus. Proc Natl Acad Sci USA 73:1260–1264
18. Fässler R (2004) Lentiviral transgene vectors. EMBO Rep 5:28–29
19. Brackett BG et al (1971) Uptake of heterologous genome by mammalian spermatozoa and its transfer to Ova through fertilization. Proc Natl Acad Sci USA 68:353–357
20. Lavitrano M et al (2003) Sperm mediated gene transfer in pig: selection of donor boars and optimization of DNA uptake. Mol Reprod Dev 64:284–291
21. Sperandio S et al (1996) Sperm-mediated DNA transfer in bovine and swine species. Anim Biotechnol 7:59–77
22. Cong L et al (2011) Reducing blood glucose level in TIDM mice by orally administering the silk glands of transgenic hIGF-I silkworms. Biochem Biophys Res Commun 410:721–725
23. Li Y et al (2011) Expression of the hIGF-I gene driven by the Fhx/P25 promoter in the silk glands of germline silkworm and transformed BmN cells. Biotechnol Lett 33: 489–494
24. Sin FY et al (2000) Electroporation of salmon sperm for gene transfer: efficiency, reliability, and fate of transgene. Mol Reprod Dev 56: 285–288
25. Campos VF et al (2011) Exogenous DNA uptake by South American catfish (Rhamdia

quelen) spermatozoa after seminal plasma removal. Anim Reprod Sci 126:136–141
26. Collares T et al (2011) Transgene transmission in chickens by sperm-mediated gene transfer after seminal plasma removal and exogenous DNA treated with dimethylsulfoxide or N, N-dimethylacetamide. J Biosci 36:613–620
27. Wang H et al (2001) Expression of porcine growth hormone gene in transgenic rabbits as reported by green fluorescent protein. Anim Biotechnol 12:101–110
28. Zhao Y et al (2011) Spontaneous uptake of exogenous DNA by goat spermatozoa and selection of donor bucks for sperm-mediated gene transfer. Mol Biol Rep. doi:10.1007/s11033-011-1019-4
29. Wang L et al (2011) Association of goat (Capra hircus) CD4 gene exon 6 polymorphisms with ability of sperm internalizing exogenous DNA. Mol Biol Rep 38:1621–1628
30. Feitosa WB et al (2010) Exogenous DNA uptake by bovine spermatozoa does not induce DNA fragmentation. Theriogenology 74:563–568
31. Francolini M et al (1993) Evidence for nuclear internalization of exogenous DNA into mammalian sperm cells. Mol Reprod Dev 34:133–139
32. Magnano A et al (1998) Sperm/DNA interaction: integration of foreign DNA sequences in the mouse sperm genome. J Reprod Immunol 41:187–196
33. Maione B et al (1997) Activation of endogenous nucleases in mature sperm cells upon interaction with exogenous DNA. DNA Cell Biol 16:1087–1097
34. Webster NL et al (2005) Multi-transgenic pigs expressing three fluorescent proteins produced with high efficiency by sperm mediated gene transfer. Mol Reprod Dev 72:68–76
35. Lavitrano M et al (1998) Sperm-mediated gene transfer. In: Cid-Arregui A, Garcìa-Carrancà A (eds) Microinjection and transgenesis: strategies and protocols. Springer, Heidelberg, pp 229–254
36. Whittingham DG (1971) Culture of mouse ova. J Reprod Fertil Suppl 14:7–21
37. Montoliu L (1997) GENERATION OF TRANSGENIC MICE. A laboratory manual. http://www.cnb.uam.es/~montoliu/transgenic.html.
38. Lavitrano M et al (1997) The interaction of sperm cells with exogenous DNA: a role of CD4 and major histocompatibility complex class II molecules. Exp Cell Res 233:56–62
39. Fantinati P et al (2005) Laparoscopic insemination technique with low numbers of spermatozoa in superovulated prepuberal gilts for biotechnological application. Theriogenology 63:806–817
40. Perry AC et al (1999) Mammalian transgenesis by intracytoplasmic sperm injection. Science 284:1180–1183
41. García-Vázqueza FA et al (2009) Effect of sperm treatment on efficiency of EGFP-expressing porcine embryos produced by ICSI-SMGT. Theriogenology 72:506–518

Chapter 45

Phenotypic Assessment of Male Fertility Status in Transgenic Animal Models

David M. de Kretser and Liza O'Donnell

Abstract

This chapter describes the approach to define the cause of male infertility in a genetically modified male mouse. It provides a guide to the establishment of the infertility status and whether it is due to the failure of mating or due to abnormalities of the sperm output, motility, and morphology. Further assessments define the nature of the spermatogenic defects and their severity and are designed to determine the pathogenic mechanisms involved.

Key words: Spermatogenesis, Male infertility, Disordered sperm motility and morphology, Hormonal and genetic defects, Light and electron microscopy methods

1. Introduction

Transgenic mice are being used increasingly to understand the many factors that control fertility in males. It is not possible in a chapter of this length to provide details of the process of spermatogenesis and the maturation and transport of sperm. Readers should consult several texts that cover these topics (1–3). This chapter is designed to provide the reader with a framework to assess a male mouse that has been unable to achieve a pregnancy despite evidence that successful copulation has been achieved. It should be taken as the entry point into a more detailed assessment of the defective process identified by the analysis provided herein. As such, the reader is referred to more detailed information in a recent review that provides specific examples of identification of mutations in the mouse genome resulting in defects that have led to the identification of novel physiological processes controlling spermatogenesis and sperm function (1).

Several approaches have been undertaken to characterise the functions of spermatogenesis genes such as the targeted deletion of a gene known to play a role in spermatogenesis or sperm transport. Alternatively, random mutations are induced in the whole genome of mice using ENU mutagenesis (4). Screening is undertaken by the assessment of fertility as described below, and then the males are bred with their sisters to produce males homozygous for the gene of interest (5). The assessment of these mice or those with a targeted deletion of a candidate gene is identical.

1.1. Overview of Spermatogenesis

Spermatogenesis is the process by which the germ cells, effectively the stem cells in the testis progress to the production and release of sperm. It takes place within the epithelium of the seminiferous tubules of the testis and culminates in the production of mature spermatozoa (Fig. 1). The other component of the seminiferous tubule epithelium is Sertoli cells, named after the man who first described them. These cells, also called the supporting or sustentacular cells, run radially from the base of the tubule to the lumen. Where they meet each other, in the basal one-third of the epithelium, they form specialised tight junction complexes that effectively block intercellular transport. Thus the Sertoli cells, in addition to supporting and nurturing the germ cells, are also involved in fluid secretion and the transfer of metabolites into and out of the tubules.

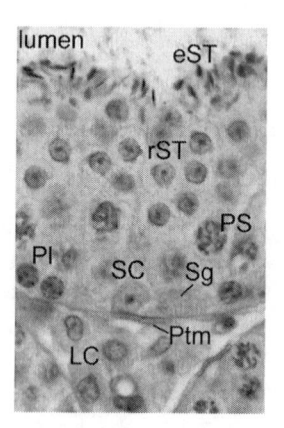

Fig. 1. Light micrograph of a seminiferous tubule and intertubular tissue with containing the stereoidogenic Leydig cells (LC) that produce testosterone. The seminiferous tubules are surrounded by peritubular myoid cells (Ptm) and a basement membrane. Within the tubule, the epithelium consists of the basal layer of spermatogonia (Sg), interspersed between the cytoplasm of the bases of the Sertoli cells (SC). The intermediate layer is formed by the primary spermatocytes that show features consistent with the process of meiosis, namely the pairing and later separation of homologous chromosomes during the first meiotic prophase. These cells divide to form secondary spermatocytes that divide again to form round spermatids. The latter do not undergo any further division but undergo the complex steps of spermiogenesis wherein a round cell is transformed into a spermatozoon. This particular micrograph shows a stage VII tubule in a wild-type mouse, with the following cell types being visible: type A spermatogonia (Sg), preleptotene spermatocytes (Pl), pachytene spermatocytes (PS), step 7 round spermatids (rST), step 16 elongated spermatids (eST).

Sperm develop from spermatogonial stem cells that undergo a series of mitotic and meiotic cell divisions followed by a complex process wherein the round spermatids cease dividing and transform into spermatozoa. This latter process is termed spermiogenesis, which will be described in a separate section of this chapter.

Spermatogenesis can be subdivided further into (1) spermatogonial renewal by mitotic division and differentiation and (2) progression through meiosis as the spermatogonia transform to primary spermatocytes that undergo the stages of the first meiotic division. The daughter cells from this division are called secondary spermatocytes, which progress through a further cell division to give rise to round spermatids, which do not divide further but commence the process of spermiogenesis.

Spermiogenesis is a complex process wherein the round spermatid is transformed into a sperm through a reorganisation of the cellular components (Fig. 2). These changes occur together but can be described separately.

1. Changes in the nucleus: The nucleus migrates toward a large vesicle, the acrosomal vesicle, which is formed from the Golgi complex. The acrosomal vesicle then spreads to form a "cap" over the pole of the nucleus that lies adjacent to the cell membrane. This acrosomal cap contains substances required to penetrate the cumulus complex and ovum during fertilisation. During spermiogenesis the centrally located nucleus and acrosomal cap of the round spermatid move eccentrically, so that part of the nuclear membrane comes into juxtaposition with the cell membrane. The acrosome finally lies in a narrow space between the cell membrane and the nuclear membrane and stains positively with the Periodic acid Schiff's (PAS) technique (*see* Subheading 3.4). This technique detects glycoproteins, and the developmental stages of the acrosome that can be visualised by this histochemical reaction are used to define the series of stages of spermiogenesis and are useful in characterising abnormalities in spermatogenesis.

2. Condensation and compaction of the spermatid nucleus: There is a progressive decrease in the volume of the nucleus as the nucleus changes shape in forming the head of the sperm. These changes are caused by compaction of the DNA and the replacement of histones by protamines such that the final volume of the sperm head is about 5% of the original nuclear volume of the round spermatid at the commencement of spermiogenesis.

3. Development of the sperm tail: This commences by the development of a microtubular structure (called the axial filament) from one of the pair of centrioles that is found in the round spermatid cytoplasm. That centriole is called the distal centriole to distinguish it from the other, called the proximal centriole. As the axial filament elongates, it together with the centrioles,

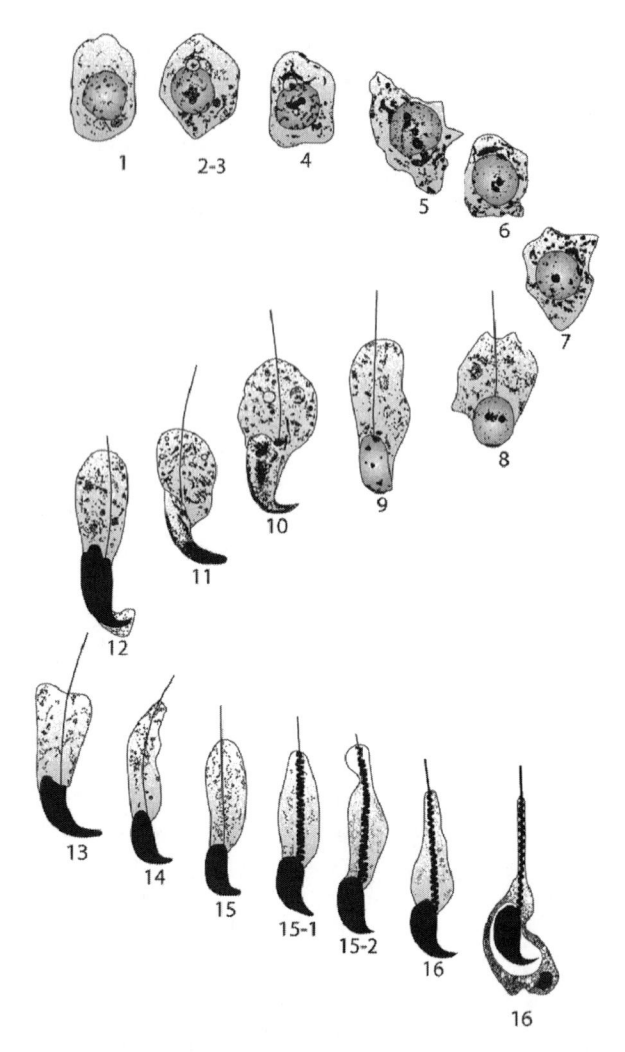

Fig. 2. The major morphological changes that a spermatid undergoes during spermiogenesis in the mouse. Mouse spermiogenesis can be divided into 16 discrete steps (indicated by the numbers) based on morphological criteria.

"migrates" so that the centriolar complex lodges in an indented region of the condensing nucleus at the pole opposite to that covered by the acrosome. This articulation with the nucleus that forms the head of the sperm is called the connecting piece. As the axial filament elongates, the central core of nine doublet microtubules surrounding two single central microtubules is progressively surrounded by additional components such as the nine outer dense fibres, one of which lies adjacent to the doublets, in the region that forms the mid-piece of the mature sperm. In the principal piece that lies distal to the mid-piece, the ribs that are disposed circumferentially around the axial filament join to two of the outer dense fibres, forming a "sheath" around the core. These structures finally disappear around the terminal region of the tail where the axial filament is only surrounded by the cell membrane.

In the final stages of spermiogenesis, two striking events occur. The mitochondria of the spermatid, which have until this stage been placed in close proximity to the cell membrane, aggregate around the mid-piece to form a helical chain, thereby bringing the source of ATP into close contact with the contractile apparatus. At this stage as well, there is a process by which the excess cytoplasm of the spermatid is "pulled off" the spermatid by the Sertoli cells during the process of "spermiation"; the shedding of testicular spermatozoa from the epithelium.

Much of the detail of the components of the spermatid/sperm can only be visualised by electron microscopy. Since the tail is responsible for generating motility, if this is compromised or abnormal, electron microscopic examination should be performed and requires someone with experience in this field.

As a spermatogonium proceeds through the germ cell stages described above, the cells progress from the basal region of the seminiferous tubule towards the lumen. Thus the spermatogonia are always in contact with the base of the tubule, whereas the primary spermatocytes are in the mid-region of the epithelium and the spermatids lie closest to the lumen (Fig. 1) although at certain stages of spermiogenesis, the developing spermatids migrate towards the basal compartment of the seminiferous epithelium.

In this chapter, we give an overview of the approach to examining fertility and spermatogenesis in transgenic male mice. We then provide methods for quantitative assessment of sperm and the preparation of testis tissue for histological and electron microscopic analysis.

1.2. Examination of Fertility and Spermatogenic Phenotype

1.2.1. Fertility Assessment

To confirm male infertility, the mice suspected of having a genetically determined infertile status are caged with normal, sexually mature female mice for a period of at least 5 days and examined each morning for a copulatory plug indicative of successful mating. Successful mating provides an indicator that the male's libido is sufficient and his penis adequate to successfully mate. The copulatory plug also indicates that the prostate and seminal vesicles have functioned to produce an ejaculate.

The outcome of this successful mating is assessed by following the females to determine the successful establishment of pregnancy and the ability of the pregnancy to proceed to parturition over the normal period of gestation. At birth, it is important to determine the number of offspring produced by that specific mating compared to mating of normal wild-type mice of the same strains. A reduced number of offspring would indicate subfertility, whereas the absence of offspring would require further matings and killing the mice at 3 days after the detection of the copulatory plug to determine if implantation sites can be identified. The latter would indicate that fertilisation and early embryonic development had occurred but foetal loss later in pregnancy was the cause of the infertility rather than a genetic defect preventing fertilisation of the ovum. Should the mating result in no offspring or a reduced number, there is a need to further assess the fertility status of the affected males.

1.2.2. Initial Assessment of Infertile Males

Further assessment of infertile males requires the euthanasia of sexually mature mice at 7 weeks of age. At this time, blood samples from each mouse should be collected by cardiac puncture, in which a large gauge needle attached to a 5 ml syringe is inserted into the heart under deep anaesthesia. These will be used to measure serum follicle stimulating hormone (FSH) and luteinizing hormone (LH), which stimulate the testis to produce sperm, and the male steroid hormone, testosterone (T). Measurement of these parameters requires access to specific and sensitive radioimmunoassays or similar techniques (6, 7). Because of the variability of serum testosterone measurements, the use of seminal vesicle weight can be used as a bioassay for testosterone (see below).

Following euthanasia, it is critical that the general features of the mice be recorded together with an assessment of organ weights for comparison with wild-type controls. In this way genetic lesions that affect processes other than fertility will be identified. Normal organ weights together with normal histological evaluation would suggest that the defect was restricted to the male reproductive tract.

Such an outcome would then lead to a detailed analysis of male fertility status.

In euthanized mice, the in situ layout of the male reproductive tract is assessed to determine if the testes are scrotally placed or can be easily brought into the scrotum from the inguinal canal. The gross morphological features of the testes, the vascular patterns on the testicular capsule, and the features of the epididymis and vas deferens are noted together with their relationship to the seminal vesicles and bladder. Finally the features of the penis, and the ano-genital distance should be noted (any shortening may indicate androgen deficiency).

It is essential that the weight of the testes and the epididymides be determined together and then separately. The weight of the accessory glands, the seminal vesicle, and the prostate should also be recorded. The former is easier to accurately determine and, since it is an androgen-dependent organ, its weight is a biomarker of androgen levels, complementing the measurement of serum testosterone.

Given that the specific gravity of the testis is close to 1, the weight essentially provides a volume measurement. Recording these parameters provides insight into the type of spermatogenic lesion that might be expected. For instance if testis weight is profoundly decreased, it indicates that there is a marked decrease in all germ cell types, whereas if the decrease in weight is small, the lesion is likely to only involve the late stages of spermiogenesis.

1.2.3. Histological Examination of the Testis

A careful histological examination of the testis can reveal much about the phenotype, the causes of infertility and the possible function of the mutant gene (see Note 1 and Table 1) (1). A comparison should be made between several (at least 3) wild-type and mutant sexually mature mice (~7 weeks). If an age-related phenotype is suspected, older mice between 6 months to 1 year should be examined. For many phenotypes it can be helpful to assess the timing of onset of the defect by examining earlier time points during spermatogenic development (see Note 1 and Table 1) (1).

Testis sections should be prepared and stained as described (see Subheadings 3.3–3.6). For initial assessments, PAS staining of the testis is very informative. The preservation of morphology and good staining is essential for the examination of spermatogenesis. Multiple sections of the testis should be examined, as sometimes lesions are focal in distribution. A low magnification assessment should be made of the size of the seminiferous tubules, whether they have a lumen, and whether the inter-tubular regions of the testis show the presence of Leydig cells, the cells responsible for the production of testosterone and estradiol.

Higher magnification examination of the seminiferous epithelium should establish the presence of each germ cell type (see Note 2). The appearance of the nuclei of Sertoli cells should be noted as the adult cell, following exposure to FSH, has a well-developed nucleolus. In mice that have a major loss of germ cells, the basement membrane and the peritubular myoid cell layer are thickened and fibrotic, the latter being well demonstrated by the use of the Masson Trichrome staining technique (see Subheading 3.5).

In some mice sexual maturation is delayed or may not occur due to a defect in the hypothalamus or pituitary gland. In such mice, testis weight is dramatically reduced as the primordial germ cells or the spermatogonial precursors, called gonocytes do not develop and divide, remaining embedded in a seminiferous cord that lacks a lumen. The latter is due to the failure of the blood-testis barrier to form and the inability of the Sertoli cells to produce seminiferous tubule fluid (8).

1.2.4. Assessment of the Epididymis

In situ examination of the epididymis on each side must include determination of appearance and the identification of abnormal dilatations suggestive of obstruction. This is especially important if an obstruction is suspected, such as in cases displaying the presence of normal spermatogenesis and epididymis together with the total failure of conception. In such instances, the course of the vas at its distal end should be carefully assessed.

The vas deferens should also be carefully assessed to determine that it is present, looks normal, and takes a normal course to the posterior aspect of the bladder where the seminal vesicles are juxtaposed.

Table 1
Overview of some common spermatogenic defects, their potential causes and some suggestions for further analysis

Testis histology	Potential causes	Further analyses
Small testis, mature spermatids present, perhaps sperm in epididymis	• Reduced Sertoli cell numbers due to defective neonatal proliferation • Defect in ability of Sertoli cells to support germ cell development • Defects in a gene involved at multiple points in germ cell development • Germ cell apoptosis	• Stereological determination of Sertoli cell numbers • Stereological determination of germ cells per Sertoli cell • Careful analysis of sites of germ cell arrest • TUNEL staining for determination of apoptosis in situ
Small testis, mature spermatids present, Sertoli cell only (SCO) tubules adjacent to normal tubules	• Focal spermatogenic lesions may arise due to defects in foetal Sertoli cell proliferation or failure of germ cells to colonise specific regions of epithelium	• Analysis of foetal and early neonatal testis histology • Immunolocalisation of protein of interest
Spermatocytes present, round and elongated spermatids absent	• Defect in spermatocytes (arrest at a particular stage of meiosis, apoptosis or inability to undergo division) • Defect in Sertoli cell ability to support meiosis	• Assessment of where meiosis arrests (e.g., zygotene or pachytene) • TUNEL analysis for spermatocyte apoptosis • Look for evidence of meiotic division (metaphase and anaphase spermatocytes)
No mature spermatids but round spermatids present	• Premature detachment of round spermatids due to defects in adhesion • Apoptosis of round spermatids	• Look for presence of detached round spermatids in epididymis • TUNEL
Multinucleated germ cells or "strings" of dead cells	• Abnormal cytokinesis during meiosis (causes multinucleated spermatids) • Apoptosis of germ cells connected by intercellular bridges	• Assessment of meiotic division, e.g., spindle by tubulin immunostaining • TUNEL

Sertoli cells and spermatogonia only	• Failure of spermatogonial proliferation/entry into meiosis • Failure of Sertoli cells to support spermatogonial development/meiosis initiation	• Immunostaining for markers of differentiated spermatogonia • Investigation of whether Sertoli cells mature and/or form intercellular tight junctions

Mature elongated spermatid nucei at the base of Sertoli cells in stages VII–XI[a]	• Failure of mature sperm to be released during spermiation (often seen in testes with hypospermatogenesis)	• Analysis of site of spermiation failure:
		  – Failure to initiate = basal spermatids in stage VII
		  – Failure to be released = basal spermatids in late stage VIII–XI

Sertoli cells only	• Defect in spermatogonial stem cell renewal or Sertoli cell function	• Compare histology of WT and mutant at earlier time points (e.g., late foetal, days 6–20 pp)
Enlarged tubule lumens	• May be related to disturbed efferent ductule function, where efferent ducts fail to resorb testicular fluid, resulting in fluid build up in testis. Can cause spermatogenic arrest • Enlarged lumens can also be caused by a massive and acute loss of germ cells (not usually seen in transgenic mice, more common in toxicant models)	• Examine tubule lumen diameter in younger mice (e.g., 20–50 dpp); if enlarged early would suggest defect in efferent ductule development • Examine efferent ductule histology
Abnormal sperm morphology	• Abnormal acrosome formation • Abnormal head shape due to defective manchette • Defect arising during the elongation phase of spermiogenesis • Defect during spermiation; failure to remove cytoplasm	• Analysis of acrosome development—PAS staining, electron microscopy (EM) • Analysis of manchette—α-tubulin immunostaining, EM • Careful examination of morphology during spermiogenesis • Careful examination of morphology during spermiation (stages VII–VIII[a])
Abnormal sperm motility	• Defect in sperm tail development/function	• Assessment of axoneme structure by EM

[a]For classification of specific stages of mouse spermatogenesis see ref. 3

1.2.5. Further Analyses of Spermatogenesis

In many cases, the use of stereology to quantify the number of germ cells and Sertoli cells per testis is invaluable in pinpointing the specific lesions to spermatogenesis in mutant mice. This technique involves the use of random sampling procedures to enable the unbiased determination of cell numbers (9) and requires the precise identification of germ cells within different spermatogenic stages (3). The details of the technique will not be discussed here, but see ref. 3 for review and (10–14) for some examples of studies using stereology to assess spermatogenesis in mutant mice. Calculating ratios of sequential germ cell populations allows the quantitation of the transition of germ cells through specific stages of spermatogenesis (14). These ratios should be made between cell populations expressed as hourly production rates, i.e., the number of a particular germ cell per testis, divided by the duration in hours of the stage(s) in which it is present. This takes into account the fact that different germ cell types exist for different durations within the seminiferous epithelium. Using this approach, for example, the ratio of round spermatids in stages I–III to pachytene spermatocytes in stages IX–XI can be used as a measurement of the efficiency of the meiotic divisions. Expressing germ cell numbers per Sertoli cell is useful when Sertoli cell numbers are different between wild-type and mutant mice, or when one wishes to examine the ability of Sertoli cells to support different germ cells (14).

Once the site of disruption to spermatogenesis has been identified, more information can be sought to pinpoint the particular spermatogenic/cellular processes that may be impaired (see Note 1) (1).

If the mutated gene is known, it is extremely helpful to know precisely where and when both the mRNA and protein are expressed in normal and mutant mice in order to understand the likely cellular processes in which it is involved. The posttranscriptional regulation of germ cell mRNAs is well known, with many mRNAs being transcribed and then stored until the protein is translated at a later stage of germ cell development (15). Hence, there is often disparity between mRNA and protein expression. An examination of the ontological expression pattern of a gene during postnatal development of spermatogenesis can be very helpful in understanding when it is first transcribed (1). Online databases with microarray information can be used to determine in which cell type(s) the gene is expressed (16, 17). If antibodies are available then a careful analysis of the immunolocalisation pattern of the gene or protein of interest is invaluable. Because of the complex arrangement of the seminiferous epithelium and the need to understand the stage-specific pattern of protein localization, it is essential that germ cell and Sertoli cell nuclear morphology be clearly visible, either by haematoxylin counterstaining for light immunohistochemistry, or DAPI-DNA staining for fluorescence microscopy.

2. Materials

1. Phosphate-buffered saline (PBS); 0.01 M phosphate, 0.154 M NaCl, pH 7.4.
2. 0.05% Triton X-100 in PBS.
3. Tissue homogeniser.
4. Neubauer haemocytometer (0.1 mm depth, Weber Scientific, England).
5. MT6 capacitation medium: 1 ml *Stock A* + 1 ml *Stock B* + 0.1 ml *Stock C* + 7.9 ml water + 40 mg BSA. *Stock A* (10× solution, can be stored for 3 months at 4°C): 125 mM NaCl (7.280 g/100 ml) + 2.7 mM KCl (0.2 g/100 ml) + 0.5 mM $MgCl_2 \cdot 6H_2O$ (0.1 g/100 ml) + 0.36 mM $NaH_2PO_4 \cdot 2H_2O$ (0.056 g/100 ml) + 6.4 mM Glucose (1 g/100 ml). *Stock B* (10× solution, can be stored for 2 weeks at 4°C): 0.11 mM $NaHCO_3$ (2.106 g/100 ml) + Phenol red (0.010 g/100 ml). *Stock C* (100× solution, can be stored for 3 months at 4°C) 1.7 mM $CaCl_2 \cdot 2H_2O$ (0.252 g/100 ml).
6. Chamber slide (Hamilton Thorne Research, 2× cell 80 μm).
7. Bouin's fixative: 75 ml aqueous picric acid, 25 ml 40% formaldehyde, 5 ml glacial acetic acid.
8. Ethanol: 70, 90, and 100%.
9. Xylene-based solvent.
10. Paraffin wax.
11. 0.5% Periodic acid in H_2O.
12. Schiff's Reagent. This can be purchased commercially or prepared as follows: Dissolve 1 g of basic fuchsin in 200 ml boiling water. Stir well and cool to 50°C. Filter and add 20 ml 10% HCl, cool to RT. Add 1.5 g potassium metabisulphite. Shake vigorously until crystals dissolve and leave standing at RT overnight. Extract colour by adding 1 g charcoal, shake, let stand for 5 min and filter. Repeat charcoal step until solution is colourless.
13. Haematoxylin.
14. Scott's tap water.
15. DPX mounting medium for coverslips.
16. Weigert's Iron Haematoxylin. Mix equal parts of Solution A (5 g haematoxylin dissolved in 500 ml absolute ethanol) and Solution B (20 ml 30% aqueous ferric chloride, 5 ml concentrated hydrochloric acid, 500 ml distilled water) immediately before use.
17. Ponceau–Acid Fuschin: 1% ponceau in MilliQ H_2O (2 parts) + 1% acid fuschin in MilliQ H_2O (1 part).

18. 1% Phosphomolybdic acid in MilliQ H_2O.
19. 1% Light green: 10% light green in 1% acetic acid.
20. 1% acetic acid in MilliQ H_2O.
21. Superfix: Dissolve 0.1 g picric acid in 243 ml 0.2 M sodium cacodylate buffer, filter, then add 166 ml 12% paraformaldehyde (4% final concentration). Add 100 ml 25% glutaraldehyde (5% final concentration), and immediately after the addition and before use, adjust pH to 7.2–7.4.
22. 0.1 M Cacodylate Buffer: 21.4 g Sodium Cacodylate Trihydrate in 1 L MilliQ H_2O, pH 7.4.
23. 2% OsO_4 in 0.1 M Cacodylate buffer: Add equal quantities of 4% $OsO_4(aq)$ and 0.2 M Sodium Cacodylate buffer (42.8 g Sodium Cacodylate Trihydrate in 1 L MilliQ H_2O, adjust pH to 7.4).
24. 2% Uranyl acetate in MilliQ H_2O.
25. Absolute ethanol.
26. Propylene oxide solvent.
27. Araldite Epon (kit).

3. Methods

3.1. Quantitative Assessment of Sperm: Elongated Spermatid Content

Overview. It is difficult to accurately determine whether germ cell numbers are reduced by a simple histological analysis. A simple method to quantify spermatogenesis is the measurement of elongated spermatid content. This technique is based on the principle that beyond ~step 17 of spermatid development, the nuclei of the elongated spermatids are resistant to destruction by detergents (18). If the duration of these stages of spermatogenesis is known, since each stage of germ cell development takes a specific length of time, the daily sperm output can be calculated as shown in the case of rat (18). The following method can be applied to the measurement of the Elongated Spermatid Content of testis or epididymis.

1. For the determination of the number of elongated spermatids in the testes, weigh and decapsulate one testis, and add to an appropriate volume of 0.05% Triton X-100 in PBS. We suggest a 2:1 ratio of buffer (ml) to testis weight (g), but smaller or larger volumes can be used if sperm are less or more concentrated.
2. For epididymal tissue, use a ratio of 2:1–4:1 buffer to organ weight. Chop the epididymal tissue with small scissors prior to buffer addition to facilitate homogenisation.

3. Homogenise tissue in a tissue homogeniser for ~20 s (longer for epididymis) until the tissue is evenly dispersed. Homogenates can be stored for 2–3 days at 4°C prior to counting.

4. Dilute an aliquot of homogenate in PBS and load into a cover-slipped Neubauer haemocytometer. Since sperm heads will readily settle in solution, it is essential that all homogenates and dilutions be well vortexed prior to use.

5. Once the sperm have been loaded into the haemocytometer, allow the sperm heads to settle for ~2 min prior to counting.

6. Count five squares from the four corners and the centre of each chamber of the haemocytometer; dilute the homogenate accordingly so that approximately 100 sperm are counted in each chamber (therefore roughly 20 sperm per square). When each square is counted, sperm landing on the left and bottom edges are excluded (exclusion boundary) and those landing on the top and right edges are included.

7. Add up the number of sperm counted in the five squares to calculate the total per chamber, and calculate the average per chamber by adding both totals together and dividing by 2.

8. Multiply the average by the dilution factor (e.g., 100) to calculate the average sperm count per chamber.

9. Calculate the *elongated spermatid content* as follows: Average sperm count per chamber × 50,000 × Z (Z = no. mls homogenisation buffer + tissue weight (g)).

3.2. Quantitative Assessment of Sperm: Assessment of Motility

Overview. The assessment of motility of sperm derived from the caput epididymis is complex as the sperm are not active immediately but become activated on dilution in medium. Further, sperm released from the head of the epididymis into media swim poorly and in circles whereas those released from the tail of the epididymis, the major storage area, swim progressively forward on release into a physiological medium. Thus any preliminary assessment should use sperm taken from the tail of the epididymis and the percentage acquiring motility should be noted and where possible the pattern of motility should be assessed and video-taped (see Note 3).

1. To assess motility, dilute sperm in a medium that permits capacitation, e.g., MT6 (19) such that their concentration is about 10^6 and 10^7 per ml.

2. Place sperm in a droplet on a chamber slide with a depth of about 80–100 μm. The use of such a chamber avoids the need to use a coverslip which can impair motility. The dilution is important to reduce the concentration of decapacitation factors.

3. Evaluate the presence of progressive forward movement at 60 and 120 min (see Note 4).

3.3. Preparation of Testis Tissue for Histological and Electron Microscopic Analysis: Fixation, Embedding for Light Microscopy

1. To fix whole testis following removal from the mouse, gently prick the tough tunica albuginea in multiple places with a 22 g needle to facilitate the penetration of the Bouin's fixative into the parenchyma of the testis and place it in the fixative for 5 h at room temperature.
2. Transfer the testis to 70% ethanol until processing can be commenced.
3. Transfer the testis into the following solutions with a duration of 2 h in each: 70% Ethanol, 90% Ethanol, 100% Ethanol, 100% Ethanol, 50:50 Ethanol:Xylene-based solvent, Xylene-based solvent, Xylene-based solvent, Paraffin wax (60°C), Paraffin wax (60°C).
4. While the tissue is at 60°C, remove it from processing cassette and place it into a histology mould containing melted paraffin wax. Solidify on a cold plate until the paraffin wax block can be removed from the mould.

3.4. Preparation of Testis Tissue for Histological and Electron Microscopic Analysis: PAS Staining

1. Dewax sections in xylene (2×5 min), then take sections through 100, 90, 70% ethanol (2 min each), then wash briefly in distilled water.
2. Immerse in 0.5% Periodic Acid for 10 min.
3. Wash in distilled water for 5 min.
4. Immerse in Schiff's Reagent for 20 min.
5. Wash in warm running tap water for 10 min.
6. Immerse in Haematoxylin for 60 s.
7. Wash in running tap water for 5 min.
8. Immerse in Scott's Tap Water for 60 s.
9. Wash in running tap water for 5 min.
10. Dehydrate sections by taking through changes of 100% ethanol (3×5 min), clear to remove ethanol in two changes of xylene (2×5 min) and mount with coverslips using DPX.

3.5. Preparation of Testis Tissue for Histological and Electron Microscopic Analysis: Masson Trichrome Staining

1. Dewax sections and wash briefly in distilled water.
2. If tissue is not yet fixed in Bouin's fixative, post fix in Bouin's for 60 min at 60°C
3. Wash in running tap water for 10 min.
4. Immerse in Weigert's iron haematoxylin for 2 min.
5. Wash in running tap water until excess stain has run clear (approx. 10–20 s).
6. Immerse in Ponceau–Acid Fuschin for 5 min.
7. Wash in running tap water until excess stain has run clear (approx. 10–20 s).
8. Immerse in 1% phosphomolybdic acid for 3 min.

9. Wash in running tap water until excess stain has run clear (approx. 10–20 s).
10. Immerse in 1% light green for 2 min.
11. Wash in running tap water until excess stain has run clear (approx. 10–20 s).
12. Immerse in 1% acetic acid for 1 min.
13. Dehydrate, clear, and mount.

3.6. Preparation of Testis Tissue for Histological and Electron Microscopic Analysis: Processing Testis for Transmission Electron Microscopy

1. Gently prick the testis as above with a fine needle to assist penetration of the fixative into the central regions of the testis.
2. Transfer the testes for fixation into 5% Superfix, and hold overnight at 4°C.
3. After fixation, wash specimen in 0.1 M Cacodylate Buffer 3×10 min.
4. Post fix in 2% OsO_4/0.1 M Cacodylate Buffer for 2 h at room temperature.
5. Wash in 0.1 M Cacodylate buffer 3×10 min.
6. Wash in MilliQ water 3×10 min.
7. En-block stain tissue in 2% Uranyl Acetate for 2 h at room temperature.
8. Wash in MilliQ water 3×10 min.
9. Dehydrate in 50, 75, 95% ethanol solutions for 15 min each.
10. Transfer to Absolute Ethanol twice for 15 min.
11. Transfer to "Dry" Absolute Ethanol (i.e., ethanol with a molecular sieve in it to absorb any water contaminant) twice for 15 min.
12. Transfer to Propylene Oxide solvent 2×15 min.
13. Transfer to 75:25 Propylene Oxide/Araldite Epon and hold for 1 h on roller.
14. Leave in 50:50 Propylene Oxide/Araldite Epon on roller overnight.
15. Transfer to 100% Araldite Epon resin 3×2 h.
16. Embed tissue into EM moulds with Araldite Epon and polymerise at 60°C 48 h.

4. Notes

1. Some common phenotypes and suggestions for further analysis: A comprehensive overview of the many possible spermatogenic phenotypes and their potential causes is beyond the scope of this chapter, and we refer the reader to an excellent recent review (1).

Table 2
Useful protein markers of testicular cell populations

Cell type	Marker
Spermatogonia	*Undifferentiated*: Plzf (24), Gpr125 (25) *Differentiated*: c-Kit (26)
Sertoli cells	Sox9, SF1, see ref. 27 *Immature*: Cytokeratin, Anti-Mullerian Hormone, see ref. 28 *Mature*: P27kip, Gata1, see ref. 28
Meiosis	Sycp1 (29), Sycp3 (30)
Haploid spermatids	CREM (31), protamines 1&2, transition proteins 1&2, see ref. 32
Leydig cells	P450SCC, see ref. 27

However for some guidance on common phenotypes, potential causes and suggestions for further analyses (Table 1).

2. Markers of germ cells and Sertoli cells: Analysis of spermatogenesis is best approached by the careful examination of the specific germ cell types and the stages in which they are present. However, in some cases, it may be useful to also utilise immunohistochemistry or PCR analysis of markers of specific cell types (Table 2). If measurement of such markers is attempted using PCR or western blotting, it is critical that changes in testicular cell composition between wild-type and mutant animals be taken into account when presenting such data. For example, if a particular mutation is associated with a loss of all spermatids, then RNA/protein from earlier germ cells and Sertoli cells will have a greater contribution to total testicular RNA/protein levels. Accordingly, if one was to measure a Sertoli cell-specific protein in this instance, there would be an apparent significant increase, despite no actual change in Sertoli cell numbers and/or expression of this protein.

3. Assessment of epididymal sperm motility: It should be noted that sperm isolated from the epididymis are fragile cells that can be damaged by shearing forces thus should be suspended in media that are carefully constituted paying particular attention to pH, osmolariy, etc. Contact with blood must be avoided, and shearing forces during centrifugation should not exceed $400 \times g$ (1).

4. The occurrence of hyperactivation can easily be detected due to the dramatic change with an exaggeration of the flagellar beat wavelength and the presence of figure of eight motility patterns. More detailed assessments involve video recording

and the use of computer-assisted sperm analysis and are require specialised equipment and skills (19, 20). Further analysis of the mechanisms involves measurements of calcium ion flux, as hyperactivation is dependent on an influx of calcium (21–23). A further test that can be performed is the assessment of the capacity of sperm to undergo hyper-activated motility, which is accepted as a measure of the sperm's ability to undergo capacitation, a process that normally takes place when sperm enter the female genital tract. Incubation of the sperm from the tail of the epididymis in MT6 medium enables this response to be evaluated. Failure to undergo capacitation, when taken together with no other testicular or epididymal abnormality can thus define a lesion that impairs sperm capacitation.

Acknowledgments

LO'D is supported by an NHMRC Australia Program Grant #494802 and both LO'D and DMdeK are supported by the Victorian Government's Operational Infrastructure Support Program.

References

1. Borg CL et al (2010) Phenotyping male infertility in the mouse: how to get the most out of a 'non-performer'. Hum Reprod Update 16: 205–224
2. Kerr JB, de Kretser DM (2010) Functional morphology of the testis. In: Jameson JL, De Groot LJ (eds) Endocrinology, vol 6. Saunders Elsevie, Philadelphia, pp 2440–2468
3. Russell LD et al (1990) Histological and histopathological evaluation of the testis. Cache River Press, Clearwater
4. Jamsai D, O'Bryan MK (2010) Genome-wide ENU mutagenesis for the discovery of novel male fertility regulators. Syst Biol Reprod Med 56:246–259
5. Jamsai D, O'Bryan MK (2011) Mouse models in male fertility research. Asian J Androl 13: 139–151
6. Lee VW et al (1975) Variations in serum FSH, LH and testosterone levels in male rats from birth to sexual maturity. J Reprod Fertil 42: 121–126
7. Buzzard JJ et al (2004) Changes in circulating and testicular levels of inhibin A and B and activin A during postnatal development in the rat. Endocrinology 145:3532–3541
8. Jegou B et al (1982) Seminiferous tubule fluid and interstitial fluid production. I. Effects of age and hormonal regulation in immature rats. Biol Reprod 27:590–595
9. Wreford NG (1995) Theory and practice of stereological techniques applied to the estimation of cell number and nuclear volume in the testis. Microsc Res Tech 32:423–436
10. De Gendt K et al (2004) A Sertoli cell-selective knockout of the androgen receptor causes spermatogenic arrest in meiosis. Proc Natl Acad Sci USA 101:1327–1332
11. Robertson KM et al (2002) The phenotype of the aromatase knockout mouse reveals dietary phytoestrogens impact significantly on testis function. Endocrinology 143:2913–2921
12. O'Shaughnessy PJ et al (2010) Direct action through the sertoli cells is essential for androgen stimulation of spermatogenesis. Endocrinology 151:2343–2348
13. Lim P et al (2008) Oestradiol-induced spermatogenesis requires a functional androgen receptor. Reprod Fertil Dev 20:861–870
14. Wreford NG et al (2001) Analysis of the testicular phenotype of the Follicle-Stimulating Hormone β-subunit knockout and the activin type II receptor knockout mice by stereological analysis. Endocrinology 142:2916–2920
15. Iguchi N et al (2006) Expression profiling reveals meiotic male germ cell mRNAs that are

15. [continued] translationally up- and down-regulated. Proc Natl Acad Sci USA 103:7712–7717
16. Lardenois A et al. (2010) GermOnline 4.0 is a genomics gateway for germline development, meiosis and the mitotic cell cycle. Database (Oxford) 2010, baq030.
17. Chalmel F et al (2007) The conserved transcriptome in human and rodent male gametogenesis. Proc Natl Acad Sci USA 104:8346–8351
18. Robb GW et al (1978) Daily sperm production and epididymal sperm reserves of pubertal and adult rats. J Reprod Fertil 54:103–107
19. O'Bryan MK et al (2008) Sox8 is a critical regulator of adult Sertoli cell function and male fertility. Dev Biol 316:359–370
20. Miki K et al (2004) Glyceraldehyde 3-phosphate dehydrogenase-S, a sperm-specific glycolytic enzyme, is required for sperm motility and male fertility. Proc Natl Acad Sci USA 101:16501–16506
21. Suarez SS et al (1993) Intracellular calcium increases with hyperactivation in intact, moving hamster sperm and oscillates with the flagellar beat cycle. Proc Natl Acad Sci USA 90: 4660–4664
22. Ho HC et al (2002) Hyperactivated motility of bull sperm is triggered at the axoneme by Ca^{2+} and not cAMP. Dev Biol 250:208–217
23. Harper CV et al (2004) Stimulation of human spermatozoa with progesterone gradients to simulate approach to the oocyte. Induction of $(Ca(2+))(i)$ oscillations and cyclical transitions in flagellar beating. J Biol Chem 279:46315–46325
24. Buaas FW et al (2004) Plzf is required in adult male germ cells for stem cell self-renewal. Nat Genet 36:647–652
25. Seandel M et al (2008) Niche players: spermatogonial progenitors marked by GPR125. Cell Cycle 7:135–140
26. Prabhu SM et al (2006) Expression of c-Kit receptor mRNA and protein in the developing, adult and irradiated rodent testis. Reproduction 131:489–499
27. Barrionuevo F et al (2009) Testis cord differentiation after the sex determination stage is independent of Sox9 but fails in the combined absence of Sox9 and Sox8. Dev Biol 327: 301–312
28. Welsh M et al (2009) Androgen action via testicular peritubular myoid cells is essential for male fertility. FASEB J 23:4218–4230
29. Vidal F et al (1998) Cre expression in primary spermatocytes: a tool for genetic engineering of the germ line. Mol Reprod Dev 51:274–280
30. Yuan L et al (2000) The murine SCP3 gene is required for synaptonemal complex assembly, chromosome synapsis, and male fertility. Mol Cell 5:73–83
31. Nantel F et al (1996) Spermiogenesis deficiency and germ-cell apoptosis in CREM-mutant mice. Nature 380:159–162
32. Hermo L et al (2010) Surfing the wave, cycle, life history, and genes/proteins expressed by testicular germ cells. Part 2: changes in spermatid organelles associated with development of spermatozoa. Microsc Res Tech 73:279–319

Index

A

A23187 114–116
Acrosome 30, 33, 35, 36, 44–47, 49, 92–94,
   176, 180, 184, 238, 251, 258, 302–305, 327, 485,
   487, 533, 534, 539
Acrosome reaction 21, 92, 94, 103, 113–118, 181, 217
Algorithm 29, 30, 32–34, 40, 47, 171, 407
Alkaline comet assay 126, 138
Aneuploidy 167–173, 247, 248, 426
Aniline blue 425–435
Annexin V 257–261
Antibody 51–59, 124, 306, 309, 310, 312–320,
   327–328, 334, 344, 345, 431, 432, 460, 461, 463,
   464, 466, 468, 470, 472–474, 482, 485–486, 540
Antigen 51–54, 270, 309–311, 315–318, 328
   localization 464–467
   retrieval 311, 313, 315, 318, 328, 334, 344
Antioxidant 122, 123, 351–354, 357, 358, 360
   capacity 122, 123, 352–353, 363–374
Anti-sperm antibody (ASA) testing 51–59
   direct-immunobead test (D-IBT) 51–55
   mixed antilobulin reaction (MAR) test 51–55
   sperm immobilization (SIT) test 52, 53, 55–58
APO-DIRECT™ 125, 127, 129–131
Apoptosis 122, 123, 126, 227–228, 261, 426, 430, 538
ART. *See* Assisted reproductive technologies (ART)
ASA. *See* Anti-sperm antibody (ASA)
Assisted reproductive technologies (ART) 29, 46, 73,
   122, 123, 215, 241, 242, 274, 275
Azoospermia factor (AZF) gene deletions
   AZFa 188–192, 194, 197–205
   AZFb 188–194, 197–205
   AZFc 188–194, 197–203

B

BDMA. *See* Benzyldimethylamine (BDMA)
Benzyldimethylamine (BDMA) 329, 339
Biotin 310, 312, 314, 319, 326, 328, 429–432
Bisulfite sequencing 452–456
Borderline normal sperm 10, 28, 32–34
Bouin's fixative 301, 311, 312, 317, 318, 482, 541, 544

C

Cacodylate buffer 336, 542, 545
Calcium ionophore 114–116, 181
Capacitation 21, 62, 63, 78, 79, 85, 103, 104,
   108, 113–118, 123, 181, 217–219, 241, 269, 270,
   277, 527, 541, 543, 547
CASA. *See* Computer-aided sperm analysis (CASA)
Caspase-3 .... 257–258, 428, 430, 432, 433
Chaotropic agent 380
Chemiluminescence 352, 354, 363–374
ChIP-Chip 463, 468–471, 474
Chromatin 126, 139, 148, 152, 161, 247, 248, 251,
   263, 270, 272, 274, 303, 305, 326, 380, 425–435,
   437, 451, 460–465, 467–474, 521
Classical microdeletions 188, 189, 194, 197, 200,
   202, 204, 205
CNV. *See* Copy number variation (CNV)
Collidine buffer 336
Colloidal gold 326, 328, 334, 344
Comet assay 137–145
Computer-aided sperm analysis (CASA) 77–86, 94, 223
Copy number variation (CNV) 399–401, 403,
   404, 406–408
Cryopreservation
   sperm 209–215, 261, 288–289, 363, 480
   testicular tissue 210–213
Cytoplasmic retention 430, 431, 433

D

Decondensation 103, 104, 138, 139, 169, 172,
   434, 435, 460, 462–465, 467, 472–473
Density gradient 6, 62, 71, 86, 114, 115, 117,
   172, 217–225, 228, 237, 249–250, 258–260, 265,
   271, 352, 359, 412, 420, 437–438, 467, 472
Deoxyribonucleic acid (DNA) 50, 121, 137, 148, 169,
   176, 188, 218, 248, 257, 263, 270, 303, 352, 363,
   379, 386, 400, 426, 437, 451, 459, 505, 519, 531
   content 280–285, 287, 291–292
   damage 121–125, 127, 128, 130, 135, 137–145,
   152, 153, 218, 248, 274, 275, 352, 363, 364
   extraction 379–384, 470

Douglas T. Carrell and Kenneth I. Aston (eds.), *Spermatogenesis: Methods and Protocols*, Methods in Molecular Biology, vol. 927,
DOI 10.1007/978-1-62703-038-0, © Springer Science+Business Media, LLC 2013

550 SPERMATOGENESIS: METHODS AND PROTOCOLS
Index

Deoxyribonucleic acid (DNA) (*cont.*)
   fragmentation
      comet 126, 137, 138, 145, 149
      direct test 148
      indirect test 148
      SCSA 126, 148–150, 152, 155
      TUNEL 121–135, 148, 152, 271
   methylation 451–457
Deoxyuridine triphosphate (dUTP) 123–127, 129, 430
Deparaffinization 310, 313
DFI. *See* DNA fragmentation index (DFI)
Diagnostic testing 92
Dichotomous key 29
Disordered sperm morphology 529
Disordered sperm motility 529
Dithiothreitol (DTT) 139, 140, 142, 144, 148, 153, 168, 169, 172, 380–383, 389–391, 393, 414, 418, 429, 430, 432, 434, 435, 439, 440, 442, 446, 447, 449, 462
DNA fragmentation index (DFI) 126, 148–155, 159–161
Double strand breaks 137
2D-PAGE. *See* Two-dimensional polyacrylamide gel electrophoresis (2D-PAGE)
DTT. *See* Dithiothreitol (DTT)
dUTP. *See* Deoxyuridine triphosphate (dUTP)

E

Electron microscopy 322, 324, 328–333, 335, 340, 343, 345, 347, 535, 539, 544–545
Electrophoresis 126, 137, 140–144, 195, 199, 202, 271–275, 381, 412, 413, 417–419, 445–450
Electrophoretic sperm separation 270, 275
Electrostatic charge 269–277
Embedding 264, 300, 301, 310, 312–313, 317, 323, 325, 326, 328–329, 332–333, 337–340, 344–346, 484–485, 495, 537, 544, 545
Endonuclease digestion 440, 469
Eosin-nigrosin staining 13–17, 239
Epigenetics 379–384, 454, 461
Eponate 328–329, 333, 336, 339, 345
N-ethyl *N*-nitrosourea (ENU) mutagenesis 532

F

Fertility 28–30, 122, 126, 152–155, 159, 176, 189, 200, 243, 244, 247, 263, 293, 351, 352, 398, 426, 490, 503, 523, 525, 531–547
Fertilizing ability 51–52, 218
F12GFB medium 505–514
FISH. *See* Fluorescence in situ hybridization (FISH)
FITC. *See* Fluorescein isothiocyanate (FITC)
Fixation 40, 41, 127, 128, 168, 169, 311–313, 317, 327–328, 331, 332, 338, 345, 429, 544
Flow cytometer 127, 157, 158, 161, 178, 281, 284

Fluorescein isothiocyanate (FITC) 114, 115, 117, 124, 127, 129, 133, 134, 175–177, 180, 181, 183, 184
Fluorescence in situ hybridization (FISH) 167–173, 432
Fluorophores 124, 170, 171, 173, 310, 316, 401

G

Gamete interaction 101
GC. *See* Germ cell (GC)
Genetic defects 46, 535–536
Genetics 46, 121, 122, 125, 127, 155, 201, 274, 279, 293, 379–384, 397–409, 452, 490, 495, 496, 499, 505, 519, 520, 535–536
Genome-wide association (GWA) 404–406, 408, 409
Genome-wide association study (GWAS) 402–404
Genomics 188–195, 197–202, 204, 380, 381, 384, 386, 392, 393, 398–402, 406, 407, 451–455, 471
Genotyping 403, 404, 406, 407, 524
Germ cell (GC) 32, 34, 115, 189, 200, 202, 272, 273, 299, 303–304, 314, 322, 324, 327, 330, 332, 337, 452, 479, 489–499, 504, 509–511, 514, 515, 532, 535–540, 542, 546
Germline stem cell (GS cell) 504–511, 513–517
Glutaraldehyde 300, 301, 327–328, 331, 334–336, 338, 345, 346, 542
GS cell. *See* Germline stem cell (GS cell)
Guanidinium thiocyanate 386
GWA. *See* Genome-wide association (GWA)
GWAS. *See* Genome-wide association study (GWAS)

H

HBDR. *See* Homozygosity by descent region (HBDR)
Hematoxylin 40, 42, 47, 300–303, 311, 314, 320, 464
Hemizona assay (HZA) 55, 91–101
Hemizona index (HZI) 93–94, 100, 101
Hemocytometer 5, 7, 10–11, 66, 104–106, 109, 140, 388, 463, 508, 515, 524, 525
Histochemistry 322, 327, 330, 334–325
Histone localization 460–461
Histone retention 270
Histones 138, 148, 149, 264, 270, 305, 380, 426, 437, 459–474, 533
Hoechst 13, 117, 184, 239, 281–285, 387, 388, 391, 392, 482, 486
Homozygosity by descent region (HBDR) 407
Hormonal defects 529
Horseradish peroxidase (HRP) 310, 312, 314, 316, 319, 364, 366–368, 371–374
HOS test. *See* Hypo-osmotic swelling (HOS) test
HRP. *See* Horseradish peroxidase (HRP)
Human spermatozoa 22, 61, 77, 79, 85, 92, 93, 218, 222, 242, 324, 385, 472
Human sperm biomarkers 425, 426
Hyaluronic acid 263–267

SPERMATOGENESIS: METHODS AND PROTOCOLS
Index | 551

Hyperactivation 62, 63, 77–86, 113, 123, 271, 546–547
Hypo-osmotic swelling (HOS) test 14, 16–18, 22–25
Hypotonic solution 21
HZA. *See* Hemizona assay (HZA)
HZI. *See* Hemizona index (HZI)

I

ICSI. *See* Intracytoplasmic sperm injection (ICSI)
In vitro fertilization (IVF) 59, 61, 68, 91–96, 101, 104, 152, 167, 218, 222, 223, 227, 229, 230, 232, 234, 235, 237, 239, 258, 259, 273–275, 286, 289, 293, 426, 522–524
In vitro spermatogenesis 479–488
Imidazole 428–430, 433
Immunobead test 52–54, 58
Immunocytochemistry 322, 326, 328, 330, 334, 344, 345, 432, 433, 460
Immunohistochemistry 301, 309–320, 485, 540
IMSI. *See* Intracytoplasmic morphologically-selected sperm injection (IMSI)
Infertility 21, 51, 52, 78, 91, 93, 103, 122, 125, 130, 138, 152, 188, 189, 201, 209, 227, 248, 257, 263, 351, 353, 361, 363, 397–398, 409, 426, 535–537
Intracytoplasmic morphologically-selected sperm injection (IMSI) 248, 250–252
Intracytoplasmic sperm injection (ICSI) 14, 18, 22, 46, 61, 68, 91, 94, 104, 123, 201, 214, 218, 222, 223, 227–229, 232, 234, 235, 239, 247, 248, 252, 253, 257–261, 263–267, 270, 275, 526

J

JC-1 .... 175–177, 180–184

K

Kinematics 78–80, 84–86, 274
Kinetics 92, 153, 202, 204, 299, 300, 367
Knockout serum replacement (KSR) 480, 481
KSR. *See* Knockout serum replacement (KSR)

L

LC-MS/MS. *See* Liquid chromatography followed by tandem MS (LC-MS/MS)
Lentivirus 505, 507, 508, 511–516
Leptotene 300, 303
Light microscopy 309, 310, 316, 325, 328–331, 337, 340, 544
Liquid chromatography followed by tandem MS (LC-MS/MS) 411–413, 415, 419–420
Liquid nitrogen 96, 109, 210–214, 289, 482, 485, 487, 506, 508, 509
LIS. *See* Lithium 3,5-diiodosalicylate (LIS)
Lithium 3,5-diiodosalicylate (LIS) 139, 140, 142, 144, 429, 432, 434, 435

Luminol 352–354, 356, 358, 359, 364–366, 370, 373
Luminometer 352–356, 364–368, 370–375

M

MACS. *See* Magnetic-activated cell sorting (MACS)
Magnetic-activated cell sorting (MACS) 258–261
MALDI-MS. *See* Matrix assisted laser desorption ionization MS (MALDI-MS)
Male factor infertility 103, 152, 209, 257, 263, 361, 397
Male germ cell transplantation 489–499
Male infertility 21, 103, 125, 138, 188, 201, 248, 257, 351, 353, 363, 397, 398, 409, 426, 535
MAR test. *See* Mixed antiglobulin reaction (MAR) test
Mass spectrometry 105, 259–261, 408, 411–413, 419, 421, 482
Matrix assisted laser desorption ionization MS (MALDI-MS) 411, 413
Mature sperm 200, 264, 269–271, 459–474, 534, 539
MeDIP. *See* Methylated DNA immunoprecipitation (MeDIP)
Meiotic errors 165
Messenger RNA (mRNA) 385–396, 540
Methylated DNA immunoprecipitation (MeDIP) 452
Microarray 386, 394, 400–401, 404, 406, 468, 471–474
Micrococcal nuclease digestion 462, 467–468, 473
Microinjection 248, 253, 490, 496, 519, 520
Microslicer 330
Mitochondria 35, 45, 49, 122, 123, 175, 176, 180, 182, 184, 251, 326, 535
Mitochondrial membrane potential 181–182
Mixed antiglobulin reaction (MAR) test 51–55
Motile sperm organelle morphology examination (MSOME) 248, 250–253
Motility 13, 18, 62, 64–74, 77–86, 92, 106, 115, 118, 215, 223, 228, 232, 237, 242, 253, 254, 258, 266, 463, 527, 535, 543, 546
Motility index 74, 353, 354
Mouse 81, 82, 133, 152, 300, 302–306, 366, 398, 437–443, 452, 464, 466, 467, 471, 473, 480, 482, 483, 487, 492–495, 497–499, 504, 505, 519, 522–524, 526, 531, 536, 544
mRNA. *See* Messenger RNA (mRNA)
MSOME. *See* Motile sperm organelle morphology examination (MSOME)
Mucus penetration 103

N

Next generation sequencing 401–402, 452
Nuclear halo 440
Nuclear matrix 437, 438, 440, 521
Nuclear proteins 138, 305, 380, 412, 445, 468, 471

O

Oocyte 61, 77, 91–101, 104, 113, 121, 125, 217, 247, 252, 253, 257, 263, 266, 270, 519, 520, 524, 526
Orange G6 .... 41, 42
Organ culture 479–488
Osmium tetroxide 328, 335, 344–346
Oxidative stress 122, 123, 155, 351–361, 363, 364
Oxygen radicals 363, 364

P

PAF. See Platelet-activating factor (PAF)
Papanicolaou stain 31, 42, 45, 49
PAS staining. See Periodic acid schiff's (PAS) staining
Paternal contribution 386, 426
Peanut agglutinin (PNA) 175, 176, 180, 181, 184
Pentoxifylline 230, 232, 234, 235, 239, 241–244
Percoll 218, 219, 223, 224, 414–416, 420, 462, 472
Perfusion 300, 301, 317, 323–325, 331, 334, 336, 337, 346
Periodic acid schiff's (PAS)
  staining 301, 303, 304, 484, 485, 533, 537, 539, 544
Pig 53, 56–58, 152, 154, 293, 520
Pisum sativum agglutinin (PSA) 114–117, 176
Plasmid 452, 455, 507, 514, 515, 526
Platelet-activating factor (PAF) 241–244
PNA. See Peanut agglutinin (PNA)
Polyacrylamide gel electrophoresis separation
  in the presence of SDS
  (SDS-PAGE) 412, 413, 418, 421, 445
Polychrome EA50 .... 41, 42
Polymerase chain reaction (PCR) multiplex
  assays 188, 189, 194, 196–198, 202–205
P1/P2 ratio 449
Predetermination of sex 279, 293
Preleptotene 300, 303, 304, 532
Progesterone 114–116
Progressive motility 59, 61, 62, 68, 71, 77, 117, 231, 237, 266, 274, 288, 289, 522, 543
Pronuclear injection 519
Propidium iodide (PI) 13, 116, 124, 127, 129, 130, 134, 175–178, 180–184, 282, 283, 465
Protamines 138, 148, 149, 271, 305, 380, 412, 426, 437, 438, 445–450, 460, 461, 464, 466, 470
Proteinase K 379–382, 468, 471
Proteome 411–421
Proteomics 411, 415–417
PSA. See Pisum sativum agglutinin (PSA)
Puresperm gradient 224, 266, 388, 395, 472

Q

Qiacube 386, 387, 389–391, 395

R

Reactive oxygen species (ROS) 50, 122, 123, 155, 218, 224, 351–361, 363–365, 373
Recovery 108, 116, 117, 150–151, 214, 227–239, 271, 274, 275, 277, 286, 369, 370, 373, 385, 386, 389, 460–461, 471, 497, 506, 509
Ribonucleic acid (RNA) extraction 385–387, 389–391, 395
RNA isolation 379, 380, 385, 386, 390, 391
RNeasy Minelute kit 387
RNeasy Mini kit 387, 389, 390
ROS. See Reactive oxygen species (ROS)

S

Salt storage solution 92, 95, 96, 100, 101
Scanning electron microscopy (SEM) 321–347
SCSA. See Sperm chromatin structure assay (SCSA)
SDS-PAGE. See Polyacrylamide gel electrophoresis
  separation in the presence of SDS (SDS-PAGE)
Sectioning 107, 264, 299, 302, 312–314, 316, 318, 320, 329, 331, 333, 339–342, 407, 493, 507
SEM. See Scanning electron microscopy (SEM)
Semen 3, 15, 22, 27, 39, 53, 63, 73, 86, 92, 103, 115, 121, 139, 153, 169, 178, 209, 220, 227, 241, 249, 259, 265, 273, 283, 321, 351, 366, 380, 417, 425, 447, 467, 520
Semen analysis 22, 27, 28, 42, 70, 73, 103, 121, 153, 211, 274, 359
Seminal plasma 17, 31, 50, 52, 54, 58, 62, 63, 84, 86, 106, 114, 122, 123, 128, 131, 217, 218, 222, 225, 241, 273, 338, 353, 357, 359, 363–375, 415
Seminiferous epithelium cycle 300–303
Seminiferous tubule 228, 232–234, 299–307, 322, 324, 325, 334–337, 483, 484, 490, 494, 496, 497, 532, 535, 537
Sequence tagged sites (STSs) 188
Sex-sorted sperm 282, 288, 292, 293
Shotgun bisulfite sequencing 453, 455–457
Silane-coated colloidal silica 217, 223
Single nucleotide polymorphism (SNP) 399–401, 403, 404, 406–408
Single strand breaks 122, 144, 443
Small noncoding RNA (sncRNA) 385–396
SMGT. See Sperm-mediated gene transfer (SMGT)
sncRNA. See Small noncoding RNA (sncRNA)
SNP. See Single nucleotide polymorphism (SNP)
Sperm 4, 14, 21, 28, 42, 51, 61, 73, 77, 91, 103, 113, 121, 137, 148, 167, 175, 186, 209, 217, 227, 241, 247, 257, 263, 279, 299, 321, 351, 371, 379, 385, 411, 425, 437, 445, 452, 459, 480, 489, 519, 530
  aspiration 228, 231, 267, 443

banking 207
chromatin 139, 148, 152, 247, 248, 263, 270, 272, 274, 426, 430–431, 434, 460–464, 467–473, 521
classification 28–30, 32–34, 148, 155, 460
concentration 3–12, 15, 17, 18, 47, 61, 68–71, 78, 84, 92, 99, 106, 128, 158, 169, 172, 231, 232, 237, 244, 261, 265, 277, 283, 286, 288, 425, 426, 430, 431, 447, 463, 467, 525
count 7, 9, 106, 109, 128, 162, 163, 171–173, 353, 354, 359, 383, 543
DNA 148, 280, 282, 291, 380, 381, 437, 438, 442, 451–457, 463, 465, 468, 470, 521–527
    damage 121–123, 125, 137–145, 152, 274, 363
    fragmentation 121–135, 148, 149, 152, 155, 258
    integrity 431–432
extraction 232–234, 379–384, 386, 438
function 13, 21, 28, 32, 51, 52, 61, 62, 68, 91–101, 103, 123, 217, 220, 221, 224, 242, 248, 322, 352, 397, 426, 531
membrane integrity 13–14, 21–25
motility 10, 15, 18, 54, 56, 61–73, 77–86, 92, 101, 117, 122, 176, 209, 212–215, 220, 231, 232, 235, 237, 241–244, 253, 254, 258, 266, 267, 271, 274, 275, 288–290, 354, 425, 426, 430, 522, 525, 539, 543, 546, 547
nuclear halo 440, 441
nuclear matrix 438, 440, 441
nuclei 148, 152, 160, 170–172, 235, 248, 280–282, 292, 434, 437–443, 462, 463, 467, 472, 527
plasmalemma 270
progressive motility 61, 62, 68, 71, 77, 117, 237, 274, 288, 522
selection 215, 232, 247, 248, 250–252, 254, 257–261, 263–267, 269–277, 522, 523, 525, 526
sorting 157, 281, 283, 285, 286, 290, 291, 293
viability 22, 58, 176–181, 214, 239, 257, 258, 274, 289
vitality 13–18, 127, 525
washing 58, 68, 148, 163, 172, 429, 462, 525, 527
Spermatids 299, 303–306, 380, 420, 485, 486, 533, 535, 540, 542
Spermatocytes 200, 238, 299, 300, 303, 304, 306, 323, 532, 533, 535, 538, 540
Spermatogenesis 189, 201, 264, 299–307, 309–320, 322, 397, 398, 459, 479–490, 496, 498, 499, 503, 532–535, 537, 540, 542
Spermatogonia 200, 299, 303–307, 323, 479, 490, 509, 510, 515, 532, 533, 535, 537, 546
Spermatogonial stem cells (SSCs) 490, 503–517, 533, 539
Spermatozoa 3, 13, 21, 32, 40, 61, 77, 92, 114, 121, 178, 209, 217, 227, 241, 257, 299, 321, 351, 379, 385, 412, 426, 438, 461, 521, 530

Sperm chromatin structure assay
    (SCSA) 125–127, 147–163
Spermiation 28, 50, 300, 303, 535, 539
Sperm-mediated gene transfer (SMGT) 519–527
Sperm morphology
    strict morphology 28, 30, 32, 39–50, 98
    tybergerg sperm morphology 423
Sperm-oocyte penetration 94
SSCs. *See* Spermatogonial stem cells (SSCs)
Stages 50, 93, 200, 232, 264, 299–307, 340, 459, 533, 535, 536, 539, 540, 542, 546
Staging 299–307
Streptavidin 310, 312, 314, 316, 317, 326, 328
Strict morphology 28, 39–50, 431
Structural variation 399
STSs. *See* Sequence tagged sites (STSs)
SYBR-14 177–184

T

TAC. *See* Total antioxidant capacity (TAC)
TEM. *See* Transmission electron microscopy (TEM)
Terminal deoxynucleotidyl transferase dUTP
    nick end labeling (TUNEL) 121–135, 148, 271, 538
TESA. *See* Testicular sperm aspiration (TESA)
TESE. *See* Testicular sperm extraction (TESE)
Testes 227, 306, 312, 313, 317, 322, 331, 332, 336, 337, 380, 480, 483, 485, 492, 496, 498, 504, 536, 542, 545
Testicular sperm aspiration (TESA) 228, 232
Testicular sperm extraction (TESE) 205, 228, 232, 234
Testicular tissue grafting 489–499
Testis 200, 227–239, 304, 310, 312–313, 317, 322, 336, 337, 398, 480–490, 492–495, 497–499, 504, 532, 535–538, 540, 542, 544–545
Testis development 103, 490
Tissue culture 108, 484, 485, 491, 492, 495, 506, 516
Total antioxidant capacity (TAC) 352–354, 357–358, 360, 361, 364, 366, 370, 371, 374
Transgene 505, 516, 520, 521
Transgenesis 520
Transgenic 487, 490, 498, 503–517, 519, 520, 522, 524, 526, 531, 535
Transgenic animals 490, 519, 520, 522, 531–547
Transmission electron microscopy
    (TEM) 321–323, 326, 327, 330–333, 340, 345, 346, 545
Trolox 353, 357, 358, 360, 366, 368–370, 372
TUNEL. *See* Terminal deoxynucleotidyl transferase dUTP nick end labeling (TUNEL)
Two-dimensional polyacrylamide gel electrophoresis
    (2D-PAGE) 412–415, 417–419, 421
Tybergerg sperm morphology 423

U

Ultramicrotome 330, 333, 340
Ultrastructural 321–347
   histochemistry 322, 327, 330, 334–335
   immunocytochemistry 322, 326, 334
Uranyl acetate 329, 333, 335, 343, 344, 542, 545

V

Viability 17, 175, 228, 239, 257, 421
Vibratome 330, 334, 345, 346

W

Wedge biopsy 232, 234
WHO. *See* World Health Organization (WHO)

Whole genome 399, 401, 402, 471, 474, 532
World Health Organization (WHO) 9, 14, 27–36, 39, 48, 49, 51, 54, 55, 62, 94, 153, 224

X

X- and Y-sperm, mammals 279, 280, 291
Xenografting 489, 490

Y

Y chromosome microdeletions 187–205

Z

Zeta potential 270, 271
Zona pellucida 61, 77, 91–94, 96, 98, 101, 103, 104, 107, 113, 264

Printed by Publishers' Graphics LLC
AMZ20121008.19.19.135